Oxidations in Organic Chemistry

Oxidations in Organic Chemistry

Miloš Hudlický
Virginia Polytechnic Institute and State University

ACS Monograph 186

American Chemical Society
Washington, DC   1990

Library of Congress Cataloging-in-Publication Data
Hudlický, Miloš, 1919–
  Oxidations in organic chemistry/Miloš Hudlický.
    p.   cm.—(ACS monograph, ISSN 0065-7719; 186)
  Includes bibliographical references.
  ISBN 0-8412-1780-7 (cloth)
  ISBN 0-8412-1781-5 (pbk.)
  1. Oxidation. 2. Chemistry, Organic.
  I. Title. II. Series: Monograph series (American Chemical Society); no. 186.
  QD281.09H83   1990
  547'.23—dc20   90-34564
                                                                CIP

The paper used in this publication meets the minimum requirements of American National Standard for Information Sciences—Permanence of Paper for Printed Library Materials, ANSI Z39.48—1984.

∞

Copyright © 1990

American Chemical Society

All Rights Reserved. The copyright owner consents that reprographic copies of the chapters may be made for personal or internal use or for the personal or internal use of specific clients. This consent is given on the condition, however, that the copier pay the stated per-copy fee through the Copyright Clearance Center, Inc., 27 Congress Street, Salem, MA 01970, for copying beyond that permitted by Sections 107 or 108 of the U.S. Copyright Law. This consent does not extend to copying or transmission by any means—graphic or electronic—for any other purpose, such as for general distribution, for advertising or promotional purposes, for creating a new collective work, for resale, or for information storage and retrieval systems. The copying fee for each chapter is a base fee of $4.00 plus $0.50 per copy of the chapter. Please report your copying to the Copyright Clearance Center with this code: for the clothbound book, 0-8412-1780-7/90/$04.00 + .50; and for the paperback book, 0-8412-1781-5/90/$04.00 + .50.

The citation of trade names and/or names of manufacturers in this publication is not to be construed as an endorsement or as approval by ACS of the commercial products or services referenced herein; nor should the mere reference herein to any drawing, specification, chemical process, or other data be regarded as a license or as a conveyance of any right or permission to the holder, reader, or any other person or corporation, to manufacture, reproduce, use, or sell any patented invention or copyrighted work that may in any way be related thereto. Registered names, trademarks, etc., used in this publication, even without specific indication thereof, are not to be considered unprotected by law.

PRINTED IN THE UNITED STATES OF AMERICA

Third printing 1997

To my dear wife Alena

FOREWORD

ACS MONOGRAPH SERIES was started by arrangement with the interallied Conference of Pure and Applied Chemistry, which met in London and Brussels in July 1919, when the American Chemical Society undertook the production and publication of Scientific and Technological Monographs on chemical subjects. At the same time it was agreed that the National Research Council, in cooperation with the American Chemical Society and the American Physical Society, should undertake the production and publication of Critical Tables of Chemical and Physical Constants. The American Chemical Society and the National Research Council mutually agreed to care for these two fields of chemical progress.

The Council of the American Chemical Society, acting through its Committee on National Policy, appointed editors and associates to select authors of competent authority in their respective fields and to consider critically the manuscripts submitted. The first Monograph appeared in 1921. Since 1944 the Scientific and Technological Monographs have been combined in the Series.

These Monographs are intended to serve two principal purposes: first, to make available to chemists a thorough treatment of a selected area in a form usable by persons working in more or less unrelated fields so that they may correlate their own work with a larger area of physical science; and second, to stimulate further research in the specific field treated. To implement this purpose, the authors of Monographs give extended references to the literature.

Publisher's Acknowledgment

The American Chemical Society thanks Robert J. Alaimo of Norwich Eaton Pharmaceuticals, Inc., and chairman of the Division of Chemical Health and Safety of the American Chemical Society for thoroughly evaluating the safety aspects of this book, for writing the safety advisory that precedes the preparative procedures, and for advising us in the presentation of safety information.

ABOUT THE AUTHOR

MILOŠ HUDLICKÝ, a native of Czechoslovakia, obtained his Ph.D. from the Technical University in Prague, Czechoslovakia. After having spent the year 1948 at the Ohio State University as a UNESCO postdoctoral fellow, he taught as an assistant professor and, later, as an associate professor at the Technical University in Prague until 1958. He then worked as a research associate in the Research Institute of Pharmacy and Biochemistry in Prague. After the Russian occupation of Czechoslovakia in 1968, he moved to the United States, where he was offered a professorship at Virginia Polytechnic Institute and State University.

His field of interest is organofluorine chemistry. He has written 65 research papers, 17 review papers, 28 patents, and 12 books (5 in English). His chefs-d'oeuvre are *Chemistry of Organic Fluorine Compounds* and *Reductions in Organic Chemistry,* published in 1976 and 1984, respectively.

CONTENTS

Preface xv
Disclaimer xvii
The System of the Book xix

1. Oxidation Agents 1
   Air, Oxygen, Ozone, and Electrolysis 1
   Hydrogen Peroxide and Its Derivatives 6
   Organic Peroxy Acids 10
   Derivatives of Group 1 Elements 14
   Derivatives of Group 2 Elements 16
   Derivatives of Group 3 Elements 17
   Derivatives of Group 4 Elements 17
   Derivatives of Group 5 Elements 18
   Derivatives of Group 6 Elements 20
   Derivatives of Group 7 Elements 26
   Derivatives of Group 8 Elements 37
   Organic Oxidants 39
   Chiral Oxidants 44
   Biochemical (Microbial) Oxidations 45

2. Dehydrogenations 47
   Introduction of Double Bonds 47
   Dehydrogenative Aromatizations 50
   Dehydrogenative Couplings 52

3. Oxidations 57
   Alkanes and Cycloalkanes 57
   Alkenes and Cycloalkenes 60
     Epoxidation 60
     Formation of Dioxetanes 64
     Formation of Ozonides 65
     Hydroxylation 67
     Conversion into Halohydrins, Amino Hydroxy Compounds,
       and Esters 73
     Conversion into Ketones 75
     Oxidative Cleavage of Double Bonds 77
     Oxidations at Allylic Positions 84
   Dienes 87
   Alkynes (Acetylenes) 90
   Aromatic Compounds 92
     Hydroxylation of Aromatic Rings 92

Oxidation of Aromatic Compounds to Quinones 94
Oxidative Cleavage of the Benzene Ring...................... 96
Oxidation of Side Chains of Carbocyclic and Heterocyclic
   Aromatic Compounds... 99
Halogen Derivatives and Tosylates.............................. 109
Alcohols.. 114
   Dehydrogenation and Oxidation of Primary Alcohols
      to Aldehydes.. 114
   Oxidation of Primary Alcohols to Carboxylic Acids 127
   Oxidation of Primary Alcohols to Esters 131
   Oxidation of Secondary Alcohols to Ketones 132
   Oxidation of Secondary Alcohols to α-Hydroxy
      Hydroperoxides... 149
   Oxidation of Secondary Alcohols to Carboxylic Acids 150
   Oxidation of Tertiary Alcohols................................ 150
   Oxidation of Unsaturated Alcohols at Multiple Bonds 151
   Oxidation of Diols 155
   Cleavage of Vicinal Diols to Carbonyl Compounds 159
Phenols.. 163
Ethers 168
   Oxidation of Unsaturated Ethers at Multiple Bonds 170
   Oxidation of Epoxides (Oxiranes) 173
Aldehydes.. 174
   Oxidation of Aldehydes Having Other Functionalities 182
   Oxidation of Aldehyde Derivatives............................ 184
Ketones 186
   Oxidation of Ketones to Esters (Baeyer–Villiger Reaction)... 186
   Baeyer–Villiger Oxidation of Functionalized Ketones 191
   Hydroxylation of Ketones 196
   Oxidation of Ketones to α-Dicarbonyl Compounds 199
   Oxidation of Ketones to Carboxylic Acids 202
   Oxidation of Unsaturated Ketones............................ 212
   Oxidation of Hydroxy Ketones to Diketones 215
   Oxidation of Dicarbonyl Compounds to Carboxylic Acids 218
   Oxidation of Ketone Derivatives.............................. 219
Carboxylic Acids and Their Derivatives 222
   Oxidation of Carboxylic Acids and Their Esters.............. 222
   Oxidation of Amides, Hydrazides, and Nitriles............... 229
Nitrogen Compounds... 230
   Oxidation of Nitro Compounds 230
   Oxidation of Nitroso Compounds............................. 231
   Oxidation of Hydroxylamines 231
   Oxidation of Azo Compounds and Azides 232
   Oxidation of Hydrazo Compounds............................ 233

  Oxidation of Primary Amines at Nitrogen 234
  Oxidation of Secondary and Tertiary Amines at Nitrogen 236
  Dehydrogenation of Primary Amines to Imines and
    Aldehydes or Ketones 239
  Dehydrogenation of Secondary Amines to Imines............. 240
  Dehydrogenation of Primary Amines to Nitriles.............. 241
  Oxidation of the Carbon Chain of Amines..................... 242
  Oxidation of Aromatic Amines to Quinones................... 246
 Phosphorus and Arsenic Compounds 248
 Sulfur Compounds.. 250
  Oxidation of Thiols (Mercaptans)............................. 250
  Oxidation of Sulfides 252
  Oxidation of Thioacetals (Mercaptals and Mercaptoles) 259
  Oxidation of Sulfoxides... 262
  Oxidation of Sulfones... 263
  Oxidation of Disulfides 263
 Selenides 265
 Iodo Compounds 266
 Boron Compounds 267
 Silicon and Tin Compounds...................................... 270
 Organomagnesium and Organomercury Compounds............. 271

4. Preparative Procedures.. 273
 1. Preparation of Anhydrous Hydrogen Peroxide in Ether 273
 2. Preparation of 90% Hydrogen Peroxide and of 90%
   Peroxyacetic Acid... 273
 3. Preparation of the Jones Reagent............................ 273
 4. Preparation of Chromium Trioxide–Pyridine Complex,
   $CrO_3 \cdot 2C_5H_5N$ (Collins Reagent) 274
 5. Preparation of Pyridinium Chlorochromate Adsorbed
   on Alumina.. 274
 6. Preparation of Tetrabutylammonium Chromate 274
 7. Preparation of Active Manganese Dioxide 274
 8. Dehydrogenation by 2,3-Dichloro-5,6-dicyano-*p*-
   benzoquinone (DDQ) 275
 9. Dehydrogenation via Selenoxides............................. 275
 10. Oxygenation with Singlet Oxygen............................. 275
 11. Ozonization 276
 12. Electrolytic Decarboxylative Coupling of Carboxylic Acids 276
 13. Oxidation with Hydrogen Peroxide........................... 277
 14. Oxidation with *tert*-Butyl Hydroperoxide..................... 277
 15. Oxidation with Potassium Peroxysulfate...................... 278
 16. Oxidation with Organic Peroxy Acids 278
 17. Oxidation with Cupric Acetate................................ 279

18. Oxidation with Silver Oxide (In Situ) 279
19. Oxidation with Mercuric Oxide 280
20. Oxidation with Thallium Trinitrate 280
21. Oxidation with Ceric Ammonium Nitrate 281
22. Oxidation with Lead Tetraacetate 281
23. Oxidation with Nitric Acid 281
24. Oxidation with Ammonium Nitrate 282
25. Oxidation with Lead Nitrate 282
26. Oxidation with Sulfur 282
27. Oxidation with Selenium Dioxide 283
28. Oxidation with Chromium Trioxide 283
29. Oxidation with Chromium Trioxide–Pyridine Complex 283
30. Oxidation with Chromium Trioxide–3,5-Dimethylpyrazole Complex 284
31. Oxidation with Pyridinium Chlorochromate (PCC) 284
32. Oxidation with Pyridinium Chlorochromate Adsorbed on Alumina 284
33. Oxidation with Tetrabutylammonium Chromate 285
34. Oxidation with Sodium Dichromate 285
35. Oxidation with Sodium Hypochlorite 285
36. Oxidation with Sodium Chlorite 285
37. Oxidation with Sodium Hypobromite 286
38. Oxidation with N-Bromoacetamide 286
39. Oxidation with Acetyl Hypoiodite 286
40. Oxidation with Periodic Acid 287
41. Oxidation with Sodium Periodate 287
42. Oxidation with Alumina-Supported Sodium Periodate 287
43. Oxidation with Sodium Periodate and Potassium Permanganate 288
44. Oxidation with Manganese Dioxide 288
45. Oxidation with Potassium Permanganate 289
46. Oxidation with Potassium-Permanganate-Coated Molecular Sieves 290
47. Oxidation with Tetrabutylammonium Permanganate (Purple Benzene) 290
48. Oxidation with Ferric Chloride 290
49. Oxidation with Osmium Tetroxide 290
50. Oxidation with Ruthenium Tetroxide 291
51. Oxidation with Diethyl Azodicarboxylate 291
52. Oxidation with Potassium Nitrosodisulfonate (Fremy Salt) 292
53. Oxidation with p-Nitrosodimethylaniline 292
54. Sommelet Oxidation 292
55. Oxidation with Dimethyl Sulfide and N-Chlorosuccinimide 293

56. Oxidation with Dimethyl Sulfoxide 293
  57. Biochemical Oxidation 294
Correlation Tables 295
Abbreviations and Generic Names 319
References 321
Bibliography... 359

INDEXES

Author Index... 363
Subject Index 375

PREFACE

FOUR YEARS after my book *Reductions in Organic Chemistry* was published by Ellis Horwood in Chichester, England, I submitted its pendant, *Oxidations in Organic Chemistry*. The present monograph is intended to become an aid to a bench chemist. That means that its scope is experimental rather than theoretical, and emphasis is laid on the preparative aspects and synthetic usefulness of individual reactions rather than on the historical developments and mechanisms. This is the only way to compress a critical survey of countless oxidations into a book of reasonable size. The criteria for inclusion of oxidation reactions were simplicity of the reactions, clarity of their descriptions, availability of the oxidants, and yields of the products. The last aspect is somewhat problematic, because analogous compounds may give widely varied results.

To compare the results of oxidations by various oxidants, fairly simple types of compounds are shown to illustrate reaction conditions. After all, these compounds were selected by the researchers who tried to advertise their new methods. More sophisticated examples were avoided, especially because of lack of space.

The systematic literature coverage, initiated by screening *Organic Syntheses,* Theilheimer's *Synthetic Methods,* Harrison and Harrison's *Compendium of Synthetic Methods, Methods in Organic Synthesis, Synthetic Pathways,* and my own files, includes original papers in some 50 major chemical journals through the end of 1986 (with a few more recent additions). Quotations of patents and cyclopedias were avoided, because these sources are generally not very readily available in chemical libraries.

Acknowledgments

It is my pleasure to express my gratitude to my wife, Alena, without whose efficient support, loving care, and understanding this book would not have been written. I gladly acknowledge the help of my son, Tomáš Hudlický, for his critical comments after reading the manuscript. I also appreciate very much the secretarial help of Angela Miller, Wanda Ritter, and Vera Good for typing the text and handling the ChemDraw software; the permission of the Department of Chemistry of Virginia Polytechnic Institute and State University to use the word-processing facilities; and the permission of A. Bader, Aldrich Chemical Company, Inc., to quote physical constants of many reagents from the Aldrich 1988–1989 catalog.

Finally, I would like to give credit to my reviewers, M. S. Newman and J. F. Wolfe, for helping me improve the quality of my book, and to the ACS Books Department editors, Paula M. Bérard, Robin Giroux, and especially A. Maureen Rouhi, for the meticulous, yet humane, editorial work.

MILOŠ HUDLICKÝ
Virginia Polytechnic Institute
 and State University
Blacksburg, VA 24061–0699
July 19, 1989

DISCLAIMER

THIS BOOK CONTAINS information and materials involving chemicals and chemical reactions, and included within some of the chapters are certain procedures for preparing oxidation agents and performing oxidation reactions. These materials and procedures have been compiled carefully, and introductory safety statements precede the preparative procedures section of this book. There are also indications of the preparations and reactants that should be noted as particularly hazardous. The running notice at the bottom of each page alerts each reader to the fact that oxidations are strong reactions. The safety information is included within this book by the American Chemical Society as a precaution to the readers.

The materials, safety information, and procedures contained in this book are believed to be reliable. This information and these procedures should serve only as a starting point for laboratory practices, and they do not purport to specify minimal legal standards or to represent the policy of the American Chemical Society. No warranty, guarantee, or representation is made by the American Chemical Society as to the accuracy or specificity of the information contained herein, and the American Chemical Society assumes no responsibility in connection therewith. The added safety information is intended to provide basic guidelines for safe practices. Therefore, it cannot be assumed that all necessary warnings and precautionary measures are contained in this document and that other or additional information and measures may not be required. Users of this book and the procedures contained herein should consult the primary literature and other sources of safe laboratory practices for more exhaustive information.

THE SYSTEM OF THE BOOK
(HOW TO USE THIS BOOK)

IN ADDITION to the subject index with entries of several aspects, such as oxidation with . . . , oxidation of . . . , or . . . by oxidation of, the desired oxidation can be located in the following ways.

The first part of the book surveys oxidizing agents and their scopes and limitations in the oxidation of different functional groups. The arrangement is indicated in the table of contents: first, oxygen and ozone; then, hydrogen peroxide and its inorganic and organic derivatives; then, oxidants according to the sequence in the periodic table; and, finally, organic oxidants, chiral agents, and microorganisms.

The second larger part of the book lists oxidations of individual functional groups and types of compounds. The system is remotely reminiscent of Beilstein's *Handbuch der Organischen Chemie* (with certain modifications), and the oxidations of organic compounds are discussed according to the following order: alkanes and cycloalkanes; alkenes and cycloalkenes; aromatic (including heterocyclic aromatic) compounds; halogen derivatives; alcohols; ethers; aldehydes and ketones and their derivatives; carboxylic acids and their derivatives; nitrogen compounds; phosphorus and arsenic compounds; sulfur compounds; selenides; iodo compounds; and compounds of boron, silicon, tin, magnesium, and mercury.

If two or more functional groups are present, the oxidation of either group is described in the place where the functional group higher up in the system is discussed. For example, the oxidation of hydroxy aldehydes to oxo aldehydes or hydroxy acids is dealt with in the section on aldehydes (not alcohols). Sometimes, especially if one functional group is far enough from and unaffected by the other, such oxidation may be described in the place where the lower functional group is reviewed. For example, the epoxidation of oleic acid is discussed under "Alkenes" rather than "Carboxylic Acids." The discussion of the oxidation affecting a particular functional group precedes that of the oxidation in a more distant part of the molecule. Thus the oxidation of propene to its epoxide is described prior to the oxidation of propene to allyl hydroperoxide or acrolein.

The relative importance of individual oxidants is shown in different fonts: the most common oxidants are printed in **boldface** type, and the less frequently used ones are set in *italic* type. Normal type is used for agents with which there is not yet enough laboratory experience. No distinction is made between yields of products as found by gas–liquid chromatography, by derivatization, or by isolation.

Frequent cross-references should help locate the searched-for reaction. Despite all these provisions, there will certainly be reactions that cannot be found in this book, simply because this monograph contains only a few parts per million of all oxidations described in the chemical literature.

Chapter 1
Oxidation Agents

AIR, OXYGEN, OZONE, AND ELECTROLYSIS

Air, the cheapest oxidant, is used only rarely without irradiation and without catalysts. Examples of oxidations by air alone are the conversion of aldehydes into carboxylic acids (autoxidation) and the oxidation of acyloins to α-diketones. Usually, exposure to light, irradiation with ultraviolet light, or catalysts are needed. Under such circumstances, dehydrogenative coupling in benzylic positions takes place at very mild conditions [1]. In the presence of catalysts, terminal acetylenes are coupled to give diacetylenes [2], and anthracene is oxidized to anthraquinone [3]. Alcohols are converted into aldehydes or ketones with limited amounts of air [4, 5, 6, 7]. Air oxidizes esters to keto esters [8], thiols to disulfides [9], and sulfoxides to sulfones [10]. In the presence of mercuric bromide and under irradiation, methylene groups in allylic and benzylic positions are oxidized to carbonyls [11].

Oxygen, O_2, exists in two states. Stable **ground-state oxygen (triplet oxygen)** has two odd electrons with parallel spins. It behaves like a diradical and is paramagnetic: $\uparrow \cdot O{:}O \cdot \uparrow$. In **excited-state oxygen (singlet oxygen),** the two odd electrons possess antiparallel spins: $\uparrow \cdot O{:}O \cdot \downarrow$. Such a molecule is unstable, with a half-life of 10^{-6} s, and is diamagnetic. Each form reacts differently with organic molecules.

Singlet oxygen [12] is formed by irradiation of gaseous oxygen with ultraviolet light in the presence of sensitizers that absorb the light of the particular wavelength: benzophenone, methylene blue, rose bengal, chlorophyll, and others. It reacts with organic compounds in the same way as singlet oxygen generated chemically by several reactions, such as the treatment of hydrogen peroxide in alkaline medium with sodium hypochlorite [13, 14, 15], calcium hypochlorite [14], bromine [16], or organic peroxy acids [16]. Singlet oxygen can be also obtained by the decomposition of aromatic endoperoxides, such as 9,10-diphenylanthracene endoperoxide [17] or rubrene peroxide [18], and by the decomposition of triphenyl phosphite ozonide [19, 20] (equations 1–4).

[13]

$$H_2O_2 + NaOH + NaOCl: \quad \overset{-}{O}-O-H \quad \overset{+}{Cl} \longrightarrow O_2 + HCl \qquad 1$$

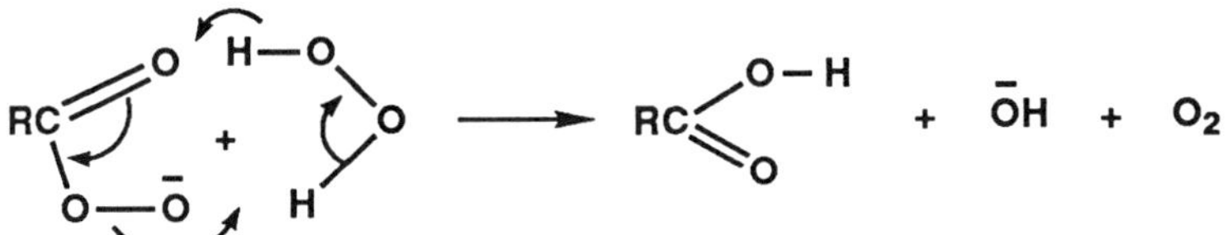

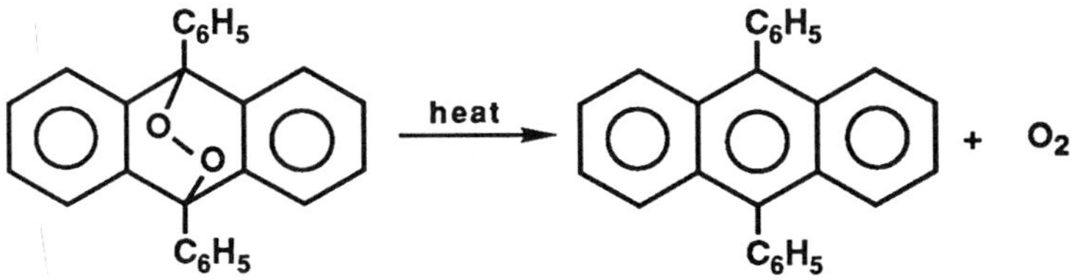

[16]

(reaction scheme showing RC(=O)O−O⁻ + H−O−O−H peroxide intermediate → RC(=O)O−H + ⁻OH + O₂) 2

[17]

(9,10-diphenylanthracene endoperoxide) $\xrightarrow{\text{heat}}$ (9,10-diphenylanthracene) + O_2 3

[19]

$$(C_6H_5O)_3P \xrightarrow{O_3} (C_6H_5O)_3P\begin{pmatrix}O-O\\O-O\end{pmatrix} \longrightarrow (C_6H_5O)_3PO + O_2 \qquad 4$$

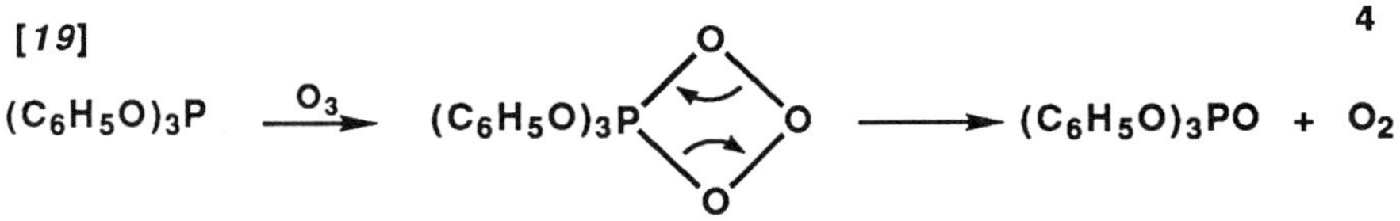

A reaction peculiar to singlet oxygen is the formation of hydroperoxides in the α positions with respect to double bonds [14, 15, 19, 21, 22, 23, 24, 25, 26] or aromatic rings [23, 27, 28, 29, 30] (equation 5).

[15]

(ene reaction of (CH₃)₂C=C(CH₃)CH₂H with O₂, 54–63%, giving allylic hydroperoxide) 5

$$ArCH_2R \xrightarrow{O_2} ArCHR\!-\!OOH$$

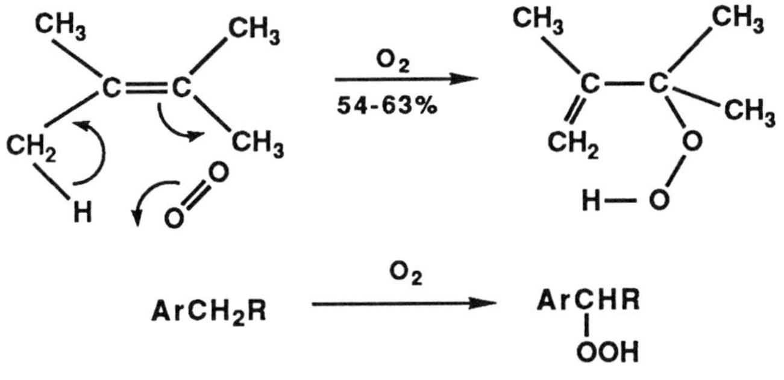

Instead of the fully accepted concerted mechanism via an **ene reaction**, a mechanism with perepoxides and dioxetanes as intermediates for hydroperoxides and carbonyl products, respectively, was proposed [21] (equation 6).

Another type of oxidation typical of singlet oxygen is the formation of oxides or **endooxides** from conjugated dienes by 1,4-addition (Diels–Alder reaction) [13, 15, 17, 19, 24, 31, 32, 33, 34, 35] (equation 7).

[21]   6

[13]   7

A less frequent reaction is the formation of ***dioxetanes*** from compounds containing double bonds [20, 36, 37, 38, 39, 40, 41, 42]. Such compounds are not stable and disintegrate to dicarbonyl compounds (equation 8).

[39]   8

Other oxidations with singlet oxygen are conversions of alkenes into epoxides [43], of secondary alcohols into ketones via alcohol hydroperoxides [44, 45] (equation 9) and the oxidative degradation of tertiary amines to secondary amines [46] (equation 10).

[44]   9

[46]   10

Ground-state oxygen alone rarely oxidizes organic compounds. A classical example is the autoxidation of benzaldehyde to benzoic acid, a usually undesirable reaction that takes place even in the absence of light. Other examples of autoxidation without illumination are oxidations at the α positions with respect to aromatic rings or at tertiary carbons [47, 48, 49, 50] and the formation of alkyl hydroperoxides from alkyl dichloroboranes [51]. Some oxidations take place when a compound is treated with oxygen in the presence of bases [9, 52, 53].

Because oxidations with oxygen are free-radical reactions, free radicals should be good initiators. Indeed, in the presence of hydrogen bromide at high enough temperatures, lower molecular weight alkanes are oxidized to alcohols, ketones, or acids [54]. Much more practical are oxidations catalyzed by transition metals, such as platinum [5, 6, 55, 56], or, more often, metal oxides and salts, especially salts soluble in organic solvents (acetates, acetylacetonates, etc.). The favored catalysts are vanadium pentoxide [3] and chlorides or acetates of copper [2, 57, 58, 59, 60, 61, 62, 63, 64, 65, 66], iron [67], cobalt [68, 69], palladium [60, 70], rhodium [10], iridium [10], and platinum [5, 6, 56, 57].

Oxidations by oxygen and catalysts are used for the conversion of alkanes into alcohols, ketones, or acids [54]; for the epoxidation of alkenes [43]; for the formation of alkenyl hydroperoxides [22]; for the conversion of terminal alkenes into methyl ketones [60, 65]; for the coupling of terminal acetylenes [2, 59, 66]; for the oxidation of aromatic compounds to quinones [3] or carboxylic acids [68]; for the dehydrogenation of alcohols to aldehydes [4, 55, 56] or ketones [56, 57, 62, 70]; for the conversion of alcohols [56, 69], aldehydes [5, 6, 63], and ketones [52, 67] into carboxylic acids; and for the oxidation of primary amines to nitriles [64], of thiols to disulfides [9] or sulfonic acids [53], of sulfoxides to sulfones [10], and of alkyl dichloroboranes to alkyl hydroperoxides [51].

Ozone, O_3 (O=O–O:), a blue gas or a dark blue liquid (bp −106, −116, or −125 °C, depending on the source of data), is used in a mixture with oxygen. Such mixtures are commercially available but are usually prepared in the laboratories. For microscale ozonization that is suitable for handling 0.01–0.1 mL of solutions, a microozonizer is assembled by using an electrical vacuum tester as a source of high-voltage electricity [71].

To prepare larger amounts of ozone, oxygen is passed through a silent electrical discharge with a potential drop of 11,000–15,000 V in an especially designed apparatus [72, 73]. Although the conversion of oxygen into ozone was reported to be as high as 6–8%, the commonly reached concentration of ozone in oxygen is 2–3%. Some compounds tend to decompose ozone, for example, dilute sodium hydroxide and phosphorus pentoxide [74].

Ozonizations are carried out by passing ozone-containing oxygen through solutions of organic compounds in solvents that do not react with ozone and are liquid at low temperatures. Cooling with dry-ice–acetone baths (–78 °C) is frequently needed to prevent the decomposition of ozonides, some of which are unstable at room temperature. The most common solvents are pentane, cyclohexane, dichloromethane, chloroform, methanol, acetic acid, and ethyl acetate.

To avoid an excess of ozone, the gas is metered by using conventional flow meters. Otherwise, the introduction of ozone is maintained until ozone appears in the exit gas. The presence of ozone is easily recognized by the starch–iodine reaction. The exit gas is passed through an aqueous solution containing 5% of alkaline iodide, 5% of sulfuric acid, and a few drops of a starch solution. Iodine, which is liberated by ozone, forms a dark blue color with starch [71]. Another sign of the end of ozonization is the light-blue tinge of the ozonized solution.

By far the most common reaction of ozone is that with alkenes, which first add ozone to form unstable molozonides. The molozonides break down into two separate species, which recombine to form true ozonides [75, 76] (equation 11).

Ozonides are rarely isolated [75, 76, 77, 78, 79]. These substances tend to decompose, sometimes violently, on heating and must, therefore, be handled with *utmost safety precautions* (safety goggles or face shield, protective shield, and work in the hood). In most instances, ozonides are worked up in the same solutions in which they have been prepared. Depending on the desired final products, ozonide cleavage is done by reductive or oxidative methods. Reductions of ozonides to *aldehydes* are performed by catalytic hydrogenation over palladium on carbon or other supports [80, 81, 82, 83], platinum oxide [84], or Raney nickel [85] and often by reduction with zinc in acetic acid [72, 81, 86, 87]. Other reducing agents are triphenylphosphine [88], trimethyl phosphite [89], dimethyl sulfide (DMS) [90, 91, 92], and sodium iodide [93]. Lithium aluminum hydride [94, 95] and sodium borohydride [95, 96] convert *ozonides into alcohols.*

The oxidative cleavage of *ozonides to carboxylic acids* is achieved with hydrogen peroxide in sodium hydroxide solution [97], in formic acid

[98, 99, 100], or in acetic acid [81, 101, 102, 103] and, less often, with silver oxide and nitric acid [77].

If only one molecule of a carboxylic acid is expected from the ozonolysis, boiling the ozonide with acetic acid is sufficient, because the ozonide contains one atom of oxygen necessary for such oxidation.

In addition to the ozonolysis of alkenes and a few aromatic compounds [93, 104], ozone oxidizes other groups. Thus saturated hydrocarbons containing tertiary hydrogen atoms are converted into tertiary alcohols [105, 106], and some alkenes are transformed into epoxides [107] or α,β-unsaturated ketones [108]. Benzene rings are oxidized to carboxylic groups [109]; ethers [110] and aldehyde acetals [111] to esters; aldehydes to peroxy acids [112]; sulfides to sulfoxides and sulfones [113]; phosphines and phosphites to phosphine oxides and phosphates, respectively [113]; and organomercury compounds to ketones or carboxylic acids [114].

Electrolytic oxidation takes place at the anode of an electrolytic cell, which may or may not contain a diaphragm to separate cathodic and anodic spaces. The anode must be made of a metal that resists oxidation, such as lead [115], nickel [116], or, most frequently, platinum. The anode is usually in the shape of a cylinder made of a wire gauze [117, 118]. The possible electrolytes are dilute sulfuric acid [117] or sodium methoxide prepared in situ from methanol and sodium [118]. The direct-current voltage is 3–100 V, the current density is 10–20 A/dm^2, and the electrolysis temperature is 20–80 °C.

Electrooxidation is rarely encountered. Examples are the epoxidation of alkenes [119]; the conversion of alkenes into ketones [120]; the oxidation of aromatic rings and side chains to carboxylic acids [117]; the oxidation of primary alcohols to carboxylic acids [116], of secondary alcohols to ketones [121], and of vicinal diols to carboxylic acids [122]; the hydroxylation or dehydrogenative coupling of phenols [115]; and, especially, the Kolbe synthesis of hydrocarbons. The Kolbe reaction cannot be performed in any other way and gives moderate to high yields of products resulting from the one-electron oxidation of carboxylate anions, decarboxylation, and subsequent coupling of the remaining free radicals [118, 123] (*see* equation 469).

HYDROGEN PEROXIDE AND ITS DERIVATIVES

Hydrogen peroxide, H_2O_2, is commercially available in aqueous solutions of 30 or 90% concentration. The 30% hydrogen peroxide is a colorless liquid (d 1.110) and is stabilized against decomposition, which occurs in the presence of traces of iron, copper, aluminum, platinum, and other transition metals.

Under reduced pressure, water can be distilled off from 30% hydrogen peroxide by using a column, and a solution of up to 90% hydrogen peroxide can thus be obtained. All the all-glass apparatus for such a distillation must be scrupulously clean and free of organic materials and transition metals and their compounds. The 90% hydrogen peroxide is a clear colorless liquid (mp −11 °C, bp 140 °C dec, d^{18} 1.393, n_D^{20} 1.3998 [*124*]).

Although 90% hydrogen peroxide is stable at 30 °C (the decomposition rate is 1%/year), it decomposes slowly at higher temperatures and rapidly with boiling at 140 °C. All *safety precautions* should be taken when working with highly concentrated hydrogen peroxide.

Practically anhydrous hydrogen peroxide as a solution in *tert*-butyl or *tert*-amyl alcohol is prepared by azeotropic removal of water by distillation of a mixture of 35% hydrogen peroxide in the alcohol at 50 °C and 50 mm of Hg until no more water distills. A solution containing 12% of hydrogen peroxide in *tert*-amyl alcohol is thus obtained [*125*].

A solution of anhydrous hydrogen peroxide in ether (~2.5 M) is obtained by repeated extraction of 30% aqueous hydrogen peroxide with ether and drying of the ether extracts over anhydrous magnesium sulfate. The hydrogen peroxide content is determined by iodometric titration [*126*].

Molecular complexes can be prepared from some tertiary amines and 90 or 30% solutions of hydrogen peroxide. 1,4-Diazabicyclooctane (Dabco) forms an adduct with two molecules of hydrogen peroxide. The adduct consists of hygroscopic crystals melting at 112 °C (dec), is stable for a limited time at room temperature, and can be used as a source of hydrogen peroxide in oxidations [*127*].

Whereas 90% hydrogen peroxide is difficult to obtain commercially, 30% hydrogen peroxide is commonly available under the trade names of Perhydrol or Superoxol. The 30% hydrogen peroxide does not mix with nonpolar organic compounds. To increase its solubility, ethanol, *tert*-butyl alcohol, trifluoroethanol [*128*], tetrahydrofuran, acetonitrile, formic acid, and, especially, acetic acid are used. When formic or acetic acid is used, the reacting species is the corresponding peroxy acid. Under such conditions, the products of oxidation by hydrogen peroxide resemble those obtained with peroxy acids. Different products may result from oxidation by hydrogen peroxide in alkaline medium, such as in pyridine, ammonia, and, most often, sodium or potassium hydroxide.

Transition metal catalysts not only increase the reaction rate but may also affect the outcome of the oxidation, especially the stereochemistry of the products. Whereas hydrogen peroxide alone in acetonitrile oxidizes alkenes to epoxides [*129*], osmic acid catalyzes *syn* hydroxylation [*130*], and tungstic acid catalyzes *anti* hydroxylation [*131*]. The most frequently used catalysts are titanium trichloride [*132*], vanadium pentoxide [*133, 134*], sodium vanadate [*135*], selenium dioxide [*125*], chromium trioxide [*134*], ammonium molybdate [*136*], tungsten trioxide [*131*], tungstic acid [*137*], sodium tungstate [*135, 138*], ferric chloride [*87*], palladium or palladium acetate [*139,140*], and osmium tetroxide (osmic acid) [*130, 141*].

The domain of oxidations with hydrogen peroxide includes the epoxidation [*129, 133, 142, 143, 144, 145, 146, 147, 148, 149, 150, 151*] and hydroxylation [*130, 131, 134, 137, 152*] of double bonds and the oxidation of primary amines to nitroso compounds [*153*], of secondary amines to hydroxylamines [*154*], of tertiary amines to amine oxides [*155, 156, 157, 158, 159, 160, 161, 162*], of sulfides to sulfoxides or sulfones [*132, 135, 163, 164, 165, 166*], of selenides to selenoxides [*167, 168, 169*], of aldehydes to acids [*125, 170*], of aromatic aldehydes to phenols [*171, 172, 173*], and of carboxylic acids to peroxy acids [*174, 175, 176*]. Hydrogen peroxide also oxidizes organic compounds to hydroperoxides or peroxides [*177, 178, 179, 180, 181, 182*], alkylboranes to alcohols [*183, 184, 185*], thiols to disulfides [*186*], and disulfides to sulfonic acids [*187*]. It is also used for the **Baeyer–Villiger reaction** [*128*], oxidation to quinones [*139*], and oxidative degradations [*87, 134, 188, 189*].

Sodium peroxide, Na_2O_2, a pale-yellow solid (mp 460 °C), is used very rarely. Some examples are the conversion of aldoximes into carboxylic acids [*190*] and the oxidation of ketones to esters (lactones) [*128, 191*].

Potassium superoxide, KO_2, when in a phase-transfer system with Aliquat 336 (methyltrioctylammonium chloride), oxidizes ketones containing α-hydrogens to acids at room temperature [*192*].

Sodium perborate, $NaBO_3 \cdot 4H_2O$ (mp 60 °C dec), is used for oxidations of primary aromatic amines to azo compounds [*193*] or nitro compounds [*194*] and of sulfides to sulfoxides and sulfones [*194*]. This reagent does not affect alcohols and only slightly affects alkenes [*194*].

Sodium persulfate ($Na_2S_2O_8$), potassium persulfate ($K_2S_2O_8$), and ammonium persulfate [$(NH_4)_2S_2O_8$], when applied in the presence of sulfuric acid, act as H_2SO_5 or $KHSO_5$ [*195*]. Alkaline persulfates are also used for the reoxidation of high-valency metal salts such as K_2RuO_4 [*196*]. A reaction peculiar to persulfates is the **Elbs persulfate oxidation** of phenols to *p*- or *o*-dihydroxy aromatic compounds [*197*]. Aliphatic primary amines are dehydrogenated to imines [*198*], and aromatic primary amines are oxidized to nitroso compounds [*199*].

Persulfuric acid (sulfomonoperacid, peroxymonosulfuric acid, or Caro acid), H_2SO_5, is prepared in situ either from hydrogen peroxide [*200*], potassium persulfate ($K_2S_2O_8$), or ammonium persulfate [$(NH_4)_2S_2O_8$] [*200*]. The reagent converts aldehydes into esters (in the presence of alcohols) [*200*]; ketones into lactones or esters [*201*]; primary aromatic amines into azoxy [*202*], nitroso [*195, 202*], or nitro [*195,203*] compounds; and iodo compounds into iodoxy compounds [*204*]. Explosions have occurred after the addition of concentrated peroxymonosulfuric acid to some secondary and tertiary alcohols [*see* reference 1157]. The use of persulfuric acid is very limited, because organic peroxy acids are now preferred.

Potassium persulfate, KHSO$_5$, when used in the mixture **2KHSO$_5$·KHSO$_4$·K$_2$SO$_4$** (**Oxone,** a registered trade mark of E. I. du Pont de Nemours and Company), is a very versatile although not too common oxidant [*205*]. Good yields are obtained in the oxidation of alcohols to ketones and esters [*205*], of amines to nitroso compounds [*205*], of mercaptans to sulfonic acids [*205*], of sulfides to sulfoxides [*206*] and sulfones [*205, 207, 208*], of phosphines to phosphine oxides [*205*], and of iodo compounds to iodoxy compounds [*205*]. Selective oxidations of sulfides to sulfones in the presence of double bonds and carbonyl groups can also be carried out [*207*]. Because the reagent is practically insoluble in organic solvents, a phase-transfer technique using tetrabutylammonium bromide is advantageous [*208, 209*].

Dioxiranes, prepared from acetone and other aliphatic ketones by treatment with Oxone, can accomplish oxidations that are usually not achieved by Oxone itself [*210, 211*]. Dioxiranes can be isolated by vacuum codistillation with the respective ketones [*210*], or else, they may be formed in situ and applied in the same reaction vessel [*210, 211*]. Examples of the applications of dioxiranes are epoxidations [*210*] and the oxidation of primary amines to nitro compounds [*211*], of tertiary amines to amine oxides [*210*], and of sulfides to sulfoxides [*210*] (equation 12).

[*210*]   R\R'C=O →(Oxone, NaHCO$_3$, H$_2$O)→ R\R'C(O-O)   12

tert*-Butyl hydroperoxide, (CH$_3$)$_3$COOH,** bp 33–34 °C at 17 mm of Hg [*212*], is prepared from *tert*-butyl hydrogen sulfate and 27% hydrogen peroxide [*212*] and is commercially available as a 70 or 90% solution containing water and *tert*-butyl alcohol. Anhydrous *tert*-butyl hydroperoxide is obtained from the 70% aqueous solution by azeotropic distillation with toluene [*213*]. Anhydrous, as well as highly concentrated, *tert*-butyl hydroperoxide must be ***handled with utmost care, because it may decompose violently in the presence of strong acids and some transition metals, especially manganese, iron, and cobalt [*213, 214*].

The bulk of oxidations with *tert*-butyl hydroperoxide consists of epoxidations of alkenes in the presence of transition metals [*147, 215, 216, 217, 218*]. In this way, α,β-unsaturated aldehydes [*219*] and ketones [*220*] are selectively oxidized to epoxides without the involvement of the carbonyl function. Other applications of *tert*-butyl hydroperoxide such as the oxidation of lactams to imides [*225*], of tertiary amines to amine oxides [*226, 227*], of phosphites to phosphates [*228*], and of sulfides to sulfoxides [*224*] are rare. In the presence of a chiral compound, enantioselective epoxidations of alcohols are successfully accomplished with moderate to high enantiomeric excesses [*221, 222, 223*].

Sharpless reagent is the most popular of the chiral oxidants. It is a mixture of tetraisopropoxytitanium, diethyl (R,R)- or (S,S)-tartrate, water, and *tert*-butyl hydroperoxide in the molar ratio 1:2:1:1 [*224*].

Benzoyl peroxide (dibenzoyl peroxide), $(C_6H_5COO)_2$ (mp 104–106 °C dec), and ***p*-nitrobenzoyl peroxide (p-$O_2NC_6H_4COO)_2$** (mp 156 °C dec), which is synthesized from *p*-nitrobenzoyl chloride and sodium peroxide [*229*], are rarely used as oxidants, and if so, they do not offer appreciable advantages over other organic oxidation agents. The *anti* addition of benzoyl groups to double bonds and the benzoxylation of aromatic rings are achieved in the presence of iodine [*230*], and alcohols are oxidized to carbonyl compounds in the presence of nickel dibromide [*231*].

Diisopropyl peroxydicarbonate, $[(CH_3)_2CHOCOO]_2$, is a rather exotic reagent with limited availability and use. The treatment of toluene with diisopropyl peroxydicarbonate yields a mixture of all three tolyl isopropyl carbonates [*232*].

***tert*-Butyl peroxyacetate, $(CH_3)_3COOCOCH_3$** [*233*], and, better still, ***tert*-butyl peroxybenzoate, $(CH_3)_3COOCOC_6H_5$** (bp 75–76 °C at 0.2 mm of Hg) [*233, 234*], in the presence of cuprous salts, are used for the acetoxylation and benzoxylation of alkenes in the α positions with respect to the double bonds. Treatment of a Grignard compound with *tert*-butyl peroxybenzoate gives a *tert*-butyl ether, which decomposes to a hydroxy compound on heating [*235*].

Bis(trimethylsilyl) peroxide, $(CH_3)_3SiOOSi(CH_3)_3$, is prepared from trimethylsilyl chloride, 1,4-diaza[2,2,2]bicyclooctane, and Dabco's complex with 2 mol of hydrogen peroxide [*127*]. It is used alone [*228*] or in the presence of catalysts such as pyridinium dichromate [*236*]; trimethylsilyl trifluoromethanesulfonate, $CF_3SO_3Si(CH_3)_3$ [*228, 237*]; or tris(triphenylphosphine)ruthenium dichloride, $[(C_6H_5)_3P]_3RuCl_2$ [*236*]. This reagent oxidizes primary alcohols to aldehydes (in preference to the oxidation of secondary alcohols to ketones [*236*]), ketones to esters or lactones (***Baeyer–Villiger reaction***) [*238*], and nucleoside phosphites to phosphates [*228*]. All these oxidations require anhydrous conditions.

ORGANIC PEROXY ACIDS

Peroxyformic acid (performic acid), HCO_3H, is always prepared in situ from 25–30% hydrogen peroxide and 88 or 98–100% formic acid. At 26.5 °C, the conversion of formic acid into peroxyformic acid requires approximately 1 h (as proven by analysis of peroxyformic acid and hydrogen peroxide contents [*239*]). The preparation of peroxyformic acid is carried out in the presence of the compounds to be oxidized, most often alkenes and double-bond-containing carboxylic acids. After an induction period of about 5 min, the reaction becomes very exothermic (and may even become explosive) and requires cooling, which is best accomplished with ice and water. The temperature is maintained at 40 °C, and after a few hours, the reaction of peroxyformic acid is finished. In the case of alkenes or other unsaturated compounds, the reaction products are formyl esters of diols. These products are not isolated but are heated with sodium or potassium hydroxide to give vicinal diols with *anti* stereochemistry [*101, 240, 241, 242, 243, 244*] (equation 13).

$$HCO_2H + H_2O_2 \longrightarrow HCO_3H \qquad 13$$

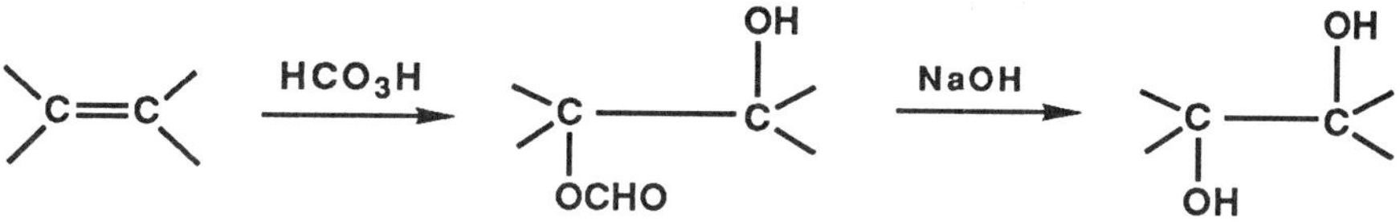

$$\text{>C=C<} \xrightarrow{HCO_3H} \text{>C(OCHO)-C<(OH)} \xrightarrow{NaOH} \text{>C(OH)-C<(OH)}$$

Sometimes the formyl ester of the diol is directly converted into a ketone [*245, 246*] (equation 14).

[*246*] 14

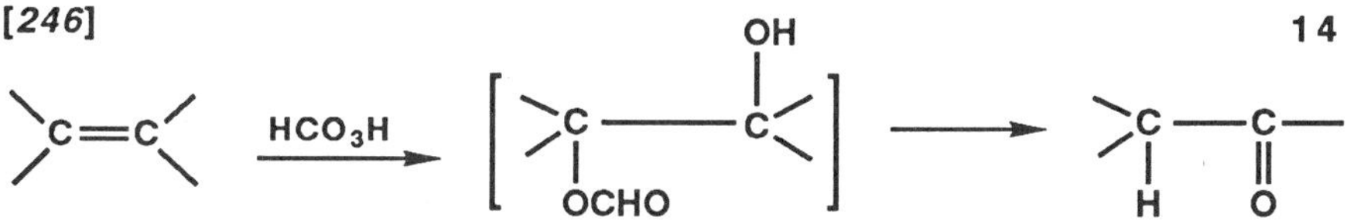

$$\text{>C=C<} \xrightarrow{HCO_3H} \left[\text{>C(OCHO)-C<(OH)} \right] \longrightarrow \text{>C(H)-C(=O)-}$$

A much rarer application of performic acid is the transformation of 2- or 4-dialkylaminoperhalopyridines into either amine oxides or *N,N*-dialkylhydroxylamines [*247, 248*] (equation 15).

[*247, 248*] 15

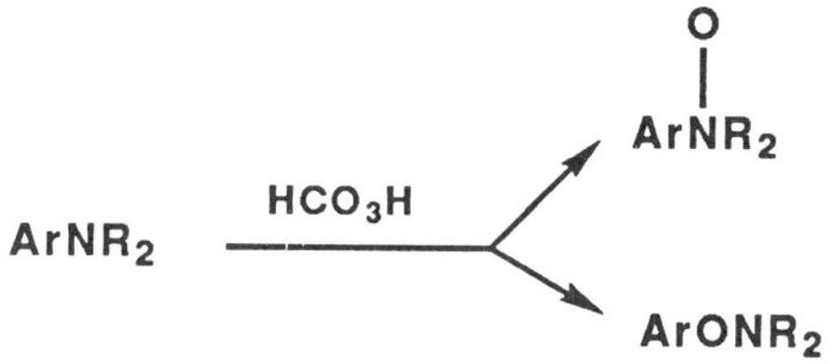

$$ArNR_2 \xrightarrow{HCO_3H} \begin{array}{c} ArN(\rightarrow O)R_2 \\ \text{or} \\ ArONR_2 \end{array}$$

Peroxyacetic acid (peracetic acid), CH_3CO_3H, can be formed in situ from 30 or 90% hydrogen peroxide, usually in the presence of catalytic amounts of sulfuric or perchloric acid and sometimes also in the presence of acetic anhydride. The reaction of hydrogen peroxide with acetic acid to form peroxyacetic acid is reversible. The presence of substrate to be oxidized drives the oxidation to completion at moderate temperatures (25–70 °C). External cooling is frequently necessary, because the reaction is strongly exothermic.

Peroxyacetic acid is commercially available as 40–45% solutions in acetic acid. Such a reagent is prepared by adding 1 mol of 90% hydrogen peroxide to a cooled mixture of 1.5 mol of glacial acetic acid and 1% of sulfuric acid (based on the total weight). After 4 h, the reaction mixture contains 45% of peroxyacetic acid and 6% of hydrogen peroxide [*249*].

Reactions with such peroxyacetic acid solutions are carried out without solvents or in solutions in acetic acid, dichloromethane [*250, 251, 252*], chloroform [*148, 253, 254*], ethyl acetate [*225, 255*], or the dimethyl ether of ethylene glycol (dimethyl cellosolve) [*256*].

The most important applications of peroxyacetic acid are the epoxidation [*250, 251, 252, 254, 257, 258*] and *anti* hydroxylation of double bonds [*241, 252*]; the **Dakin reaction** of aldehydes [*259*]; the **Baeyer–Villiger reaction** of ketones [*148, 254, 258, 260, 261, 262*]; the oxidation of primary amines to nitroso [*153*] or nitrocompounds [*253*], of tertiary amines to amine oxides [*158, 263*], of sulfides to sulfoxides and sulfones [*264, 265*], and of iodo compounds to iodoso or iodoxy compounds [*266, 267*]; the degradation of alkynes [*268*] and diketones [*269, 270, 271*] to carboxylic acids; and the oxidative opening of aromatic rings to aromatic dicarboxylic acids [*256, 272, 271, 272, 273, 274*]. Occasionally, peroxyacetic acid is used for the dehydrogenation [*275*] and oxidation of aromatic compounds to quinones [*249*], of alcohols to ketones [*276*], of aldehyde acetals to carboxylic acids [*277*], and of lactams to imides [*225, 255*]. The last two reactions are carried out in the presence of manganese salts. The oxidation of alcohols to ketones is catalyzed by chromium trioxide, and the role of peroxyacetic acid is to reoxidize the trivalent chromium [*276*].

Peroxytrifluoroacetic acid (trifluoroperacetic acid), CF_3CO_3H, is prepared from 90% hydrogen peroxide, trifluoroacetic acid, and, usually, trifluoroacetic anhydride. Trifluoroacetic acid is a very strong acid so that no catalyst is needed for its conversion into peroxytrifluoroacetic acid. A peroxy acid of lower concentration can also be obtained by using 30% hydrogen peroxide [*278, 279*]. Pure peroxytrifluoroacetic acid results from the addition of the calculated amount of trifluoroacetic anhydride to 90% hydrogen peroxide without solvent [*280*] or in dichloromethane [*281*]. Cooling of the reaction mixture is necessary. The amounts of peroxy acid and unreacted hydrogen peroxide are determined by iodometric and cerimetric titrations, respectively [*239*].

Peroxytrifluoroacetic acid is usually applied as its solution in trifluoroacetic acid. Other solvents are 1,2-dichloroethane [*280, 282*] and, especially, dichloromethane. To neutralize trifluoroacetic acid, sodium carbonate [*253, 283*] or disodium hydrogen phosphate [*284*] is used.

The scope of peroxytrifluoroacetic acid oxidations resembles that of peroxyacetic acid and other peracids, although sometimes the outcome may be different [*285*]. The advantage of peroxytrifluoroacetic acid is that it reacts faster than the other peroxy acids: peroxyacetic acid, peroxybenzoic acid, *m*-chloroperoxybenzoic acid [*286*], and peroxyformic acid [*285*].

Peroxytrifluoroacetic acid is used for the epoxidation [*283, 287*] and *anti* hydroxylation [*285, 288*] of alkenes; the **Baeyer–Villiger oxidation** of ketones [*280, 282, 284, 289*]; and especially the oxidation of primary amines [*253, 278, 281, 285, 290*] and nitroso compounds [*278, 291*] to nitro compounds, of tertiary amines to amine oxides [*158, 248*], and of sulfides to sulfoxides and sulfones [*279*]. In the presence of cobalt salts, benzene is converted into phenyl trifluoroacetate [*292*].

Peroxybenzoic acid (perbenzoic acid), $C_6H_5CO_3H$ (mp 41–43 °C), is commercially available. It is prepared from dibenzoyl peroxide according to equation 16 [*293*] and is obtained as a solution in chloroform. The peroxyacid content of the solution can be determined iodometrically [*293*].

[*293*]   16
$(C_6H_5COO)_2 \xrightarrow{CH_3ONa} C_6H_5CO_2CH_3 + C_6H_5CO_3Na \xrightarrow{H_2SO_4} C_6H_5CO_3H$

Solid peroxybenzoic acid is obtained by evaporation of the chloroform under reduced pressure. Better procedures for the preparation of peroxybenzoic acid are the reaction of benzoyl chloride with sodium peroxide [*294*] and the treatment of benzoic acid with 70% hydrogen peroxide in the presence of methanesulfonic acid [*176*]. Peroxybenzoic acid can also be prepared in situ from benzaldehyde and oxygen [*295*] or ozone [*112*].

Oxidations with peroxybenzoic acid are carried out in solutions in dichloromethane, chloroform, benzene, ether, or ethyl acetate at or below room temperature and include epoxidation of double bonds [*295, 296, 297, 298, 299, 300, 301*], oxidation of benzaldehydes to carboxylic acids or phenols [*302*], the *Baeyer–Villiger reaction* of ketones [*303, 304, 305, 306, 307*], and oxidation of sulfides to sulfoxides [*308, 309*]. Peroxybenzoic acid is also used for the *anti* hydroxylation of double bonds [*310*], the oxidation of pyrrolidines to pyrrolidones [*311*] and of pyrroles to succinimides [*311*], and the preparation of azoxy compounds from azo compounds [*312*].

***m*-Chloroperoxybenzoic acid (*m*-chloroperbenzoic acid), *m*-$ClC_6H_4CO_3H$**, is commercially available in 80–85% purity (mp 92–94 °C dec). It is prepared from *m*-chlorobenzoyl chloride and sodium peroxide [*294*] or sodium hydrogen peroxide [*313*].

Oxidations with *m*-chloroperoxybenzoic acid are carried out in solutions in hexane, dichloromethane, chloroform, methanol, or tetrahydrofuran at temperatures ranging from −78 to 40 °C. The applications of *m*-chloroperoxybenzoic acid are epoxidation [*287, 314, 315, 316*]; the *Baeyer–Villiger reaction* [*286, 315, 317, 318*]; and the oxidation of primary amines to nitro compounds [*319*], of tertiary amines to amine oxides [*320*], of sulfides to sulfoxides [*321, 322, 323, 324*], and of selenides to selenones [*325*]. Secondary alcohols are oxidized to ketones in the presence of hydrogen chloride [*326*], and acetals are oxidized to esters with boron trifluoride etherate as a catalyst [*327*]. The addition of potassium fluoride to reaction mixtures facilitates product isolation, because both *m*-chlorobenzoic acid and the unreacted *m*-chloroperoxybenzoic acid are precipitated as solids from the solutions in dichloromethane [*315*]. The results of oxidations by *m*-chloroperoxybenzoic acid resemble those of other peroxyacids, although some differences are found in the regiochemistry [*286*] and the stereochemistry of the products [*287*].

p-Nitroperoxybenzoic acid, p-O$_2$NC$_6$H$_4$CO$_3$H (mp 155–156 °C), is prepared from *p*-nitrobenzaldehyde and ozone [*112*] or from *p*-nitrobenzoyl chloride and sodium peroxide [*229*]. In addition to its use for the stereoselective epoxidations of allylic alcohols [*328*], *p*-nitroperoxybenzoic acid is also used for the highly regio- and stereoselective hydroxylation of hydrocarbons at tertiary carbons [*329*].

Peroxyphthalic acid, o-C$_6$H$_4$(CO$_2$H)CO$_3$H, is obtained from phthalic anhydride and hydrogen peroxide [*330*] or sodium hydrogen peroxide [*331*] and is applied usually in ethereal solutions to epoxidations [*330, 332*], the *Baeyer–Villiger reaction* [*333, 334*], and oxidations of sulfides to sulfoxides [*163, 335*] and sulfones [*336, 337*]. The oxidation of sulfides to sulfones takes precedence over epoxidation, as evidenced by the fact that unsaturated ketone thioacetals are oxidized to unsaturated disulfones [*337*].

Peroxymaleic acid, cis-HO$_2$CCH=CHCO$_3$H, is prepared from 90% hydrogen peroxide and maleic anhydride [*338*], and **peroxydichloromaleic acid, cis-HO$_2$CCCl=CClCO$_3$H,** from 90% hydrogen peroxide and dichloromaleic anhydride [*339*]. Both oxidizing agents are used in solutions in dichloromethane. Peroxymaleic acid oxidizes alkenes to epoxides [*338*], ketones to esters [*338*], primary aromatic amines to nitro compounds [*338*], and sulfoxides to sulfones [*338*]. Peroxydichloromaleic acid oxidizes 3-chloropyridazine to its 1-oxide [*339*].

Peroxypropionic acid (C$_2$H$_5$CO$_3$H) [*340*], **peroxylauric acid (C$_{11}$H$_{23}$CO$_3$H)** [*174*], and other aliphatic peroxy acids [*341*] are used rarely, and the results do not differ appreciably from those obtained from more common peracids. The same is true of **peroxypentafluorobenzoic acid, C$_6$F$_5$CO$_3$H,** which is obtained from pentafluorobenzaldehyde and ozone and is used for epoxidations; the Baeyer–Villiger reaction; and the preparation of amine oxides, sulfoxides, and sulfones [*342*].

Benzeneperoxyseleninic acid, C$_6$H$_5$Se(O)OOH, is prepared in situ from benzeneseleninic acid and 30% hydrogen peroxide in tetrahydrofuran or methylene chloride. This reagent oxidizes cyclic ketones to lactones, sometimes with better results than those with other oxidants [*343*].

DERIVATIVES OF GROUP 1 ELEMENTS

Elemental copper is used as a dehydrogenation catalyst for the conversion of alcohols into aldehydes or ketones [*344*]. Its efficiency is enhanced when it is applied on large-surface-area supports and in the presence of silver [*345*].

Copper bronze (finely ground copper, which may lose its activity on aging) is a hydrogenation–dehydrogenation catalyst in the ***Guerbet reaction***, in which a primary aliphatic alcohol, in the presence of sodium and copper, is dehydrogenated to the corresponding aldehyde. The aldehyde undergoes aldol condensation and subsequent dehydration, and the resulting α,β-unsaturated aldehyde is reduced by the hydrogen formed to a saturated alcohol [*346*]. The reaction requires temperatures around 300 °C and autoclaves that will stand the autogenous pressure (50–60 atm). A classical example is the preparation of 2-ethylhexanol from butanol [*346*].

Cuprous chloride, CuCl (mp 430 °C), when complexed with amines such as pyridine or phenanthroline, catalyzes the oxidation of alcohols to aldehydes and ketones by air [*347*]. In the presence of palladium dichloride, $PdCl_2$, in aqueous dimethylformamide, terminal alkenes are converted by oxygen into methyl ketones [*348*].

Copper oxide, CuO, is an oxidant for the conversion of alcohols into aldehydes or ketones [*349*] and for the transformation of hydrazo compounds into azo compounds [*350*].

Copper sulfate, $CuSO_4 \cdot 5H_2O$, is used for the oxidative coupling of terminal acetylenes [*58*]; for the conversion of α-hydroxy ketones (acyloins) into α-diketones [*351, 352*]; and, in cooperation with potassium peroxydisulfate, for the selective oxidation of methyl groups on benzene rings to aldehyde groups [*353*].

Copper chromite, $CuCr_2O_4$, and mixtures of cupric oxide with chromium sesquioxide and special additives (the ***Adkins*** catalyst), dehydrogenate primary alcohols to aldehydes [*354, 355*] and secondary alcohols to ketones [*354, 355, 356*].

Copper acetate, $Cu(OCOCH_3)_2$ or **$Cu(OCOCH_3)_2 \cdot H_2O$**, resembles copper sulfate in its oxidizing properties and is used for the oxidative coupling of terminal acetylenes [*58, 357*] and for the conversion of acyloins into α-diketones [*358, 359*]. Its presence favorably affects the acetoxylation of toluenes to benzyl acetates by sodium persulfate [*360*].

Copper carbonate, basic, $CuCO_3 \cdot Cu(OH)_2$, causes the hydroxylation of benzoic acids exclusively in the *ortho* position to produce salicylic acids [*361*].

Silver, Ag, (as well as **silver oxide, Ag_2O**, and **silver nitrate, $AgNO_3$**), is used as a catalyst in the oxidation of aromatic aldehydes to carboxylic acids in the presence of alkalies (***Cannizzaro reaction***) [*362, 363, 364*].

Silver oxide (argentous oxide), Ag_2O, is a brown-black powder. It is available commercially or can be prepared by treatment of a solution of silver nitrate with a solution of potassium hydroxide. The resulting precipitate is decanted and washed with distilled water until no nitrate ions are detectable in the supernatant liquid. Ultimately, the precipitate, hydrated silver oxide, is filtered with suction; washed with water, alcohol, and ether; and dried in the air. If the oxidations are carried out in an aqueous solution, silver oxide (hydroxide) is prepared in situ from silver nitrate and sodium or potassium hydroxide.

The domain of oxidations with silver oxide includes the conversion of aldehydes into acids [63, 206, 362, 365, 366, 367] and of hydroxy aromatic compounds into quinones [171, 368, 369]. Less frequently, silver oxide is used for the oxidation of aldehyde and ketone hydrazones to diazo compounds [370, 371], of hydrazo compounds to azo compounds [372], and of hydroxylamines to nitroso compounds [373] or nitroxyls [374] and for the dehydrogenation of CH–NH bonds to –C=N– [375]. Similar results with **silver carbonate** are obtained in oxidations of alcohols to ketones [376] or acids [377] and of hydroxylamines to nitroso compounds [378].

Argentic oxide, AgO, which is prepared by the electrolysis of silver nitrate in nitric acid [379], converts aromatic methyl homologues into aldehydes [380], primary alcohols into aldehydes [380] or acids [381], aldehydes into acids [382], primary amines into aldehydes and nitriles [381], and phosphines into phosphine oxides [381].

DERIVATIVES OF GROUP 2 ELEMENTS

Mercuric bromide, $HgBr_2$ (mp 236 °C), oxidizes methylene groups that are adjacent to double bonds or aromatic rings to carbonyls [11].

Mercuric oxide, HgO (yellow modification or the less reactive red modification), resembles silver oxide in its oxidizing properties. This reagent transforms phenols and hydroquinones into quinones [383, 384] and is used especially for the conversion of hydrazones into diazo compounds [385, 386, 387, 388, 389, 390, 391, 392]. Dihydrazones of α-diketones furnish acetylenes [393, 394, 395, 396]. N-Aminopiperidines are dehydrogenated to tetrazenes [397] or converted into hydrocarbons [398].

Mercuric acetate, $Hg(OCOCH_3)_2$ (mp 179–182 °C), is used for dehydrogenations, resulting in the introduction of double bonds into the α positions with respect to conjugated systems [399, 400], and for the dehydrogenation of amines to imines [401, 402]. The reagent can also demethylate tertiary amines to secondary amines [403] and introduce acetoxyl groups into the α positions with respect to double bonds [404].

Mercurous trifluoroacetate, $HgOCOCF_3$, converts ketone monohydrazones into acetylenes [405].

Mercuric trifluoroacetate, $Hg(OCOCF_3)_2$, parallels the actions of mercuric oxide. It oxidizes hydroquinones to quinones [383] and hydrazones to diazo compounds [406].

DERIVATIVES OF GROUP 3 ELEMENTS

Aluminum chloride (anhydrous), AlCl$_3$, is used successfully for the dehydrogenative cyclization of aromatic polynuclear compounds [407, 408].

Thallium trinitrate, Tl(NO$_3$)$_3$·3H$_2$O (mp 102–105 °C), oxidizes phenols and dihydroxy aromatic compounds to quinones [409]; acetylenes to α-hydroxy ketones, α-diketones, or carboxylic acids [413]; and methyl ketones to α-keto acids [414].

Thallium monoacetate, TlOCOCH$_3$ (hygroscopic), behaves similarly to silver acetate or silver benzoate. In the presence of iodine, it accomplishes stereoselective hydroxylations of alkenes [410].

Thallium triacetate, Tl(OCOCH$_3$)$_3$·1.5H$_2$O (mp 182 °C), like the monoacetate, is used for the stereoselective acetoxylation of alkenes [411] and for oxidations of alkenes to epoxides [412].

All thallium compounds are highly toxic and must be handled with utmost care.

DERIVATIVES OF GROUP 4 ELEMENTS

Ammonium cerium nitrate (ceric ammonium nitrate), (NH$_4$)$_2$Ce(NO$_3$)$_6$, a newcomer to the arena of oxidants, is useful for the acetoxylation of aromatic side chains in benzylic positions [415, 416] and for the oxidation of methylene or methyl groups that are adjacent to aromatic rings to carbonyl groups [238, 415, 417]. The reagent also oxidizes alcohols to aldehydes [418, 419, 420, 421] and phenols to quinones [422, 423], cleaves vicinal diols to ketones and α-hydroxy ketones to acids [424, 425], and converts diaryl sulfides into sulfoxides [426]. A specialty of ammonium cerium nitrate is the oxidative recovery of carbonyl compounds from their oximes and semicarbazones [422, 427] and of carboxylic acids from their hydrazides [428] under mild conditions.

Ammonium cerium sulfate (ceric ammonium sulfate), (NH$_4$)$_2$Ce(SO$_4$)$_4$ ·2(NH$_4$)$_2$SO$_4$·2H$_2$O, converts aromatic polynuclear hydrocarbons into quinones [429].

Lead dioxide (lead superoxide), PbO$_2$, and red lead, Pb$_3$O$_4$, are used only to a limited extent. PbO$_2$ is used mainly for the oxidation of phenols to quinones [368, 430, 431] and of aromatic methyl groups to carboxyls [432]. Red lead, Pb$_3$O$_4$, in acetic acid oxidizes α-hydroxy acids to carbonyl compounds [143, 433] and cleaves vicinal diols to carbonyl compounds [433]. In this respect, red lead acts as lead tetraacetate in situ [434]. Lead oxide, Pb$_3$O$_4$, is also used in the preparation of lead tetraacetate and tetrakis(trifluoroacetate) [435].

Lead tetraacetate, $Pb(OCOCH_3)_4$, a white hygroscopic crystalline compound, is easily hydrolyzed with water to lead dioxide. Therefore, oxidations have to be carried out in strictly anhydrous solvents such as benzene, dioxane, and glacial acetic acid, usually at room temperature or at slightly elevated temperatures. In reactions of lead tetraacetate, acetoxy radicals are the reactive species. The addition of acetoxy radicals across double bonds produces vicinal diacetates [*436*], and their attack at the benzylic positions and at the α positions with respect to carbonyl groups produces benzylic acetates [*434, 437*] and α-acetoxy ketones [*438, 439*], respectively. Lead tetraacetate also causes one-electron oxidations resulting in the formation of quinones from phenols [*369, 440*] and from aromatic amino compounds with *p*-amino groups [*441*]. Primary alcohols are dehydrogenated to aldehydes [*442*], and primary amines are converted into nitriles [*443, 444*]. Ketone hydrazones are transformed into diazo compounds [*445*]. A very important application of lead tetraacetate is the cleavage of the carbon chains in vicinal diols in nonaqueous media [*446, 447, 448*].

Lead tetrakis(trifluoroacetate), $Pb(OCOCF_3)_4$, which is prepared from red lead oxide (Pb_3O_4), trifluoroacetic acid, and trifluoroacetic anhydride [*435*], has very limited applications. It converts alkylbenzenes into benzylic trifluoroacetates [*435*] and benzene and its derivatives into phenyl trifluoroacetates [*435, 449*].

DERIVATIVES OF GROUP 5 ELEMENTS

Nitrous anhydride (nitrogen sesquioxide), N_2O_3, is obtained by the reduction of nitric acid by heating with starch or arsenic oxide, As_2O_3 [*450*]. This reagent is used for the oxidation of a methylene group flanked by two carbonyl or carboxyl groups to a carbonyl group [*450*].

Nitrous acid, HNO_2, is stable only at low temperatures. It is prepared in situ from sodium or potassium nitrite and dilute sulfuric or hydrochloric acid. This reagent is used for the oxidation of *p*-aminophenols to quinones [*451*].

Alkyl nitrites, RONO, such as ethyl nitrite, C_2H_5ONO; butyl nitrite, C_4H_9ONO; and amyl (pentyl) nitrite, $C_5H_{11}ONO$, are commercially available. These reagents are obtained from alkaline nitrites, alcohols, and sulfuric acid and act like nitrous anhydride. Methylene groups that are sufficiently activated by two adjacent aromatic rings or two carbonyl or carboxyl groups are converted into carbonyl groups via nitroso and isonitroso compounds (oximes) [*452*].

Dinitrogen tetroxide, N_2O_4, a light-brown gas or liquid (bp 21.5–22.5 °C), is commercially available and is prepared from nitrososulfuric acid and potassium nitrate [*453*] or, more simply, by the thermal decomposition of lead nitrate [*454*]. Dinitrogen tetroxide is utilized for the oxidation of perfluoroalkyl iodides to perfluorocarboxylic acids [*455*] and of benzylic alcohols to ketones [*456*]. A mixture of predominantly dinitrogen tetroxide and nitrous anhydride, N_2O_3, produced by heating fuming nitric acid, concentrated sulfuric acid, and arsenous oxide, is applied to a clean-cut transformation of hydroquinones into quinones [*457*].

Nitric acid, HNO_3, is available as concentrated (68%) nitric acid (*d* 1.41, azeotropic mixture with water, bp 120.5 °C), fuming (100%) nitric acid (*d* 1.52), and, rarely, red fuming nitric acid, containing nitrogen oxides. For many oxidations, dilute nitric acid (approximately 35%) obtained by dilution with water in a 1:1 ratio is employed. Most of the oxidations are carried out at atmospheric pressure in glass equipment or in porcelain dishes. If higher temperatures are needed, either sealed glass tubes or stainless steel or glass-lined autoclaves must be used.

Nitric acid is a very strong but not very selective oxidant. The advantages of its use are simple and usually clean-cut isolation of the products. The obnoxious fumes generated during the oxidations are a disadvantage. The main applications of nitric acid are the dehydrogenation (aromatization) of dihydropyridines [*458*]; the degradation of aromatic rings to carboxylic acids [*459*]; the oxidation of aromatic side chains to carboxyls [*460, 461, 462, 463, 464*]; the oxidation of alcohols [*465, 466, 467, 468, 469*], aldehydes [*467,470*], ketones [*471*], and esters [*472*] to carboxylic acids; and the oxidation of aromatic amines to quinones [*473*], thiols to sulfonic acids [*474*], and iodo compounds to iodoso compounds [*475*]. In combination with silver oxide, nitric acid is used for the oxidative cleavage of ozonides to carboxylic acids [*77*].

Ammonium nitrate, NH_4NO_3, oxidizes acyloins to α-diketones in the presence of catalytic amounts of cupric acetate [*476*].

Lead nitrate, $Pb(NO_3)_2$ (mp 470 °C dec), converts benzylic halides into the corresponding carbonyl compounds [*477*].

Vanadium pentoxide, V_2O_5 (mp 690 °C), is one of the strongest catalysts used for dehydrogenations [*478*] and oxidations with air or oxygen [*3, 479*], especially in the gas phase and at very high temperatures. It also catalyzes the hydroxylation of alkenes with aqueous hydrogen peroxide [*134*].

Vanadyl acetylacetonate, $VO(acac)_2$, catalyzes stereoselective epoxidations of unsaturated alcohols [*216, 328*].

Arsenic pentoxide, As_2O_5, functions as an oxidant in the *Skraup synthesis* of quinolines [*480*].

Bismuth sesquioxide, Bi_2O_3, in the presence of acetic acid, oxidizes acyloins to α-diketones in high yields [*481*].

Sodium bismuthate, $NaBiO_3$, cleaves vicinal diols to dicarbonyl compounds [*482, 483*] but offers no advantage over the more common oxidants of vicinal diols, such as lead tetraacetate and periodic acid.

Potassium nitrosodisulfonate (Fremy salt), ·ON(SO₃K)₂, is a free radical capable of converting phenols [487, 488, 489] and aromatic amines [490] into quinones.

DERIVATIVES OF GROUP 6 ELEMENTS

Potassium hydroxide, KOH, is applied to the conversion of fatty alcohols into salts of fatty acids (having the same number of carbon atoms) [484] and to the hydroxylation of aromatic compounds of low electron density, such as nitrobenzene [485] and pyridine [486].

Sulfur, S_x, is a dehydrogenating agent that converts compounds containing six-membered rings into their aromatic analogues. In the reaction, sulfur is reduced to hydrogen sulfide [491, 492]. Such dehydrogenations occur at lower temperatures compared with those by selenium [493] or noble metals [494, 495, 496, 497].

The most useful application of sulfur is the **Willgerodt reaction**—the conversion of ketones into acids having the same number of carbon atoms [498]. The conversion is accomplished by heating ketones with sulfur and ammonium sulfide [499, 500, 501, 502] at 160–210 °C under pressure. The **Kindler modification** avoids operation under pressure by using sulfur and high-boiling secondary amines such as morpholine, instead of ammonium sulfide [503, 504]. Such a reaction carried out at reflux gives the thiomorpholides, which are subsequently hydrolyzed to acids.

A peculiar oxido-reductive reaction takes place when *p*-aminotoluene is heated with sodium polysulfide; the product is *p*-aminobenzaldehyde [505].

Sulfuric acid, H_2SO_4, rarely acts as an oxidant. It can be used for the dehydrogenative coupling of aromatic rings [395]. Equally rare is the oxidation of a sulfide to a sulfoxide with **sulfuryl chloride, SO_2Cl_2** [506].

Selenium, Se (its red modification), has been successfully employed for dehydrogenations of six-membered rings to aromatic systems. Compared with sulfur, selenium requires higher temperatures. Consequently, side reactions such as rearrangements occur [493].

Selenium dioxide, SeO_2 (mp 315 °C, sublimes), and **selenious acid, H_2SeO_3,** which is obtained by the evaporation of an aqueous solution of SeO_2 [507, 508], are very selective oxidants. They are capable of mild dehydrogenation to form double bonds [375] and can oxidize alkenes and acetylenes to vicinal dicarbonyl compounds [509, 510] and allylic ethers to aldehydes [511]. The most important applications are conversions of alkenes into allylic alcohols [512]; of benzylic, methyl, or methylene groups into carbonyl groups [513, 514, 515]; and of carbonyl compounds into α-dicarbonyl compounds [*507, 508, 516, 517, 518, 519, 520*]. Other oxidations, such as the conversion of sulfides into sulfoxides, are exceptions [*521*].

Selenium dioxide oxidations may be accomplished by heating with neat substrates [*519*], but more often, they are carried out in solvents such as water, *tert*-butyl alcohol, ethanol, dioxane, acetic acid, and acetic anhydride. The solvent used often affects the outcome of the reaction. The use of acetic acid or acetic anhydride favors oxidation to acetates (and thence to alcohols), whereas in water or dioxane, carbonyl compounds are usually formed. An unpleasant feature of oxidations with selenium dioxide is the formation of red colloidal selenium, which cannot always be separated by distillation but can be removed by treatment of the product with potassium cyanide [*522*], mercury [*523*], or deactivated Raney nickel [*524*].

Benzeneseleninic anhydride, $C_6H_5Se(O)O(O)SeC_6H_5$, which is prepared in situ from diphenyldiselenide and *tert*-butyl hydroperoxide, is used for the oxidation of alcohols to aldehydes or ketones [*525*]. This reagent is a suitable dehydrogenating agent for the introduction of double bonds α to carbonyl groups [*526*] and the regeneration of ketones from their oximes, semicarbazones, and phenylhydrazones [*527*].

Molybdenum hexafluoride, MoF_6 (mp 17 °C, bp 35 °C) [*528*], and **molybdenum oxychloride, $MoOCl_3$,** which is prepared by the partial hydrolysis of molybdenum pentachloride [*528*], are suitable for the recovery of ketones from their dimethylhydrazones and tosylhydrazones under very mild conditions [*528*].

Molybdenum hexacarbonyl, $Mo(CO)_6$ [*328*], and **molybdenum acetylacetonate, $MoO_2(acac)_2$** [*530*], catalyze the stereoselective epoxidation of allylic alcohols with *tert*-butyl hydroperoxide.

Oxodiperoxymolybdenum–pyridine–hexamethylphosphoric triamide, $MoO_5 \cdot C_5H_5N \cdot HMPA$ (MoOPH), which is prepared from molybdenum trioxide, MoO_3 [*531,532*], hydroxylates the enolates of ketones and esters in the α position with respect to the carbonyl groups [*531, 532, 533*] and, in the presence of mercuric acetate, converts acetylenes into α-dicarbonyl compounds [*534*].

Tungsten hexafluoride, WF_6 (mp 2.5 °C, bp 17 °C) [*529*], resembles MoF_6 and $MoOCl_3$ in its oxidative properties. However, less exotic reagents such as ozone, ammonium cerium sulfate, or zinc dichromate may be used for a similar purpose.

Chromium trioxide [chromium(VI) oxide, chromic acid, or chromic anhydride], CrO_3, (dark red crystals, mp 195 °C), is one of the most powerful and universal oxidants. It is applied in solutions in acetic acid, dilute sulfuric acid, a mixture of acetic acid and dilute sulfuric acid, dilute sulfuric acid and acetone (**Jones reagent**), acetic anhydride and acetic acid (**Fieser reagent**) [*535*], water, water and ether [*536, 537, 538*], dichloromethane [*539*], benzene [*540*], pyridine [*541*], dimethylformamide (DMF) [*542*], and hexamethylphosphoric triamide (HMPA) [*543*]. It provides 1.5 equivalents of oxygen per mole.

Oxidations with chromic oxide encompass hydroxylation of methylene [*544*] and methine [*544, 545, 546*] groups; conversion of methyl groups into formyl groups [*539, 547, 548, 549*] or carboxylic groups [*550, 551*] and of methylene groups into carbonyls [*275, 552, 553, 554, 555*]; oxidation of aromatic hydrocarbons [*556, 557, 558*] and phenols [*559*] to quinones, of primary halides to aldehydes [*540*], and of secondary halides to ketones [*560, 561*]; epoxidation of alkenes [*562, 563, 564*]; and oxidation of alkenes to ketones [*565, 566*] and to carboxylic acids [*567, 568, 569*].

The main applications of oxidation with chromium trioxide are transformations of primary alcohols into aldehydes [*184, 537, 538, 543, 570, 571, 572, 573*] or, rarely, into carboxylic acids [*184, 574*], and of secondary alcohols into ketones [*406, 536, 542, 543, 575, 576, 577, 578, 579, 580, 581, 582, 583, 584*]. **Jones reagent** is especially successful for such oxidations. It is prepared by diluting with water a solution of 267 g of chromium trioxide in a mixture of 230 mL of concentrated sulfuric acid and 400 mL of water to 1 L to form an 8 N CrO_3 solution [*565, 572, 579, 581, 585, 586*]. Other oxidations with chromic oxide include the cleavage of carbon–carbon bonds to give carbonyl compounds or carboxylic acids [*482, 566, 567, 569, 580, 587, 588*], the conversion of sulfides into sulfoxides [*541*] and sulfones [*589*], and the transformation of alkyl silyl ethers into ketones or carboxylic acids [*590*].

Chromium trioxide–pyridine complex or dipyridine chromium(VI) oxide (Collins reagent), $CrO_3·2C_5H_5N$, is commercially available. It forms red crystals and is prepared by adding chromic oxide to an excess of anhydrous pyridine cooled below 20 °C. Efficient stirring is essential to prevent local overheating, which might result in flashes and in flaming. (This overheating would occur if pyridine were added to chromic oxide [*591, 592, 593, 594*]). The complex is practically insoluble in carbon tetrachloride, benzene, and ether [*591*] and sparingly soluble in chloroform (4.5 g/100 mL), pyridine (6.1 g/100 mL), and *cis*-1,2-dichloroethylene (7.1 g/100 mL). The best solvent is dichloromethane (12.5 g/100 mL) [*595*]. A solution of dipyridine chromium(VI) oxide in dichloromethane is conveniently prepared by adding 1 mol of chromic oxide to a solution of 2 mol of pyridine in dichloromethane [*596*]. The complex may also be formed on the surface of silica gel by adding pyridine to silica gel impregnated with chromic oxide [*597*].

Oxidations with dipyridine chromium(VI) oxide are carried out (usually at room temperature) in solutions in dichloromethane. Primary alcohols, including allylic and benzylic alcohols, are oxidized to aldehydes [*594, 595, 596, 597, 598, 599, 600*] and secondary alcohols to ketones [*592, 595,

597, 599]. Methylene groups next to double [*593*] or triple [*601*] bonds are converted into carbonyl groups to yield α,β-unsaturated ketones.

Chromium trioxide–3,5-dimethylpyrazole complex, $CrO_3 \cdot C_5H_8N_2$, is prepared in situ by adding 1 mol of 3,5-dimethylpyrazole to a suspension of 1 mol of chromium trioxide in dichloromethane at room temperature. The scope and limitations of this oxidant are similar to those of the chromium trioxide–pyridine complex: the oxidation of primary alcohols to aldehydes and of secondary alcohols to ketones at room temperature [*602*].

Pyridinium chlorochromate (PCC), $C_5H_5N \cdot HCl \cdot CrO_3$, mp 205 °C (dec), which is commercially available, is prepared by adding chromium trioxide to a solution of pyridine in hydrochloric acid [*603*]. The preparation is easier than that of the chromium trioxide–pyridine complex.

Oxidations by pyridinium chlorochromate resemble those by dipyridine chromium(VI) oxide, both in scope and the mild conditions required. At room temperature, primary alcohols give aldehydes [*604, 605*], secondary alcohols afford ketones [*605*], allylic and benzylic methylene groups are oxidized to carbonyl groups [*606, 607*], enol ethers are converted into esters [*608*] or lactones [*609*], trimethylsilyl ethers of diphenols are transformed into quinones [*610*], and alkylboranes are converted into aldehydes [*611*].

4-Dimethylaminopyridinium chlorochromate, $4\text{-}(CH_3)_2NC_5H_4NH \cdot CrO_3Cl$, is a yellow-orange solid prepared by the addition of 1 mol of 4-dimethylaminopyridine to a solution of 1 mol of chromium trioxide in dilute hydrochloric acid. The reagent is light sensitive [*530*]. For the oxidation of allylic and benzylic alcohols to aldehydes in dichloromethane at room temperatures, 4–6 mol of 4-dimethylaminopyridinium chlorochromate is used. Saturated alcohols are not oxidized [*530*].

Poly(vinylpyridinium) chlorochromate (PVPCC) is obtained as a bright-orange solid by adding chromium trioxide and concentrated hydrochloric acid to an aqueous suspension of cross-linked poly(vinylpyridine) (PVP) resin (50–100 mesh). Oxidations of primary and secondary alcohols to carbonyl compounds are carried out on heating with a slight excess of the resin at 75–80 °C. The reaction is best carried out in cyclohexane [*612*].

Potassium chromate, K_2CrO_4, is used very rarely. It converts hydrazomethane into azomethane [*613*]. In the presence of hexamethylphosphoramide and crown ether, it transforms allylic and benzylic bromides into α,β-unsaturated aldehydes [*614*], and in the presence of dilute sulfuric acid, it oxidizes chlorohydroquinone to chloro-*p*-benzoquinone [*615*].

Silver chromate, Ag_2CrO_4, a reddish-brown precipitate prepared by adding an aqueous solution of silver nitrate to an aqueous solution of potassium chromate, is used with iodine for the preparation of α-iodoketones from alkenes [*616*].

Tetrabutylammonium chromate, $(C_4H_9)_4NHCrO_4$, is obtained as a yellow-orange precipitate when aqueous solutions of chromium trioxide and tetrabutylammonium chloride are mixed. After having been dried, the reagent is used in organic solvents such as dichloromethane [*617*] and chloroform [*618*] for the oxidation of alcohols to carbonyl compounds. The reagent can be prepared and used in situ in a mixture of water and chloroform [*618*].

Tetrabutylammonium chlorochromate, $(C_4H_9)_4NCrO_3Cl$, is formed by adding a solution of chromium trioxide in concentrated hydrochloric acid to a solution of tetrabutylammonium hydrogen sulfate in 5 N hydrochloric acid at 0 °C. The dry compound is used in chloroform for the oxidation of alcohols to carbonyl compounds and the oxidative coupling of thiols to disulfides [*619*].

Sodium dichromate, $Na_2Cr_2O_7$, is applied not only in solutions in sulfuric acid but also in solutions in perchloric acid [*620, 621*], acetic acid [*622, 623, 624, 625, 626, 627, 628*], benzene, and dimethyl sulfoxide [*629*]. In the presence of tetrabutylammonium bisulfate, dichloromethane, used as a water-insoluble solvent, extracts the product from the aqueous phase (phase transfer) [*630*].

Sodium dichromate hydroxylates tertiary carbons [*620*] and oxidizes methylene groups to carbonyls [*622, 623, 625, 626, 631*]; methyl and methylene groups, especially as side chains in aromatic compounds, to carboxylic groups [*624, 632, 633, 634, 635*]; and benzene rings to quinones [*630, 636, 637*] or carboxylic acids [*638*]. The reagent is often used for the conversion of primary alcohols into aldehydes [*629, 630, 639*] or, less frequently, into carboxylic acids or their esters [*640*]; of secondary alcohols into ketones [*621, 629, 630, 641, 642, 643, 644*]; of phenylhydroxylamine into nitrosobenzene [*645*]; and of alkylboranes into carbonyl compounds [*646*].

Potassium dichromate, $K_2Cr_2O_7$, is applied under similar conditions as its sodium analogue to oxidize benzene rings to quinones [*647, 648, 649, 650*], methylene groups adjacent to aromatic rings to carbonyls [*514*], primary alcohols to aldehydes [*651, 652, 653*], secondary alcohols to ketones [*644, 652, 654, 655*], and aldehydes to acids [*656*]. Phenylhydroxylamine is transformed into nitrosobenzene [*657*], and an aromatic nitroso compound, into a nitro compound [*658*].

Sodium dichromate and potassium dichromate possess almost identical oxidative properties. Both reagents furnish an equivalent of three oxygen atoms, as the oxidation state of chromium changes from hexavalent to trivalent, through tetravalent and pentavalent chromium as intermediates. The molecular weights of both reagents are almost identical, because sodium dichromate crystallizes with two molecules of water of crystallization. In the presence of dilute sulfuric acid, the most common solvent, chromic acid, H_2CrO_4, is liberated. Standard 10% solutions of CrO_3 are known as a **Kiliani mixture** or a **Beckmann mixture**. The Kiliani mixture is obtained by dissolving 60 g of $Na_2Cr_2O_7 \cdot 2H_2O$ in dilute sulfuric acid, prepared by adding 80 g of concentrated sulfuric acid to 270 mL of water (not the other way around). The Beckmann mixture is prepared by dissolving 60 g of $K_2Cr_2O_7$ in dilute sulfuric acid (80 g of concentrated sulfuric acid and 270 mL of water).

Silver dichromate–pyridine complex (tetrapyridine silver dichromate), $Ag_2Cr_2O_7 \cdot 4C_5H_5N$, an orange-yellow solid prepared by adding a warm solution of potassium dichromate to a solution of silver nitrate in water and pyridine, oxidizes allylic and benzylic alcohols to aldehydes and acyloins to α-diketones [659].

Zinc dichromate trihydrate, $ZnCr_2O_7 \cdot 3H_2O$, is obtained as an orange-red solid by adding zinc carbonate to a cold solution of chromium trioxide in dilute sulfuric acid [660]. The applications are oxidations of acetylenes to α-diketones, of aromatic hydrocarbons to quinones, of alcohols to aldehydes, and of ethers to esters and the oxidative regeneration of carbonyl compounds from their oximes [660].

Pyridinium dichromate, $(C_5H_5NH)_2Cr_2O_7$, a stable bright-orange solid (mp 144–146 °C), is prepared by adding pyridine to a cooled solution of chromium trioxide in water at a temperature lower than 30 °C and by diluting the mixture with acetone [603]. The reagent is applied at room temperature in solutions in dichloromethane or dimethylformamide (DMF). In dichloromethane solutions, primary alcohols are oxidized to aldehydes and secondary alcohols to ketones [603]. In DMF solutions, allylic alcohols are oxidized to α,β-unsaturated aldehydes or ketones. Nonconjugated aldehydes are converted into carboxylic acids, when an excess of the reagent is used. Acid- and base-sensitive functional groups are not endangered [603].

α,β-Unsaturated aldehydes and trimethylsilyl cyanide give trimethylsilyl cyanohydrins, which are transformed by pyridinium dichromate into α-keto nitriles and ultimately give unsaturated γ-lactones [661]. In the presence of trimethylsilyl peroxide, primary alcohols are oxidized in preference to secondary alcohols in a ratio of 20:1 [236].

Utmost care must be taken in preparing and handling pyridinium dichromate, because explosions have occurred even in aqueous media **[662].**

3-Carboxypyridinium dichromate and 4-carboxypyridinium dichromate, $(HO_2CC_6H_4NH)_2Cr_2O_7$, are prepared by adding nicotinic and isonicotinic acid, respectively, to aqueous solutions of chromium trioxide at 0–5 °C followed by acetone [663]. The orange-yellow solids (mp 215–217 and 250–253 °C, respectively) are applied in dichloromethane or in dichloromethane–pyridine solutions at room temperature to oxidize alcohols to carbonyl compounds, thiols to disulfides, and aromatic rings to quinones [663].

Quinolinium chlorochromate, $C_9H_7N \cdot HCl \cdot CrO_3$, is obtained as a yellowish-brown solid by addition of quinoline to a stirred mixture of chromic acid and concentrated hydrochloric acid in water. When applied in dichloromethane, this reagent oxidizes primary alcohols in preference to secondary alcohols [664].

All the pyridine- and pyridine-analogue-containing hexavalent chromium compounds excel by giving good to very high yields under very mild conditions, usually at temperatures not higher than room temperature. In addition, these compounds are sometimes selective oxidants. The disadvantage is the necessity of often using a large excess of the reagent.

Chromyl chloride, $Cr_2O_2Cl_2$, a dark-red liquid (mp −96.5 °C, bp 117 °C, d 1.911), is prepared from chromium trioxide or sodium dichromate, hydrochloric acid, and sulfuric acid [665]. The reagent is used in solutions in carbon disulfide, dichloromethane, acetone, tert-butyl alcohol, and pyridine. Oxidations with chromyl chloride are often complicated by side reactions and do not always give satisfactory yields. The mechanism of the oxidation with chromyl chloride, the *Etard* reaction, is probably of free-radical nature [666]. Complexes of chromyl chloride with the compounds to be oxidized have been isolated [666, 667, 668].

The reaction of chromyl chloride with alkenes gives epoxides, chlorohydrins, chloroketones, and ketones [668, 667, 668, 669, 670] or aldehydes (in the presence of zinc) [671, 672]. Benzene homologues are oxidized to aldehydes [667, 668] or ketones [666, 668, 673]. Primary alcohols are converted into aldehydes [674, 675], and trimethylsilyl ethers of enols are transformed into α-hydroxy ketones [676].

***tert*-Butyl chromate, $(tert\text{-}C_4H_9O)_2CrO_4$**, results from the addition of chromium trioxide to ice-cooled *tert*-butyl alcohol and is used in benzene or petroleum ether solutions [677]. This reagent oxidizes primary alcohols to aldehydes at low temperatures [678]; but its applications are rare.

Acetyl chromate, $(CH_3CO)_2CrO_4$; trichloroacetyl chromate, $(CCl_3CO)_2CrO_4$; trifluoroacetyl chromate, $(CF_3CO)_2CrO_4$; and benzoyl chromate, $(C_6H_5CO)_2CrO_4$, are prepared from chromium trioxide and the respective anhydrides [679, 680]. Trifluoroacetyl chromate reacts with many solvents and with silicone stopcock grease.

Acyl chromates are used for epoxidation [679] and for the conversion of alkenes into aldehydes [680].

DERIVATIVES OF GROUP 7 ELEMENTS

Chlorine, Cl_2, a greenish-yellow gas (mp −101 °C, bp −34.1 °C) delivered in steel containers under a pressure of 6 atm at 21 °C, is extremely corrosive and toxic. The use of good hoods is imperative for any work involving chlorine. Applications of chlorine gas for oxidation purposes are rare; rather, its compounds such as sodium, potassium, and calcium hypochlorites; *tert*-butyl hypochlorite; sodium chlorite; and sodium, potassium, silver, and barium chlorates are used. Chlorine is soluble in dichloromethane, chloroform, and carbon tetrachloride and is frequently applied in water, in which it is only sparingly soluble.

Oxidations by chlorine are limited to only few types of compounds. In organic solvents and pyridine [*681*] or hexamethylphosphoramide (HMPA) [*682*] as cosolvents, primary alcohols are oxidized to aldehydes and secondary alcohols to ketones [*681, 682*]. Secondary alcohols are oxidized in preference to primary alcohols [*681*]. Many oxidations with chlorine are carried out in aqueous media and involve sulfur-containing compounds. Mercaptans [*683*], alkyl thiolcarboxylates [*683*], thiocyanates [*684*], isothioureas [*684*], disulfides [*685*], and sulfinic acids [*686*] are transformed into sulfonyl chlorides. The chlorination of dimethyl disulfide in acetic anhydride yields methanesulfinyl chloride [*687*].

Sodium hypochlorite, NaOCl, is stable only as a very dilute aqueous solution (5.25%) and is commercially available under the name Clorox or bleach. This reagent can be prepared in situ by introducing chlorine into an aqueous solution of sodium hydroxide [*688*]. In the presence of tetrabutylammonium hydrogen sulfate, it is transformed into tetrabutylammonium hypochlorite, which is soluble in organic solvents [*689, 690*].

Sodium hypochlorite is used for the epoxidation of double bonds [*689, 691*]; for the oxidation of primary alcohols to aldehydes [*692*], of secondary alcohols to ketones [*693*], and of primary amines to carbonyl compounds [*692*]; for the conversion of benzylic halides into acids or ketones [*690*]; for the oxidation of aromatic rings to quinones [*694*] and of sulfides to sulfones [*695*]; and, especially, for the degradation of methyl ketones to carboxylic acids with one less carbon atom [*688, 696, 697, 698, 699*] and of α-amino acids to aldehydes with one less carbon [*700*]. Sodium hypochlorite is also used for the reoxidation of low-valence ruthenium compounds to ruthenium tetroxide in oxidations by ruthenium trichloride [*701*].

Potassium hypochlorite, KOCl, acts like sodium hypochlorite. This reagent is prepared in situ from potassium hydroxide and chlorine and oxidizes primary alcohols to aldehydes [*702*] and methyl ketones to carboxylic acids [*703*].

Calcium hypochlorite, Ca(OCl)$_2$, is a solid compound and is used for the epoxidation of double bonds [*704*] and the oxidation of alcohols to carbonyl compounds [*705*].

***tert*-Butyl hypochlorite, *tert*-C$_4$H$_9$OCl,** is obtained when chlorine is passed through a solution of *tert*-butyl alcohol in aqueous sodium hydroxide. The reagent is a liquid (bp 77–78 °C at 760 mm of Hg, d_{20}^{20} 0.910, n_D^{20} 1.403 [*706*]) and effects the dehydrogenation of amines to imines [*707*] and of alcohols to carbonyl compounds [*708, 709*] and the oxidation of sulfides to sulfoxides [*710*] and of selenides to selenoxides [*711*].

Sodium chlorite, NaClO$_2$, which is commercially available, oxidizes aldehydes to carboxylic acids at room temperature [*69, 712*]. The byproducts, hypochlorous acid and chlorine dioxide, must be removed during the oxidation to prevent side reactions, especially when unsaturated aldehydes are involved. As scavenger of the byproducts, hydrogen peroxide [*712*], sulfuric acid [*69, 712*], dimethyl sulfoxide [*712*], or resorcinol [*69*] is used. Another application of sodium chlorite is the oxidation of aldehyde–bisulfite compounds to carboxylic acids in the presence of dimethyl sulfoxide and acetic anhydride [*713*]. Oxidations of aldehydes to acids with sodium chlorite are very selective and much cheaper than those with silver oxide.

Sodium chlorate, $NaClO_3$; potassium chlorate, $KClO_3$; silver chlorate, $AgClO_3$; and barium chlorate, $Ba(ClO_3)_2$, oxidize organic compounds only in the presence of catalysts, usually osmium tetroxide [*310, 714, 715, 716, 718*] or vanadium pentoxide [*716, 718*]. Because such oxidations do not occur without catalysts, it is likely that the real oxidants are osmium tetroxide and vanadium pentoxide, respectively, and that the function of the chlorates is reoxidation.

Such oxidants cause the *syn* hydroxylation of double bonds [*310, 714, 715, 716*] and convert triple bonds into vicinal carbonyls [*717*]. Other applications, such as the conversion of furfural into fumaric acid [*716, 718*] or the oxidation of hydroquinone to quinone [*719*], are rare.

Organochlorine compounds with chlorine bonded to nitrogen give different oxidation products, depending on the nature of the organic molecule.

***N*-Chlorosuccinimide, $C_4H_4ClNO_2$** (mp 144-146 °C), either alone [*709*] or in the presence of dimethyl sulfide [*720, 721*], dehydrogenates alcohols to carbonyl compounds.

Trichloroisocyanuric acid, $C_3Cl_3N_3O_3$ (mp 249–251 °C), converts ethers into esters [*722*], and **sodium *N*-chloro-*p*-toluenesulfonamide (Chloramine-T), *p*-$CH_3C_6H_4SO_2NClNa·3H_2O$** (mp 167–170 °C), in the presence of osmium tetroxide, transforms alkenes into vicinal hydroxy sulfonamido compounds [*723*].

Bromine, Br_2, a dark brown liquid (mp −7.3 °C, bp 59.5 °C, *d* 3.1), forms crystalline molecular complexes with **dioxane, $C_4H_8O_2·Br_2$** (mp 60–61 °C) [*724*]; **pyridine, $C_5H_5N·Br_2$** (mp 62–63 °C) [*725*]; and **1,4-diazabicyclo[2,2,2]octane (Dabco), $C_6H_{12}N_2·2Br_2$ (Dabco·$2Br_2$)** (mp 155–160 °C dec), that are very stable to light and moist air [*726, 727*]. Bromine can also be generated in situ from sodium bromate and sodium bromide in acidic solutions [*728*]. It is applied in aqueous media [*729, 730*] or organic solvents, such as dichloromethane [*682*], carbon tetrachloride [*731, 732*], ethanol [*733*], acetonitrile [*726*], or hexamethylphosphoric triamide (HMPA) [*682*].

Bromine dehydrogenates alcohols to carbonyl compounds [*682, 726, 729, 734*] (secondary alcohols in preference to primary alcohols [*682*]) and hydrazo compounds to azo compounds [*733*] and oxidizes sulfides to sulfoxides [727, 728] and disulfides to sulfonic acids [730]. Benzyl alkyl ethers are degraded to benzaldehyde [731].

Sodium hypobromite, NaOBr, is prepared in situ from bromine and aqueous sodium hydroxide under cooling. Its almost exclusive application is the degradation of methyl ketones to acids with one less carbon [635, 735, 736, 737]. Such degradation is achieved not only with methyl ketones but also with other alkyl ketones [103, 160]. Other oxidations, such as the conversion of hydroxylamines into nitroso compounds, are rare [738].

Sodium bromite, $NaBrO_2 \cdot H_2O$, a crystalline compound, oxidizes secondary alcohols to ketones [739] and sulfides to sulfoxides [739]. Primary alcohols are transformed into esters [739].

Sodium bromate, $NaBrO_3$, a white crystalline compound, converts acyloins into α-diketones under forcing conditions [740]. More often, this reagent is used as a reoxidant of ammonium cerium nitrate [421], cerium sulfate [741], or ruthenium trichloride [741] in oxidations of alcohols to aldehydes [421] or carboxylic acids [741].

Potassium bromate, $KBrO_3$, and hydrobromic acid are used in the in situ generation of bromine for the oxidation of primary alcohols to esters [742].

Organobromine compounds most frequently used for dehydrogenations and oxidations are N-bromoacetamide, $CH_3CONHBr$ (mp 108–109 °C), and N-bromosuccinimide, $C_4H_4BrNO_2$ (mp 180–183 °C), both of which are commercially available.

***N*-Bromoacetamide** is successfully applied to the oxidation of secondary alcohols in preference to allylic primary alcohols [743].

***N*-Bromosuccinimide** is used in the dehydrogenation of hydrazo compounds to azo compounds [744] and in the oxidative degradation of α-hydroxy acids to aldehydes or ketones [745]. This reagent also oxidizes alkyl trimethylsilyl ethers to esters or ketones [744].

In aqueous solutions, N-bromosuccinimide converts alkenes into bromohydrins or, ultimately, into epoxides (after the treatment of bromohydrins with alkali [746] or when the oxidation is done in alkaline media).

***N*-Bromobenzophenoneimine, $(C_6H_5)_2C=NBr$** (mp 38.5 °C) [747], which is prepared from benzophenoneimine, bromine, and alkalies, is used for the conversion of alcohols into carbonyl compounds [748].

Iodine, I_2 (dark-violet crystals, mp 113.5 °C), sublimes, and its vapor is violet. It is suitable for the oxidation of thiols and thiol-group-containing compounds to the corresponding disulfides [749].

Sodium hypoiodite, NaOI, which is used for the iodoform reaction [750], is prepared in situ from iodine and sodium hydroxide.

Iodine and silver oxide oxidize alkenes to epoxides [751], whereas **iodine and silver chromate** convert alkenes into α-iodoketones [610].

Potassium hypoiodite, KOI, which is prepared in situ from iodine and potassium hydroxide, oxidizes methyl ketones (and compounds that are oxidizable to methyl ketones) to carboxylic acids. The byproduct of the reaction, iodoform, is a heavy yellow crystalline compound possessing a unique smell and is easily identifiable by its melting point (120-123 °C). Because of this conspicuous byproduct, the reaction has long been used for the identification of methyl ketones [*750*]. With the advent of NMR spectroscopy, in which the COCH$_3$ group gives a characteristic singlet at 2.1 ppm, the importance of the iodoform reaction (***Lieben test***) has decreased considerably. Not only methyl ketones but also some cyclic ketones are oxidized to carboxylic acids by potassium hypoiodite [*752*].

Sodium iodate, NaIO$_3$, is used rarely; its applications are limited to the oxidation of aromatic polyhydroxy compounds to quinones [*753, 754*].

Periodic acid (metaperiodic acid), HIO$_4$, and its dihydrate (**paraperiodic acid, H$_5$IO$_6$,** mp 122 °C) are commercially available. Both reagents are applied usually in aqueous or aqueous–alcoholic solutions and only rarely in anhydrous tetrahydrofuran [*755*], aqueous dioxane [*756*], acetone [*567*], or acetic acid [*757, 758*].

The oxidative cleavage of carbon–carbon bonds in vicinal diols [*756, 759*] is a reaction widely used in saccharide chemistry. Besides its application in this reaction, periodic acid achieves the oxidative coupling [*757*] or oxidation to quinones [*758*] of polynuclear aromatic hydrocarbons, the oxidation of methyl groups in aromatic compounds to carbonyl groups [*760*], the conversion of epoxides into dicarbonyl compounds [*761*], and the oxidative cleavage of trimethylsilyl ethers of acyloins to carboxylic acids [*755*].

Sodium periodate (sodium metaperiodate), NaIO$_4$ (mp 300 °C dec), which is commercially available, is applied mainly in aqueous or aqueous–alcoholic solutions. Like the free periodic acid, sodium periodate cleaves vicinal diols to carbonyl compounds [*762*]. This reaction is especially useful in connection with potassium permanganate [*763, 764*] or osmium tetroxide [*765*]. Such mixed oxidants oxidize alkenes to carbonyl compounds or carboxylic acids, evidently by way of vicinal diols as intermediates. Sulfides are transformed by sodium periodate into sulfoxides [*322, 323, 766, 767, 768, 769, 770, 771, 772*], and selenides are converted into selenoxides [*773*]. Sodium periodate is also a reoxidant of lower valency ruthenium in oxidations with ruthenium tetroxide [*567, 774*].

Potassium periodate, KIO$_4$, when used in dilute sulfuric acid, acts like periodic acid in degrading α-hydroxy acids to aldehydes [*775*].

Tetrabutylammonium periodate, (C$_4$H$_9$)$_4$NIO$_4$ (mp 175 °C dec), which is usually prepared in situ from tetrabutylammonium hydrogen sulfate and sodium periodate [*776*], is useful in two-phase systems, because it dissolves in chloroform and other organic solvents. Its scope and limitations resemble those of periodic acid and alkaline periodates: the oxidation of carboxylic acids [*777*] and α-hydroxy acids [*776, 778*] to aldehydes with one less carbon, the oxidation of sulfides to sulfoxides [*776,778*], and the oxidation of benzyl-type bromides to aldehydes [*778*].

Acetyl hypoiodite (iodine acetate), CH$_3$COOI, is prepared by treatment of silver acetate in acetic acid with iodine at room temperature [*779, 780*]. The reagent cleaves vicinal diols to dicarbonyl compounds [*779*] and degrades tertiary alcohols to ketones [*780*]. Acetyl hypoiodite is also an intermediate in the reaction of alkenes with the so-called "**Simonini complex,**" an addition product of **iodine** with 2 mol of **silver acetate** [*781, 782, 783*].

Benzoyl hypoiodite, C$_6$H$_5$COOI, a similar complex with similar applications, is formed from **iodine** and 2 mol of **silver benzoate** [*783, 784*]. Silver *p*-chloro- and 3,5-dinitrobenzoates react analogously [*783*] (equation 17).

[*783*]   17

$$2\ C_6H_5CO_2Ag + I_2 \longrightarrow C_6H_5CO_2Ag \cdot C_6H_5CO_2I + AgI$$

Both complexes are used in the hydroxylation of double bonds via diacetates or dibenzoates of vicinal diols. The reaction is stereospecific. In anhydrous medium (the ***Prévost reaction*** [*783*]), the reaction takes place in the *anti* mode. In the presence of water (the ***Woodward modification*** [*783*]), the reaction results in a *syn* addition. The mechanisms of both reactions are shown in the section Hydroxylation of Alkenes and Cycloalkenes in Chapter 3 (*see* equation 78).

Iodine triacetate, I(OCOCH$_3$)$_3$, which is prepared from iodine trichloride and silver acetate in acetic acid at room temperature, cleaves vicinal diols to dialdehydes [*779*].

Iodine tris(trifluoroacetate), I(OCOCF$_3$)$_3$, forms epoxides from certain alkenes [*785*].

Several oxidants are derivatives of iodoaromatic compounds containing trivalent and pentavalent iodine.

Iodobenzene dichloride, C$_6$H$_5$ICl$_2$, oxidizes selenides to selenoxides [*773*].

Iodosobenzene, C$_6$H$_5$IO (mp 210 °C, explodes), which is prepared by way of iodobenzene diacetate [*266*], is used for the conversion of acetylenes into α-dicarbonyl compounds [*786, 787*], of alcohols into aldehydes or ketones [*788*], and of aldehydes into acids [*788*]. This oxidant is also applied to the preparation of carbonyl compounds from vicinal diols [*789*], the syntheses of α-hydroxy ketones [*790*], and the oxidation of sulfides to sulfoxides [*791*]. Similar reactions can be accomplished by ***p*-iodosotoluene** or ***o*-iodosobenzoic acid** [*792*].

Iodobenzene diacetate, C$_6$H$_5$I(OCOCH$_3$)$_2$ (mp 163–165 °C) (Aldrich's product), is obtained by treatment of iodobenzene with peroxyacetic acid [*266*] or with 30% hydrogen peroxide and acetic anhydride at 40 °C

[793]. The reagent hydroxylates double bonds via diacetates [789], hydroxylates esters in the α position with respect to the carboxylic group [794], cleaves vicinal diols to carbonyl compounds [789], and oxidizes sulfides to sulfoxides [795].

Pentafluoroiodobenzene bis(trifluoroacetate), $C_6F_5I(OCOCF_3)_2$ (mp 119–120 °C), a product of the treatment of iodopentafluorobenzene with trifluoroacetic anhydride and fuming nitric acid [796], converts acetylenes into carboxylic acids [797].

Iodoxybenzene, $C_6H_5IO_2$ (mp 230 °C, explodes), is prepared by the oxidation of iodobenzene with peroxyacetic acid [267]. This reagent cleaves vicinal diols to aldehydes [798] and, in the presence of α-dipyridyl diselenide, converts alkenes into α-carbonyl alkenes (α,β-unsaturated ketones) [799].

***m*-Iodoxybenzoic acid, m-$O_2IC_6H_4CO_2H$,** in connection with phenylseleninic anhydride, is used for the dehydrogenation of ketones to α,β-unsaturated ketones [526].

Dess–Martin "periodinane" (mp 138–140 °C), which is derived from *o*-iodobenzoic acid, accomplishes the oxidation of primary and secondary alcohols under very mild conditions and in high yields [800, 801] (equation 18).

[*801*]   18

Manganic sulfate, $Mn_2(SO_4)_3$, is obtained by the oxidation of manganese sulfate (tetrahydrate) in 6 M sulfuric acid by aqueous potassium permanganate at 0–25 °C. This oxidant oxidizes aromatic hydrocarbons to quinones [802].

Manganic acetate (manganese triacetate), $Mn(OCOCH_3)_3$, is prepared by refluxing a solution of manganese acetate tetrahydrate in acetic acid with potassium permanganate [803]. This oxidant hydroxylates benzylic methylene groups [416] and forms lactones from terminal alkenes [803, 804] (*see* equation 88).

Manganese dioxide, MnO_2, is a commercially available, dark-brown powder prepared by the simultaneous addition of a 42.5% aqueous solution of manganese sulfate tetrahydrate and a 40% solution of sodium hydroxide to a hot stirred 14% aqueous solution of potassium permanganate. The brown precipitate is collected and washed with water until the washings are colorless, dried at 100–120 °C, and ground to a fine powder before use

[805]. Its activity can be modified by changing the particle size, by using different drying temperatures (220–280 °C), and by aftertreatment with 15% nitric acid and drying at 220–250 °C [806]. A different procedure involves the addition of a slight excess of a concentrated aqueous solution of potassium permanganate to a stirred solution of manganese sulfate at 90 °C; filtration of the precipitate; washing of the solid with hot water, methanol, and ether; and drying at 120–130 °C to constant weight [807]. Many more modifications are described in the literature [808]. The removal of water from the hydrated manganese dioxide can also be achieved by azeotropic distillation with benzene [809].

The differences in the activities and the results of oxidations with manganese dioxide are due to differences in methods of preparation [808, 810] and solvents used [808]. The most frequently used solvents are petroleum ether [541, 805, 806, 808], hexane [382, 806, 811], cyclohexane [812], benzene [808, 813, 814, 815, 816], dichloromethane [817, 818], chloroform [807, 808, 811, 813, 816], carbon tetrachloride [805, 808], ether [806, 808, 811, 813, 819, 820], acetone [821, 822], acetic acid [823], and water [813].

Manganese dioxide is a reagent of choice for the oxidation of allylic [382, 805, 806, 807, 808, 809, 810, 815, 819, 821, 824] and benzylic [382, 806, 808, 809, 811, 815, 816] alcohols to aldehydes or ketones. Saturated alcohols do not always give good yields [807, 808, 813, 815]. The oxidation of aldehydes to acids is exceptional [813].

In addition, manganese dioxide oxidizes benzylic methyl or methylene to carbonyl [814] and cleaves the carbon bonds of vicinal diols [817, 822]. It converts amines into imines [811, 825]; tertiary amines into secondary amines [812], formamides [826, 827, 828], or ketones; aromatic primary amines [813, 825] and hydrazo compounds [825] into azo compounds; hydroxylamines into nitroso [816] or nitro compounds [823]; hydrazones into diazo compounds [820]; phosphines into phosphine oxides [813]; thiols into disulfides [816]; and sulfides into sulfoxides [541].

Potassium manganate, K_2MnO_4, purple crystals similar to potassium permanganate [829], forms a dark-green aqueous solution [830]. It is prepared from potassium permanganate and potassium hydroxide at 100 °C [830] or 120–140 °C [829]. Its use is limited to the *syn* hydroxylation of double bonds [829] and the hydroxylation of tertiary carbons in branched carboxylic acids [830, 831]. It offers no advantages over potassium permanganate.

Barium manganate, $BaMnO_4$, is commercially available. The darkblue crystals are obtained from aqueous solutions of barium chloride and potassium permanganate [832, 833]. It oxidizes alcohols, especially benzylic alcohols, to carbonyl compounds [832, 833]; hydroquinone to quinone [833]; benzylamines to benzaldehydes [833]; aromatic amines to azo compounds [833]; and phosphines to phosphine oxides [833].

Sodium permanganate monohydrate, NaMnO$_4$·H$_2$O, which is commercially available, is used for the oxidation of alkenes to carboxylic acids [834] and of alcohols to carbonyl compounds [835], the conversion of sulfinic acids into sulfonic acids [836], and the selective oxidation of sulfoxides to sulfones (sulfides are not oxidized with sodium permanganate in dioxane solutions) [837].

Potassium permanganate, KMnO$_4$, dark-purple crystals with a metallic luster (d 2.7, solubility: 6.6% in cold water and 22% in boiling water), is soluble in acetone without reacting with the solvent [838]. It is applied most frequently in aqueous solutions but can be used in the presence of organic solvents such as petroleum ether [839], benzene [840, 841, 842, 843, 844, 845, 846, 847], dichloromethane [848, 849], tert-butyl alcohol [850, 851, 852, 853], acetone [296, 838, 853, 854, 855, 856, 857, 858, 859], acetic acid [843, 857], acetic anhydride [860, 861], and pyridine [855, 862]. The solubilization of potassium permanganate in organic solvents can be achieved by crown ethers [841] or by phase-transfer agents, such as quaternary ammonium salts that form quaternary permanganates soluble in solvents immiscible with water [843, 845, 848] (purple benzene [841]). Oxidations in organic solvents can also be carried out with potassium permanganate adsorbed on molecular sieves [847, 863], bentonite [849], montmorillonite [863], or copper sulfate pentahydrate [844, 849].

Potassium permanganate furnishes three oxidation equivalents per mole (three oxygens per 2 mol). Its reduction to manganese dioxide liberates potassium hydroxide. Therefore, most oxidations with potassium permanganate take place in strongly alkaline medium. Some oxidations occur at or require lower pH, which can be attained by using buffers such as carbon dioxide [858, 864], sodium bicarbonate [856], potassium carbonate [852], aluminum oxide [865], disodium hydrogen phosphate [839], magnesium nitrate [866], magnesium sulfate [853, 856, 859, 867], zinc sulfate [850, 851], calcium sulfate (better than zinc sulfate) [851], or sulfuric acid [846, 868].

The results of some oxidations with potassium permanganate differ depending on the pH of the reaction. For example, stearolic acid gives 9,10-diketostearic acid at pH 7–7.5 (achieved with carbon dioxide) and azelaic acid on treatment at pH 12 [864]. In some reactions, potassium permanganate is used as a catalyst for oxidation with other oxidants, such as sodium periodate. Thus alkenes are cleaved to carbonyl compounds or acids via vicinal diols obtained by hydroxylation with potassium permanganate, followed by cleavage by sodium periodate [763, 852].

The intensity of oxidations by potassium permanganate can be regulated further by concentration of the oxidant in water or water–organic solvent systems and by temperature variations (from –5 °C to refluxing temperature). Usually, a slight excess of the reagent over the stoichiometric amount required for the oxidation is used. Frequently, potassium permanganate or its solution is added to the organic substrate until the oxidant ceases to react. The excess of potassium permanganate is difficult to notice, because the reaction mixture is dark owing to the presence of the darkbrown manganese dioxide. If a drop of the dark mixture is placed on a filter paper, an excess of potassium permanganate is revealed by a purple ring surrounding the brown spot of manganese dioxide. An excess of potassium permanganate is easily destroyed by adding methanol or ethanol to the mixture until the test for the presence of potassium permanganate is negative.

The workup of reaction mixtures after oxidations with potassium permanganate starts with the removal of the manganese dioxide, either by centrifugation or by filtration. The microcrystalline manganese dioxide easily clogs the pores of filters, and the filtrations are slow. In addition, colloidal solutions of manganese dioxide are sometimes formed. Both these drawbacks can be handled by filtering the reaction mixture with suction over a layer of activated charcoal. Because of the large surface area of manganese dioxide, some products are obstinately retained in this byproduct. Intensive washing of the dioxide is necessary to remove the adsorbed products.

All these difficulties can be avoided when the manganese dioxide is reduced to manganese sulfate, which is water soluble and almost colorless. Such reductions are easily accomplished by passing a stream of sulfur dioxide through the reaction mixture until the complete disappearance of the manganese dioxide [*838, 868, 869*].

The spectrum of applications of potassium permanganate is very broad. This reagent is used for dehydrogenative coupling [*870*], hydroxylates tertiary carbons to form hydroxy compounds [*830, 831*], hydroxylates double bonds to form vicinal diols [*101, 296, 858, 871*], oxidizes alkenes to α-diketones [*860, 861*], cleaves double bonds to form carbonyl compounds [*840, 842, 852*] or carboxylic acids [*763, 841, 843, 845, 852, 869, 872, 873, 874*], and converts acetylenes into dicarbonyl compounds [*848, 856, 864*] or carboxylic acids [*843, 864*]. Aromatic rings are degraded to carboxylic acids [*875, 876*], and side chains in aromatic compounds are oxidized to ketones [*866, 877*] or carboxylic acids [*503, 878, 879, 880, 881, 882, 883*]. Primary alcohols [*884*] and aldehydes [*749, 868, 885*] are converted into carboxylic acids, secondary alcohols into ketones [*749, 839, 844, 863, 865, 886, 887*], ketones into keto acids [*888, 889, 890*] or acids [*889, 891*], ethers into esters [*855*], and amines into amides [*854, 855*] or imines [*851*]. Aromatic amines are oxidized to nitro compounds [*738, 859, 892*], aliphatic nitro compounds to ketones [*862, 867*], sulfides to sulfones [*846*], selenides to selenones [*325*], and iodo compounds to iodoso compounds [*893*].

Cupric permanganate, $Cu(MnO_4)_2 \cdot 8H_2O$, is commercially available and oxidizes primary alcohols and aldehydes to acids, secondary alcohols to ketones, and sulfides to sulfones in very high yields [*894*].

Bis(2,2′-bipyridyl)copper(II) permanganate (BBCP) (mp 105–110 °C) is prepared as a crystalline product on cooling the dark-purple solution obtained by adding an aqueous solution of potassium permanganate to an aqueous solution of 2 mol of bis(2,2'-bipyridyl)copper(II) chloride [*895*]. The compound is used in dry dichloromethane and is more selective than potassium permanganate. It oxidizes allylic and benzylic alcohols to aldehydes [*895*], acyloins to α-diketones [*895*], primary aromatic amines to azo compounds [*895*], and thiols to disulfides [*896*]. Similar results are obtained with purple crystalline **bis(2,2'-bipyridyl)silver permanganate** [*897*].

Magnesium permanganate, $Mg(MnO_4)_2$, and **zinc permanganate, $Zn(MnO_4)_2$**, are much more reactive than potassium permanganate and may cause ignition when added to alcohols, tetrahydrofuran, acetone, and acetic acid. They are used safely when supported on silica gel in dichloromethane or chloroform at room or refluxing temperature [*898*]. Their specialties are the conversion of acetylenes into α-diketones, of cyclic ethers into lactones, and of amines into amides [*898*].

Tetrabutylammonium permanganate, $(C_4H_9)_4NMnO_4$ (mp 120–121 °C dec), is prepared by adding a slight excess of a concentrated aqueous solution of tetrabutylammonium bromide to a stirred aqueous solution of potassium permanganate. The purple precipitate is filtered, washed with water, and dried in vacuo [*899*]. The reagent is applied in solutions in dichloromethane, chloroform, acetone, or pyridine at room temperature and oxidizes alkenes to diols, dialdehydes [*900*], or acids [*899*] and aromatic aldehydes to acids [*899*]. Tetrabutylammonium permanganate is to be *handled with caution,* because ignition of the crystalline material during weighing has been reported [*901*].

Benzyltriethylammonium permanganate, $C_6H_5CH_2N(C_2H_5)_3MnO_4$ (violet crystals, mp 127-129 °C dec), is obtained by the dropwise addition of an aqueous solution of benzyltriethylammonium chloride to an aqueous solution of potassium permanganate. The crystals are filtered and dried at 40 °C at 0.2 mm of Hg [*902*]. The reagent is used in dichloromethane or acetic acid for the hydroxylation of some tertiary carbons and the conversion of methylene in hydrocarbons and ethers into carbonyl [*902*], of aldehydes into carboxylic acids [*903*], and of amines into amides [*904*]. Like tetrabutylammonium permanganate, benzyltriethylammonium permanganate should be *handled with due precaution.*

Triphenylmethylphosphonium permanganate, $(C_6H_5)_3PCH_3MnO_4$, a violet solid, is obtained by adding an aqueous solution of potassium permanganate to an aqueous solution of triphenylmethylphosphonium bromide. The crystals are filtered and dried in vacuo over phosphorus pentoxide. The compound can be stored for several weeks in the dark but may decompose explosively at 70 °C. An *explosion* may also occur when organic compounds are added to the solid oxidant. Oxidations of alkenes to vicinal diols are carried out by adding solutions of alkenes in dichloromethane to a solution prepared by adding, portionwise, triphenylmethylphosphonium permanganate to dichloromethane at −70 °C [*905*].

DERIVATIVES OF GROUP 8 ELEMENTS

Ferric chloride, FeCl$_3$ or FeCl$_3$·6H$_2$O, is a fairly selective reagent capable mainly of various dehydrogenations: of alcohols to aldehydes or ketones [*906*], of hydroquinones to quinones [*907*], of diamino [*908*] and amino hydroxy aromatic compounds [*909*] to quinones, and of α-amino ketones to α-diketones [*910*]. The reagent is also used for the dehydrogenative coupling of ketones in the α positions with respect to carbonyl groups [*911*] and the coupling of naphthols [*912*] or of aromatic *o*-diamines [*913*] and of thiols to disulfides [*907*].

Ferric sulfate, Fe$_2$(SO$_4$)$_3$, either as such [*914*] or prepared in situ from **ferric oxide and sulfuric acid** [*915*], is used for similar intra- and intermolecular dehydrogenations.

Potassium ferrate, K$_2$FeO$_4$, which is prepared from ferric nitrate and alkaline hypochlorites [*916*], is a selective oxidant for the conversion of primary alcohols into aldehydes (not acids), of secondary alcohols into ketones, and of primary amines into aldehydes [*917, 918*].

Potassium ferricyanide [potassium hexacyanoferrate(III)], K$_3$Fe(CN)$_6$, in the presence of a base, dehydrogenates hydroaromatic compounds to aromatic compounds [*919*] and can cause dehydrogenative cyclizations [*920*]. The reagent is used for the conversion of acid hydrazides into aldehydes [*921*], of sterically hindered phenols into phenoxy radicals [*922, 923*], and of primary amines into nitriles [*924*]. Tertiary amines are demethylated to secondary amines [*925, 926*].

Raney nickel, Ni(R), effects the dehydrogenation of secondary alcohols to ketones in excellent yields at moderate temperatures in the presence [*927*] or absence [*927, 928*] of hydrogen acceptors.

Nickel peroxide, an undefined black oxide of nickel, is prepared from nickel sulfate hexahydrate by oxidation in alkaline medium with an ozone–oxygen mixture [*929*] or with sodium hypochlorite [*930, 931, 932, 933*]. Its main applications are the oxidation of aromatic side chains to carboxyls [*933*], of allylic and benzylic alcohols to aldehydes in organic solvents [*929, 932*] or to acids in aqueous alkaline solutions [*929, 930, 932*], and of aldehydes to acids [*934*]; the conversion of aldehyde or ketone hydrazones into diazo compounds [*935*]; the dehydrogenative coupling of ketones in the α positions with respect to carbonyl groups [*931*]; and the dehydrogenation of primary amines to nitriles or azo compounds [*936*].

Ruthenium tetroxide, RuO$_4$, a reddish-orange substance (mp 25.5 °C, bp 108 °C dec) that is 60 times more soluble in carbon tetrachloride than in water [*937*], is prepared usually in situ from black ruthenium dioxide, **RuO$_2$ or RuO$_2$·XH$_2$O,** or ruthenium trichloride, **RuCl$_3$ or RuCl$_3$·3H$_2$O** (mp 500 °C dec), by oxidation with sodium hypochlorite [*701, 938, 939*], sodium bromate [*940*], periodic acid [*939*], sodium periodate [*774*], or peroxyacetic acid [*939*]. This reagent actually acts as a catalyst, because it is reoxidized after the reaction with organic compounds by one of the previously mentioned oxidants. All work with ruthenium tetroxide should be done in hoods, because the reaction mixtures generate an obnoxious smell reminiscent of ozone. A thorough study of oxidation by ruthenium tetroxide generated in situ from ruthenium dioxide or ruthenium trichloride hydrates shows especially the importance of using hydrates containing no less than 20% of water. Ruthenium dioxide hydrate with less than 10% of water did not work [*see 1207*].

The applications of ruthenium tetroxide range from the common types of oxidations, such as those of alkenes, alcohols, and aldehydes to carboxylic acids [*701, 774, 939, 940*]; of secondary alcohols to ketones [*701, 940, 941*]; of aldehydes to acids (in poor yields) [*940*]; of aromatic hydrocarbons to quinones [*942, 943*] or acids [*701, 774, 941*]; and of sulfides to sulfoxides and sulfones [*942*], to specific ones like the oxidation of acetylenes to vicinal dicarbonyl compounds [*938*], of ethers to esters [*940*], of cyclic imines to lactams [*944*], and of lactams to imides [*940*].

Sodium ruthenate, Na$_2$RuO$_4$, is prepared in situ from ruthenium tetroxide (in solution in carbon tetrachloride) and 1 M sodium hydroxide by shaking for 2 h at room temperature. The reagent remains in the aqueous layer, which acquires bright-orange color [*937*]. It oxidizes primary alcohols to carboxylic acids and secondary alcohols to ketones and is comparable with but stronger than potassium ferrate [*937*].

Potassium ruthenate, K$_2$RuO$_4$, is prepared in situ from ruthenium trichloride and aqueous persulfate. The reagent catalyzes persulfate oxidations of primary alcohols to acids, secondary to ketones, and primary amines to nitriles or acids at room temperature in high yields [*196*].

Palladium, Pd, and **platinum, Pt,** usually on activated carbon or asbestos, are catalysts for the dehydrogenation of hydroaromatic and some heterocyclic compounds to aromatic compounds. The reaction takes place at high temperatures (300–350 °C), and consequently, side reactions such as rearrangements often take place [*495, 496, 497, 945, 946, 947, 948*]. The catalytic dehydrogenations played a very important role in the elucidation of terpene and alkaloid structures. Because spectroscopic methods, especially NMR spectroscopy, can help to determine structures much more reliably, catalytic dehydrogenation over palladium and platinum are rare nowadays.

Osmium tetroxide (osmic acid), OsO$_4$ (white crystals, mp 39.5–41 °C, bp 130 °C), is used rarely in stoichiometric amounts, because it is expensive and very toxic. It is applied in solutions in ether [949], dioxane [482], or pyridine [950] and reacts with alkenes to form five-membered cyclic esters of vicinal diols, which, on hydrolysis, afford vicinal diols. The result is the *syn* addition of two hydroxylic groups across a double bond [482, 949, 950]. In the presence of chiral amines such as quinuclidine, enantioselective addition with an enantiomeric excess of 26–82% of one enantiomer has been accomplished [951]. Most of the *syn* hydroxylations with osmium tetroxide are carried out with only catalytic amounts in the presence of oxidants that regenerate osmium tetroxide, such as hydrogen peroxide [130], sodium chlorate [715,716], potassium chlorate [715], silver chlorate, barium chlorate [310], and amine oxides [952]. In the presence of sodium perchlorate [953], sodium periodate [765], and an excess of hydrogen peroxide [141], alkenes are oxidized to carbonyl compounds after cleavage of the carbon–carbon bond of the vicinal diols formed primarily. Osmium tetroxide oxidizes sulfoxides (but not sulfides) to sulfones [837].

ORGANIC OXIDANTS

Carbon tetrachloride, CCl_4, in the presence of potassium hydroxide at 25–80 °C, transforms primary alcohols into carboxylic acids, methyl ketones into acids with the same number of carbons or with one less carbon in the chain, and aryl methyl sulfones into arenesulfonic acids [954].

Hexamine (hexamethylenetetramine or methenamine), $C_6H_{12}N_4$ (white crystals, mp 280 °C, sublimes), reacts with reactive (allylic or benzylic) halides to form salts, which, on steam distillation or refluxing with water, yield aldehydes [955, 956] (***Sommelet reaction***). Aromatic aldehydes can be obtained from benzyl chlorides on treatment with formaldehyde and hydrochloric acid and subsequent refluxing with hexamethylenetetramine [957].

Chloral (trichloroacetaldehyde), CCl_3CHO, oxidizes secondary alcohols to ketones in the presence of very active aluminum oxide (Woelm). This reaction seems to be superior to the ***Oppenauer oxidation,*** because it takes place at room temperature or at only slightly elevated temperatures [958, 959, 960].

Acetone, cyclohexanone, benzophenone, cinnamaldehyde, and other carbonyl compounds are hydrogen acceptors in the ***Oppenauer oxidation*** of alcohols to carbonyl compounds. The reaction is catalyzed by Raney nickel [961], aluminum alkoxides [962], tris(isopropoxide), or tris(*tert*-butoxide) as bases soluble in organic solvents [963, 964]. These dehydrogenations of alcohols to aldehydes and ketones require refluxing or distillations and have given way to dimethyl sulfoxide oxidations, which take place at room temperature.

Mesitylglyoxal, 3-nitromesitylglyoxal, 3,5-dinitromesitylglyoxal, and **3,5-di-*tert*-butyl-*o*-benzoquinone** are used for the conversion of primary amines with the amino groups on secondary carbons into ketones via *Schiff bases* at 0–23 °C in high yields (transamination) [965] (equation 19).

[965]   19

$$\text{>CHNH}_2 + \text{OCHCO-Ar} \longrightarrow \text{>CHN=CHCO-Ar} \longrightarrow$$

$$\text{>C=NCH}_2\text{CO-Ar} \xrightarrow{H_2O} \text{>C=O} + H_2\text{NCH}_2\text{CO-Ar}$$

Chloranil (tetrachloro-*p*-benzoquinone), 2,3-dichloro-5,6-dicyano-*p*-benzoquinone (DDQ), and other high-potential quinones are good hydrogen acceptors in dehydrogenations forming double bonds adjacent to conjugated double bonds [966] or aromatic rings [967, 968], in aromatizations of six-membered alicyclic compounds [969, 970], in dehydrogenative coupling [971, 972], in oxidations of allylic alcohols to α,β-unsaturated carbonyl compounds [973, 974, 975], and in oxidations of phenols to quinones [971]. Some of these dehydrogenations take place at room temperature, whereas some reactions require refluxing in ether, benzene, xylene, methanol, or dioxane.

The sodium salt of 2,6-dichlorophenolindophenol, *N*-(3,5-dichloro-4-hydroxyphenyl)-*p*-benzoquinone imine **(Tillman reagent)**, is used especially in steroid chemistry for the conversion of α-hydroxy ketones into α-hydroxy acids and α-keto acids [976] (equation 20).

[976]   NaO-(Cl,Cl-C6H2)-N=C6H4=O   20

$$RCOCH_2OH \longrightarrow RCHCO_2H \longrightarrow RCOCO_2H$$
$$\!\!|$$
$$\!\!OH$$

Diethyl azodicarboxylate, $C_2H_5O_2CN=NCO_2C_2H_5$ (bp 106 °C at 13 mm of Hg), is a strong hydrogen acceptor in dehydrogenations and oxidations. It is converted into diethyl hydrazodicarboxylate and reacts usually at room temperature, although some of its reactions require refluxing in benzene. The reagent converts alcohols into aldehydes or ketones, thiols into disulfides, aromatic amines and hydrazo compounds into azo compounds [977], hydroxylamines into nitroso compounds [978], and sulfides into sulfoxides [*979*]. Diethyl azodicarboxylate is very useful for the demethylation of tertiary amines to secondary amines [*980*].

Triphenylmethyl tetrafluoroborate, $(C_6H_5)_3C^+BF_4^-$, oxidizes secondary alcohols in preference to primary alcohols [*981*].

Diphenylpicrylhydrazyl, which is prepared from N,N-diphenyl-N'-(2,4,6-trinitrophenyl)hydrazine by oxidation with lead dioxide [*982*], is used for the dehydrogenation of 1,4-dihydronaphthalene to naphthalene and of hydrazobenzene to azobenzene [*983*]. Simpler and cheaper compounds are, however, available for such purposes (equation 21).

[*982*] 21

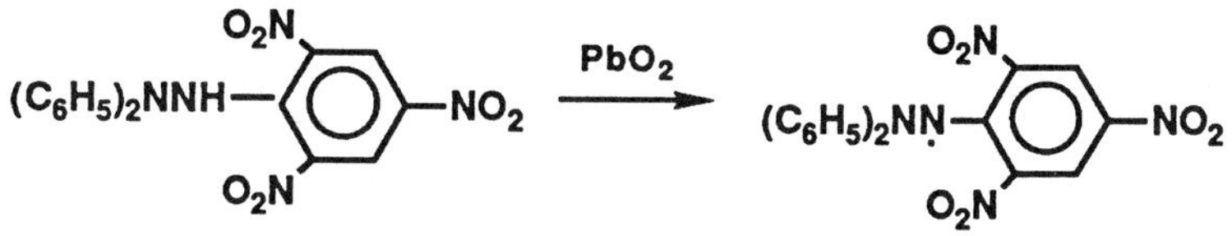

$$(C_6H_5)_2NNH-\underset{O_2N}{\underset{|}{\overset{O_2N}{\overset{|}{C_6H_3}}}}-NO_2 \xrightarrow{PbO_2} (C_6H_5)_2N\dot{N}-\underset{O_2N}{\underset{|}{\overset{O_2N}{\overset{|}{C_6H_3}}}}-NO_2$$

Nitrosobenzene, C_6H_5NO, which is obtained by the oxidation of phenylhydroxylamine, and ***p*-nitrosodimethylaniline,** $p\text{-}(CH_3)_2NC_6H_4NO$, which is easily prepared by the nitrosation of dimethylaniline, are fairly specific oxidizing agents for the preparation of aromatic aldehydes from benzyl halides or tosylates and of α-dicarbonyl compounds from from α-halo ketones [*984, 985*]. Also, a methylene group flanked by two carbonyls can be oxidized to a carbonyl group by nitrosodimethylaniline [*986*]. Pyridine is frequently used to form quaternary pyridinium salts from reactive halides prior to their oxidation to aromatic aldehydes, α-ketoaldehydes, or α-diketones [*984*] (equations 22 and 23).

[*984*] $ArCH_2X \xrightarrow{C_5H_5N} ArCH_2-\overset{+}{N}C_5H_5 \; \overset{-}{X} \xrightarrow{\overset{-}{O}H}$ 22

X = Cl, Br, I, or TosO

$$\longrightarrow Ar\overset{..}{C}H-\overset{+}{N}C_5H_5 \xrightarrow[-H_2O]{-C_5H_5N} \underset{R-N-O}{ArCH} \xrightarrow{H_2O} \underset{RNHOH}{ArCHO}$$
$$\uparrow$$
$$R-\ddot{N}=O$$

[*984*] $RCOCH_2X \xrightarrow{C_5H_5N} RCOCH_2-\overset{+}{N}C_5H_5 \; \overset{-}{X} \xrightarrow{\overset{-}{O}H}$ 23

$$\longrightarrow RCO\overset{..}{C}H-\overset{+}{N}C_5H_5 \xrightarrow[-H_2O]{-C_5H_5N} RCOCH=NR'$$
$$\underset{R'-\ddot{N}=O}{\curvearrowleft}$$
$$\downarrow H_2O$$
$$RCOCHO + R'NHOH$$

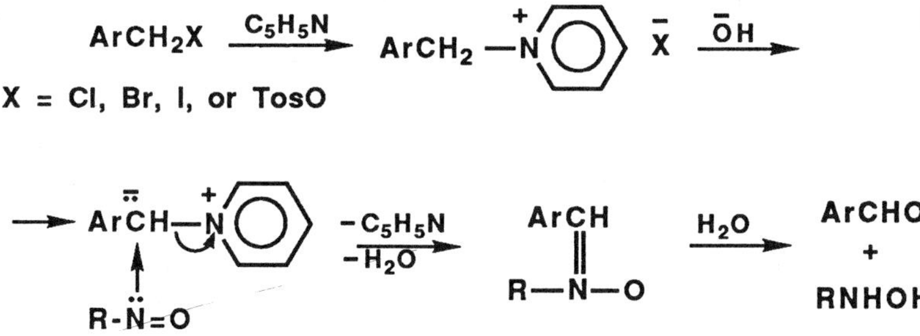

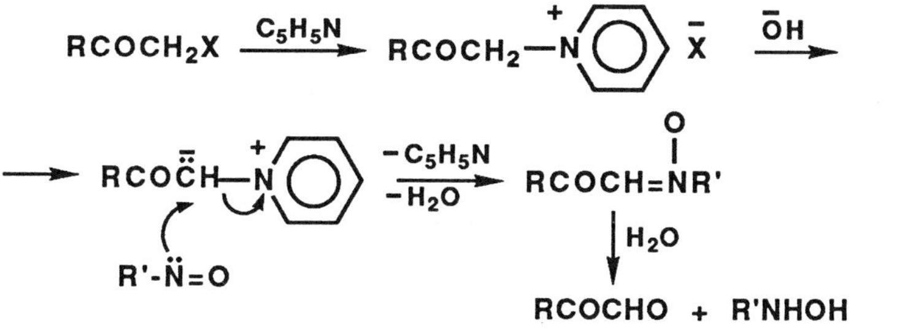

Nitro compounds in alkaline medium are used for rather specific oxidations. **2-Nitropropane** in the form of its sodium salt converts benzylic halides into aromatic aldehydes under fairly mild conditions [987]. **Nitrobenzene** transforms isoeugenol [988] and similar aromatic compounds with side chains that contain double bonds conjugated with the aromatic rings into aldehydes [988, 989]. In the *Skraup synthesis,* nitrobenzene dehydrogenates the dihydroquinoline intermediate to quinoline.

In the presence of alkalies and sulfur, *p*-nitrotoluene gives *p*-aminobenzaldehyde. The nitro group oxidizes the methyl group intramolecularly [990].

Trimethylamine oxide, $(CH_3)_3NO$, or its dihydrate, $(CH_3)_3NO \cdot 2H_2O$ (mp 97–100 °C), oxidizes boranes to alcohols [991, 992] and alkyl iodides to aldehydes [993] under gentle conditions. **N-Methylmorpholine oxide, $O(CH_2CH_2)_2NO$,** is used as a reoxidant in hydroxylations of alkenes with osmium tetroxide [952] and for the oxidation of nucleoside phosphites to nucleoside phosphates in an anhydrous medium [228]. **Pyridine oxide** hydroxylates aromatic rings to phenols [994].

1-Oxo-2,2,6,6-tetramethylpiperidinium chloride, an exotic reagent prepared from chlorine and 2,2,6,6-tetramethyl-1-piperidinyloxyl (a free-radical Tempo), converts diols into acyloins or lactones [995] (equation 24).

[995]   (structure) $\xrightarrow{Cl_2}$ (structure) $\bar{Cl}$ 24

Triphenyl phosphite ozonide, $(C_6H_5O)_3PO_3$, which is obtained by ozonization of triphenyl phosphite in dichloromethane solution at −78 °C, decomposes to triphenyl phosphate and singlet oxygen [178]. As an oxidant, it converts the trimethylsilyl ethers of enols of α,β-unsaturated ketones into unsaturated hydroxy ketones [996].

Dimethyl sulfide and chlorine or, better still, **dimethyl sulfide and N-chlorosuccinimide,** form a system capable of the selective dehydrogenation of alcohols to aldehydes or ketones. The intermediates, such as $(CH_3)_2S^+Cl\ Cl^-$, react with bases according to the scheme in equation 25 [720].

[720]   (succinimide structure)$N-\overset{+}{S}(CH_3)_2\bar{Cl}$ $\xrightarrow[-C_4H_5NO_2]{RCH_2OH}$ 25

$\xrightarrow[-HCl]{Et_3N}$ $(CH_3)_2\overset{+}{S}-O-CHR$ $\longrightarrow$ $(CH_3)_2S$ + $O=CHR$
                                        $|$
                                        $H \curvearrowleft NEt_3$ + $Et_3\overset{+}{N}H\bar{Cl}$

Dimethyl sulfoxide (DMSO), $(CH_3)_2SO$, is a versatile reagent for the oxidation of alcohols to carbonyl compounds under gentle conditions. In addition to the previously mentioned dehydrogenations, it is capable of other oxidations: acetylenes to α-diketones [997], alkyl halides to aldehydes [998, 999], tosyl esters to aldehydes [1000], methylene groups adjacent to carbonyl groups to carbonyls [1001, 1002], α-halocarbonyl compounds to α-dicarbonyl compounds [1003, 1004, 1005], aldehydes to acids [1006], and phosphine sulfides and selenides to phosphine oxides [1007].

The main domain of oxidation with dimethyl sulfoxide is the conversion of primary alcohols into aldehydes and of secondary alcohols into ketones. These reactions are accomplished under very mild conditions, sometimes at temperatures well below 0 °C. The reactions require the presence of acid catalysts such as acetic anhydride [713, 1008, 1009], trifluoroacetic acid [1010], trifluoroacetic anhydride [1011, 1012, 1013], trifluoromethanesulfonic acid [1014], phosphoric acid [1015, 1016], phosphorus pentoxide [1006, 1017], hydrobromic acid [1001], sulfur trioxide [1018], chlorine [1019, 1020], N-bromosuccinimide [997], carbonyl chloride (phosgene) [1021], and oxalyl chloride (the **Swern oxidation**) [1022, 1023, 1024]. Dimethyl sulfoxide also converts sufficiently reactive halogen derivatives into aldehydes or ketones [998, 999] and epoxides to α-hydroxy ketones at −78 °C [1014].

The mechanism of the reactions of DMSO with alcohols is shown in equation 26 and in a few more examples. Dicyclohexylcarbodiimide is sometimes used in addition to the acid catalysts (equation 26).

[*1021*]   **26**

$$RR'CHOH + ClCOCl \xrightarrow{-HCl} RR'CH-O-CO \xrightarrow{-\bar{Cl}}$$

with $(CH_3)_2S=O$ attacking, Cl leaving

$$RR'CH-O-CO \atop {}^+(CH_3)_2S\text{———}O \xrightarrow{-CO_2} \xrightarrow{Et_3N} \xrightarrow{Et_3N} RR'C-O \atop H \quad S(CH_3)_2^+$$

$$\longrightarrow RR'C=O + S(CH_3)_2 + (C_2H_5)_3\overset{+}{N}H$$

Phenylselenyl chloride, C_6H_5SeCl, and phenylselenyl bromide, C_6H_5SeBr, in connection with oxidizing agents such as hydrogen peroxide or sodium periodate, are used for the conversion of aldehydes, ketones, and esters into their α,β-unsaturated analogues. The key intermediate is alkyl phenyl selenoxide, which decomposes via a five-membered transition state [167] (equation 27).

[167]

RCH$_2$—CH$_2$COX $\xrightarrow[\text{or PhSeBr}]{\text{PhSeCl}}$ RCH$_2$—CHCOX(SeC$_6$H$_5$) $\xrightarrow[\text{THF}]{\text{30\% H}_2\text{O}_2}$ 27

RCH—CHCOX with H and Se(=O)C$_6$H$_5$ ⟶ RCH=CHCOX + C$_6$H$_5$SeOH

CHIRAL OXIDANTS

The ambition of synthetic organic chemists is not satisfied by inventing oxidations that are diastereospecific, that is, reactions that yield only, or predominantly, one of the two possible diastereomers, for example, either the *cis* or the *trans* product. The predominant formation of one enantiomer is desirable whenever it is possible. Such ***enantioselective oxidations*** take place when microorganisms or enzymes are involved. Only recently have such oxidations been accomplished by purely chemical means. Enantioselective oxidations require the presence of a chiral component and a metal catalyst that forms a complex with the chiral reagent. Some metallic compounds used are vanadyl acetylacetonate, VO(acac)$_2$ [216]; molybdenyl acetylacetonate, MoO$_2$(acac)$_2$ [1025]; and especially titanium tetraisopropoxide, Ti[OCH(CH$_3$)$_2$]$_4$. **Sharpless reagent,** which gives the best results, is a mixture of 1 mol of titanium isopropoxide, 1–2 mol of *tert*-butyl hydroperoxide as the oxidant, and 2 mol of (*R,R*)- or (*S,S*)-diethyl tartrate as the chiral component [221, 222, 223, 224, 1025, 1026, 1027, 1028]. The addition of 1 mol of water to this mixture is claimed to give better results in aqueous media [224, 1029]. *N*-Methyl- and *N*-ethylephedrine [1030], quinine and quinidine acetates [951], and *N*-phenylcamphoryl hydroxamic acid [1031] have also been used as chiral reagents.

(+)-**(2*R*,8*aS*)- or (−)-(2*S*,8*aR*)-camphorylsulfonyloxaziridine,** another chiral reagent, contains oxygen that reacts with the enolates of esters to form chiral α-hydroxy esters [1032] (equation 28).

[1032] camphorylsulfonimine $\xrightarrow{\text{Oxone}}$ camphorylsulfonyloxaziridine 28

The main applications of chiral oxidations are the epoxidation of double bonds, especially in allylic alcohols [*221, 222, 223, 1026, 1027, 1028, 1030, 1031*], the hydroxylation of double bonds [*951, 1033*] and of the α position with respect to an ester group [*1032*], and the oxidation of non-symmetrical sulfides to chiral sulfoxides [*224, 1025, 1029*]. Optical yields (enantiomeric excesses [ee]) are frequently over 95%.

BIOCHEMICAL (MICROBIAL) OXIDATIONS

The oxidations accomplished by microorganisms or enzymes excel in regiospecificity, stereospecificity, and enantioselectivity. Although the yields of such oxidations are sometimes fair and even low, optical purity (enantiomeric excess) is usually very high and frequently 100%. The spectrum of microbial and enzymatic oxidations is unbelievably broad; many different positions in steroidal rings can be hydroxylated by different microorganisms, and usually, only one diastereomer is formed. From achiral molecules, optically active compounds are generated.

Microbial oxidations occur under very mild conditions, usually around 37 °C, and in dilute solutions, for example, 1 g of a substance in 6 L of water. They are very slow and often take days. The isolation of products from the reaction media is done by extraction with proper solvents. Out of countless examples described in the literature, only a very narrow selection is presented in this chapter to show the versatility of microbial and enzymatic oxidations.

A single enzyme is sometimes capable of many various oxidations. In the presence of NADH (reduced nicotinamide adenine dinucleotide), **cyclohexanone oxygenase** from *Acinetobacter* NCIB9871 converts aldehydes into acids, formates of alcohols, and alcohols; ketones into esters (***Baeyer–Villiger reaction***); phenylboronic acids into phenols; sulfides into optically active sulfoxides; and selenides into selenoxides [*1034*]. **Horse liver alcohol dehydrogenase** oxidizes primary alcohols to acids (esters) [*1035*] and secondary alcohols to ketones [*1036*]. **Horseradish peroxidase** accomplishes the dehydrogenative coupling [*1037*] and oxidation of phenols to quinones [*1038*]. **Mushroom polyphenol oxidase** hydroxylates phenols and oxidizes them to quinones [*1039*].

The following is an alphabetical list of microorganisms capable of various oxidations: ***Acetobacter* spp.** [*1040, 1041*], ***Acinetobacter* spp.** [*1034, 1042, 1043*], ***Arthrobacter* spp.** [*1044*], ***Aspergillus* spp.** [*1045, 1046, 1047, 1048, 1049, 1050*], ***Bacillus megaterium*** [*1051*], ***Calonectria* spp.** [*575, 1052*], ***Cellulomonas* spp.** [*1053*], ***Cladiosporium* spp.** [*1054*], ***Corynebacterium* spp.** [*1055, 1056, 1057, 1058*], ***Cunninghamella* spp.** [*1059, 1060, 1061*], ***Curvularia* spp.** [*1062, 1063, 1064*], ***Cylindrocarbon* spp.** [*1065*], ***Fusarium* spp.** [*1066*], ***Gibberella* spp.** [*1067*], ***Heliostylum* spp.**

[*1059*], **Helminthosporium spp.** [*1054, 1068*], **Hygroporus spp.** [*1069*], **Mortierella spp.** [*1068*], **Nocardia spp.** [*1070, 1071*], **Penicillium spp.** [*1065, 1072*], **Pseudomonas spp.** [*92, 1073, 1074*], **Rhizopus spp.** [*1059, 1075, 1076, 1077, 1078, 1079*], **Sporotrichum spp.** [*1080, 1081, 1082, 1083, 1084*], **Streptomyces spp.** [*1050, 1085*], **Torulopsis spp.** [*1086, 1087*], and **Saccharomyces spp. (yeast)** [*1088*].

Among the various dehydrogenations and oxidations achieved by microorganisms are dehydrogenation to form double bonds [*1053, 1056, 1057, 1065, 1071*]; dehydrogenative coupling [*1037*]; oxidation of methyl groups to primary alcoholic groups [*1087*] or carboxylic groups [*1053, 1071, 1086, 1087*]; oxidation of $-CH_2X$, in which X is a halogen, to carboxylic groups [*1086*]; hydroxylation of methylene [*575, 1039, 1045, 1050, 1051, 1054, 1060, 1061, 1062, 1064, 1066, 1067, 1075, 1076, 1077, 1078, 1079, 1080, 1081, 1082, 1083, 1084*] and methine [*1059, 1062, 1083, 1089*] groups; oxidation of methylene groups to carbonyls [*1070, 1080, 1081*]; epoxidation of double bonds [*1055, 1063*]; hydroxylation of double bonds [*1073*]; conversion of alkenes into ketones [*1069*]; oxidation of primary alcohols to carboxylic acids [*1035, 1044*], of secondary alcohols to ketones [*1040, 1041, 1056, 1059, 1088*], of aldehydes to carboxylic acids, [*1034*], and of ketones to esters (Baeyer–Villiger reaction) [*1034, 1042, 1043, 1049, 1065, 1072*]; hydroxylation of aromatic rings [*1039*]; and oxidation of phenols to quinones [*1038, 1039*], of phenylboronic acids to phenols [*1034*], of primary amines to nitro compounds [*1085*], of sulfides to optically active sulfoxides [*1034, 1046, 1047, 1048, 1052, 1058*], of sulfoxides to sulfones [*1034, 1048*], and of selenides to selenoxides [*1034*].

Chapter 2
Dehydrogenations

DEHYDROGENATIONS, which involve the elimination of hydrogen from organic molecules, lead to compounds containing double bonds, multiple bonds, or aromatic rings. For practical reasons, only the formation of carbon–carbon double bonds, of carbon–nitrogen double bonds in cyclic amines, and of aromatic rings (both carbocyclic and heterocyclic) will be discussed in this chapter. The conversion of alcohols into aldehydes and ketones and of amines into imines and nitriles will be discussed in the chapter Oxidations (Chapter 3).

In addition to the reactions just mentioned, dehydrogenations also give rise to coupled products (intermolecular coupling) and to cyclic structures from open-chain compounds (intramolecular dehydrogenation). Examples are the formation of diphenyl from benzene (intermolecular coupling) and of toluene from heptane (intramolecular dehydrogenation). However, similar dehydrogenative doublings of compounds, such as the formation of hydrazo compounds from amines or of disulfides from thiols, will be treated in Oxidations (Chapter 3).

INTRODUCTION OF DOUBLE BONDS

Many dehydrogenations resulting in the *formation of carbon–carbon double bonds* are known in steroids and occur always in the α positions with respect to double bonds or carbonyl groups. Thus ergosterol or ergosterol acetate gives dehydroergosterol or its acetate in 40 or 37% yield, respectively, by refluxing in acetic acid with *mercuric acetate* [399, 400] (equation 29).

3β-Acetoxylanost-8-ene is converted into 3β-acetoxylanosta-7,9(11)-diene in 92% yield on treatment with acetic acid and 30% *hydrogen peroxide* for 45 h at 22 °C [275].

When testosterone acetate is refluxed with *2,3-dichloro-5,6-dicyano-p-benzoquinone* and *p*-toluenesulfonic acid in benzene for 5 h, a 60% yield of 6-dehydrotestosterone acetate is obtained [966]. The treatment of androst-4-ene-3,17-dione in dioxane with the same quinone and gaseous hydrogen chloride gives a 72% yield of androsta-4,6-diene-3,17-dione [966] (equation 30).

[*399*] 29

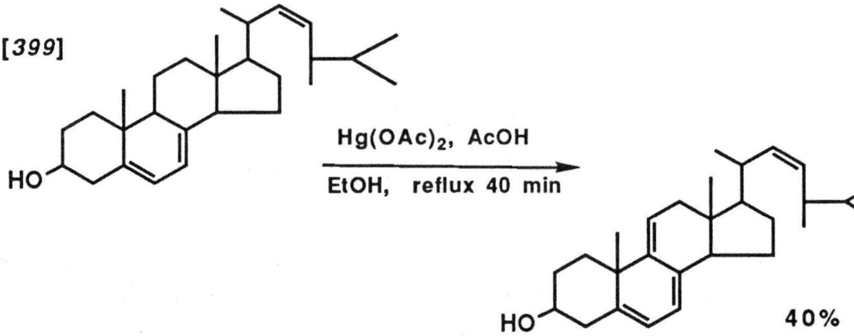
Hg(OAc)2, AcOH
EtOH, reflux 40 min
HO
HO
40%

[*966*] 30

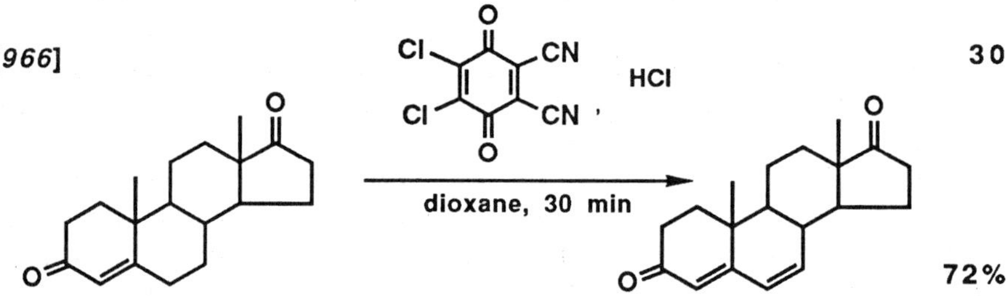
Cl
Cl
CN
CN
O
O
HCl
dioxane, 30 min
72%

Both cholestan-3-one and cholestan-3-ol yield 1,4-cholestadien-3-one on treatment with iodoxybenzene or *m*-iodoxybenzoic acid and a catalytic amount of benzeneseleninic anhydride, which is generated in situ from diphenyldiselenide [*526*]. The treatment of cholestan-3-one with phenylselenyl chloride and hydrogen peroxide gives 1-cholesten-3-one [*167*] (equations 31 and 32).

Similar dehydrogenations are often accomplished *biochemically*. On treatment with *Corynebacterium simplex,* 19-nortestosterone is transformed into estradiol [*1056*]. Many adrenocortical hormones are converted

[*526*] 31

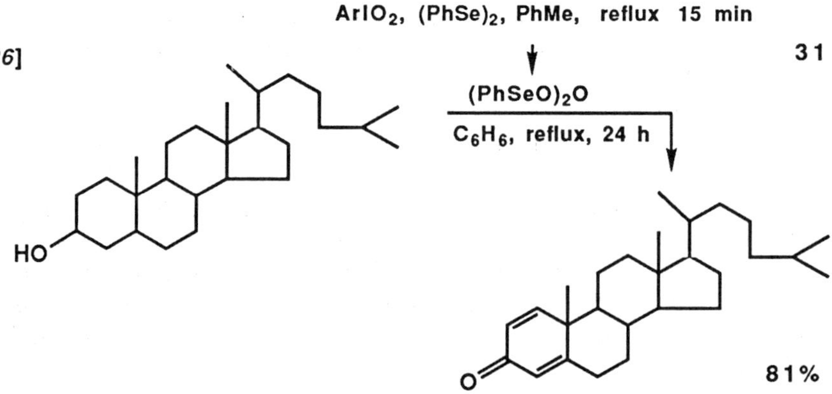
ArIO2, (PhSe)2, PhMe, reflux 15 min
(PhSeO)2O
C6H6, reflux, 24 h
HO
O
81%

[167]   32

                              1. PhSeCl,
                                 AcOEt,
                                 1 h
                              ────────────→
                              2. 30% H$_2$O$_2$
                                 THF, 35 °C,
                                 1 h
                                                                                                    96%

into their 1,2-dehydro analogues by the same microorganism [1057] (equation 33).

[1056]   33

                           Corynebacterium simplex
                                ATCC 6946
                              ────────────→
                                  30 °C
                                                                                                    75%

A classical example of the dehydrogenative formation of a carbon–carbon double bond conjugated with an aromatic ring is the dehydrogenation of ethylbenzene to styrene at 500–600 °C over a complex *catalyst* containing oxides of zinc and chromium [1090] or at 625 °C over vanadium pentoxide on alumina [478].

Several dehydrogenations forming double bonds conjugated with aromatic rings are achieved by heating with *quinones*. Acenaphthene gives a 75% yield of acenaphthylene when refluxed with chloranil for 16 h in xylene at 140 °C [967], and *p,p'*-dimethoxybibenzyl affords an 83–85% yield of *p,p'*-dimethoxystilbene when refluxed with 2,3-dichloro-5,6-dicyano-*p*-benzoquinone in dioxane at 105 °C [968].

Double bonds between carbon and nitrogen in secondary and tertiary cyclic amines are formed on treatment with *mercuric acetate* [401, 402, 1091], *silver oxide* [375], or tert-*butyl hypochlorite* [707]. tert-Butyl hypochlorite converts methyl pyrrolidine-2-carboxylate into methyl 1-pyrroline-2-carboxylate in 71% yield [707], and mercuric acetate transforms 2-*tert*-butylpiperidine into 2-*tert*-butyl-1-piperideine in 75% yield [1091]. Tertiary cyclic amines are dehydrogenated by mercuric acetate via quaternary salts [401,402]. 1-Methyl-2-ethylpiperidine yields ultimately 1-methyl-2-ethyl-2-piperideine [402] and quinolizidine $\Delta^{1(10)}$-dehydroquinolizidine [401] (equation 34).

[401] Hg(OAc)$_2$, 5% aqueous AcOH, 100 °C, 1.5 h → (iminium) 59% ClO$_4^-$ → 40% NaOH → 34, 68%

DEHYDROGENATIVE AROMATIZATIONS

Compounds containing six-membered rings are usually dehydrogenated all the way to carbocyclic and heterocyclic aromatic compounds. The driving force for such a deep dehydrogenation is the resonance energy liberated during the process: 36 kcal for benzene, 61 kcal for naphthalene, 84 kcal for anthracene, 92 kcal for phenanthrene, and 23 kcal for pyridine. Such "aromatizations" are easier if one or especially two double bonds are already present in a ring. The tendency for the aromatization is so strong that the formation of aromatic compounds occurs even when it requires the elimination of quaternary groups [495] and intramolecular rearrangements [493].

Dehydrogenative aromatizations are effected by heating hydroaromatic compounds with catalytic amounts of *platinum* [496, 945, 946] or, more often, *palladium* [408, 494, 496, 497, 945, 946, 948, 1092], usually supported on activated charcoal, asbestos, or other large-surface-area materials. Such **catalytic dehydrogenations** require temperatures of 250–350 °C and are carried out by refluxing the compounds with the metals or by passing their vapors over the metals. **Chemical dehydrogenations** are achieved by heating the compounds with elemental *sulfur* at 230–270 °C [491, 492] or red *selenium* at 280–350 °C [493].

During dehydrogenations, quaternary groups such as carboxyl or methyl are eliminated [495]. Other side chains are preserved [495], and if they are unsaturated, they may be hydrogenated [1093]. Alcohols are dehydrated, reduced, or rearranged [1092]. Ketones in six-membered rings are converted into phenols or hydrocarbons [496, 947], whereas those in side chains are left unchanged. However, if the ketone groups are adjacent to six-membered aromatic rings, they are hydrogenated [494,1092]. Cyclic structures other than six-membered rings are unchanged or rearranged [497].

The yields of dehydrogenations often leave much to be desired, but in many cases, the purpose of such reactions is not to prepare but, rather, to identify natural products such as terpenes or alkaloids by converting them into known aromatic compounds. With the advent of spectroscopic techniques, dehydrogenative aromatizations have lost much of their importance. Selected examples of both catalytic and chemical dehydrogenations are shown in equations 35–40 [*492, 493, 494, 495, 496, 497*].

[*492*]

35

[*494*] 36

[*496*] 37

[*495*] 38

[*497*]

39

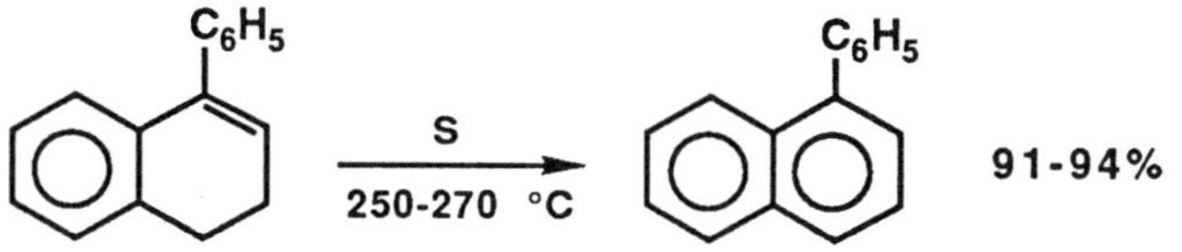

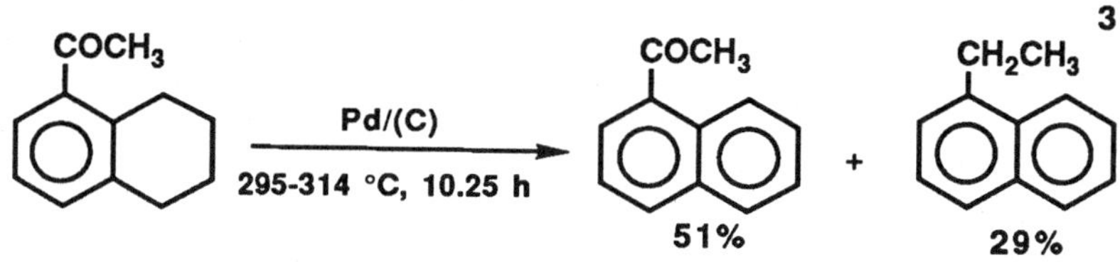

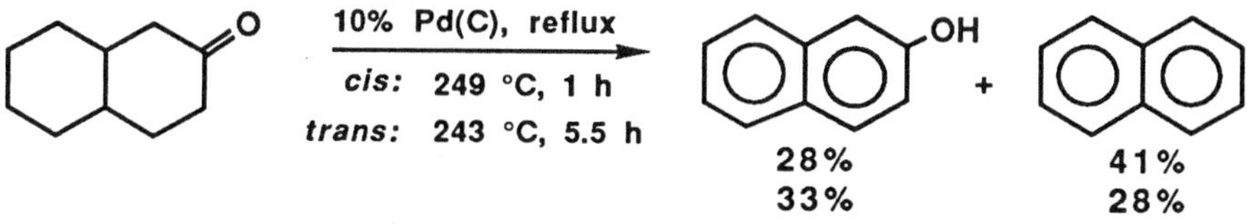

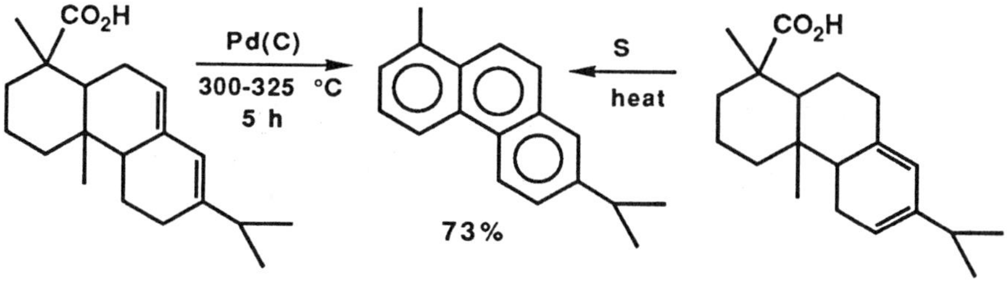

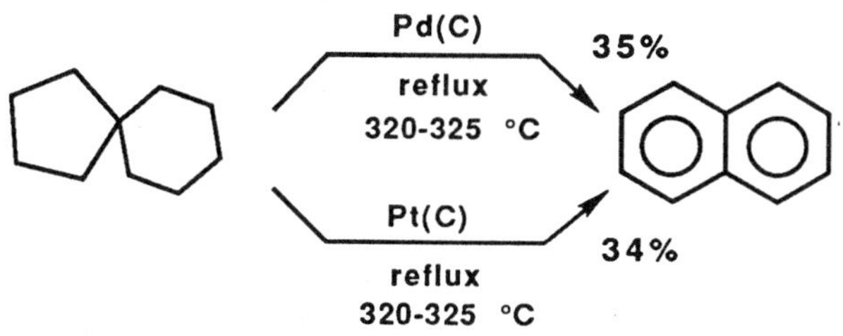

[493]

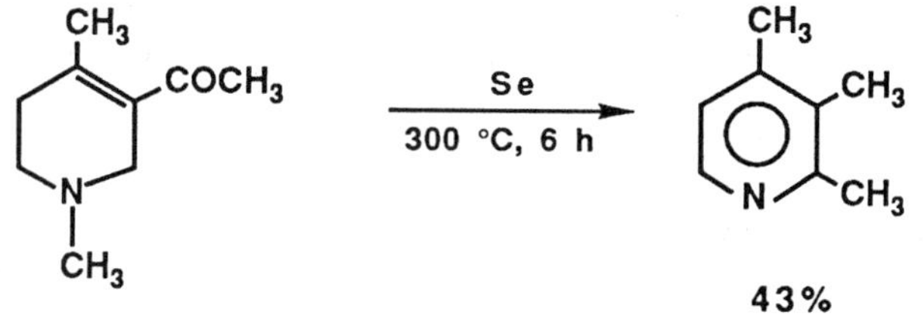

$$\underset{300\ °C,\ 6\ h}{\xrightarrow{Se}}$$

43%

40

Some aromatizations are accomplished with agents other than those previously mentioned. Thus tetralin is converted quantitatively into naphthalene by refluxing with *2,3-dichloro-5,6-dicyano-p-benzoquinone* [970], and 1,2,3,4-tetrahydrocarbazoles are transformed into carbazoles in 50–95% yields by refluxing with *chloranil* in xylene [969]. Anhydrous *aluminum chloride* in refluxing carbon disulfide is used in the preparation of coronene [408].

Dehydrogenations are very easy in **six-membered nitrogen-containing compounds** that already contain one or two double bonds in the heterocyclic ring. Such dehydrogenations are carried out by *nitric acid* [458], *arsenic oxide* (As_2O_5) [480], *manganese dioxide* [825], *potassium ferricyanide* [919], *nitrobenzene*, and others (equations 41 and 42).

[458]   41

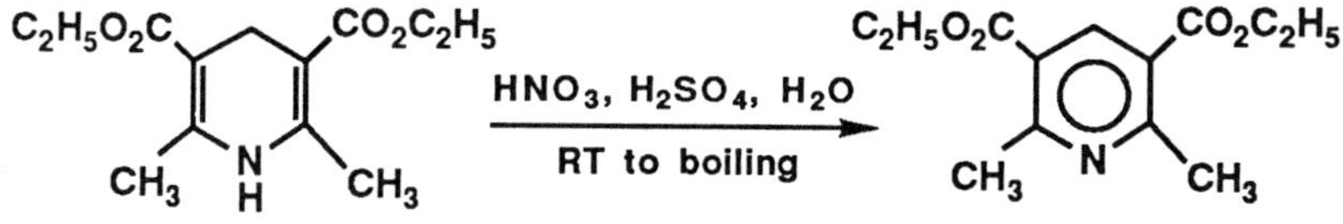

$$\xrightarrow[\text{RT to boiling}]{HNO_3,\ H_2SO_4,\ H_2O}$$

58-65%

[825]

$$\xrightarrow[81\ °C]{MnO_2,\ C_6H_6}$$

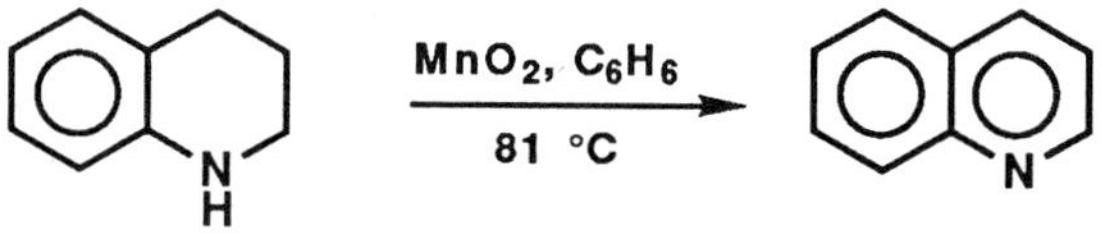

79%   42

DEHYDROGENATIVE COUPLINGS

Some compounds are capable of coupling after the initial abstraction of a hydrogen atom from one carbon, usually by a free-radical process. Such abstraction of hydrogen occurs easily when the remaining free radical is stabilized by resonance.

Thus benzene is converted into biphenyl in 54% yield when its vapors are passed at 720 °C through an iron tube [1094], and pyrene furnishes

1,1'-bipyrene in 70–75% yield when heated with a solution of *periodic acid* in acetic acid at 48–51 °C [*757*]. Coupling takes place easily at benzylic positions. Diarylmethanes give tetraarylethanes on treatment with *oxygen* in alcoholic potassium hydroxide [*1*] and on oxidation with *manganese dioxide* [*814*] or *potassium permanganate* [*870*] (equation 43).

[*870*] NaNH$_2$, NH$_3$; KMnO$_4$, -33 °C → 42% **43**

[*870*] (C$_6$H$_5$)$_2$CH$_2$ KNH$_2$, NH$_3$; KMnO$_4$, -33 °C → 69% (C$_6$H$_5$)$_2$CHCH(C$_6$H$_5$)$_2$

[*814*] MnO$_2$, C$_6$H$_6$, Ph$_2$, 211 °C → 81%

Coupling in *para* or *ortho* positions is easily accomplished with phenols [*912, 1037, 1095*] and aromatic amines [*915*]. β-Naphthol, on refluxing with anhydrous *ferric chloride* in ether, is converted in 60% yield into α,α'-bi-β-naphthol [*912*], and 2,4-bis(*tert*-butyl)-5-methylphenol, on heating with *chromic acid*, yields the corresponding "dimer" [*1095*] (equation 44).

[*1095*] K$_2$Cr$_2$O$_7$, AcOH, H$_2$SO$_4$, H$_2$O, 50-55 °C   **44**

The refluxing of 1-amino-2-methylnaphthalene with 66% *sulfuric acid* for 50 h or in the presence of *ferric oxide* for 8 h yields 4,4'-bis(1-amino-2-methyl)binaphthyl [*915*] (equation 45).

[*915*] 66% H$_2$SO$_4$, reflux 50 h → 74% **45**

Fe$_2$O$_3$, 66% H$_2$SO$_4$, RT, 24 h; reflux 8 h → 68%

If such oxidative couplings occur within the same molecule, cyclizations, usually to six-membered rings, take place. Anhydrous *aluminum chloride* proves useful in such intramolecular dehydrogenations. At 120 °C, it cyclizes benzil to 9,10-phenanthrenequinone [*407*] and bridges benzene rings in the synthesis of coronene [*408*]. At the same time, partial aromatization takes place (equations 46 and 47).

[*407*] 46

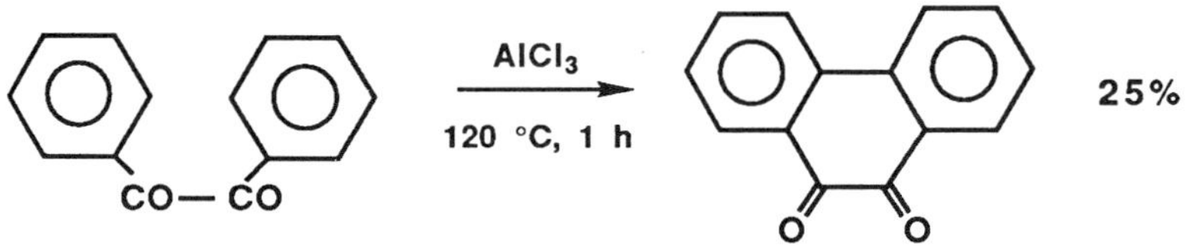

benzil $\xrightarrow{\text{AlCl}_3,\ 120\ °C,\ 1\ h}$ 9,10-phenanthrenequinone 25%

[*408*] 47

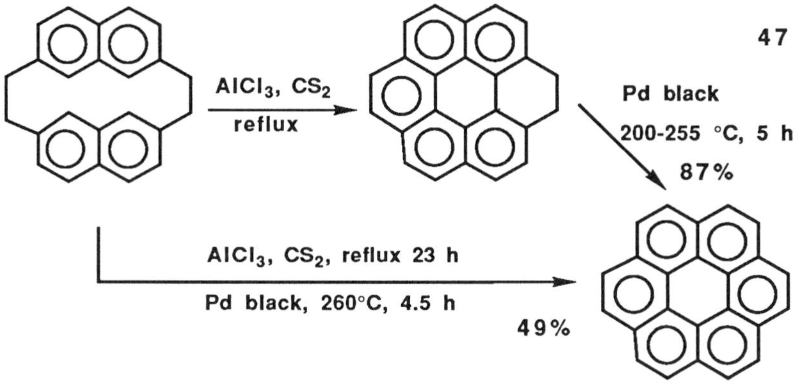

$\xrightarrow{\text{AlCl}_3,\ \text{CS}_2,\ \text{reflux}}$ $\xrightarrow[87\%]{\text{Pd black},\ 200\text{-}255\ °C,\ 5\ h}$ coronene

$\xrightarrow[49\%]{\text{AlCl}_3,\ \text{CS}_2,\ \text{reflux 23 h};\ \text{Pd black},\ 260°C,\ 4.5\ h}$

[*1037*] 48

laudanosoline methiodide $\xrightarrow{\text{1. 0.02% H}_2\text{O}_2,\ 1\ h,\ \text{Et}_3\text{N},\ \text{horseradish peroxidase};\ \text{2. HCl}}$ apomorphine methochloride 60%

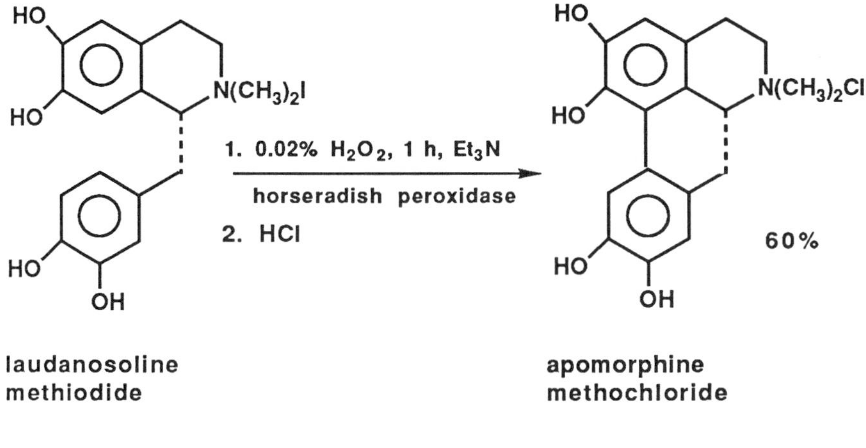

A large-scale application of intramolecular coupling combined with concomitant aromatization is the industrial conversion of heptane into toluene in 72% yield on heating at 500 °C in the presence of chromium and aluminum oxides [*1096*].

Dehydrogenative cyclizations are also accomplished with *potassium ferricyanide* [*920*] and with *hydrogen peroxide* in the presence of horseradish peroxidase [*1037*] (equation 48).

Chapter 3
Oxidations

OXIDATIONS OF INDIVIDUAL FUNCTIONAL GROUPS are organized according to the following sequence: alkanes and cycloalkanes; alkenes and cycloalkenes; alkynes (acetylenes); aromatic compounds (including heterocyclic aromatic compounds); halogen compounds; alcohols; ethers; aldehydes, ketones, and their derivatives; carboxylic acids and their derivatives; nitrogen compounds; phosphorus and arsenic compounds; sulfur compounds; selenium compounds; iodo compounds; boron compounds; silicon and tin compounds; and organomagnesium and organomercury compounds. Within the individual sections, the oxidants are arranged more or less in the same order as in Chapter 1, Oxidation Agents. The most commonly used oxidants are printed in **boldface** type, and the less frequently used ones are printed in *italic* type. The rest of the oxidants are applied only on special occasions or sporadically or have been tested only on a limited number of cases. Other items of importance, such as important compound types or reactions and safety warnings, are emphasized by printing in ***boldface italic*** type.

If a functional group characteristic of the classes of compounds just mentioned is remote enough to be affected by another functional group present in the molecule, its oxidation may be discussed at the same place as would be the oxidation of the parent compound. For example, the epoxidation of oleic acid is described in the section on alkenes. However, the oxidation of mesityl oxide, where the double bond is conjugated with the carbonyl group, is discussed under ketones. Oxidations of difunctionalized compounds are described in the sections discussing the higher functional groups. Thus, oxidations of aldols to dicarbonyl compounds will be discussed in the section on aldehydes or ketones (not alcohols). Frequent cross-references should facilitate location of the desired oxidation.

ALKANES AND CYCLOALKANES

Oxidations of alkanes and cycloalkanes are rare and not easy. The rates of oxidation with chromic acid of methyl, methylene, and methine groups differ by 2 orders of magnitude: $CH_3:CH_2:CH$ 1:65:3500 (later, rates in the ratio 1:114:7000–18,000 were found) [1097]. These rates are comparable with those of hydrogen abstraction by bromine (1:82:1640) [620].

A similar increase in reactivities in the methyl–methylene–methine series is found in the free-radical oxidations of lower alkanes with oxygen in the presence of hydrogen bromide as an initiator of the reaction. Ethane gives a 64% yield of acetic acid at 220 °C, propane gives a 72% yield of acetone at 189 °C, and isobutane gives a 69.5% yield of *tert*-butyl hydroperoxide, a 10% yield of *tert*-butyl alcohol, and a 6% yield of di-*tert*-butyl peroxide at 163 °C [54].

Terminal **methyl groups,** as well as the halomethyl groups in aliphatic halides with 15–18 carbon atoms, are oxidized biochemically to **carboxyls** (equation 49) [1086, 1087].

49

[1086]
[1087] $CH_3(CH_2)_{14}CH_2X \xrightarrow[25\ °C,\ 24\ h]{Torulopsis\ gropengiesseri} HO_2C(CH_2)_{14}CO_2H^*$

X	Yield (%)
F	9
Cl	33
Br	50
I	43

*The products were isolated as the methyl esters.

Similar microbial oxidations of terminal methyl groups accompanied by degradation and dehydrogenations of the carbon chains take place in alkyl benzenes [1053, 1071].

Methylene groups are hydroxylated to give alcoholic groups. Such **hydroxylations** are rarely achieved chemically but occur frequently in *biochemical processes*. The vast majority of such hydroxylations takes place in steroidal ketones and will be discussed in the section on ketones.

A methylene group is oxidized to a carbonyl group only if it is activated by neighboring aromatic rings, as in fluorene, or by carbonyl or carboxyl groups.

The relatively most frequent reaction is oxidation at the **methine group** giving tertiary alcohols as main products and secondary alcohols and ketones as minor products [902]. The hydroxylation occurs predominantly with retention of configuration [329, 620]. The oxidation of 3-ethylpentane with benzyltriethylammonium permanganate yields 24% of 3-ethyl-3-pentanol, 25% of 3-ethyl-2-pentanone, and 25% of 3-pentanone [902]. Under the same conditions, methylcyclohexane yields 72% of 1-methylcyclohexanol and 3% of 2-methylcyclohexanone [902].

Decalins oxidized by *p*-nitroperoxybenzoic acid in refluxing chloroform give 9-decalols with 95–96% regiospecificity and 99–100% stereospecificity (the isolated yield of *cis*-decalol is 69%) [*329*]. The distribution of the products of oxidation of diastereomeric decalins with other oxidants is shown in equation 50 [*545, 902*].

[*902, 545*]

	PhCH$_2$NEt$_3$MnO$_4$, AcOH, 3 °C, 3 days	CrO$_3$, AcOH, Ac$_2$O, <35 °C, 1 h, RT, 6 h	
	cis / trans	cis / trans	
9-decalol (OH)	67% / 37%	32% / 7%	
9,10-diol	—	5% / 3%	
1-decalone	43%	4% / 8% (combined)	
2-decalone	3%		

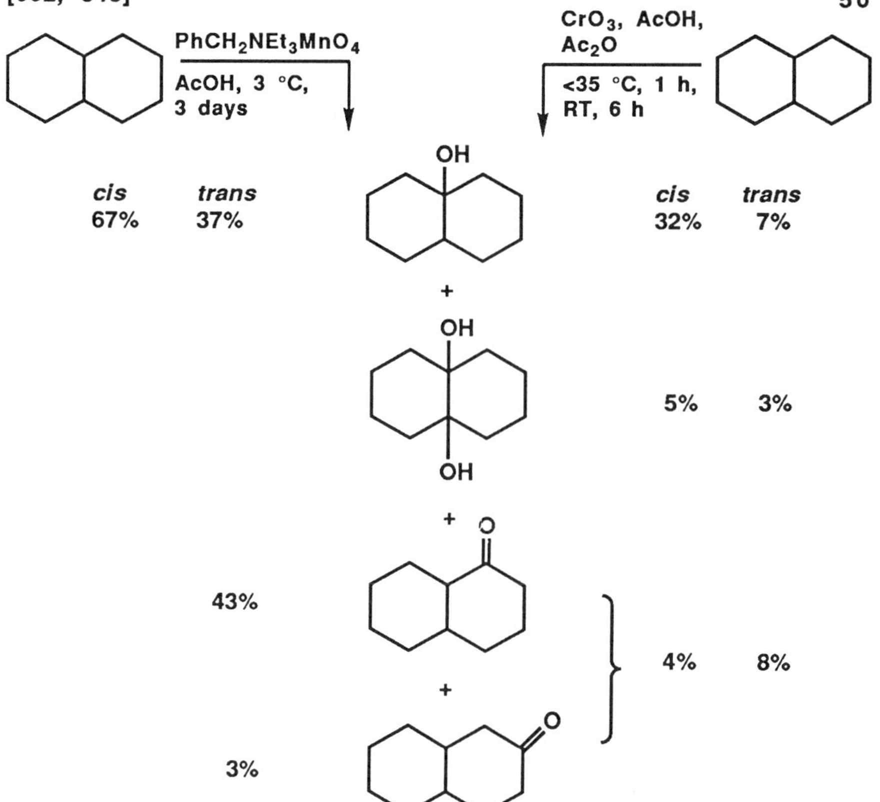

The oxidation of adamantane occurs almost exclusively at the tertiary carbon, C-1 (equation 51) [*106, 545, 546*].

[*545*] adamantane → CrO$_3$, Ac$_2$O, AcOH; 35 °C, 1 h; RT, 6 h → 71% 1-adamantanol*

[*546*] adamantane → CrO$_3$, H$_2$SO$_4$, 120–125 °C → 52%

[*106*] adamantane → O$_3$, SiO$_2$; −78 to −60 °C, 2 h; to RT, 3 h → 81–84%

51

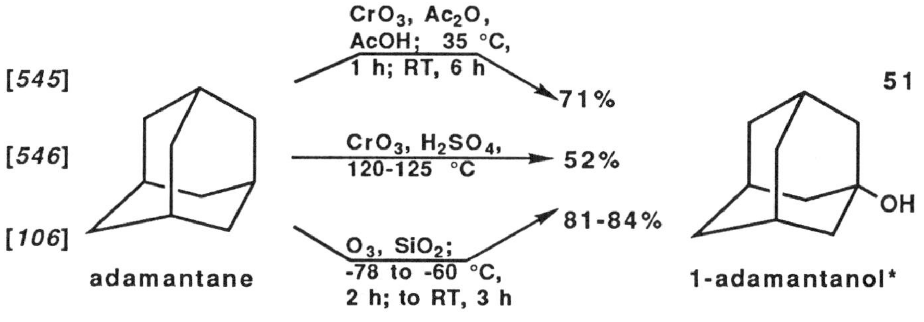

*The byproduct (7%) was 2-adamantanone.

The oxidation of hydrocarbons containing tertiary hydrogens gives **hydroperoxides** [54], sometimes in very low conversions [48] (equations 52 and 53).

[54]
$$(CH_3)_3CH \xrightarrow[163\ °C]{O_2,\ HBr\ catalyst} (CH_3)_3COOH \qquad 70\% \qquad 52$$

[48]
decalin $\xrightarrow[110\ °C,\ 24\ h]{O_2}$ decalin-H,OOH   *trans* 1.5%   53

ALKENES AND CYCLOALKENES

The oxidation of alkenes and cycloalkenes may affect the double bonds, the rest of the molecule, or both. Also included in this section are aromatic hydrocarbons containing double bonds in their side chains. Compounds containing double bonds and other functional groups, such as hydroxyl, carbonyl, or carboxyl, will be discussed in the appropriate sections, such as unsaturated alcohols, aldehydes, ketones, acids, and esters.

Epoxidation

The simplest imaginable case of oxidation of a double bond is the industrial production of ethylene oxide from ethylene and oxygen in the presence of silver oxide on alumina at 270 °C [1098].

In the laboratory, epoxidations of alkenes are usually accomplished by hydrogen peroxide or its derivatives. **Hydrogen peroxide** is applied in an alkaline medium so that it reacts as an anion (equation 54) [138, 142, 143, 146, 147, 148, 1099].

$$\rangle C=C\langle + \bar{O}-O-H \longrightarrow \rangle C=C\langle \longrightarrow \rangle C-C\langle + \bar{O}H \qquad 54$$

In the presence of a nitrile, alkaline hydrogen peroxide forms peroxycarboximidic acid, which, in the presence of an alkene, gives an epoxide and a carboxamide (equation 55) [129].

[129]
$$CH_3CN + H_2O_2 \xrightarrow[pH\ 8]{\bar{O}H} CH_3C\langle^{NH}_{OOH} \xrightarrow{\rangle C=C\langle} \rangle C-C\langle + CH_3CONH_2 \qquad 55$$

***tert*-Butyl hydroperoxide** epoxidizes double bonds either in alkaline solutions [*220*] or in the presence of transition metal compounds of titanium, vanadium, chromium, or molybdenum [*215, 216, 217, 221, 222, 223*]. In the presence of chiral reagents, such as titanium tetraisopropoxide and (*R,R*)- or (*S,S*)-diethyl tartrate, chiral epoxides are obtained (where applicable) [*223*].

The most popular derivatives of hydrogen peroxide used for epoxidation are organic **peroxy acids** (equation 56): peroxyacetic [*250, 251*], peroxytrifluoroacetic [*283, 287*], peroxybenzoic [*295, 296, 297, 300*], *m*-chloroperoxybenzoic [*315, 316*], *p*-nitroperoxybenzoic [*328*], peroxyphthalic [*330*], peroxymaleic [*338*], and peroxylauric [*174*].

$$\begin{array}{c}\diagdown\\ \diagup\end{array}C=C\begin{array}{c}\diagup\\ \diagdown\end{array} \longrightarrow \begin{array}{c}\diagdown\\ \diagup\end{array}C-C\begin{array}{c}\diagup\\ \diagdown\end{array} \quad \mathbf{56}$$

Other oxidants used for epoxidation are dimethyldioxirane [*210*], ozone [*107*], thallium triacetate [*412*], *chromic oxide* [*564*], chromyl compounds [*679*], *sodium hypochlorite* [*112, 689*], calcium hypochlorite [*704*], N-*bromosuccinimide* in water [*746*], iodine and silver oxide [*751*], iodine triacetate [*785*], electrolysis [*119*], and *microorganisms* [*1055, 1063*].

Epoxidation takes place preferentially or more rapidly at ***electron-rich*** (i.e., tetraalkylated) ***double bonds*** [*217*]. The reaction is ***stereospecific:*** *cis* alkenes give *cis* epoxides, and *trans* alkenes give *trans* oxides [*217*]. 1,2-Dimethylcyclopentene is oxidized with *peroxybenzoic acid* to 1,2-dimethylcyclopentene oxide in 85% yield [*296*], and *cis*-cyclooctene is transformed by hydrogen peroxide into *cis*-cyclooctene oxide in 60–61% yield [*1099*].

4-Vinylcyclohexene and *tert*-butyl hydroperoxide, in the presence of chromium acetylacetonate, yield exclusively 4-vinylcyclohexene oxide [*217*].

In 1,2-dimethylcyclohexa-1,4-diene, m-*chloroperoxybenzoic acid* epoxidizes only the electron-rich tetrasubstituted double bond to give a 68–78% yield of the 1,2-oxide (equation 57) [*316*].

The oxidation of isobutylene with thallium triacetate gives 82% of isobutylene oxide [*412*]. The oxidation of 1-pentene with *peroxytrifluoroacetic acid* yields 81% of propyloxirane [*283*]. The electrooxidation of

2-octene in dimethylformamide gives 73% of 1-methyl-2-pentyloxirane [*119*], and the oxidation of 1-hexadecene with *Corynebacterium equii* furnishes a 41% yield of (*R*)-1-tetradecyloxirane of 100% optical purity [*1055*]. 2-Cholestene is converted by iodine tris(trifluoroacetate) at –18 °C into 2-cholestene oxide [*785*], and the treatment of 1-mesitylstyrene with ozone in pentane at –78 °C gives 1-mesitylstyrene oxide in 71% yield [*107*]. The reaction of 1,1-dimethyl-2,2-diphenylethylene with *chromic oxide* in acetic anhydride yields 76% of the corresponding epoxide (oxirane) [*564*].

In *cis,trans*-1,5-cyclodecadiene, only the *trans* double bond is epoxidized with *peroxyacetic acid* in 80% yield.

Halogenated alkenes are transformed into halogenated epoxides by oxygen or hydrogen peroxide (equation 58) [*1100*].

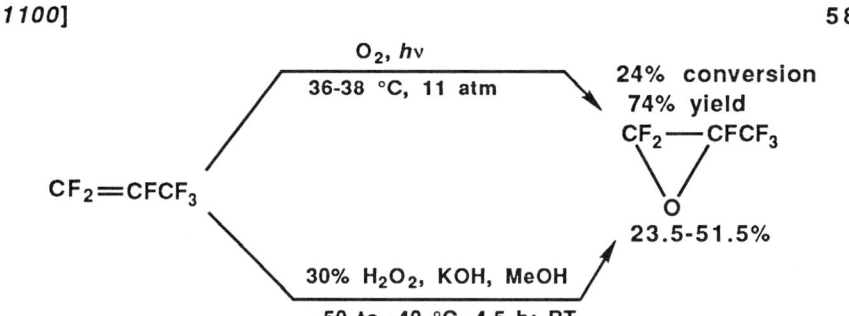

More examples of epoxidations of alkenes, cycloalkenes, and aromatic compounds having double bonds in their side chains are shown in equations 59–63 [*129, 210, 217, 251, 300, 315, 330, 338, 679, 689, 746, 751*].

Epoxidations of unsaturated alcohols, aldehydes, ketones, and acids and their derivatives are discussed in the appropriate sections.

EXAMPLES OF EPOXIDATION

$$CH_2=CHC_6H_{13} \longrightarrow \underset{CH_2-CHC_6H_{13}}{\overset{O}{\triangle}} \quad 59$$

[*679*] $CrO_2(NO_3)_2$, DMF, CH_2Cl_2, CCl_4, –78 °C, 30 min   43%

[*338*] 90% H_2O_2, (CHCO)$_2$O, CH_2Cl_2, 0 °C, 1.8 h   80%

[*217*] *t*-BuOOH; Cr, V, VO, or MoO_2 acetylacetonates; 25 °C   53%

60

cyclohexene → cyclohexene oxide

[129]	50% H_2O_2, MeCN, MeOH, NaOH (pH 9.5-10), 60 °C, 3.5 h	85%
[315]	m-ClC$_6$H$_4$CO$_3$H, KF, CH$_2$Cl$_2$, RT, overnight	>95%
[330]	30% H_2O_2, phthaloyl peroxide, Et$_2$O, RT; 0 °C	64%
[679]	CrO$_2$(NO$_3$)$_2$, C$_5$H$_5$N, CH$_2$Cl$_2$, CCl$_4$, -78 °C, 30 min	75%
[746]	NBS, H$_2$O, RT, 10 min; aqueous NaOH, 60 °C, 30 min	81%
[751]	I$_2$, Ag$_2$O, dioxane-H$_2$O (12:1), RT, 7 h	80%

$C_6H_5CH=CH_2$ → $C_6H_5CH\text{—}CH_2$ (epoxide) **61**

[129]	50% H_2O_2, MeCN, MeOH, NaOH (pH 7.5), 50 °C, 6 h	74%
[300]	BzO$_2$H, CHCl$_3$, 0 °C, 24 h	69-75%
[746]	NBS, H$_2$O, RT; 20% NaOH, 60 °C, 30 min	85%
[679]	CrO$_2$(NO$_3$)$_2$, DMF, CH$_2$Cl$_2$, -78 °C, 30 min	83%

62

trans-stilbene → trans-stilbene oxide

[251]	40% AcO$_2$H, AcONa, CH$_2$Cl$_2$, 20-35 °C, 15 h	70-75%
[210]	[2KHSO$_5$·KHSO$_4$·K$_2$SO$_4$ (Oxone) + Me$_2$CO] → Me$_2$C(O-O) dioxirane, Me$_2$CO, 22 °C, 2 h	73%
[679]	CrO$_2$(NO$_3$)$_2$, DMF, CH$_2$Cl$_2$, CCl$_4$, -78 °C, 30 min	30% + C_6H_5CHO (9%)

64   Oxidations in Organic Chemistry

[reaction scheme 63: phenanthrene → phenanthrene 9,10-epoxide]

[689]   NaOCl (pH 8-9), Bu$_4$NHSO$_4$, CHCl$_3$, RT,   90%
        15 min to 24 h
[210]   Me$_2$CO, K$_2$SO$_5$, Bu$_4$NHSO$_4$ (pH 7.5-8.5),   65%
        CH$_2$Cl$_2$, 0-10 °C, 5.5 h

Formation of Dioxetanes

The addition of one molecule of oxygen across a double bond results in the formation of dioxetanes, compounds containing two atoms of oxygen in a four-membered ring. Such compounds are formed especially from alkenes that do not possess allylic hydrogens. The addition is stereospecific, occurs in the *syn* mode [36], and takes place when **singlet oxygen,** generated chemically or photochemically, is applied. Many dioxetanes are isolated and even distilled at low temperatures. In other cases, they are assumed to be intermediates in the formation of dicarbonyl compounds.

Biadamantylene dissolved in methylene chloride at −78 °C and treated with a solution of triphenyl phosphite ozonide in the same solvent and subsequently with a mixture of methanol and pyridine gives biadamantylene dioxetane in 91% yield (equation 64). The dioxetane decomposes at 150 °C [1101].

[1101] [reaction scheme 64: biadamantylene → biadamantylene dioxetane]
        1. (PhO)$_3$PO$_3$, CH$_2$Cl$_2$, -78 °C
        2. MeOH-C$_5$H$_5$N (1:1), -78 °C
        91%

In the oxidation of indene [39] and of 1,2-diphenylcyclobutene [25] with singlet oxygen generated by irradiation of the solutions of the compounds in the presence of sensitizers, dioxetanes are the probable intermediates in the conversion of the unsaturated hydrocarbons into dialdehydes and diketones, respectively. Different products may be formed depending on the solvents used (equation 65) [25].

The stereospecific addition of singlet oxygen to *cis*- and *trans*-1,2-diethoxyethylene under irradiation with rose bengal as a sensitizer in fluorotrichloroethane at −78 °C gives *cis*- and *trans*-1,2-diethoxydioxetane [40]. Similarly, *cis*- and *trans*-α,α′-dimethoxystilbene give *cis*- and *trans*-1,2-dimethoxy-1,2-diphenyloxetane (equation 66) [38].

Tetramethoxyethylene in ether at −70 °C treated with oxygen under irradiation in the presence of zinc tetraphenylporphyrin furnishes a 94% yield of 1,1,2,2-tetramethoxydioxetane [37]. Chemically generated singlet oxygen converts substituted ketenes into α-peroxylactones (equation 67) [20].

Formation of Ozonides

The mechanism of the reaction of **ozone** with double bonds (equation 68) is very complex and still subject to arguments. An alkene and ozone may first form a π complex (**a**), which forms a σ complex (**b**), a "molozonide" (**c**), or both. The molozonide may change to a dipolar ion (**d**), which breaks down with the fission of the carbon–carbon bond to a carbonyl compound and another dipolar ion (**e**). The two species recombine to give the ultimate product, ozonide (**f**) (1,3,4-trioxolane, also known as 1,3,4-trioxacyclopentane) [76]. The temporary presence of the carbonyl compounds and the dipolar ion **e** explains formation of the "crossed" ozonides **g** and **h**.

The reaction is even more complicated when *cis–trans* isomerism is involved. Whereas *trans*-2,2,5,5-tetramethyl-3-hexene gives *trans*-di-*tert*-butylethylene ozonide, the *cis* isomer gives a mixture of 70% of *cis* ozonide and 30% of *trans* ozonide [79]. The ratios of *cis* and *trans* isomers may vary over a wide range depending mainly on the bulkiness of the alkyl groups at the double bonds. Whereas both *cis*- and *trans*-2-butene yield *cis* and *trans* ozonides in the ratio 2:3, *cis*-ethyl-*tert*-butylethylene gives *cis* and *trans* ozonides in the ratio 2:1. With the *trans* alkene, the ratio is 3:7 [76].

In the majority of cases, the ozonide structure is unimportant, because the purpose of the ozonation of double bonds is to prove structures by cleaving the chain or ring to form alcohols, aldehydes, ketones, or acids after reductive or oxidative workup of the ozonides.

When ozonides are to be isolated, a stream of 2–6% ozone in oxygen is passed through a solution of an alkene or a cycloalkene in low-boiling solvents, such as methyl chloride, pentane, hexane, chloroform, or carbon tetrachloride, at low temperatures (–78 to 5 °C) until the solution acquires a blue tinge (the color of ozone). The solvent is then evaporated at reduced pressure, and the ozonide is left as a solid, which is frequently polymeric, or a liquid, which can sometimes be distilled in vacuo. *Safety precautions* such as the use of face shields or protective shields and working in a hood *are imperative* [75].

Isobutylene gives a liquid ozonide, which is isolated in 40% yield by distillation at 43–43.5 °C at 143 mm of Hg (equation 69) [75].

[75] $(CH_3)_2C=CH_2$ $\xrightarrow[\text{distillation}]{6\% \text{ O}_3, \text{ MeCl} \atop -75 \text{ °C;}}$ $(CH_3)_2C\overset{O}{\underset{O-O}{\diagdown}}CH_2$ 69

40%

bp 43-43.5 °C at 143 mm of Hg

Similarly, 1-pentene ozonide distilling at 44–45 °C (at 32 mm of Hg) is obtained in 48% yield from 1-pentene [75]. *cis*- And *trans*-di-*tert*-butylethylene ozonides (bp 23 °C at 0.3 mm of Hg) are obtained by ozonization of *cis*- and *trans*-2,2,5,5-tetramethyl-3-hexene in pentane solutions at –75 °C in 82 and 58% yields, respectively [78, 79].

1-Dodecene and 1-tridecene give quantitative yields of the ozonides in pentane and chloroform solutions, respectively. Pentane is more suitable as a solvent than chloroform or carbon tetrachloride (equation 70) [77].

[77] $CH_2=CHR$ $\xrightarrow{O_3}$ $CH_2\overset{O}{\underset{O-O}{\diagdown}}CHR$ 70

R = $C_{10}H_{21}$ C_5H_{12}, -10 °C   99%
R = $C_{11}H_{23}$ $CHCl_3$, -5 °C   99%

Ozonides are hardly ever isolated. Rather, conversions to products of **ozonolysis** are carried out in the same reaction vessels (*see* Oxidative Cleavage of Double Bonds, page 77).

Hydroxylation

Hydroxylation, the addition of two hydroxyl groups across double bonds, converts **alkenes and cycloalkenes into vicinal diols. Stereochemically**, the addition may occur in the *syn* or the *anti* mode. In open-chain alkenes (with the exception of terminal alkenes for which stereochemistry is irrelevant), *syn* hydroxylation transforms *cis* alkenes into *erythro* (or *meso*) diols and *trans* alkenes into *threo* (or DL) diols. *anti* Hydroxylation of *cis* alkenes gives *threo* (or DL) diols, whereas *anti* hydroxylation of *trans* alkenes yields *erythro* (or *meso*) diols. *syn* Hydroxylation of cycloalkenes gives *cis* diols, whereas *anti* hydroxylation furnishes *trans* diols (Table I).

syn Hydroxylation results from the hydrolysis of five-membered cyclic

Table I. Stereochemistry of Hydroxylation of Alkenes and Cycloalkenes

Structure of Alkene	Mode of Addition	Structure of Diol
cis (symmetrical)	syn	meso
	anti	DL
trans (symmetrical)	syn	DL
	anti	meso
cis (nonsymmetrical)	syn	DL-*erythro*
	anti	DL-*threo*
trans (nonsymmetrical)	syn	DL-*threo*
	anti	DL-*erythro*
cyclic (3–7-membered ring)	syn	cis
	anti	trans

esters of osmic or manganic acid obtained by treatment of alkenes with osmic acid (osmium tetroxide) or potassium permanganate (equations 71 and 72).

syn *Hydroxylations* (equation 73) are achieved by dilute aqueous solutions (1–2%) of **potassium permanganate** at low temperatures (0–5 °C). Yields are not too high because the isolation of the water-soluble diols from dilute aqueous solutions is very difficult [296, 853]. Therefore, the alternative hydroxylation with **osmium tetroxide (osmic acid)** is preferred [949]. Because this reagent is very toxic and very expensive, usually only catalytic amounts are used in the presence of oxidizing agents such as *hydrogen peroxide* [*130, 152*], *sodium chlorate* [*714*], *potassium chlorate* [*715*], silver chlorate [*310*], barium chlorate [*310*], or amine oxides [*952*], which reoxidize hexavalent osmium to octavalent osmium.

anti *Hydroxylations* are achieved by reagents forming three-membered epoxides, which are hydrolytically cleaved by back-side nucleophilic attack (inversion) (equation 74).

$$\begin{array}{c} \diagdown \diagup \\ C \\ \| \\ C \\ \diagup \diagdown \end{array} \xrightarrow{RCO_3H} \begin{array}{c} \diagdown \diagup \\ C \\ | \\ C \\ \diagup \diagdown \end{array} O \xrightarrow{H_2O} \begin{array}{c} \diagdown \diagup \\ C \\ | \\ C \\ \diagup \diagdown \end{array} O \cdots H-O-H \rightarrow \begin{array}{c} \diagdown \diagup \\ C-O^- \\ | \\ H\overset{+}{O}-C \\ | \\ H \end{array} \rightarrow \begin{array}{c} \diagdown \diagup \\ C-OH \\ | \\ HO-C \\ \diagup \diagdown \end{array} \quad 74$$

With strong acids such as **formic acid** or **trifluoroacetic acid,** the opening of the epoxide gives the half-ester of the diol, which is easily hydrolyzed to the diol (equations 75–77) [*240, 288*].

$$\begin{array}{c} \diagdown \diagup \\ C \\ \| \\ C \\ \diagup \diagdown \end{array} \xrightarrow{HCO_3H} \begin{array}{c} \diagdown \diagup \\ C \\ | \\ C \\ \diagup \diagdown \end{array} O \xrightarrow{HCO_2H} \begin{array}{c} \diagdown \diagup \\ C-OH \\ | \\ HCOO-C \\ \diagup \diagdown \end{array} \xrightarrow{H_2O} \begin{array}{c} \diagdown \diagup \\ C-OH \\ | \\ HO-C \\ \diagup \diagdown \end{array} \quad 75$$

[*240*]

$$CH_2=CHC_{12}H_{25} \xrightarrow[\text{2. 3 N KOH/EtOH, reflux 1 h}]{\text{1. 25.6\% } H_2O_2, \text{ 98–100\% } HCO_2H, \text{ 40 °C, 24 h}} \underset{\underset{HO}{|}\quad\underset{OH}{|}}{CH_2CHC_{12}H_{25}} \quad 76$$

95% crude
69% pure

[*288*]

$$CH_3CH=CHC_2H_5 \xrightarrow[\text{2. 3\% HCl/MeOH, reflux 2 h}]{\text{1. 90\% } H_2O_2, (CF_3CO)_2O, CH_2Cl_2, CF_3CO_2H.Et_3N} \underset{74\%}{\overset{OH}{\underset{|}{CH_3CH-CHC_2H_5}}} \quad 77$$

Another kind of *anti* hydroxylation is the reaction of alkenes with **hydrogen peroxide** in the presence of vanadium pentoxide, molybdenum trioxide, selenium dioxide, and especially tungsten trioxide (tungstic acid) [*131, 144*]. Cyclohexene is thus converted into DL-*trans*-1,2-cyclohexanediol [*131*].

Stereospecific hydroxylations can also be accomplished via **silver acetates** [*781, 782*] or **silver benzoates** [*784*] and **iodine** followed by hydrolysis of the intermediate esters. Depending on the reaction conditions, either *syn* or *anti* hydroxylation (acyloxylation) can be accomplished (equations 78 and 79).

With dry reagents, *anti* addition takes place to form diesters, which yield diols on hydrolysis *(Prévost reaction)*.

$$2\ RCOOAg + I_2 \longrightarrow AgI + RCO_2Ag \cdot RCO_2I \qquad 78$$

anti addition   *syn* addition

[783]   BzOAg, I₂, C₆H₆
          ─────────────
          3 days

1. AcOAg, I₂, AcOH, 40 min
2. H₂O, 20 °C, 12 h
3. LiAlH₄, Et₂O, reflux 1 h

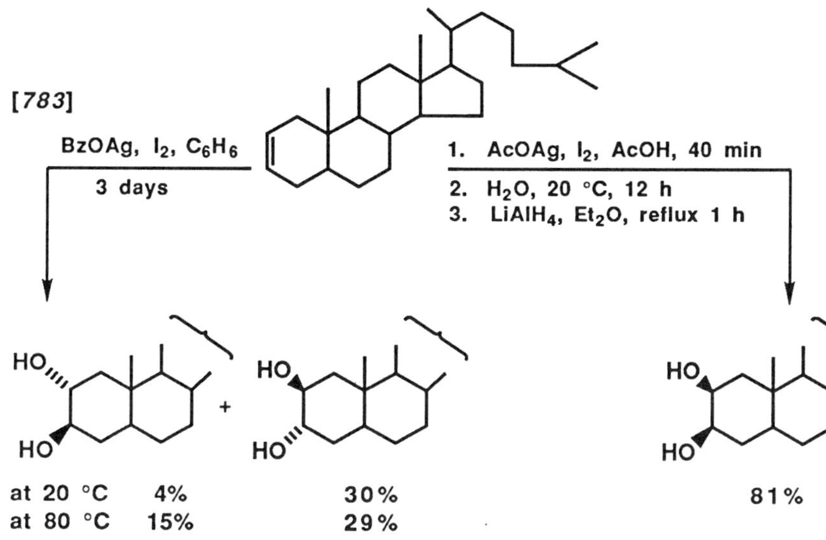

79

at 20 °C   4%   30%   81%
at 80 °C 15%   29%

If, on the other hand, aqueous acetic acid is added to the reaction mixture containing the positively charged intermediate, the product is a monoester of the diol, which, on hydrolysis, gives a diol. Such a reaction mimics a *syn* addition **(Woodward reaction)** [781, 783].

Instead of silver salts, the somewhat cheaper but more toxic thallium(I) acetate may be used [410]. Examples of *syn* and *anti* hydroxylations of cycloalkenes are shown in equations 80–82 [131, 242, 244, 410, 782, 900, 905, 952].

Double bonds in the side chains of aromatic compounds undergo hydroxylation in the same way as those in simple alkenes [784]. With some compounds, such as stilbene, **enantioselective hydroxylation** can be accomplished with *chiral compounds,* which, by complexing osmium tetroxide, form enantiomeric products in high enantiomeric excesses (equation 83) [951, 1033].

The presence in alkenes of chiral groups, such as phenyl sulfoxide [1102] or *N,S*-dimethyl-*S*-phenylsulfoximine [1103], affects the direction of the access of osmium tetroxide from just one (or predominantly one) face of the double bond. Thus predominantly one enantiomer of the *syn*-hydroxylation product is formed (equations 84 and 85).

Enantioselective hydroxylation of double bonds also occurs in *biochemical oxidation* by **Pseudomonas putida** [1073].

The same oxidizing agents and conditions used for the hydroxylation of alkenes and cycloalkenes are applied to unsaturated alcohols, ethers, aldehydes, ketones, and acids and their derivatives.

EXAMPLES OF *syn* AND *anti* HYDROXYLATIONS   80

syn hydroxylation

[782] 1. AcOAg, I₂, AcOH, RT, 4.5 h   66%
      2. AcOH, H₂O, reflux, 1 h

[952] OsO₄, (N-methylmorpholine N-oxide),   89-90%
      Me₂CO, H₂O

[410] 1. AcOTl, I₂, AcOH, 80 °C, 0.5 h   70-75%
      2. H₂O, reflux 9 h
      3. NaOH, EtOH, reflux 2 h

anti hydroxylation

[242, 1. 30%, H₂O₂, 88% HCO₂H, 40-45 °C   73-75%
 244] 2. 35% NaOH, 45 °C

[131] 30% H₂O₂, AcOH, WO₃, 50 °C,   74-77%
      5-6.5 h

[410] 1. AcOTl, I₂, AcOH, reflux 9 h   65-70%
      2. NaOH, EtOH, reflux 3 h

[905] cyclooctene → Ph₃MePMnO₄, CH₂Cl₂, -70 °C, 30 min → *cis* diol   *cis* 80%   81

[900]   82

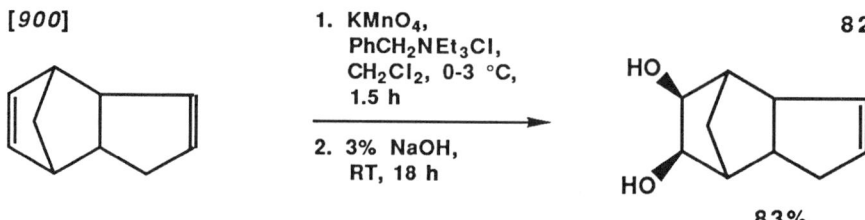

1. KMnO₄, PhCH₂NEt₃Cl, CH₂Cl₂, 0-3 °C, 1.5 h
2. 3% NaOH, RT, 18 h

83%

[951] 1. OsO$_4$, PhMe, RT, dihydroquinidine acetate   83
2. LiAlH$_4$ 85%, 82% ee

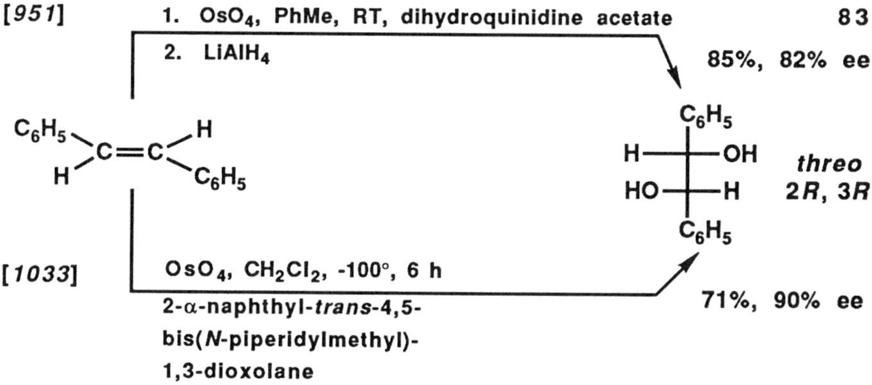

C$_6$H$_5$CH=CHC$_6$H$_5$ (cis) → threo 2R, 3R diol (C$_6$H$_5$, H, OH; HO, H, C$_6$H$_5$)

[1033] OsO$_4$, CH$_2$Cl$_2$, -100°, 6 h
2-α-naphthyl-*trans*-4,5-bis(*N*-piperidylmethyl)-1,3-dioxolane   71%, 90% ee

[1102]   84

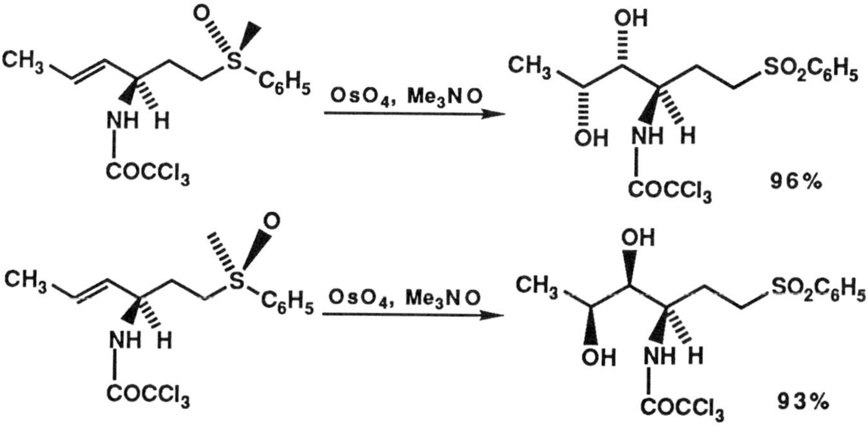

CH$_3$-CH=CH-C(NHCOCCl$_3$)(H)-CH$_2$-S(O)-C$_6$H$_5$ →(OsO$_4$, Me$_3$NO)→ CH$_3$-CH(OH)-CH(OH)-C(NHCOCCl$_3$)(H)-CH$_2$-CH$_2$-SO$_2$C$_6$H$_5$ 96%

(diastereomer) → 93%

[1103]   85

cyclohexenyl sulfoximine →(OsO$_4$)→ diol   (+) 81%
(−) 70%

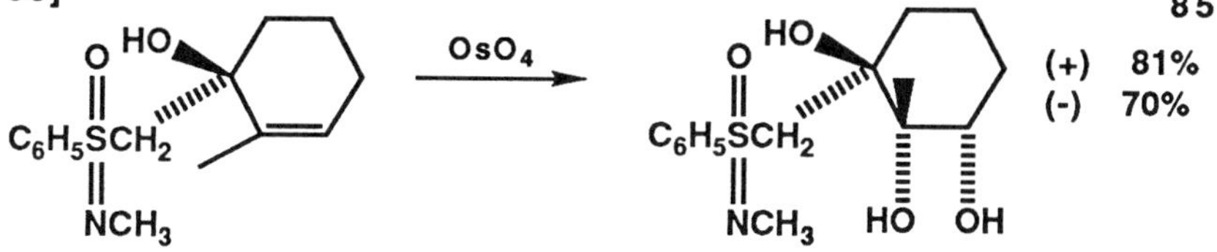

Conversion into Halohydrins, Amino Hydroxy Compounds, and Esters

The addition of halogens and hydroxyls across double bonds leads to **halohydrins,** which are useful intermediates, especially for the synthesis of epoxides. Such additions are achieved by treatment of alkenes with **N-bromoacetamide** [*1104*] or **N-bromosuccinimide** [*746*] in aqueous media and give products of *anti* addition. On heating with alkalies, bromohydrins give epoxides in good yields. Thus, styrene affords an 82% yield of styrene bromohydrin (2-bromo-1-phenylethanol) and an 85% yield of styrene oxide [746]. Cyclohexene is converted into cyclohexene bromohydrin and cyclohexene oxide in good yields (equation 86) [764, 1104].

86

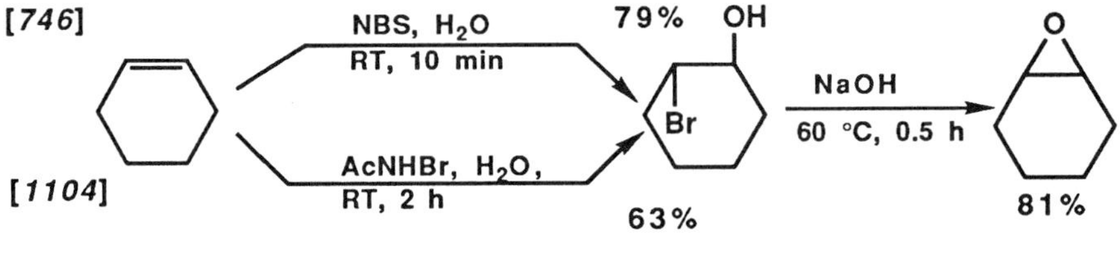

[746] cyclohexene → (NBS, H$_2$O, RT, 10 min, 79%) → 2-bromocyclohexanol (OH, Br) → (NaOH, 60 °C, 0.5 h) → cyclohexene oxide, 81%

[1104] (AcNHBr, H$_2$O, RT, 2 h, 63%)

Vicinal *amino hydroxy compounds* are prepared by treatment of alkenes with *N*-chloroamides in the presence of osmium tetroxide (equation 87) [723, 1105].

87

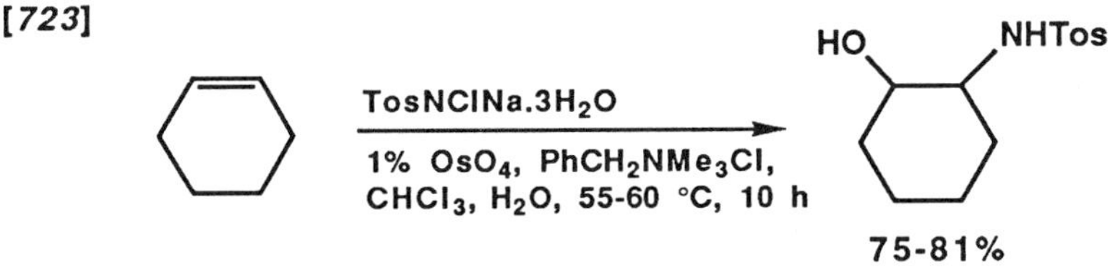

[723] cyclohexene → (TosNClNa·3H$_2$O, 1% OsO$_4$, PhCH$_2$NMe$_3$Cl, CHCl$_3$, H$_2$O, 55–60 °C, 10 h) → trans-2-(tosylamino)cyclohexanol (HO, NHTos), 75–81%

Heating alkenes and cycloalkenes with manganese triacetate yields lactones resulting from an oxidative addition of acetoxyls across the double bonds (equation 88) [803].

88

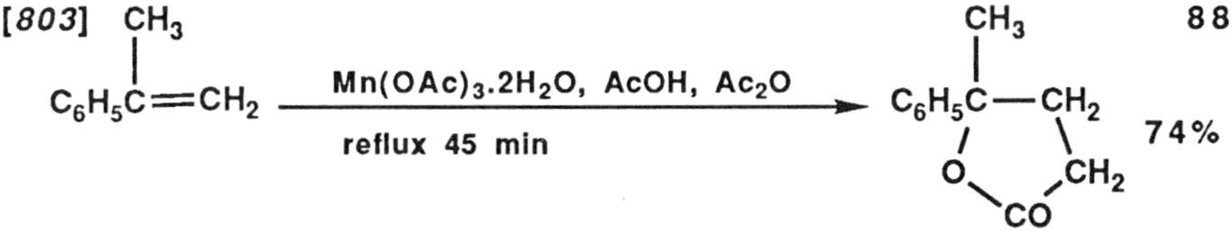

[803] C$_6$H$_5$C(CH$_3$)=CH$_2$ → (Mn(OAc)$_3$·2H$_2$O, AcOH, Ac$_2$O, reflux 45 min) → γ-butyrolactone derivative, 74%

Acyl esters of vicinal diols are obtained by the reaction of alkenes with metal carboxylates [436]. **Lead tetraacetate** in acetic acid at 70 °C converts 1,2-dihydronaphthalene to *trans*-1,2-diacetoxy-1,2,3,4-tetrahydronaphthalene in 72% yield [436]. The reaction is not always stereospecific. Cyclohexene treated with thallium triacetate gives a mixture of diastereomers in varying ratios, depending on reaction conditions, and byproducts as a result of rearrangements (equation 89) [411].

Heating stilbene with benzoyl peroxide and iodine in carbon tetrachloride at 80 °C for 48 h gives an 83% yield of the dibenzoyl ester of hydrobenzoin (1,2-dibenzoxy-1,2-diphenylethane) [230]. Esters of vicinal diols are intermediates in the preparation of diols by the Prévost and the Woodward methods (equation 78).

[411]

Tl(OAc)₃, dry AcOH, RT, a few days → 12% (OAc, OAc diacetate) + 88% (OAc, OAc bridged) + 89 (cyclopentyl-CH(OAc)₂ etc.)

Tl(OAc)₃, wet AcOH, RT → 81% + 19%, 40-55%

Conversion into Ketones

Alkenes can be oxidized to **ketones of the same chain length** by using salts of copper, palladium, and mercury as catalysts and *air, electrolysis* [*120*], *hydrogen peroxide*, or *chromium compounds* as oxidants [*60, 65, 140, 565*] (equation 90).

EXAMPLES OF OXIDATION OF ALKENES TO KETONES

$$CH_2=CHR \longrightarrow CH_3COR \quad 90$$

Ref	R	Conditions	Yield
[65]	C_8H_{17}	O_2, $PdCl_2$, CuCl, H_2O, DMF, RT, 24 h	65-73%
[60]	$C_{10}H_{21}$	O_2, $PdCl_2$, $CuCl_2 \cdot 2H_2O$, H_2O, DMF, 60-70 °C, 2.5-3.5 h	51-87%
[140]	C_6H_{13}	30% H_2O_2, Pd(OAc)$_2$, AcOH, 80 °C, 6 h	96%
[140]	$C_{10}H_{21}$	30% H_2O_2, Pd(OAc)$_2$, AcOH, 80 °C, 6 h	92%
[565]	C_6H_{13}	CrO_3, H_2SO_4, H_2O, Hg(OCOEt)$_2$ Me$_2$CO, 25 °C, 4 h	82%
[120]	C_8H_{17}	DDQ*, electrolysis, Pd(OAc)$_2$, Et$_4$NBF$_4$, Pt anode, 0.83 A/dm^2, RT	63%

*DDQ is 2,3-dichloro-5,6-dicyano-*p*-benzoquinone.

The oxidation of indene with peroxyformic acid gives a 69–81% yield of 2-indanone after hydrolysis of the intermediate, 1-formoxy-2-hydroxyindane (equation 91) [*246*].

The oxidation of 2,4,4-trimethylpentene with *chromyl chloride* followed by treatment with zinc affords a 70–78% yield of 2,4,4-trimethyl-

[246] Scheme: indene → 30% H₂O₂, 88% HCO₂H, 30-40 °C, 2 h, RT, 7 h → indanol with OCHO substituent → H₂SO₄, H₂O, reflux → indanone (69-81%)   **91**

pentanal [672]. A similar reaction of *trans*-cyclododecene with chromyl chloride followed by treatment with zinc dust in acetic acid affords cyclododecanone in 75% isolated yield [670]. The intermediates in these reactions are **α-chloro ketones,** which are prepared in 38–79% yields by adding chromyl chloride to alkenes or cycloalkenes in acetone solutions at low temperatures (−70 or −5 °C) [670].

α-Iodo ketones result from the reaction of alkenes and cycloalkenes, especially the electron-rich ones, with *silver chromate and iodine* (equations 92 and 93) [616].

[616]   **92**

$Ag_2CrO_4 + I_2 \longrightarrow$ AgI + AgO−Cr(=O)−OI ; with $C_6H_5CH=CH_2$ → AgO−Cr(=O)−O−I intermediate → $C_6H_5C-CH_2I$ cyclic intermediate with AgO−Cr=O → $C_6H_5COCH_2I$ + $AgHCrO_3$

[670] cyclohexene: CrO_2Cl_2, Me_2CO, −70 to −65 °C, 30 min; −75 °C, 1 h; 23-25 °C, 1 h → 2-chlorocyclohexanone (38%)

[616] cyclohexene: $Ag_2CrO_4, I_2, C_5H_5N, CH_2Cl_2$, 0 °C, 25 min → 2-iodocyclohexanone (60%)   **93**

A solution of *potassium permanganate* in aqueous acetone buffered with acetic acid converts internal alkenes into **α-hydroxy ketones** (equation 94) [857].

Alkenes may also be transformed into **vicinal dicarbonyl compounds.** The treatment of alkenes with selenium dioxide, although possible, does not give satisfactory yields [509]. Better results are obtained when unsaturated compounds such as oleic acid are oxidized with **potassium permanganate** buffered with acetic anhydride [861].

[857]

$$\underset{H}{\overset{C_4H_9}{>}}C=C\underset{C_4H_9}{\overset{H}{<}} \xrightarrow[25\ °C]{KMnO_4,\ Me_2CO,\ H_2O,\ AcOH} C_4H_9\underset{OH}{\overset{\overset{O}{\|}}{CH-CC_4H_9}} \quad 94$$

73%

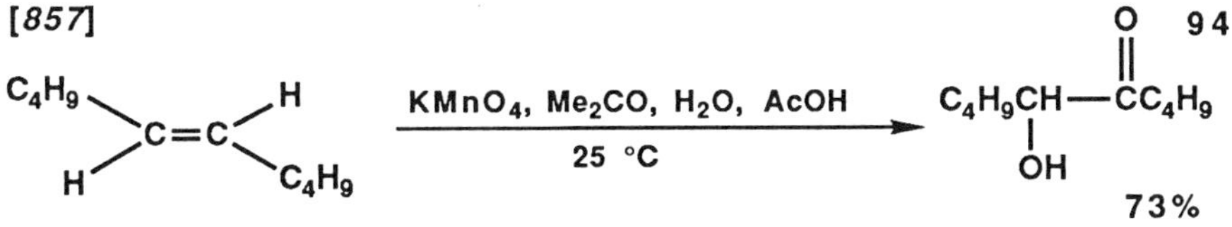

Oxidative Cleavage of Double Bonds

Double bonds are oxidatively cleaved to alcohols, aldehydes, ketones, or acids. The specific product formed depends on the structure of the alkene, that is, the presence or absence of hydrogen atoms at the carbons of the double bonds, and on the oxidants used.

The oldest and still very common cleavage is the reaction of unsaturated compounds with ozone and the subsequent treatment of the ozonides formed (equations 95 and 96). Reduction by strong reducing agents such as complex hydrides gives **alcohols** [94]. Cycloalkenes yield **diols** [82].

[94]

$$C_2H_5CH=CHC_4H_9 \xrightarrow[-42\ to\ -38\ °C,\ 6.5\ h]{O_3,\ C_5H_{12}} \left[C_2H_5CH\underset{O-O}{\overset{O}{\diagup\diagdown}}CHC_4H_9 \right] \quad 95$$

↓ LiAlH$_4$, Et$_2$O
-10 °C, 2 h

87% C$_2$H$_5$CH$_2$OH
+
87% C$_4$H$_9$CH$_2$OH

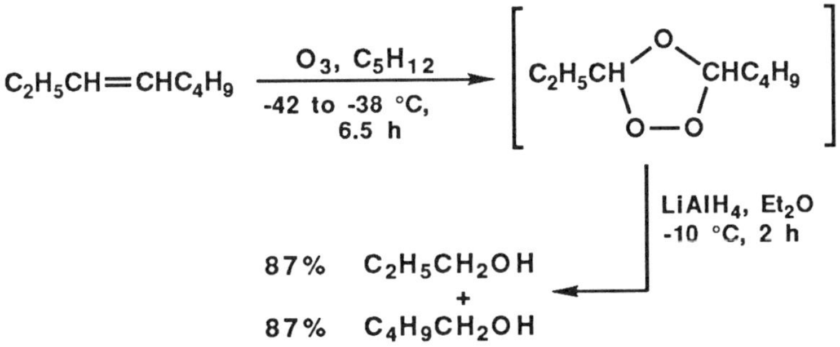

[82]

cyclohexene $\xrightarrow[-78\ to\ 20\ °C]{O_3,\ MeOH}$ [cyclohexane with OCH$_3$, OOH, CHO] $\xrightarrow[\substack{H_2/Pt,\\35-50\ °C,\\3.5-11\ atm}]{\substack{H_2/Pt,\\0-15\ °C,\\1-3.5\ atm}}$ cyclohexane-CH$_2$OH, CH$_2$OH 96

95%

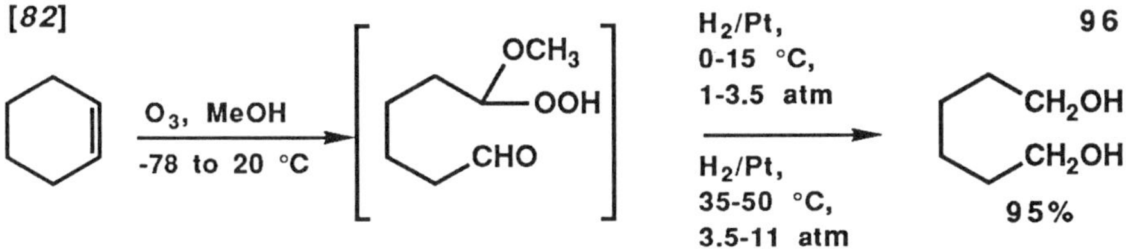

An example showing the different products formed by different reducing agents is the ozonolysis and reduction of 2-phenylskatole (3-methyl-2-phenylindole) (equation 97) [95].

A much more frequently used reaction is the cleavage of unsaturated compounds to **aldehydes** (equations 98 and 99). Alkenes and cycloalkenes that possess one or two hydrogens at the double bonds are oxidized by ozone to ozonides, which have to be reduced to prevent a subsequent oxidation to acids by the excess oxygen atom. Reductions are carried out, usually without isolation of the ozonides, by **catalytic hydrogenation over palladium catalyst** [80, 81, 1106] or Raney nickel [85] or by treatment with

[95]

97

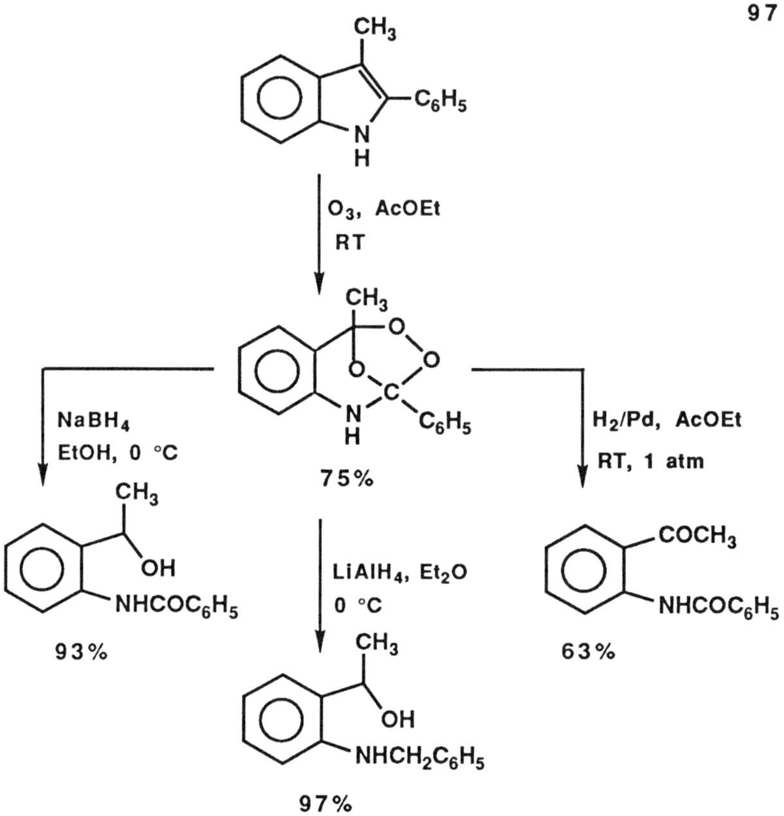

(Scheme: 3-methyl-2-phenylindole + O_3, AcOEt, RT → ozonide intermediate (75%); three branches:
- NaBH$_4$, EtOH, 0 °C → 2-(1-hydroxyethyl)-N-benzoylaniline (93%)
- LiAlH$_4$, Et$_2$O, 0 °C → 2-(1-hydroxyethyl)-N-benzylaniline (97%)
- H$_2$/Pd, AcOEt, RT, 1 atm → 2-acetyl-N-benzoylaniline (63%))

zinc and acetic acid [*81, 86, 87, 1107*], with **trimethyl phosphite** [*89*], or with **dimethyl sulfide** [*90, 91, 92, 1108*] (equations 100 and 101).

As alternatives to ozonolysis, other oxidations are used to prepare aldehydes from unsaturated compounds: treatment of alkenes with chromyl

[*91*]

$$C_6H_{13}CH=CH_2 \xrightarrow[\text{2. Me}_2\text{S, -60 °C; -10 °C, 1 h; RT, 1 h}]{\text{1. O}_3\text{, MeOH, -30 to -60 °C}} C_6H_{13}CHO$$

98

75%

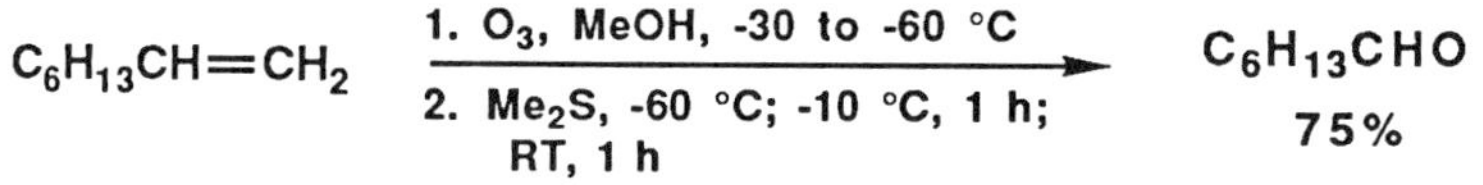

[*86*]

99

(4-hydroxy-3-methoxy-propenylbenzene → 4-hydroxy-3-methoxybenzaldehyde
- 1. O$_3$, AcOEt; 2. Zn, Et$_2$O, AcOH, cooling → 71%
- 6.3% H$_2$O$_2$ in *t*-BuOH, V$_2$O$_5$, RT, 12 h → 66%)

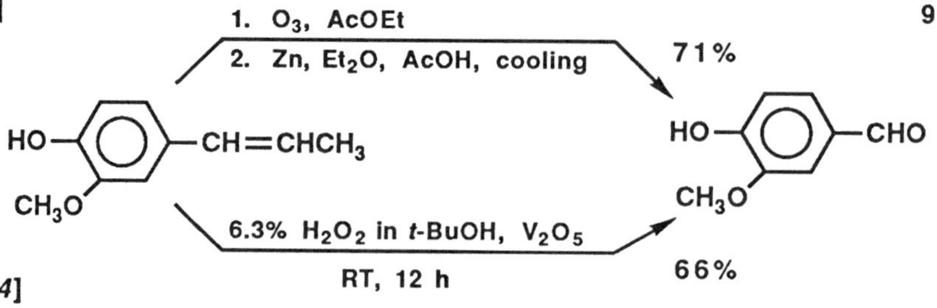

[*134*]

		100
		cyclohexane-1,2-dicarbaldehyde (CHO, CHO)
[*80, 1106*]	1. O$_3$, AcOEt, -50 to -70 °C 2. H$_2$/Pd(CaCO$_3$)	70%
[*89*]	1. O$_3$, MeOH, -65 to -70 °C 2. (MeO)$_3$P, -20 °C	85%
[*1108*]	1. O$_3$, MeOH, CH$_2$Cl$_2$, -78 °C 2. TosOH, RT, 1.5 h 3. NaHCO$_3$, RT, 15 min 4. Me$_2$S, 12 h	68-70%
[*765*]	OsO$_4$, NaIO$_4$, Et$_2$O, H$_2$O 24-26 °C, 2 h	77%
[*680*]	CrO$_2$(OCOCCl$_3$)$_2$, CCl$_4$, Me$_2$CO, 1.25 h; (CO$_2$H)$_2$, 3 h	46%

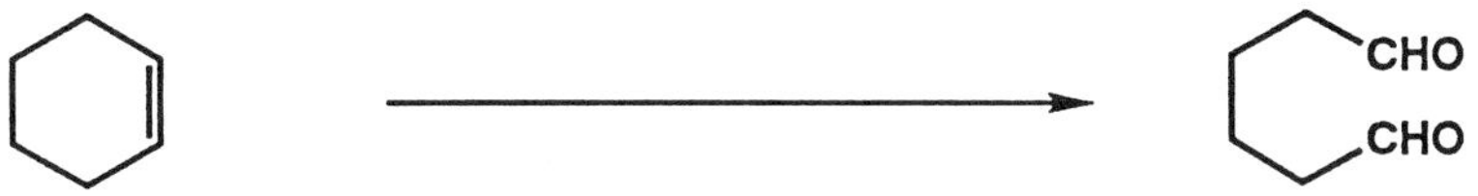

[*81*] CH$_3$(CH$_2$)$_7$CH=CH(CH$_2$)$_7$CO$_2$CH$_3$ **101**

$$\xrightarrow[-20\ °C,\ 3.25\ h]{O_3,\ MeOH}$$ CH$_3$(CH$_2$)$_7$CHO + OCH(CH$_2$)$_7$CO$_2$CH$_3$

H$_2$/Pd(C), AcOMe, AcOH, 22 °C, 1.5 h	76%	83%
Zn, AcOH, MeOH, 30-35 °C	81%	94%

acetate, trichloroacetate, or benzoate followed by reduction with oxalic acid [*680*] or their oxidation with sodium periodate [*760*], with **osmium tetroxide and sodium metaperiodate** [*765*], with osmium tetroxide and anhydrous hydrogen peroxide in ether [*141*], with anhydrous hydrogen peroxide in *tert*-butyl alcohol in the presence of vanadium pentoxide [*134*], or with **potassium permanganate** under special conditions [*834, 840, 842, 853, 895, 900*] (equations 102–104).

[*141*] **102**

$$C_6H_5CH=CHC_6H_5 \xrightarrow[Na_2SO_4,\ spontaneous\ reflux,\ RT,\ overnight]{4\ H_2O_2\ (20\%\ in\ Et_2O),\ OsO_4} C_6H_5CHO$$

79%

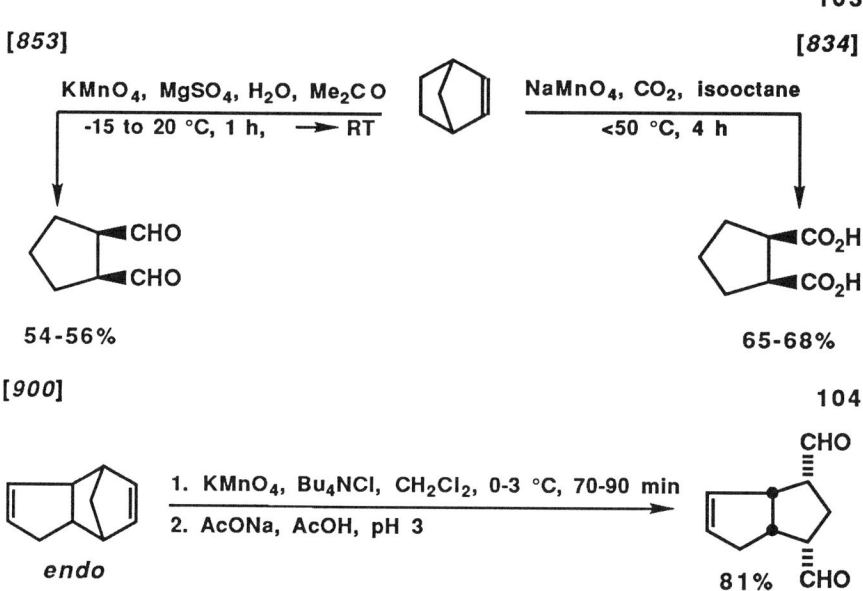

Unsaturated hydrocarbons with tetrasubstituted double bonds are cleaved to **ketones** by **ozonolysis** followed by reduction with zinc in acetic acid [1107], by *oxygen* sensitized with methylene blue under irradiation [25, 1109], by *chromium trioxide* and sulfuric acid at low temperature [564], or by sodium periodate [760] (equations 105–107).

The cleavage of double bonds in unsaturated alcohols (including saccharides) [87], phenols [86, 989], unsaturated ketones [1110], and unsaturated acids [840, 842] and esters [81] are discussed in the appropriate sections.

The oxidation of alkenes and cycloalkenes and their halogen derivatives with at least one hydrogen or halogen atom at the double bond leads to **carboxylic acids.** *Ozonolysis* usually requires the oxidative decomposition of the ozonide. The oxygen content of the ozonide is not sufficient for the formation of two molecules of acids or one dicarboxylic acid. The nonoxidative decomposition of cyclohexene ozonide gives an aldehyde–acid or its derivatives [1108]. It comes, therefore, as a surprise that carboxylic acids are claimed as products of the decomposition of ozonides by hydrogenation over the Lindlar catalyst [83] (equation 108).

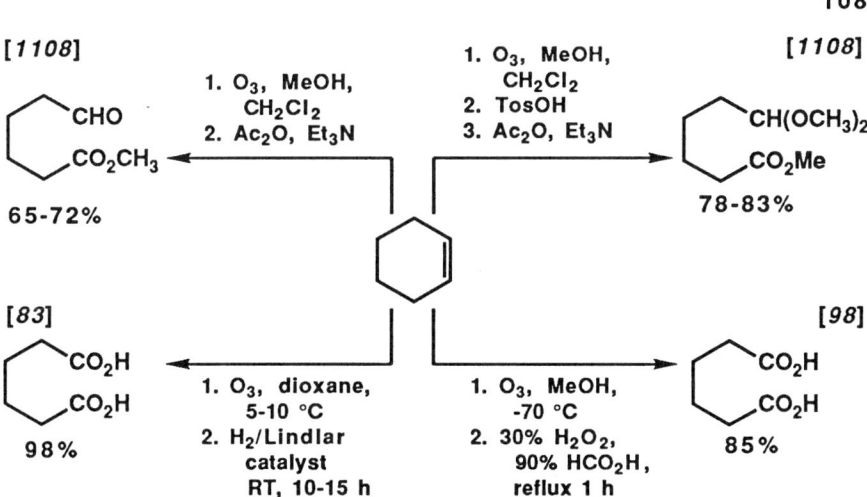

The oxidative decomposition of ozonides is further accomplished by their treatment with **hydrogen peroxide** [97] in the presence of formic acid [98, 99] or acetic acid [102, 103] or with silver oxide and nitric acid [77] (equations 109–112).

[99] $C_6H_5P(O)(CH_2CH=CH_2)_2$ $\xrightarrow[\text{2. -78 °C, 30% } H_2O_2,\text{ } HCO_2H,\text{ reflux}]{\text{1. } O_3,\text{ MeOH, -78 °C, 4 h}}$ $C_6H_5P(O)(CH_2CO_2H)_2$ **110**   83%

[103]   **111**

2-pyridyl-$CH_2CH_2CH=CCICH_3$ $\xrightarrow[\text{2. 30%, } H_2O_2,\text{ AcOH, 70 °C, 2 h}]{\text{1. } O_3,\text{ CHCl}_3,\text{ -10 °C, 8 h}}$ 2-pyridyl-$CH_2CH_2CO_2H$ 82%

[77]   **112**

$C_{10}H_{21}CH=CH_2$ $\xrightarrow[\begin{array}{l}\text{2. } Ag_2O,\text{ NaOH, 95 °C, 6 h}\\\text{3. } HNO_3,\text{ 60 °C, 1 h}\end{array}]{\text{1. } O_3,\text{ } C_5H_{12},\text{ -10 °C}}$ $C_{10}H_{21}CO_2H$ 98%

The most common and simplest cleavage of double bonds to carboxyls is achieved with **potassium permanganate** [*838, 841, 843, 852, 860, 869, 874*] or sodium permanganate [*834*] in aqueous solution (equation 113) or with tetraalkylammonium permanganate in organic solvents (equation 114) [*845, 872, 899*].

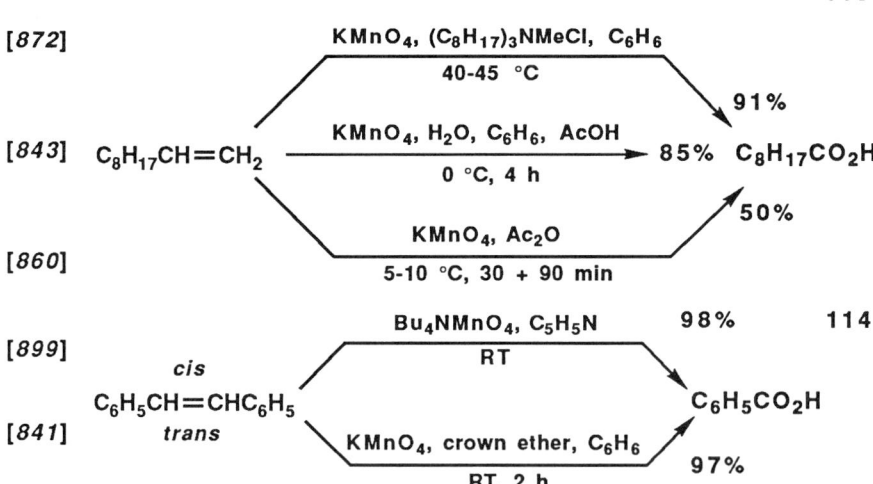

The oxidative cleavage of alkenes resulting in the formation of carboxylic acids is also accomplished with *chromium trioxide* in sulfuric acid or acetic acid [*567, 569*]; with *potassium dichromate* in sulfuric acid [*1111*]; and with **ruthenium tetroxide**, generated in situ from ruthenium trichloride and sodium periodate [774] or from ruthenium dioxide and sodium periodate [567, 774], sodium hypochlorite [939], periodic acid [939], or peroxyacetic acid [939] (equations 115–117).

[567]

[567]

[569]

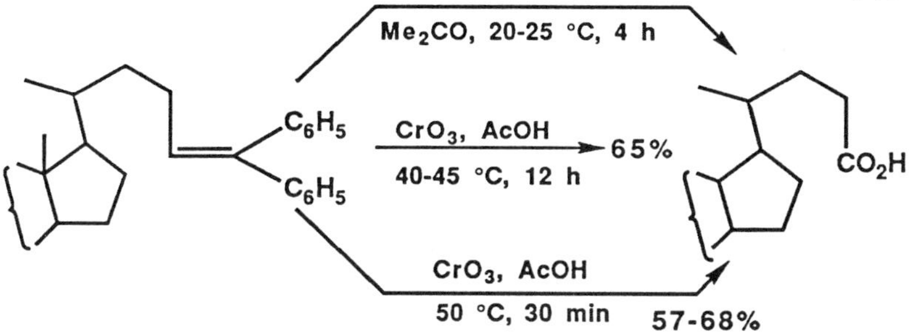

$$\text{RuO}_4, \text{NaIO}_4, \text{Me}_2\text{CO}, 20\text{-}25 \,°\text{C}, 4\text{ h} \quad 80\text{-}83\%$$

$$\text{CrO}_3, \text{AcOH}, 40\text{-}45 \,°\text{C}, 12\text{ h} \quad 65\%$$

$$\text{CrO}_3, \text{AcOH}, 50 \,°\text{C}, 30\text{ min} \quad 57\text{-}68\%$$

115

[774]

$$(E)-\text{C}_4\text{H}_9\text{CH}=\text{CHC}_4\text{H}_9 \xrightarrow[\text{H}_2\text{O, RT, 2 h}]{\text{RuCl}_3, \text{H}_2\text{O, NaIO}_4, \text{CCl}_4, \text{MeCN}} \text{C}_4\text{H}_9\text{CO}_2\text{H} \quad 88\%$$

116

[939]

$$\text{C}_8\text{F}_{17}\text{CH}=\text{CH}_2 \xrightarrow[\text{RT, 1 h}]{\text{RuO}_2, \text{AcO}_2\text{H}, \text{C}_2\text{Cl}_3\text{F}_3} \text{C}_8\text{F}_{17}\text{CO}_2\text{H} \quad 92\%$$

117

Alkenes and cycloalkenes with **at least one halogen** linked to the double bonds are oxidized in aqueous or acetone solutions with *potassium permanganate* to carboxylic acids via acyl chlorides. Such oxidations are frequently used to prepare highly fluorinated or perfluorinated carboxylic acids (equations 118–120) [838, 869, 874].

118

[838]

$$\text{CF}_3\text{CCl}=\text{CCl}_2 \xrightarrow[\substack{\text{1. KMnO}_4, \text{Me}_2\text{CO}, \text{RT, 30 min} \\ \text{2. H}_2\text{O, H}_2\text{SO}_4, \text{SO}_2}]{} \text{CF}_3\text{CO}_2\text{H} \xleftarrow[\substack{\text{1. KMnO}_4, \text{KOH}, 60 \,°\text{C, reflux 8-10 h} \\ \text{2. SO}_2, \text{H}_2\text{SO}_4}]{} \text{CF}_3\text{CCl}=\text{CClCF}_3$$

46%   83%

[869]

[838]

119

$$\begin{array}{c}\text{CF}_2\\ \text{CF} \quad \text{CF}\\ \| \quad \|\\ \text{CF} \quad \text{CF}\\ \text{CF}_2\end{array} \xrightarrow[\text{2. H}_2\text{O, H}_2\text{SO}_4, \text{SO}_2]{\text{1. KMnO}_4, \text{Me}_2\text{CO}, 20 \,°\text{C}, 30\text{ min}} \text{CF}_2(\text{CO}_2\text{H})_2 \quad 60\%$$

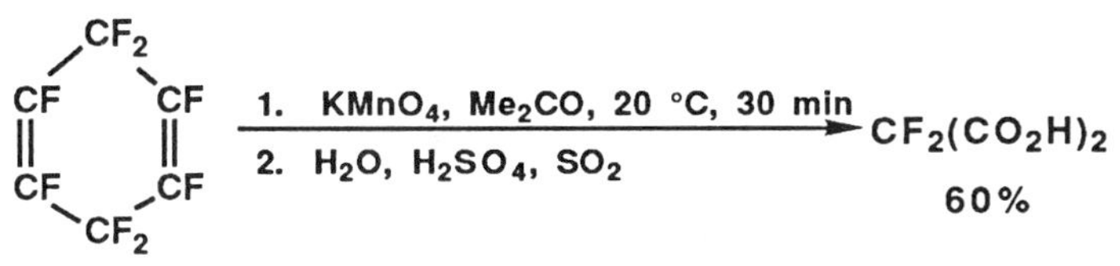

[874] 120

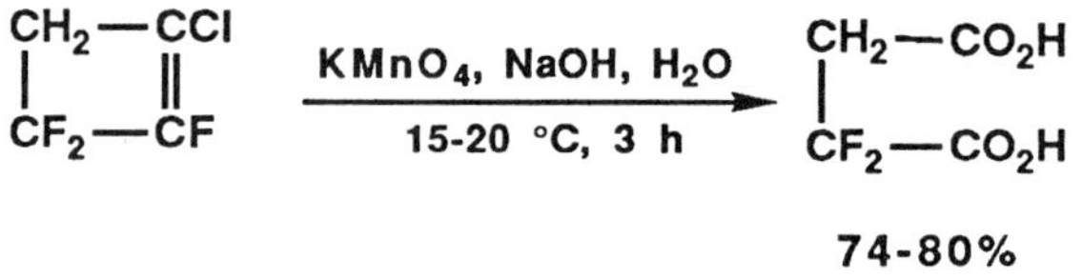

CH2—CCl
CF2—CF
KMnO4, NaOH, H2O
15-20 °C, 3 h
CH2—CO2H
CF2—CO2H
74-80%

Oxidations at Allylic Positions

Allylic positions may be oxidized to give unsaturated hydroperoxides, alcohols, esters, and carbonyl compounds.

Singlet oxygen, generated usually by the irradiation of solutions of oxygen in the presence of sensitizers, is incorporated into alkenes and forms ***hydroperoxides*** via an ***"ene"*** mechanism (equations 121 and 122) [*15, 19, 26, 1112*].

121

[1112] 82%

[19] 53%

[15] 63%

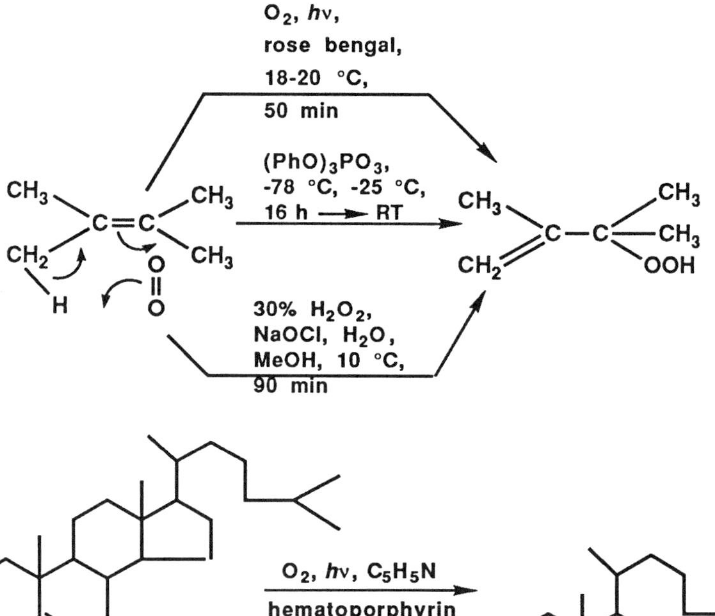

O2, hν,
rose bengal,
18-20 °C,
50 min
(PhO)3PO3,
-78 °C, -25 °C,
16 h → RT
30% H2O2,
NaOCl, H2O,
MeOH, 10 °C,
90 min
CH3
CH3
CH2
H
CH3
CH3
OOH

[26] 122

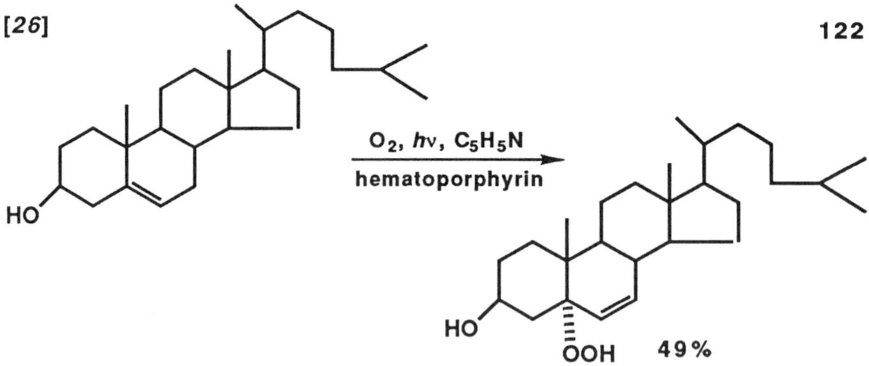

O2, hν, C5H5N
hematoporphyrin
HO
HO
OOH
49%

An alternative mechanism via ***dioxetanes*** or ***perepoxides*** has been suggested (equation 123) [*21*].

The hydroperoxides can be transformed further into alcohols or carbonyl compounds (equation 124) [*22, 23*].

Oxidations 85

[21]

123

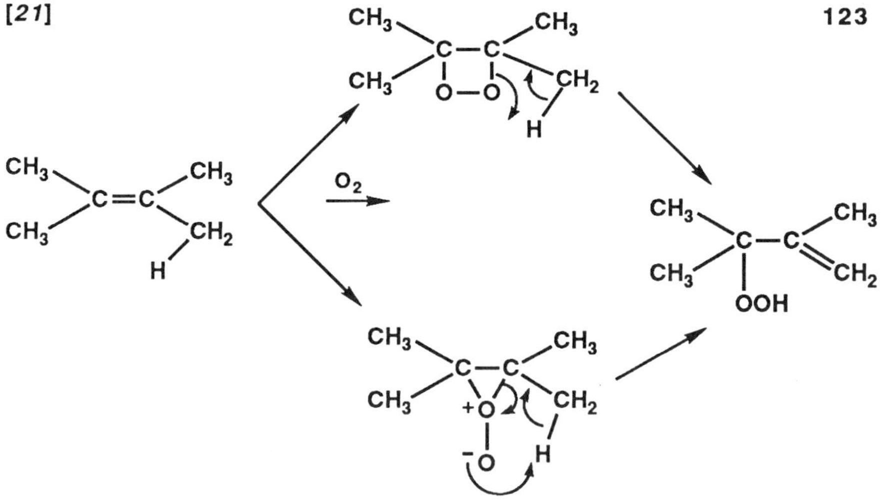

[22] O₂, hν, 15-20%, 35-40 °C, 3-4 h

[23] O₂, hν, RT, 4 h, 81%

NaOH, RT, 26 h → 82%

AcOH, HClO₄, RT, 36 h → 39%

124

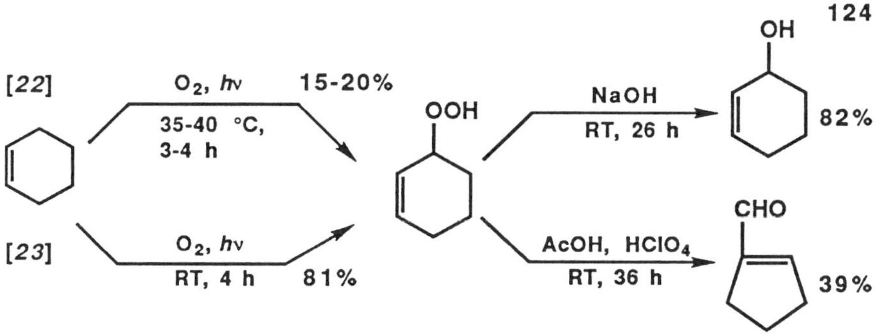

The outcome of the autoxidation can be affected by solvents (equation 125) [25].

[25]

125

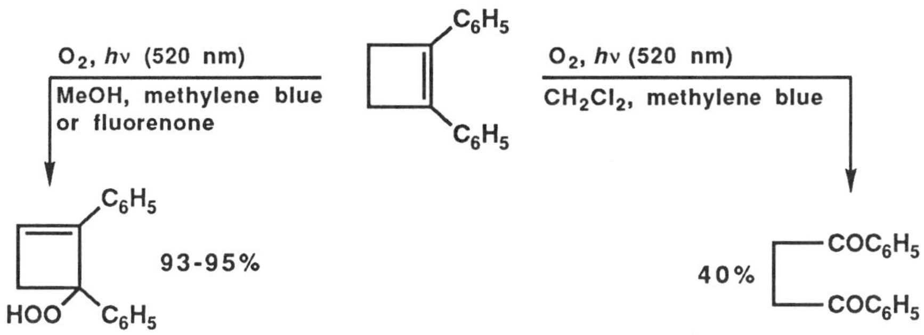

O₂, hν (520 nm), MeOH, methylene blue or fluorenone → 93-95%

O₂, hν (520 nm), CH₂Cl₂, methylene blue → 40%

Allylic alcohols are obtained from allylic hydroperoxides by treatment with *alkalies* [22] or by reduction with *sulfites* [27]. They are also formed from alkenes with **selenium dioxide** in aqueous solvents (equation 126) [*512*].

[*512*]

$SeO_2, H_2O_2, t\text{-BuOH}, H_2O$, 40-50 °C → 49-55%   **126**

Allylic esters result from the reaction of alkenes with tert-*butyl peroxyacetate* [*233*], or the more reactive *peroxybenzoate* [*233, 234, 1113*], in the presence of cuprous bromide; with mercuric acetate [*404*]; or with lead tetraacetate [*1114*] (equation 127).

[*233*] $t\text{-BuO}_2\text{COMe}$, CuBr, reflux 4 h → 54%, R=CH$_3$ **127**
[*233*] $t\text{-BuO}_2\text{COPh}$, CuBr, reflux → 50-77%, —OCOR R=C$_6$H$_5$
[*234*] $t\text{-BuO}_2\text{COPh}$, CuBr, 80-82 °C, 4 h → 71-80%, R=C$_6$H$_5$

α,β-*Unsaturated ketones* are products of the oxidation of alkenes and cycloalkenes in allylic positions. *trans*-5-Decene is converted into *trans*-5-decen-4-one in 81% yield by pyridine-α-seleninic anhydride prepared in situ from pyridine-α-diselenide and iodoxybenzene (equation 128) [*799*].

[*799*]   **128**
$(E)\text{-}C_4H_9CH=CHC_4H_9$ $\left(\text{Py-Se}\right)_2$, $PhIO_2$, C_6H_6, 80 °C, 8 h → $(E)\text{-}C_3H_7COCH=CHC_4H_9$ 81%

Cyclohexene, on irradiation in the presence of mercuric bromide, forms 2-cyclohexenone in 80% yield (equation 129) [*11*].

[*11*] $HgBr_2$, $t\text{-BuOH}$, $h\nu$ (254 nm), RT → 80%   **129**

Most allylic oxidations of methylene to carbonyl groups are carried out by **chromic oxide** [*552, 553*] or its *complexes with pyridine* [*593, 606*] and performed on steroidal esters [*552, 593, 606*] and ketones [*61, 552*] (equation 130).

The treatment of 5-cholesten-3-one with oxygen in the presence of cupric nitrate or cupric acetate in pyridine and triethylamine in methanol furnishes 5-cholesten-3,6-dione in 75% yield [*61*]. 4,4,14α-Trimethyl-Δ^8-

pregnene-20-one 3,21-diacetate is oxidized to 3β,21-diacetoxy-4,4,14α-trimethyl-Δ^8-pregnene-7,11,20-trione in 75% yield (equation 131) [552].

DIENES

Conjugated dienes (and compounds that behave like conjugated dienes in the Diels–Alder reaction) react with **singlet oxygen** to form *cyclic peroxides* as if molecular oxygen acted as a dienophile. The yields of the peroxides, prepared by *photochemical oxidation* [13, 33] or by chemical oxidations with *hydrogen peroxide and sodium hypochlorite, alkaline hydrogen peroxide and bromine, alkaline salts of peroxy acids* [14, 16], or the *ozonide of triphenyl phosphite* [19], are comparable.

A 2% solution of isoprene in dichloromethane and methanol with *methylene blue or rose bengal* as *sensitizers* gives, on irradiation under oxygen with a 500-W iodine lamp, a 50% yield of 4-methyl-1,2-dioxa-4-cyclohexene [32]. Cyclic dienes are converted into bicyclic endoperoxides (equation 132). The naturally occurring ascaridole is thus synthesized from α-terpinene (equation 133) [19, 1115].

The unsaturated endoperoxides afford **unsaturated cis diols** on re-

[13] H$_2$O$_2$, NaOCl, MeOH, -5 to -10 °C — 20% → **132**
[33] O$_2$, hν, i-PrOH, rose bengal, 6 h — 51%
[13] O$_2$, hν, MeOH, rose bengal, RT, 8 min — 35%

[1115]
O$_2$, hν, i-PrOH, chlorophyll or eosin — 40% → **133**
(PhO)$_3$PO$_3$, CH$_2$Cl$_2$, -78 °C to RT — 60%

[19] α-terpinene → ascaridole

duction with thiourea [33]. Both the oxidation and the subsequent reduction may be carried out in one step (equation 134) [33].

[33] O$_2$, hν, MeOH, CS(NH$_2$)$_2$, 0 °C, 450-W high-pressure Hg lamp with pyrex filter → 60% **134**

Just as anthracene and its homologues behave like conjugated dienes in the Diels–Alder reaction, they react with singlet oxygen to form 9,10-endoperoxides (equation 135) [16, 24].

[16] H$_2$O$_2$, KOH, H$_2$O, PhCl, Br$_2$, RT — 93% **135**
[24] O$_2$, microwave discharge, microcrystalline cellulose — 79%
[24] O$_2$, microwave discharge, silica gel — 60%

Such endoperoxides generated from anthracene and its homologues release oxygen in singlet form when heated [17, 18, 19, 1116]. This reversible reaction is typical of rubrene (tetraphenylnaphthacene), a red compound that yields a colorless endoperoxide (equation 136) [18, 19, 1116].

[18] **136**

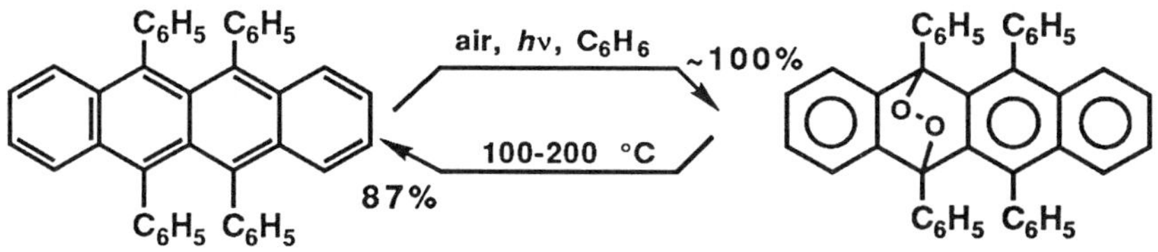

[Reaction scheme: tetraphenyl-substituted tetracene + air, hν, C₆H₆ → ~100% endoperoxide; reverse at 100–200 °C, 87%]

Not only hydrocarbon dienes but also conjugated diene–ketones, such as tetraphenylcyclopentadienone (tetracyclone), and heterocyclic dienes, such as furan derivatives, afford endoperoxides, which may undergo subsequent decompositions (equation 137) [13, 17, 19].

[19]

[13] **137**

[17]

Reagents shown:
- (PhO)$_3$PO$_3$, CH$_2$Cl$_2$, -78 °C to RT
- H$_2$O$_2$, NaOCl, dioxane, 0 °C
- 9,10-diphenylanthracene endoperoxide, CHCl$_3$ or C$_6$H$_6$, reflux 2–4 days
- −CO

Yields: 38%, 50%, 50%

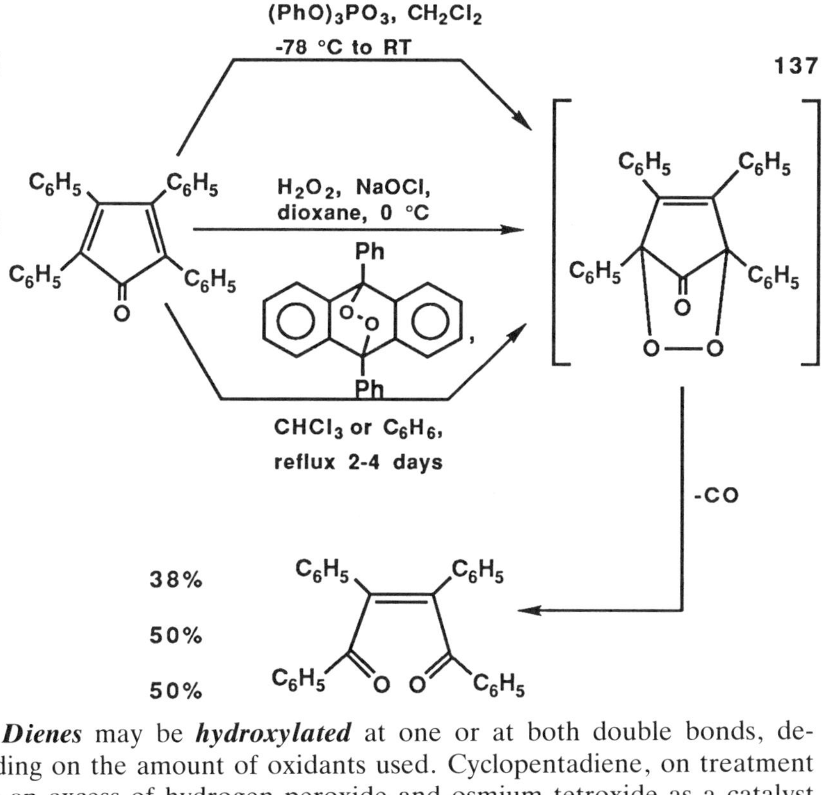

Dienes may be **hydroxylated** at one or at both double bonds, depending on the amount of oxidants used. Cyclopentadiene, on treatment with an excess of hydrogen peroxide and osmium tetroxide as a catalyst in *tert*-butyl alcohol, gives a mixture of 21% of 3,5-cyclopentenediol and 61% of 1,2,3,4-cyclopentanetetrol [152].

The cyclopentadiene dimer, when treated with potassium permanganate and triethylbenzylammonium chloride as a phase-transfer agent, furnishes an 83% yield of *exo,cis*-diol resulting from the hydroxylation of the double bond in the cyclohexene ring [900] (equation 82).

Dienes react with **ozone** at one or both double bonds to give *carbonyl compounds*. When 1,3-cyclohexadiene is dissolved in dichloromethane at −78 °C and treated with 1.5 mol of ozone in dichloromethane solution and the resulting ozonide is stirred overnight at 0 °C with dimethyl sulfide, 2-hexenedial is obtained in 67% yield. If, on the other hand, an excess of ozone is passed through a solution of 1,3-cyclohexadiene at −78 °C until a blue color appears, both double bonds are ozonolyzed, and a 70% yield of 1,4-butanediol is obtained by reducing the reaction mixture with lithium aluminum hydride (equation 138) [92]. More substituted double bonds react with ozone preferentially.

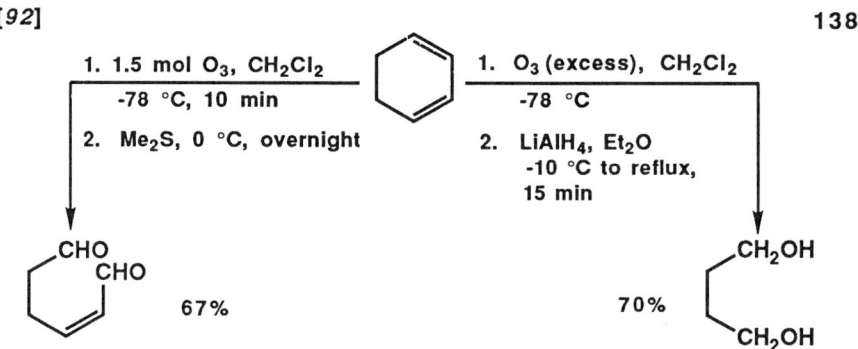

ALKYNES (ACETYLENES)

Terminal alkenes containing hydrogen linked to the triple bond undergo *oxidative coupling to diacetylenes* by *cupric salts* [58, 357] or by *oxygen* in the presence of *cuprous salts* [59, 66] (equation 139). Copper salts are solubilized by complexing with tertiary amines, most frequently pyridine [59, 357] and tetramethylethylenediamine [66]. The coupling can also be carried out with cuprous salts of acetylenes [58].

The oxidative coupling of long-chain terminal diacetylenes carried out in very dilute solutions gives macrocyclic diacetylenes and polyacetylenes

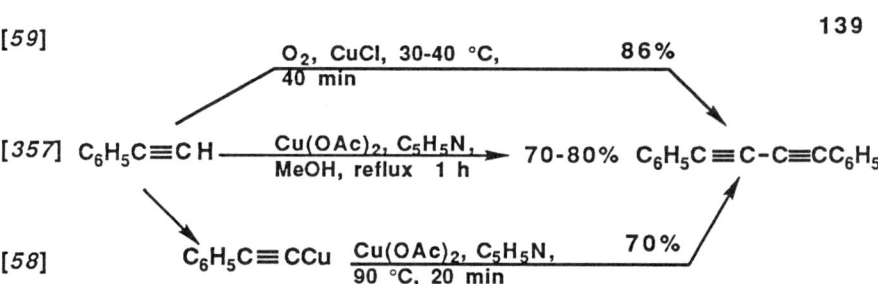

in low yields (equation 140) [58]. Oxidative coupling is frequently applied to acetylenic alcohols [2, 58].

[58] HC≡C(CH$_2$)$_{10}$C≡CH $\xrightarrow[\text{heat, 72 h}]{\text{Cu(OAc)}_2, \text{C}_5\text{H}_5\text{N}, \text{MeOH}, \text{Et}_2\text{O}}$ (CH$_2$)$_{10}$ with C≡C-C≡C 17% + cyclic dimer with two (CH$_2$)$_{10}$ and two C≡C-C≡C 30%   140

Dialkylalkynes are oxidized to ***acyloins*** or their alkyl ethers by thallium trinitrate in aqueous or aqueous–alcoholic solutions, respectively [*413*]. Only symmetrical alkynes give acceptable yields (70–90%) (equation 141).

[*413*] C$_2$H$_5$C≡CC$_2$H$_5$ $\xrightarrow[\text{MeOH, RT, 30 min}]{\text{Tl(NO}_3)_3, \text{HClO}_4}$ C$_2$H$_5$CHCOC$_2$H$_5$ with OCH$_3$ 70%   141

Alkylarylacetylenes undergo oxidative rearrangement to esters of alkylarylacetic acid (equation 142) under similar conditions [*413*].

[*413*] C$_6$H$_5$C≡CCH$_3$ $\xrightarrow[\text{reflux 2 h}]{\text{Tl(NO}_3)_3, \text{MeOH}}$ C$_6$H$_5$CHCO$_2$CH$_3$ with CH$_3$ 80%   142

Diarylacetylenes are converted in 55–90% yields into **α-*diketones*** by refluxing for 2–7 h with thallium trinitrate in glyme solutions containing perchloric acid [*413*]. Other oxidants capable of achieving the same oxidation are ozone [*84*], selenium dioxide [*509*], zinc dichromate [*660*], molybdenum peroxo complex with HMPA [*534*], **potassium permanganate in buffered solutions** [*848, 856, 864, 1117*], zinc permanganate [*898*], osmium tetroxide with potassium chlorate [*717*], ruthenium tetroxide and sodium hypochlorite or periodate [*938*], dimethyl sulfoxide and *N*-bromosuccinimide [*997*], and iodosobenzene in the presence of a ruthenium catalyst [*787*] (equation 143).

A triple bond activates the adjoining methylene group for ***oxidation to a carbonyl group.*** Thus 5-decyne furnishes a 46% yield of 5-decyn-4-one on treatment with chromium–pyridine in dichloromethane for 24 h at room temperature [*601*].

Whereas internal acetylenes are oxidized to α-diketones, terminal acetylenes give ***carboxylic acids*** with one less carbon on treatment with thallium trinitrate [*413*], potassium permanganate [*843*], iodosobenzene with tris(triphenylphosphine)ruthenium dichloride as a catalyst [*787*], or a rather rare oxidant, pentafluoroiodobenzene bis(trifluoroacetate) [*797*] (equation 144).

EXAMPLES OF OXIDATION OF ACETYLENES 143

$$C_6H_5C \equiv CC_6H_5 \longrightarrow C_6H_5COCOC_6H_5$$

Ref	Conditions	Yield
[413]	Tl(NO$_3$)$_3$, 70% HClO$_4$, glyme, reflux 3 h	85%
[509]	SeO$_2$, 280 °C	35%
[660]	ZnCr$_2$O$_7$·3H$_2$O, CCl$_4$, RT, 8 h	81%
[534]	MoO(O$_2$)$_2$, Hg(OAc)$_2$, HMPA MeOH/CH$_2$Cl$_2$ (1:9), 40 °C, 20 h	77%
[856]	Zn(MnO$_4$)$_2$/SiO$_2$, CH$_2$Cl$_2$, 40 °C, 45 min	67%
[938]	RuO$_2$, NaOCl or NaIO$_4$, CCl$_4$, RT, 1 h	83%
[997]	DMSO, NBS (2 mol), 22 °C, 24 h	98%
[787]	PhIO/(Ph$_3$P)$_3$RuCl$_2$, CH$_2$Cl$_2$, RT, 15 min	86%

$$RC \equiv CH \longrightarrow RCO_2H \qquad 144$$

Ref	Conditions	R	Yield
[413]	Tl(NO$_3$)$_3$, HClO$_4$, glyme, H$_2$O, RT, 1 h	R = C$_6$H$_{13}$	80%
[843]	KMnO$_4$, H$_2$O, C$_5$H$_{12}$, AcOH, Aliquat 336, cooling with ice, 5 h	R = C$_8$H$_{17}$	70%
[787]	PhIO/(Ph$_3$P)$_3$RuCl$_2$, CH$_2$Cl$_2$, RT, 5 min	R = C$_6$H$_5$	69%
[797]	C$_6$F$_5$I(OCOCF$_3$)$_2$, C$_6$H$_6$, H$_2$O, reflux overnight	R = C$_6$H$_5$	79%

Oxidations of acetylenic acids [101, 268, 864] and esters of acetylenic alcohols [84, 1117] to the corresponding α-dicarbonyl compounds and carboxylic acids are discussed in the section on carboxylic acids.

AROMATIC COMPOUNDS

This section will encompass the reactions of carbocyclic and heterocyclic aromatic compounds in which oxidation affects the aromatic rings and the attached side chains.

Hydroxylation of Aromatic Rings

The hydroxylation of aromatic compounds takes place on heating with solid **potassium hydroxide** of compounds with low electron density in the ring and occurs in positions prone to nucleophilic attack. Nitrobenzene is converted into *o*-nitrophenol on heating with potassium hydroxide for 2 h at 60–70 °C (yield 45%) [485]. Quinoline gives carbostyril (2-hydroxyquinoline) on heating with potassium hydroxide for 3 h at 225 °C (yield 77%) [486]. A historically important reaction is the hydroxylation of both α- and β-anthraquinonesulfonic acid by alkali fusion to yield alizarin (1,2-dihydroxyanthraquinone).

An interesting hydroxylation takes place, probably by an intramolecular transfer of oxygen, on irradiation of the 2-benzylpyridine N-oxide complex with boron trifluoride (equation 145) [994].

[994]   145

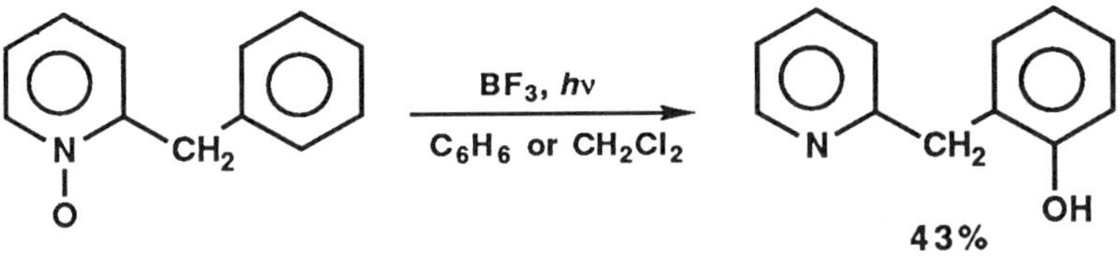

$$\text{2-benzylpyridine N-oxide} \xrightarrow[\text{C}_6\text{H}_6 \text{ or CH}_2\text{Cl}_2]{\text{BF}_3, h\nu} \text{2-(2-hydroxybenzyl)pyridine} \quad 43\%$$

The benzene rings in toluene [92, 1074], chlorobenzene [92], styrene [92], and phenylacetylene [92] undergo stereospecific *syn* hydroxylation in positions 2 and 3 with concomitant reduction to methyl-, chloro-, vinyl-, and acetylenyl-*cis*-2,3-dihydroxycyclohexa-4,6-dienes when treated with a culture of **Pseudomonas putida** (equation 146) [92, 1074].

[92]   146

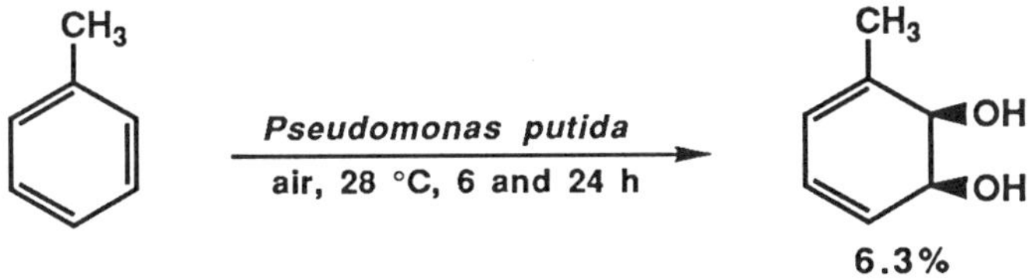

$$\text{toluene} \xrightarrow[\text{air, 28 °C, 6 and 24 h}]{\textit{Pseudomonas putida}} \text{methyl-cis-2,3-dihydroxycyclohexa-4,6-diene} \quad 6.3\%$$

Hydroxylations of aromatic rings are also achieved via transient esters that are easily hydrolyzed in situ to the corresponding phenols. In this way, anisole and diphenyl ether are oxidized by peroxytrifluoroacetic acid to predominantly *ortho*-hydroxy compounds in low yields (equation 147) [1118].

[1118]   147

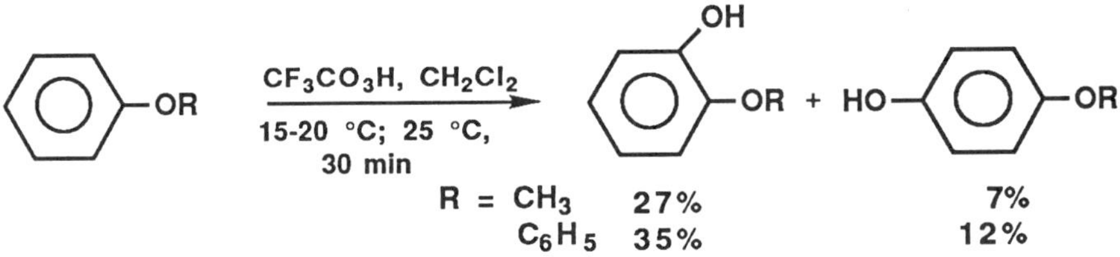

$$\text{PhOR} \xrightarrow[\substack{\text{15-20 °C; 25 °C,} \\ \text{30 min}}]{\text{CF}_3\text{CO}_3\text{H, CH}_2\text{Cl}_2} \text{2-HO-C}_6\text{H}_4\text{-OR} + \text{4-HO-C}_6\text{H}_4\text{-OR}$$

R = CH_3 27%   7%
    C_6H_5 35% 12%

The oxidation of benzoic acid by divalent copper carried out by heating cupric benzoate to 210–260 °C in high-boiling liquids furnishes 73–100% of salicylic acid [361].

Esters of phenols are obtained by treatment of aromatic hydrocarbons with peroxy acids [292], organic peroxides [1119, 1120], and lead tetraacetate or tetrakis(trifluoroacetate) [435, 449]. Benzene treated with trifluoroacetic acid, trifluoroacetic anhydride, and peroxytrifluoroacetic acid in the presence of cobaltous acetate gives a 56% yield of phenyl trifluoroacetate [*292*]. The same product is obtained on treatment of benzene with lead tetrakis(trifluoroacetate) [*435, 449*] (equation 148).

$$\text{PhH} \xrightarrow{\begin{array}{c}[435]\ \text{Pb(OCOCF}_3)_4,\ \text{RT, 35 min} \quad >45\% \\ [292]\ \text{CF}_3\text{CO}_3\text{H, CF}_3\text{CO}_2\text{H, (CF}_3\text{CO)}_2\text{O, Co(OAc)}_2,\ 0\ ^\circ\text{C, 27 h} \quad 56\% \\ [449]\ \text{Pb(OCOCF}_3)_4,\ \text{CF}_3\text{CO}_2\text{H},\ 0\ ^\circ\text{C, 20 min} \quad 76\text{-}79\%\end{array}} \text{PhOCOCF}_3 \quad 148$$

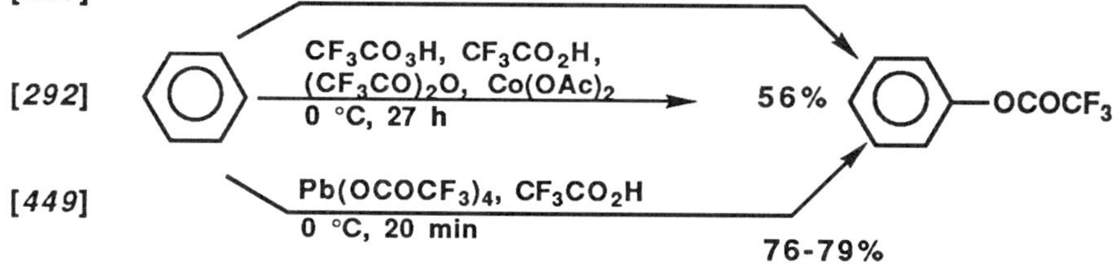

Heating benzene homologues with diisopropyl peroxydicarbonate and cupric chloride or with dibenzoyl peroxide and iodine gives isopropyloxycarbonyl or benzoyl esters of the corresponding phenols [*232, 1119, 1120*]. Unfortunately, all three isomers are usually obtained so that the synthetic uses are limited (equation 149).

149

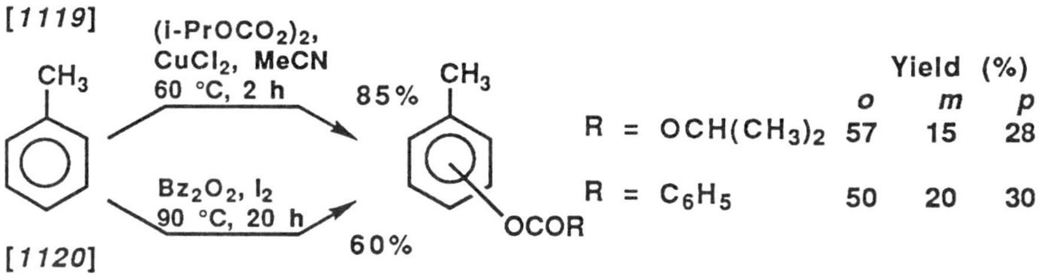

[*1119*] $(\text{i-PrOCO}_2)_2$, CuCl_2, MeCN, 60 °C, 2 h, 85%
[*1120*] Bz_2O_2, I_2, 90 °C, 20 h, 60%

Toluene → methylphenyl-OCOR

	Yield (%)		
	o	m	p
R = OCH(CH$_3$)$_2$	57	15	28
R = C$_6$H$_5$	50	20	30

Like the acyloxylation with peroxides, the treatment of benzene homologues and derivatives with lead tetrakis(trifluoroacetate) gives mixtures of *ortho, meta,* and *para* isomers, with the *para* isomers predominating [*449*].

Oxidation of Aromatic Compounds to Quinones

The oxidation of benzene to *p*-benzoquinone is impractical, because benzoquinone is obtained from other compounds [*647*]. Condensed aromatic hydrocarbons are oxidized to quinones by many reagents [*429, 758, 802*], most frequently by the compounds of hexavalent chromium [*1121*] (equation 150).

Alkylnaphthalenes are converted into alkyl-1,4-naphthoquinones with peroxyacetic acid [*139*] or chromic acid [*557*]. Under the conditions of these reactions (equation 151), the alkyl groups resist oxidation to carboxyl groups.

Oxidations

150 (1,4-naphthoquinone from naphthalene)

Ref.	Conditions	Yield
[429]	Ce(SO$_4$)$_2$·2(NH$_4$)$_2$SO$_4$·2H$_2$O, 4 N H$_2$SO$_4$, 25 °C, 6 h	90-95%
[1121]	CrO$_3$, AcOH, 0 °C, 2-3 h, RT, 3.5 days	32-35%
[802]	Mn$_2$(SO$_4$)$_3$ (from MnSO$_4$·4H$_2$O + KMnO$_4$), MeCN, H$_2$O, 25 °C, 4 h	75%
[758]	HIO$_4$, AcOH, 110 °C, 5 min, 70 °C, 30 min	70-76%

151 (2,3-dimethyl-1,4-naphthoquinone from 2,3-dimethylnaphthalene)

Ref.	Conditions	Yield
[139]	60% H$_2$O$_2$, AcOH, Pd(II) on Dowex 50w-x8, 50 °C, 8 h	51%
[557]	CrO$_3$, AcOH, 20-30 °C, 15 min, RT, 3 days	60-80%

The oxidation of ***anthracene*** to anthraquinone takes place in positions 9 and 10 [3, 249, 509, 630, 660, 663] and has industrial application in the synthesis of lake and vat dyes (equation 152).

EXAMPLES OF FORMATION OF QUINONES

152 (anthraquinone from anthracene)

Ref.	Conditions	Yield
[3]	air, V$_2$O$_5$/pumice, 420-425 °C	81%
[509]	SeO$_2$, 165-170 °C	76%
[249]	40% AcO$_2$H, AcOH, 50 °C, 6 h	70%
[630]	Na$_2$Cr$_2$O$_7$, H$_2$SO$_4$, Bu$_4$NHSO$_4$, 70 °C, 2 min	82-92%
[660]	ZnCr$_2$O$_7$·3H$_2$O, CCl$_4$, RT, 3.5 h	80%
[663]	(pyridine-CO$_2$H)$_2$Cr$_2$O$_7$, AcOH, reflux 75 h	67%

9,10-Anthraquinone is also formed from 9-methylanthracene on refluxing with sodium dichromate in acetic acid for 30 min (yields are 97% of crude product and 88% of pure product) [*1122*].

Phenanthrene is oxidized to 9,10-phenanthrenequinone [*429, 649, 650, 802*] and, to a smaller extent, to 1,4-phenanthrenequinone (yield 19%) [*802*] (equation 153).

The oxidation of ***chrysene*** (1,2-benzophenanthrene) by refluxing for 8–10 h with chromic oxide in acetic acid gives a 96–97% yield of 5,6-chrysenequinone [*556*].

		153
[*429*]	Ce(SO$_4$)$_2$.2(NH$_4$)$_2$.2H$_2$O, 4 N H$_2$SO$_4$, 50 °C, 4 h	60%
[*650*]	CrO$_3$, H$_2$SO$_4$, H$_2$O, boil 20 min	44-48%
[*649*]	K$_2$Cr$_2$O$_7$, H$_2$SO$_4$, H$_2$O, 1 h	79%
[*802*]	Mn$_2$(SO$_4$)$_3$ (from MnSO$_4$.4H$_2$O + KMnO$_4$), MeCN, H$_2$O, 25 °C, 4 h	32%

Oxidative Cleavage of the Benzene Ring

The breakdown of the benzene ring to aldehydes is extremely rare and has been achieved only by ozone. A historical and classical example is the disintegration of *o*-xylene to a mixture of glyoxal, methyl glyoxal, and diacetyl (butanedione) in the predicted ratios [*104*]. From the preparative point of view, the conversion of phenanthrene into *o,o'*-diformylbiphenyl (diphenaldehyde) [*1123*] or *o*-formylbiphenyl-*o'*-carboxylic (diphenaldehydic) acid [*1124*] is more important (equation 154).

The conversion of the benzene ring in benzene homologues to carboxylic groups is encountered rarely and has hardly any preparative value. One example involves the treatment of α-truxillic acid with ozone and subsequent oxidation with hydrogen peroxide (equation 155) [*109*].

The reagent of choice for the oxidation of phenyl groups to carboxyls seems to be ruthenium tetroxide with sodium periodate [*231, 941*] or sodium hypochlorite [*701*] as reoxidants. Phenylcyclohexane is oxidized to cyclohexanecarboxylic acid (equation 156) [*774, 941*], and β-phenylpropionic acid is transformed mainly into succinic acid (equation 157) [*701*].

The degradation of benzene rings to carboxyls is facilitated by the presence of electron-donating groups. *m*-Trifluoromethylaniline is converted into trifluoroacetic acid in 90–95% yield on heating with **sodium dichromate** and dilute sulfuric acid for 30–40 min at 70–170 °C [*638*].

Condensed polynuclear hydrocarbons and aromatic heterocycles are converted into dicarboxylic acids. Naphthalene is oxidized to phthalic an-

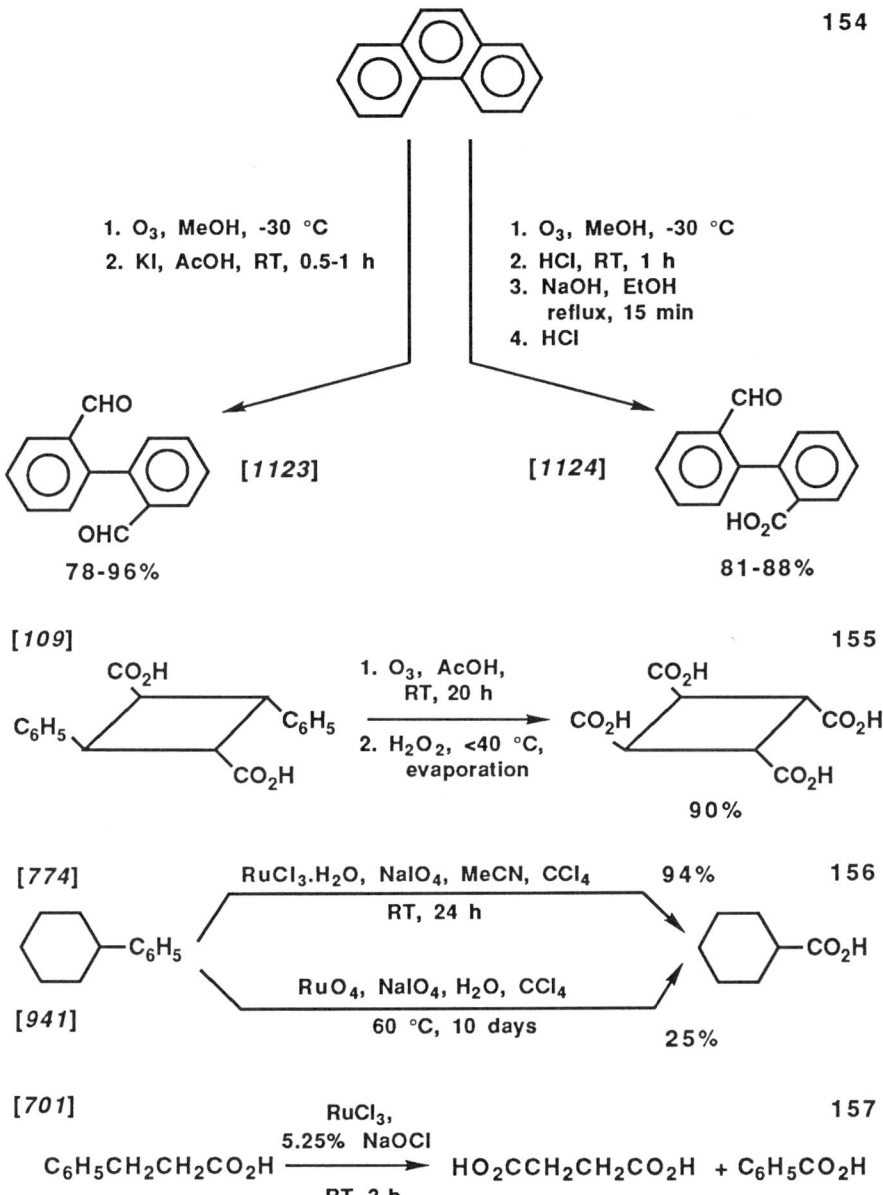

hydride by oxygen or air over vanadium pentoxide at high temperatures [479] and to phthalic acid by peroxyacetic acid [272], potassium permanganate [1125], or ruthenium tetroxide [943] (equation 158).

α-Nitronaphthalene is oxidized by ruthenium tetroxide to give 63% of 3-nitrophthalic acid and 7% of phthalic acid [943]. When oxidation is

[272] 26% AcO$_2$H, 17 days → 77% **158**
*KMnO$_4$, NaOH, reflux 2 h → 40–41%
[943] [1125]* RuO$_4$, 5.25% NaOCl, H$_2$O, CCl$_4$, RT, 60 h → 70%

Naphthalene → phthalic acid (1,2-benzenedicarboxylic acid)

carried out by potassium permanganate, the nitro-group-containing ring yields phthalonic acid [875] (equation 159).

159

1-Nitronaphthalene:
- RuO$_4$, NaOCl, H$_2$O, CCl$_4$, RT, 7 days [943] → 3-nitrophthalic acid (63%) + phthalic acid (7%)
- KMnO$_4$, NaOH, reflux 1.5 h [875] → 2-(oxoacetyl)benzoic acid / phthalonic acid (74%)

Phenanthrene is oxidized with 40% peroxyacetic acid in glyme at 100–110 °C to α,α′-diphenic acid in 65–70% yield [256]. Quinoxaline affords a 75–77% yield of pyridazine-2,3-dicarboxylic acid on refluxing for 1.5 h with potassium permanganate [876].

In heterocyclic aromatic compounds, the ring containing electron-withdrawing substituents is more resistant to oxidative degradation. Both quinoline and 8-hydroxyquinoline give quinolinic (pyridine-2,3-dicarboxylic) acid (equation 160) [117, 459].

160

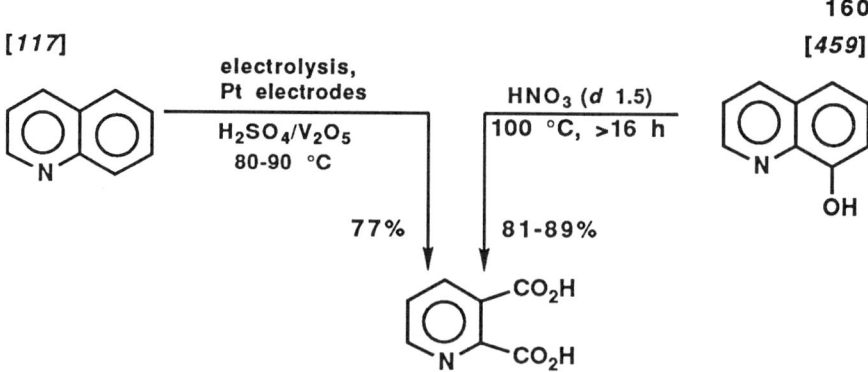

[117] Quinoline — electrolysis, Pt electrodes, H$_2$SO$_4$/V$_2$O$_5$, 80–90 °C, 77% → quinolinic acid
[459] 8-Hydroxyquinoline — HNO$_3$ (d 1.5), 100 °C, >16 h, 81–89% → quinolinic acid

Oxidation of Side Chains of Carbocyclic and Heterocyclic Aromatic Compounds

The alkyl groups attached to aromatic rings undergo several types of oxidation. The common denominator is an attack at the carbon adjacent to the ring—the benzylic carbon. This position of attack is understandable, because an intermediate resulting from such a reaction is stabilized by resonance with the ring and is formed more readily than at any other place of the chain. Only rarely does the oxidation affect more distant carbons of the chain.

FORMATION OF ARALKYL HYDROPEROXIDES

The treatment of compounds possessing primary, secondary, and especially tertiary benzylic hydrogens with oxygen, usually under irradiation, results in a free-radical chain reaction giving hydroperoxides as the final products (equation 161).

$$ArCH_3 \xrightarrow{h\nu} ArCH_2\cdot \xrightarrow{O_2} RCH_2\text{-O-O}\cdot \qquad 161$$

$$RCH_2\text{-O-O}\cdot + ArCH_3 \longrightarrow RCH_2\text{-O-O-H} + ArCH_2\cdot$$

Such reactions take place with p-xylene [28], ethylbenzene [28], and especially readily with isopropylbenzene (cumene) [29], where the intermediate free radical is stabilized not only by the aromatic ring but also by the two adjacent methyl groups. The oxidation of cumene to cumyl hydroperoxide (equation 162) followed by acid treatment is a basis for the large-scale production of phenol.

[29]   cumene $\xrightarrow[85\ °C,\ 24\ h]{O_2,\ h\nu}$ cumyl hydroperoxide   162

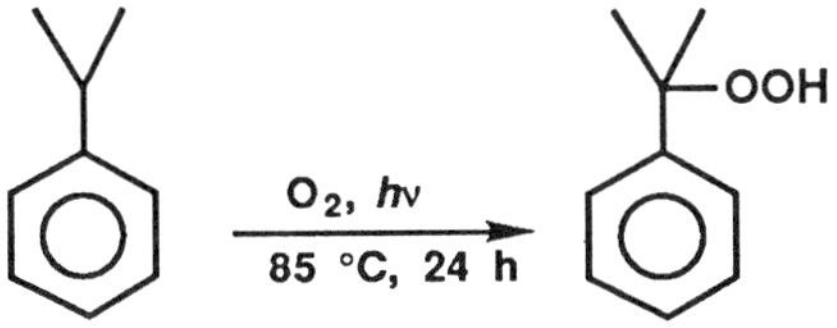

6% conversion

In compounds containing saturated rings attached to aromatic nuclei, for example, indane [27], tetralin [23, 50], and octahydroanthracene [47], oxidations take place at carbons adjacent to the aromatic rings and usually give low conversions. Only a few reactions have preparative importance (equation 163).

163

[50]   tetralin $\xrightarrow{O_2,\ 70\ °C,\ 24\text{-}48\ h}$ tetralin hydroperoxide   44–57%

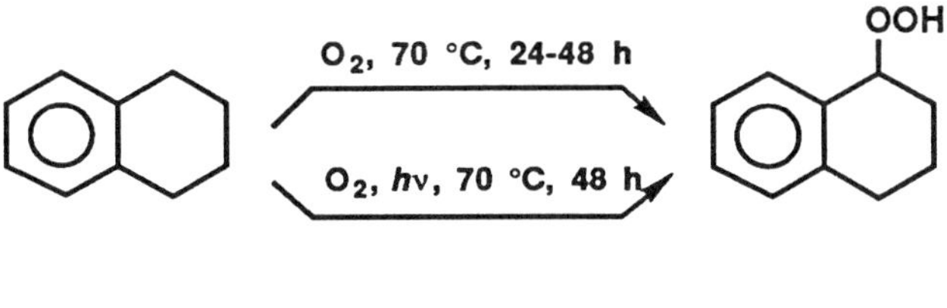

[23] $\xrightarrow{O_2,\ h\nu,\ 70\ °C,\ 48\ h}$ 66%

OXIDATION TO BENZYLIC ALCOHOLS AND ESTERS

Hydroxylation at the benzylic position is encountered very rarely. An example is the oxidation of 10-methylanthrone to 10-hydroxy-10-methylanthrone in 67% yield on treatment with 30% hydrogen peroxide in 10% sodium hydroxide in hot ethanol for 3 min (equation 164) [1126].

[1126]   164

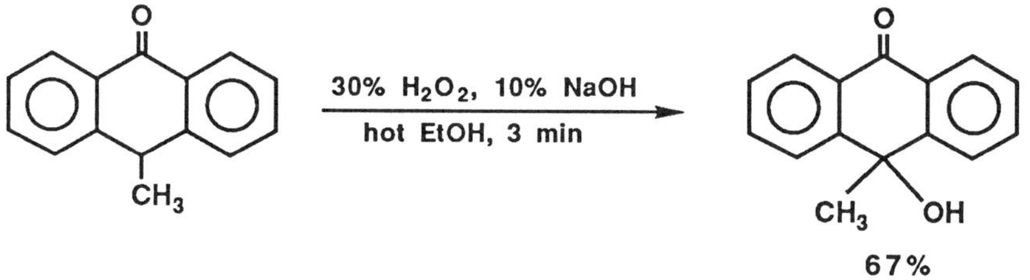

10-methylanthrone → (30% H_2O_2, 10% NaOH, hot EtOH, 3 min) → 10-hydroxy-10-methylanthrone, 67%

More frequently, the benzylic hydrogen is replaced by an acyloxy group on refluxing with ceric ammonium nitrate in 100% acetic acid [415], with **lead tetraacetate** [434, 437], or with lead tetrakis(trifluoroacetate) [435].

Oxidation with ammonium cerium nitrate has to be done in 100% acetic acid (equation 165). In 50% acetic acid, carbonyl compounds are

[415]   165

$C_6H_5CH_3$ —$(NH_4)_2Ce(NO_3)_6$, 100% AcOH→ $C_6H_5CH_2OCOCH_3$ 90%

formed. In some cases, more than one product is generated. For example, in the oxidation of p-chlorotoluene, a mixture of p-chlorobenzyl alcohol and p-chlorobenzyl acetate is obtained (equation 166) [415].

[415]   166

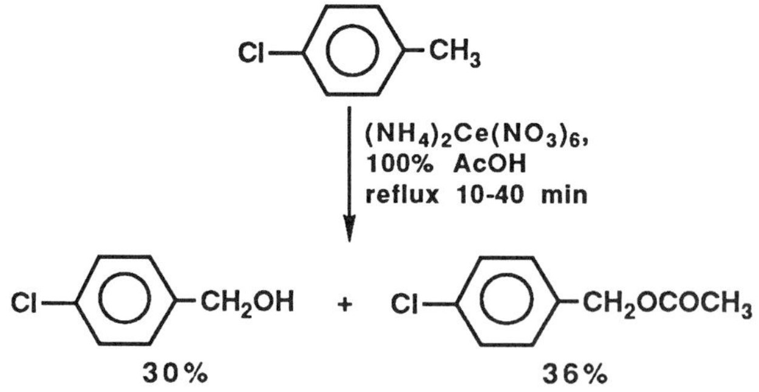

Cl–C$_6$H$_4$–CH$_3$ →($(NH_4)_2Ce(NO_3)_6$, 100% AcOH, reflux 10-40 min)→ Cl–C$_6$H$_4$–CH$_2$OH (30%) + Cl–C$_6$H$_4$–CH$_2$OCOCH$_3$ (36%)

Heating toluene with lead tetraacetate at 80 °C for 54 h gives only 25% of benzyl acetate, but ethylbenzene and anisole give 63% of α-methylbenzyl acetate after 16 h and 60% of p-methoxybenzyl acetate after 3 h at 80 °C, respectively [437]. Under similar conditions, acenaphthene yields

1-acetoxyacenaphthene (equation 167) [*434*]. Subsequent hydrolysis of the esters with sodium hydroxide yields the corresponding alcohols [*434, 435*].

[*434*]   167

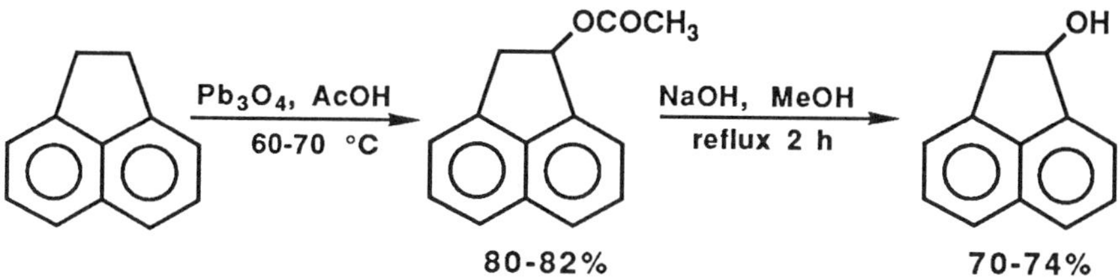

$$\text{acenaphthylene} \xrightarrow[\text{60-70 °C}]{Pb_3O_4, \ AcOH} \text{1-acetoxyacenaphthene} \xrightarrow[\text{reflux 2 h}]{NaOH, \ MeOH} \text{1-hydroxyacenaphthene}$$

80-82%   70-74%

In allylbenzene, an easy oxidation at the benzylic, and at the same time allylic, carbon would be expected. However, cinnamyl acetate is formed instead on treatment with mercuric acetate, evidently as a result of an S_N2' reaction at the double bond (equation 168) [*404*].

[*404*]   168

$$C_6H_5CH_2CH=CH_2 \xrightarrow[\text{reflux 50 h}]{Hg(OAc)_2, \ AcOH} C_6H_5CH=CHCH_2OCOCH_3$$

68%

OXIDATION TO ALDEHYDES, KETONES, AND THEIR DERIVATIVES

Methyl homologues of aromatic compounds are oxidized to **aldehydes** by silver(II) oxide (argentic oxide) [*380*], by ceric ammonium nitrate [*238, 417, 422*], by **selenium dioxide** [*513, 514, 515*], by chromyl chloride [*477, 667*], by periodic acid [*760*], and by **manganese dioxide** [*1127*] (equation 169).

169

[*417*] $(NH_4)_2Ce(NO_3)_6, \ HClO_4, \ 40 \ °C, \ 80 \ min$ 92%

[*1127*] $C_6H_5CH_3$ $\xrightarrow{MnO_2}$ C_6H_5CHO

[*380*] $4 \ AgO, \ H_3PO_4, \ -10 \ °C, \ 0.5 \ h$ 54%

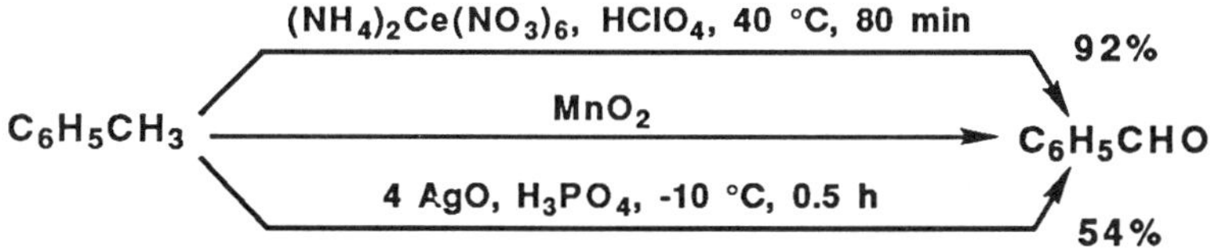

p-Nitrotoluene is oxidized by ceric ammonium nitrate [*417*] or by chromyl chloride [*667*] to *p*-nitrobenzaldehyde, whereas refluxing with aqueous alcoholic sodium polysulfide gives *p*-aminobenzaldehyde by an internal redox reaction [*505*] (equation 170).

The regiospecific oxidation of dimethylanisoles to methoxymethyl-benzaldehydes is accomplished with copper sulfate and potassium peroxydisulfate, which oxidize selectively only the methyls in the *ortho* or *para* positions with respect to the methoxy group (equation 171) [*353*].

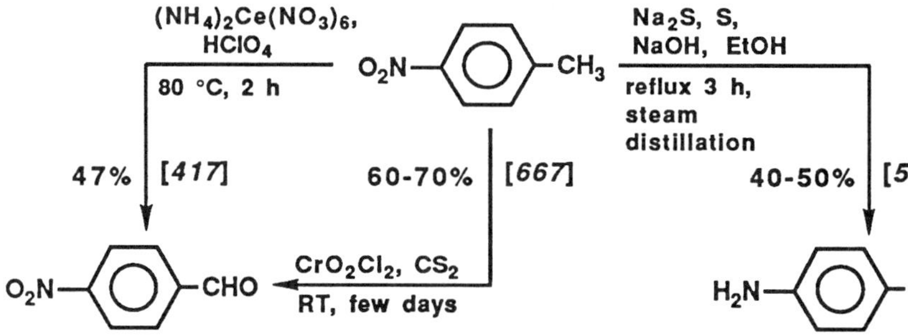

Reaction schemes

O_2N—⟨C$_6$H$_4$⟩—CH_3

- $(NH_4)_2Ce(NO_3)_6$, $HClO_4$, 80 °C, 2 h → 47% [417] → O_2N—⟨C$_6$H$_4$⟩—CHO
- CrO_2Cl_2, CS_2, RT, few days → 60–70% [667] → O_2N—⟨C$_6$H$_4$⟩—CHO
- Na_2S, S, NaOH, EtOH, reflux 3 h, steam distillation → 40–50% [505] → **170** H_2N—⟨C$_6$H$_4$⟩—CHO

[353] CH_3O—⟨C$_6$H$_3$(CH$_3$)⟩—CH_3 $\xrightarrow{CuSO_4\cdot 5H_2O,\ K_2S_2O_8,\ MeCN/H_2O\ (1:1),\ reflux\ 15\text{–}30\ min}$ CH_3O—⟨C$_6$H$_3$(CH$_3$)⟩—CHO 92% **171**

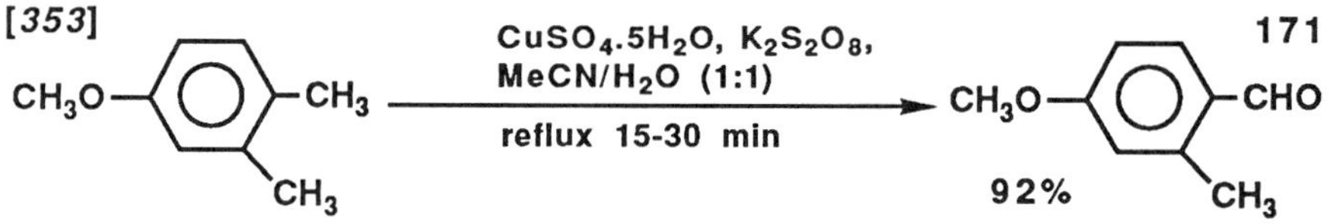

In aromatic heterocycles, the methyl groups in α positions with respect to nitrogen are oxidized especially easily. Thus quinaldine gives a 50% yield of quinoline-α-carboxaldehyde on refluxing for 1 h with selenium dioxide in aqueous dioxane [513], and 2,3,8-trimethylquinoline gives an 82% yield of 3,8-dimethylquinoline-2-carboxaldehyde on refluxing for 6 h with selenium dioxide in ethanol [515]. A similar selective oxidation is achieved when 8-ethyl-2-methylquinoline is treated with selenium dioxide or **potassium dichromate** [514]. Treatment with selenium dioxide gives 8-ethyl-2-formylquinoline, whereas treatment with potassium dichromate gives 8-acetyl-2-methylquinoline (equation 172) [514].

[514] **172**

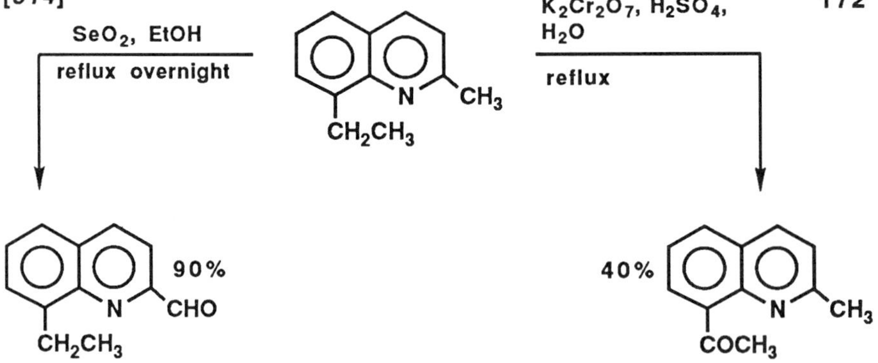

8-ethyl-2-methylquinoline

- SeO_2, EtOH, reflux overnight → 90% → 8-ethyl-2-formylquinoline
- $K_2Cr_2O_7$, H_2SO_4, H_2O, reflux → 40% → 8-acetyl-2-methylquinoline

Selectivity is also noticed in oxidations of 2,3-dimethylindole. Whereas sodium periodate cleaves the double bond and gives an 85% yield of *o*-acetamidoacetophenone, periodic acid oxidizes the methyl group adjacent to the heterocyclic nitrogen to an aldehyde group in a low yield (equation 173) [760].

[760]

HIO₄, H₂O, MeOH
ice-salt bath, 0.5 h

[Indole with 2-CH₃ and 3-CH₃ substituents]

NaIO₄, H₂O, MeOH
RT, 8 h

173

[3-methylindole-2-carbaldehyde] 10%   85%   [o-(NHCOCH₃)C₆H₄COCH₃]

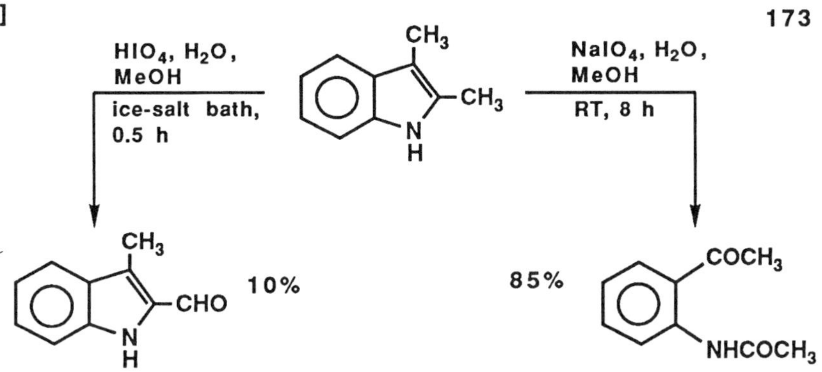

In the presence of acetic anhydride, acetic acid, and sulfuric acid, chromic acid (**chromium trioxide**) converts *o*-nitrotoluene and *p*-nitrotoluene into the corresponding ***nitrobenzaldehyde diacetates,*** from which the aldehydes are obtained on hydrolysis (equation 174) [547, 548, 549].

$$ArCH_3 \longrightarrow ArCH(OCOCH_3)_2 \longrightarrow ArCHO$$ (Ar = nitrophenyl) 174

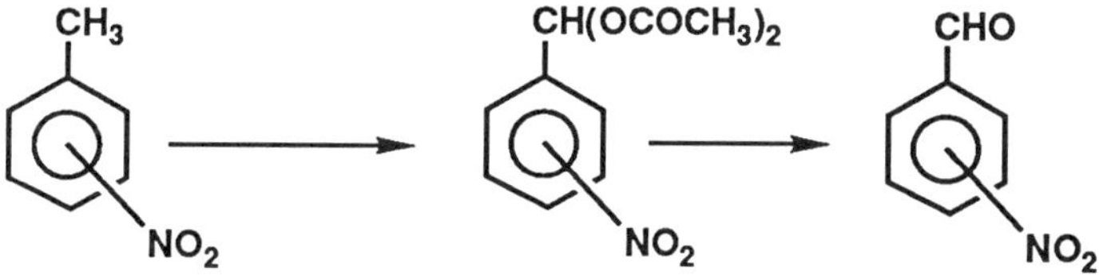

[548, 549]
ortho   CrO₃, Ac₂O, H₂SO₄   36–37%   HCl, EtOH   74%
         10 °C, 3 h   reflux 45 min

[547]
para   CrO₃, Ac₂O, AcOH,   48–54%   H₂SO₄, EtOH   89–94%
         H₂SO₄, <10 °C   reflux 30 min

Methylene groups adjacent to aromatic rings are oxidized to ***keto groups*** by oxygen with chromium sesquioxide as a catalyst [1128] or by mercuric bromide [11], ceric ammonium nitrate [380, 417, 422], selenium dioxide [509], sodium dichromate [622, 625], pyridinium chlorochromate [607], manganese dioxide [814], potassium permanganate [866, 877], and alkyl nitrites [452].

Ethylbenzene is converted into acetophenone by argentic oxide [380], by ceric ammonium nitrate [417], and by other oxidants (equation 175).

[380]   PhCH₂CH₃ —(4 AgO, H₃PO₄, −10 °C, 1 h)→ 65%   PhCOCH₃   175
[417]   —((NH₄)₂Ce(NO₃)₆, HNO₃, 70°, 1.5 h)→ 77%

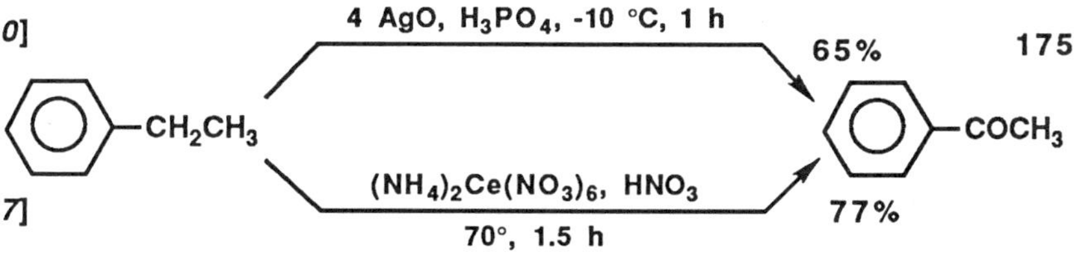

The conversion of methylene to carbonyl is especially easy when the group is flanked by two aromatic rings. Thus, diphenylmethane yields benzophenone on treatment with ceric ammonium nitrate [417, 422], selenium dioxide [509], and pyridinium chlorochromate [607]. Manganese dioxide at 125 °C converts diphenylmethane into benzophenone, whereas at 211 °C, tetraphenylethylene is formed in 81% yield [814] (equation 176).

176

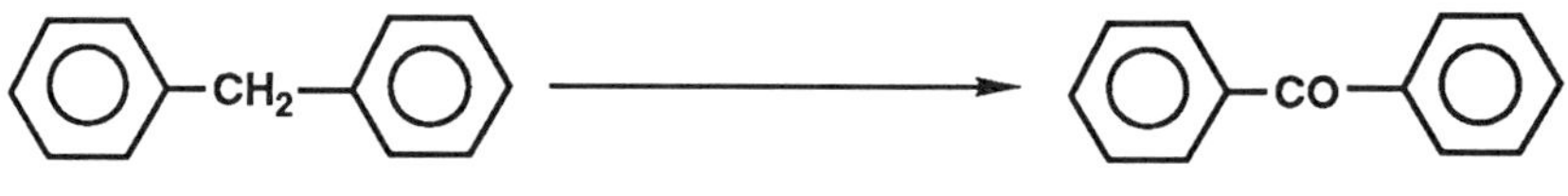

Ph–CH$_2$–Ph ⟶ Ph–CO–Ph

[417]	(NH$_4$)$_2$Ce(NO$_3$)$_6$, HNO$_3$, 90 °C, 75 min	76%
[509]	SeO$_2$, 200-210 °C, 30 min	87%
[607]	C$_5$H$_5$N.HCl.CrO$_3$, Celite*, C$_6$H$_6$, reflux 10 h	88%
[814]	MnO$_2$, 125 °C, 6 h	74%

*Celite is diatomaceous earth or infusorial earth.

Fluorene is oxidized to fluorenone in 65–70% yield by refluxing for 3 h with **sodium dichromate** in acetic acid [622], and 2-methylfluorene is converted into 2-methylfluorenone via its oxime on treatment with amyl nitrite (**pentyl nitrite**) [452]. The methylene group between two aromatic rings is oxidized in preference to the methyl group because nitrites react only with highly activated methylene groups (equation 177).

[452] **177**

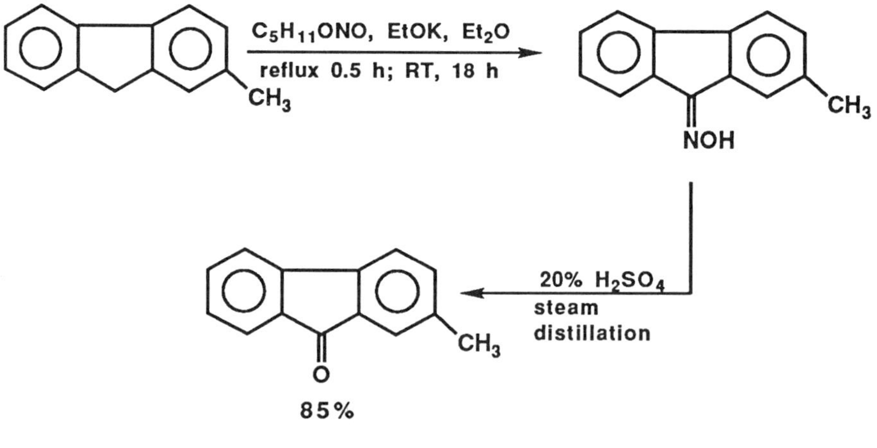

Fluorene-CH$_3$ →(C$_5$H$_{11}$ONO, EtOK, Et$_2$O; reflux 0.5 h; RT, 18 h)→ =NOH intermediate →(20% H$_2$SO$_4$, steam distillation)→ 2-methylfluorenone

85%

2-Benzylpyridine forms 2-phenacylpyridine on heating with potassium permanganate (equation 178) [877].

Indane is oxidized to 1-indanone and tetralin is converted into 1-tetralone by mercuric bromide under irradiation [11] or by argentic oxide [380] (equation 179).

[877]

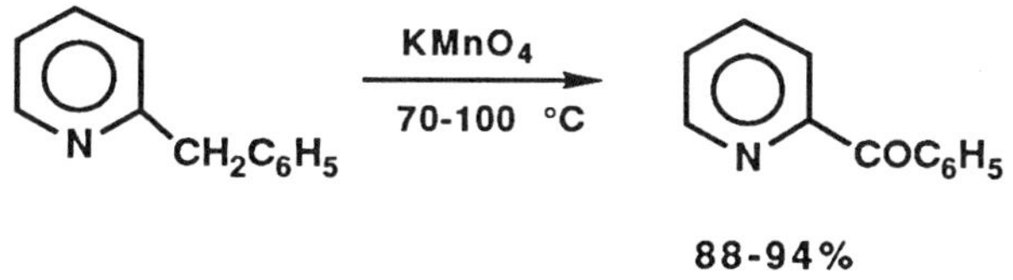

$$\text{2-PyCH}_2\text{C}_6\text{H}_5 \xrightarrow[70-100\ °C]{\text{KMnO}_4} \text{2-PyCOC}_6\text{H}_5 \quad 88\text{-}94\% \qquad 178$$

[11]

[380]

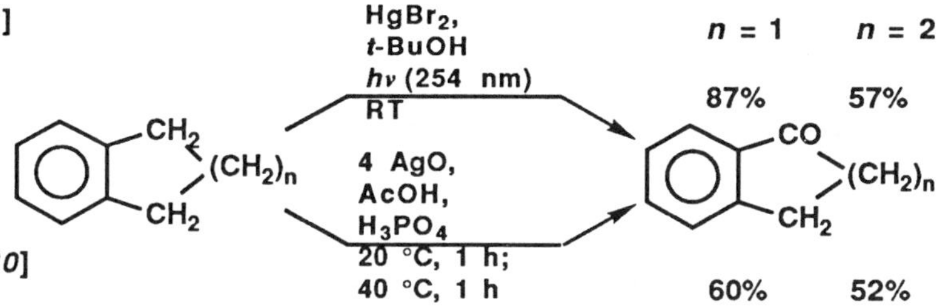

$$\text{o-C}_6\text{H}_4(\text{CH}_2)_2(\text{CH}_2)_n \xrightarrow{\substack{\text{HgBr}_2,\ t\text{-BuOH} \\ h\nu\ (254\ \text{nm}),\ \text{RT}}} \text{product} \quad (n=1:\ 87\%;\ n=2:\ 57\%)$$

$$\xrightarrow[20\ °C,\ 1\ h;\ 40\ °C,\ 1\ h]{4\ \text{AgO, AcOH, H}_3\text{PO}_4} \text{product} \quad (n=1:\ 60\%;\ n=2:\ 52\%) \qquad 179$$

In acenaphthene, both methylene groups are oxidized by sodium dichromate to give acenaphthenequinone (equation 180) [625].

[625]

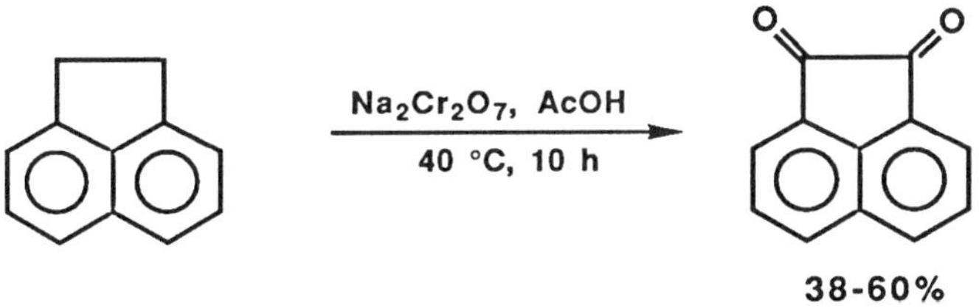

$$\text{acenaphthene} \xrightarrow[40\ °C,\ 10\ h]{\text{Na}_2\text{Cr}_2\text{O}_7,\ \text{AcOH}} \text{acenaphthenequinone} \quad 38\text{-}60\% \qquad 180$$

Oxidations with chromyl chloride (Etard reaction) of aromatic compounds with longer chains are not always suitable because the reactions result in more than one product [673].

OXIDATION TO CARBOXYLIC ACIDS

Not only methyl groups but practically any other alkyl groups attached to aromatic rings are oxidized to carboxyls by sufficiently strong oxidants, such as **nitric acid** [460, 461, 462, 463, 464, 891], **chromic acid and its derivatives** [550, 551, 624, 633, 634, 1129, 1130], and **potassium permanganate** [503, 841, 880, 881, 882, 883, 1131]. Occasionally, such oxidations have been effected by other reagents, such as ozone [68], sulfomonoperacid (Oxone) [205], sodium hypochlorite [696], and nickel dioxide [933], or by electrooxidation [117] (equation 181).

Ethylbenzene and isopropylbenzene give 80 and 75%, respectively, of benzoic acid on oxidation with 15% nitric acid at higher temperatures [462]. The disadvantages of oxidations with nitric acid are the formation of nitrated byproducts (up to 12% of nitrobenzoic acid in the oxidation of ethylbenzene) and the necessity of working in glass-lined or stainless steel autoclaves [462].

EXAMPLES OF OXIDATION OF TOLUENES TO BENZOIC ACIDS

CH_3-C$_6$H$_4$-X → CO_2H-C$_6$H$_4$-X   181

Ref	X	Reagents/Conditions	Yield
[841]	X = H	KMnO$_4$, crown ether, C$_6$H$_6$, 25 °C, 72 h	78%
[205]	X = H	2KHSO$_5$·KHSO$_4$·K$_2$SO$_4$, reflux 22 h	50%
[462]	X = H	15% HNO$_3$, 170-200 °C, 30-50 atm	85-90%
[68]	X = o-CH$_3$	O$_3$, Co(OAc)$_2$, AcOH, 115-120 °C, 1.5 h; O$_2$, 115-120 °C, 8.5 h	77%
[463]	X = o-CH$_3$	HNO$_3$ (dilute 1:2), 145-155 °C, reflux, 55 h	53-55%
[461]	X = p-F	20% HNO$_3$, 190-195 °C, 70 atm, 6 h	96%
[460]	X = p-CH$_3$, o-F	22% HNO$_3$, 200 °C, 52 atm, 15 min	79%
[882]	X = o-Cl	KMnO$_4$, H$_2$O, boil 3-4 h	76-78%
[880]	X = o-Br, m-Br, p-Br	KMnO$_4$, NaOH, H$_2$O, reflux	59% (76%)*, 63% (81%)*, 76% (83%)*
[634]	X = p-NO$_2$	Na$_2$Cr$_2$O$_7$, H$_2$SO$_4$, reflux 1 h	82-86%
[551]	X = 3,4-(NO$_2$)$_2$	CrO$_3$, H$_2$SO$_4$, 45-50 °C, 3-4 h	85-90%
[1129]	X = 2,4,6-(NO$_2$)$_3$	Na$_2$Cr$_2$O$_7$, H$_2$SO$_4$, 45-55 °C, 3-4 h	57-69%
[881]	X = HgCl	KMnO$_4$, NaOH, 95 °C, 15 min	61-74% X = OH

*The yields are based on the consumed bromotoluenes.

The partial oxidation of polyalkylated aromatic compounds is also observed. o-Xylene is oxidized to o-toluic acid by heating with ozone and oxygen at 115–120 °C in the presence of cobalt acetate in acetic acid (yield 77%) [68] or by refluxing with dilute nitric acid (1:2) (yield 53–55%) [463]. In p-cymene (p-isopropyltoluene), the isopropyl group is oxidized in preference to the methyl group to give a 51% yield of p-toluic acid on refluxing with dilute nitric acid (1:36) [464]. On the contrary, biochemical oxidation with *Nocardia* strain 107–332 converts p-cymene into p-isopropylbenzoic acid [1071].

In fairly rare cases, alkyl benzenes are converted into carboxylic acids without, or with only partial, degradation of the carbon chain. Propylbenzene is converted into benzoic acid on heating in an autoclave with an aqueous solution of sodium dichromate at 270 °C, but at 200 °C, 54% of 3-phenylpropionic acid is obtained [*632*]. Ethyl-, isopropyl-, and butylbenzene give similar results albeit in not so clear-cut reactions: isopropylbenzene yields 70% of 2-phenylpropionic acid and less than 10% of benzoic acid at 200 °C and 76% of benzoic acid and 24% of 2-phenylpropionic acid at 275 °C [*632*].

Biochemical oxidations of side chains proceed without degradation or with only limited degradation. Both ethylbenzene and butylbenzene, and even dodecylbenzene, give phenylacetic acid on incubation at 30 °C with *Nocardia* strain 107–332 (equation 182) [*1071*].

[*1071*]   182

$C_6H_5CH_2CH_3$
$C_6H_5CH_2CH_2CH_2CH_3$ $\xrightarrow{\text{Nocardia strain 107-332}}_{30\ °C}$ $C_6H_5CH_2CO_2H$
$C_6H_5CH_2CH_2C_{10}H_{21}$

Amylbenzene (pentylbenzene) is first oxidized to 5-phenylvaleric acid and later transformed, by β-degradation and dehydrogenation, to *trans*-cinnamic acid in high yields by *Cellulomonas galba* (equation 183) [*1053*].

[*1053*]   183

$C_6H_5(CH_2)_4CH_3 \longrightarrow C_6H_5(CH_2)_4CO_2H \longrightarrow C_6H_5CH=CHCO_2H$
                                         0-12%   trans
                                                                             88-100%

Homologues of *naphthalene* are oxidized to naphthalenecarboxylic acids on heating in an autoclave at 250 °C for 18 h with an aqueous solution of sodium dichromate. β-Methylnaphthalene affords β-naphthoic acid in 93% yield [*633*], and 2,3-dimethylnaphthalene, 2,3-naphthalenedicarboxylic acid in 87–93% yield [*1130*]. *Methylanthracenes* and *methylphenanthrenes* are similarly converted into the corresponding carboxylic acids in yields well over 90% [*633*].

In a derivative of acenaphthene, the five-membered ring is cleaved to form a dicarboxylic acid that yields an anhydride (equation 184) [*624*].

Alkyl groups are oxidized to carboxyls also in aromatic compounds containing functional groups, which may or may not be affected by the oxidant. Thus *p*-methylbenzophenone is transformed, on refluxing with chromium trioxide in acetic and sulfuric acid, into benzophenone-*p*-carboxylic acid (yield 62–69%) [*550*], whereas in *p*-methylacetophenone, both the methyl and the acetyl groups are converted into carboxyls (equation 185) [*696, 891*].

[624]   184

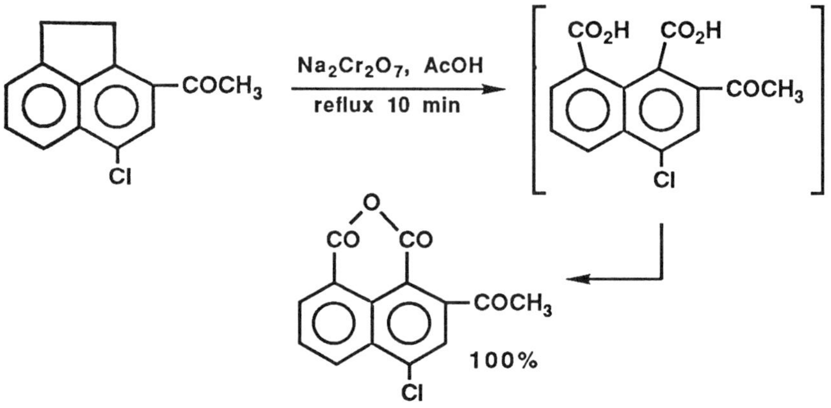

[Reaction: chloroacenaphthyl methyl ketone → Na₂Cr₂O₇, AcOH, reflux 10 min → naphthalene-1,8-dicarboxylic acid with COCH₃ and Cl substituents → cyclic anhydride product, 100%]

[696]   185

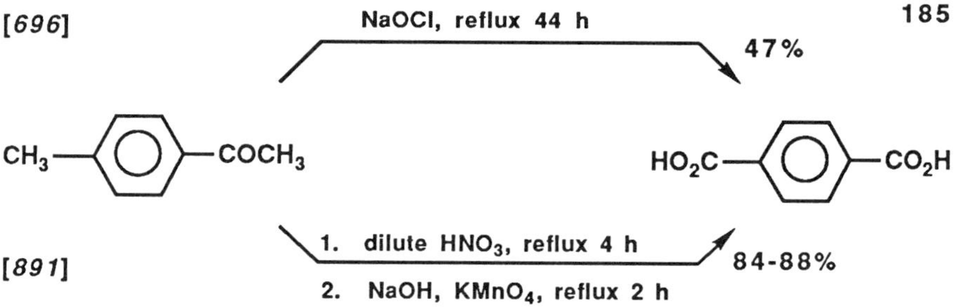

[891]

CH₃—C₆H₄—COCH₃ → HO₂C—C₆H₄—CO₂H

NaOCl, reflux 44 h, 47%

1. dilute HNO₃, reflux 4 h
2. NaOH, KMnO₄, reflux 2 h
84–88%

On treatment with potassium permanganate, the sodium salt of toluene-*o*-sulfonamide yields the corresponding carboxylic acid, which cyclizes to form saccharin (equation 186) [*1131*].

[1131]   186

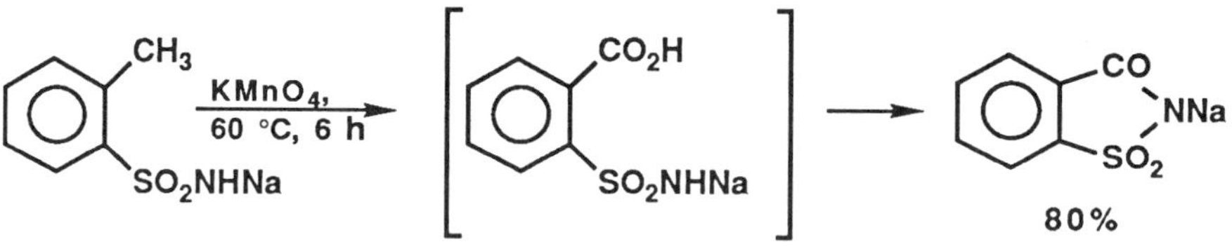

[Reaction: 2-methylbenzenesulfonamide sodium salt → KMnO₄, 60 °C, 6 h → 2-carboxybenzenesulfonamide sodium salt → saccharin sodium salt, 80%]

Homologues of pyridine are converted into the corresponding pyridinecarboxylic acids by electrooxidation [*117*], by oxidation with dilute nitric acid [*462*], or by treatment with potassium permanganate [*503, 883*] (equation 187).

2-Methylquinoline-3-carboxylic acid is oxidized quantitatively by nickel dioxide in sodium hydroxide after 12 h at 25 °C to quinoline-2,3-dicarboxylic acid [*933*].

methylpyridine → pyridinecarboxylic acid **187**

[*883*]	2-(α)-	KMnO$_4$, H$_2$O, 100 °C, 3-3.5 h	50-51%
[*117*]	3-(β)-	electrooxidation, Pb electrodes, H$_2$SO$_4$	60%
[*462*]	3-(β)-	15% HNO$_3$, 200 °C, 35-40 atm	50-60%
	4-(γ)-	10% HNO$_3$, H$_3$PO$_4$, 230 °C, 40 atm	93%
[*503*]	4-(γ)-	KMnO$_4$, H$_2$O, reflux	60-70% (50-60% pure)

HALOGEN DERIVATIVES AND TOSYLATES

In this section, only oxidations that result in the removal of halogen substituents will be discussed. Oxidations that do not affect halogens are discussed in sections describing oxidations of individual functional groups. The exception is the oxidative cleavage of haloalkenes, which leads to carboxylic acids and which is mentioned in the section Alkenes and Cycloalkenes.

Halogen compounds can be oxidized to aldehydes, ketones, or acids, depending on their structures and on the oxidants used.

Aliphatic *primary halides*—chlorides, bromides, and especially iodides—are converted into *aldehydes* by treatment with dimethyl sulfoxide [*998, 999, 1000*] or trimethylamine oxide [*993*]. The reactivity of alkyl chlorides and bromides is increased by converting them in situ to alkyl iodides by the addition of sodium iodide into the reaction mixtures [*999*] (equation 188).

A modification of the oxidation of alkyl halides to aldehydes is the transformation of the halides into *alkyl tosylates* on treatment with silver tosylate in acetonitrile at 0–5 °C followed by heating of the crude tosylates with dimethyl sulfoxide and sodium bicarbonate (equation 189) [*1000*].

Secondary bromides such as 2-bromobutane and 2-bromooctane are oxidized by dimethyl sulfoxide in the presence of sodium iodide and sodium bicarbonate after 2 h at 115 °C to 2-butanone and 2-octanone in 65 and 56% yields, respectively (equation 190) [*999*].

Allylic halides are transformed into α,β-unsaturated aldehydes or ke-

EXAMPLES OF OXIDATION OF ALKYL HALIDES TO ALDEHYDES

$$RCH_2X \longrightarrow RCHO \qquad 188$$

Ref	Conditions	Yield
[999]	R = C_3H_7, X = Br NaI, DMSO, NaHCO$_3$, 115 °C, 2 h	75%
[998]	R = C_5H_{11}, X = I DMSO, NaHCO$_3$, 150 °C, 3 min	86%
[999]	R = C_7H_{15}, X = Cl NaI, DMSO, NaHCO$_3$, 115 °C, 2 h	73%
[999]	R = C_7H_{15}, X = Br NaI, DMSO, NaHCO$_3$, 115 °C, 2 h	60%
[998]	R = C_7H_{15}, X = I DMSO, NaHCO$_3$, 148 °C, 4 min	74%
[993]	R = C_7H_{15}, X = I $(CH_3)_3NO$, CHCl$_3$, 40-50 °C, 1 h; reflux 20 min	41.5-43%

[1000]

$$C_7H_{15}CH_2X \xrightarrow[0-5\ °C]{TosOAg,\ MeCN} C_7H_{15}CH_2OTos \xrightarrow[150\ °C,\ 3\ min]{DMSO,\ NaHCO_3} C_7H_{15}CHO \qquad 189$$

X = Cl 71%
Br 74%
I 70%

[999]

$$CH_3CHBrR \xrightarrow[115\ °C,\ 2\ h]{DMSO,\ NaI,\ NaHCO_3} CH_3COR \qquad 190$$

R = C_2H_5 65%
R = C_6H_{13} 56%

tones by the **Sommelet reaction** on treatment with hexamine (hexamethylenetetramine) (equation 191) [1132].

[1132]

$$RCH=CHCH_2X \xrightarrow{C_6H_{12}N_4} RCH=CHCH_2.C_6H_{12}N_4.X \qquad 191$$

$$RCH=CHCHO \xleftarrow{\text{steam distillation or reflux with water}}$$

γ,γ-Dimethylallyl bromide and geranyl bromide are converted into the corresponding aldehydes in 75 and 82% yields, respectively, by heating at 100 °C with *sodium dichromate* in hexamethylphosphoramide in the presence of dicyclohexyl-18-crown-6 ether [614]. Similar results are achieved with polymer-supported chromic acid [540] (equation 192).

The oxidation of **benzylic halides to aromatic aldehydes and ketones** is very easy and is accomplished by many reagents. Very simple ways to prepare aromatic aldehydes from benzyl chlorides and bromides are treat-

$$(CH_3)_2C=CH(CH_2)_2C(CH_3)=CHCH_2Br \qquad \textbf{192}$$

polymer-supported H_2CrO_4, C_6H_6, reflux 2 h	$Na_2Cr_2O_7$, HMPA crown ether, 100 °C, 2 h
95% [540]	82% [614]

$$(CH_3)_2C=CH(CH_2)_2C(CH_3)=CHCHO$$

ment with **dimethyl sulfoxide** [999, 1003], refluxing with copper or lead nitrate [477], treatment with the sodium salt of 2-nitropropane [987] or with tetrabutylammonium periodate [778], and the ***Sommelet reaction***, which involves refluxing in water [1132] or in chloroform with hexamine (hexamethylenetetramine) and subsequent hydrolysis [955, 956] (equations 193–196).

193

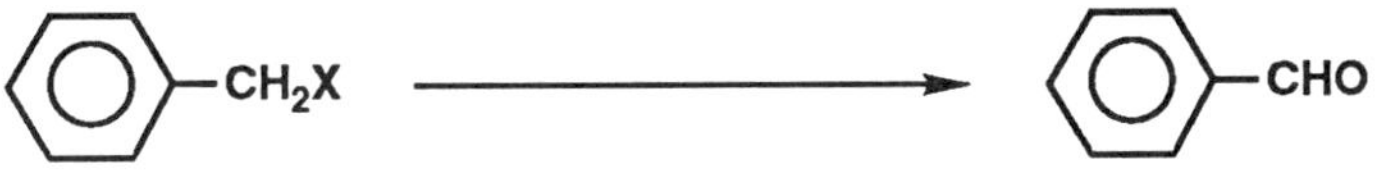

Ph–CH$_2$X ⟶ Ph–CHO

[614]	X = Cl	$K_2Cr_2O_7$, HMPA, dicyclohexano-18-crown-6 ether, 100 °C, 2 h	80%
[540]	X = Cl	H_2CrO_4 (polymer supported), C_6H_6, reflux 75 min	95%
[987]	X = Cl	Na, EtOH, Me$_2$CHNO$_2$, 80 °C, 3 h	73%
[778]	X = Br	Bu$_4$NIO$_4$, dioxane, reflux, 3 h	85%

194

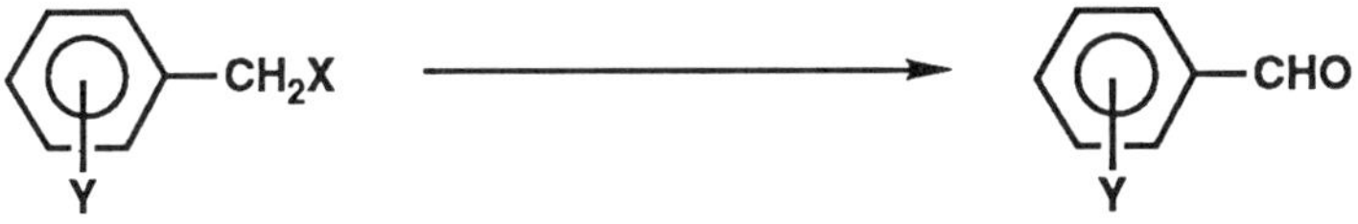

Y–C$_6$H$_4$–CH$_2$X ⟶ Y–C$_6$H$_4$–CHO

[477]	X = Br, Y = o-F	Pb(NO$_3$)$_2$, H$_2$O, reflux 5–6 h	61%
[1003]	X = Br, Y = p-NO$_2$	DMSO, MeCN, RT, 9 h	48%
[999]	X = Cl, Y = p-NO$_2$	DMSO, NaI, 120 °C, 4 h	62%

[955] **195**

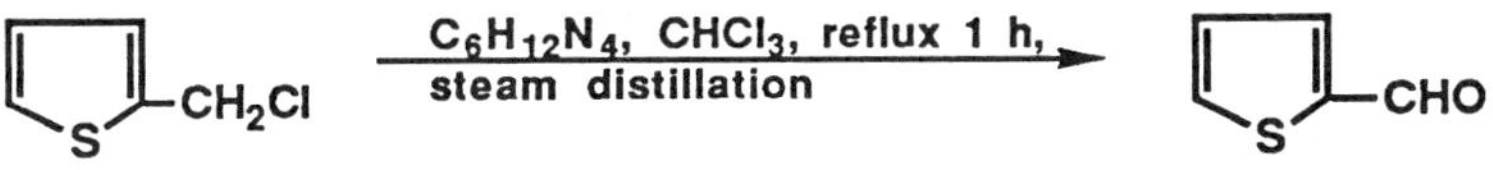

2-(chloromethyl)thiophene $\xrightarrow{\text{C}_6\text{H}_{12}\text{N}_4,\ \text{CHCl}_3,\ \text{reflux 1 h,}\ \text{steam distillation}}$ 2-thiophenecarbaldehyde

51%

[956]

196

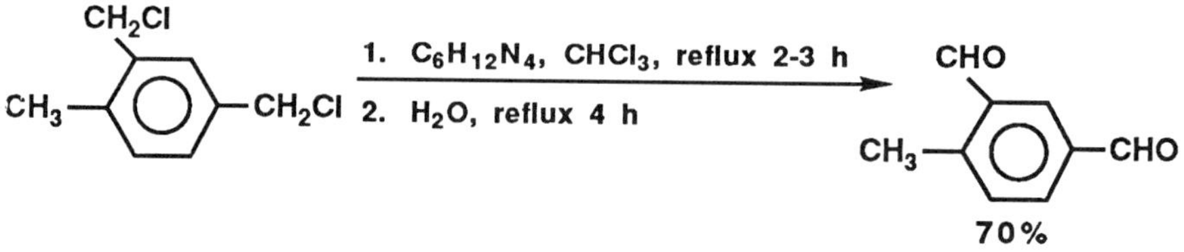

[Reaction: 4-methyl-1-(chloromethyl)...CH₂Cl benzene derivative]
1. C₆H₁₂N₄, CHCl₃, reflux 2-3 h
2. H₂O, reflux 4 h
→ 3-methyl-benzene-1,4-dicarbaldehyde (CH₃–Ar(CHO)–CHO), 70%

Benzylic halides are also converted into aldehydes on treatment with **nitroso compounds**, usually the easily prepared *p*-nitrosodimethylaniline, in the presence of pyridine with subsequent hydrolysis of the nitrones by hydrochloric acid or by hydrazine (equation 197) [*984*].

[984]

$$ArCH_2Br \xrightarrow{C_5H_5N} ArCH_2\overset{+}{N}C_5H_5\overset{-}{B}r \qquad \textbf{197}$$

↓ ON–C₆H₄–N(CH₃)₂

$$ArCHO + HONH\text{–}C_6H_4\text{–}N(CH_3)_2 \xleftarrow{H_2O} ArCH{=}\overset{\;}{N}(O)\text{–}C_6H_4\text{–}N(CH_3)_2$$

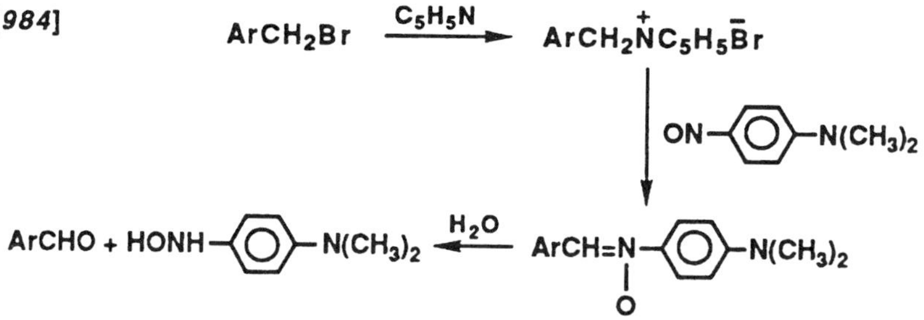

Secondary benzylic chlorides and bromides give *aromatic ketones* on oxidation with chromium trioxide [*560*] or with tetrabutylammonium hypochlorite [*690*] (equations 198 and 199).

[560]

1-chloroindane $\xrightarrow[\text{35-40 °C, 2 h}]{CrO_3, H_2O, AcOH}$ 1-indanone, 50-60%   **198**

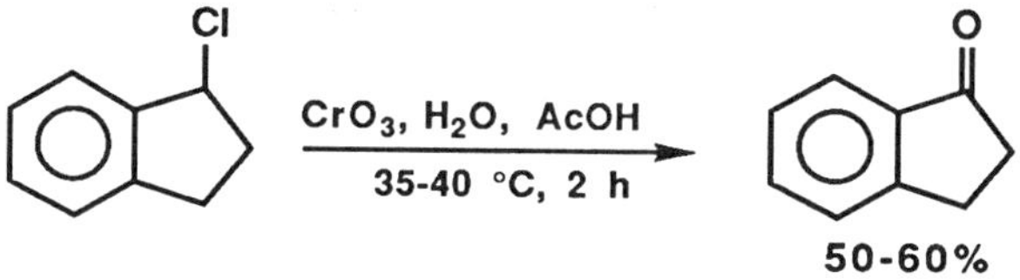

[690] $(C_6H_5)_2CHX \xrightarrow[\text{RT, 7-24 h}]{NaOCl,\ Bu_4NHSO_4,\ MeCN} (C_6H_5)_2CO$ **199**

X = Cl   89%
X = Br   90%

The oxidation of *primary alkyl halides to carboxylic acids* is accomplished on incubation of long-chain aliphatic fluorides, chlorides, bromides, and iodides with the yeast *Torulopsis gropengiesseri* [*1086, 1087*] (equation 200). The biooxidation starts at the terminal methyl groups and gives ω-halogenoalkanoic acids, which are ultimately converted to α,ω-dicarboxylic

[1086]

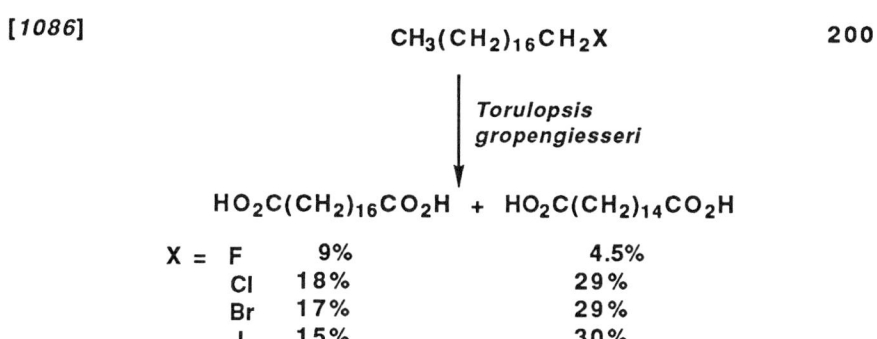

$CH_3(CH_2)_{16}CH_2X$ 200

↓ *Torulopsis gropengiesseri*

$HO_2C(CH_2)_{16}CO_2H$ + $HO_2C(CH_2)_{14}CO_2H$

X =			
	F	9%	4.5%
	Cl	18%	29%
	Br	17%	29%
	I	15%	30%

acids. In addition, dicarboxylic acids with two less carbons are produced concomitantly.

High-temperature oxidations with **nitrogen dioxide** or nitrogen dioxide and chlorine followed by treatment with water convert perfluoroalkyl hydrides and perfluoroalkyl iodides into perfluorocarboxylic acids. Perfluoroalkyl bromides and especially perfluoroalkyl chlorides do not react appreciably (equation 201) [455].

[455] 201

$C_6F_{13}CHF_2$ $\xrightarrow[\text{2. H}_2\text{O}]{\text{1. N}_2\text{O}_4, \text{Cl}_2, 600\,°\text{C}}$ $\xleftarrow[\text{2. H}_2\text{O}]{\text{1. N}_2\text{O}_4, 550\,°\text{C}}$ $C_6F_{13}CF_2I$

46%   59%
$C_6F_{13}CO_2H$

Oxidations of halogenated alkenes to carboxylic acids are discussed in a previous section, Oxidative Cleavage of Double Bonds (equations 118–120).

Perfluorinated aromatic compounds are oxidized to quinones (equations 202 and 203) [1133, 1134].

[1134] 202

[1133] (reaction scheme with perfluoroisoquinoline, 100% HNO₃, 50 °C, 1 h → 78% intermediate with NO₂ and NH; and 100% HNO₃, 100 °C, 20 h → 92% product) 203

Middle step: 100% HNO₃, 150 °C, 30 min

ALCOHOLS

Primary alcohols are oxidized to aldehydes or acids, and secondary alcohols are oxidized to ketones. Tertiary alcohols resist oxidation, unless they are dehydrated in acidic media to alkenes, which are subsequently oxidized. The conversion of alcohols into carbonyl compounds can be achieved by catalytic dehydrogenation or by chemical oxidation. Catalytic dehydrogenation is especially of advantage with primary alcohols, because it prevents overoxidation to carboxylic acids. Examples are tabulated in equations 223–227 and 265–268.

Dehydrogenation and Oxidation of Primary Alcohols to Aldehydes

Catalytic dehydrogenations of primary alcohols are achieved by passing vapors of the alcohols at 275–350 °C over a catalyst, usually supported on asbestos, silica gel, pumice, etc. Ethyl alcohol is converted into acetaldehyde in 88% yield at 93% conversion by passing it at 275 °C over a mixture of **oxides of copper, cobalt, and chromium** on asbestos [1135].

Over copper chromite on Celite (diatomaceous earth) at 300–350 °C, aliphatic alcohols with three to eight carbons are converted into aldehydes in 53–67% yields [354]. Catalytic dehydrogenations over **copper, silver, or both** [7, 345] are carried out in a current of an insufficient amount of air or oxygen at 300–380 °C and give aldehydes in yields ranging from 70 to 100%. An example of an industrial dehydrogenation is the conversion of methallyl alcohol into methacrolein [4] (equation 204).

Vapor-phase dehydrogenation of primary alcohols to aldehydes takes

[4]
$$CH_2=C(CH_3)CH_2OH \xrightarrow{Cu,\ 350\ °C,\ air} CH_2=C(CH_3)CHO$$
204
conversion 71%
yield 95%

place at 250–300 °C over **cupric oxide** in a current of helium. Yields of 90–100% are obtained with saturated and unsaturated alcohols [*349*].

The vapor-phase dehydrogenations just mentioned are applicable only to the preparation of aldehydes that tolerate such high temperatures. A catalytic oxidation of alcohols carried out by passing a current of air or oxygen through solutions of alcohols in solvents such as heptane [*56*] or ethyl acetate [*55*] in the presence of **platinum** [*55, 56*], or, better still, **platinum dioxide** [*56*], or active cobalt oxide [*1136*] is applicable even to alcohols that cannot be vaporized and that contain double bonds (equation 205) [*56*].

[*56*]
$$CH_3CH=C(CH_3)CH_2OH \xrightarrow[60\ °C,\ 2\ h]{O_2/PtO_2,\ C_7H_{16}} CH_3CH=C(CH_3)CHO$$
205
77%

Both aliphatic and aromatic alcohols, as well as unsaturated alcohols, are oxidized in the liquid phase with argentic oxide in nitric or acetic acid at temperatures from −10 through 60 °C [*380*].

Ceric ammonium nitrate in water or in 50% acetic acid oxidizes benzylic alcohols at 90 °C in very good yields [*420*]. Only catalytic amounts of the reagent and **sodium bromate** as a reoxidant are needed to convert benzyl alcohol into benzaldehyde in 90% yield on heating in acetonitrile at 80 °C [*421*]. A similar result is obtained on treatment of benzyl alcohol with **lead tetraacetate** in pyridine at room temperature for a few hours (yield 85%) [*442*].

A very simple and gentle oxidation of primary alcohols to aldehydes is their treatment in chloroform or carbon tetrachloride with a solution of *dinitrogen tetroxide*, obtained either commercially or by thermal decomposition of lead nitrate. The reaction is carried out at 0 °C through room temperature and gives high yields (91–98%) of benzaldehydes (equation 206) [*454*].

[*454*]
$$p\text{-}ClC_6H_4CH_2OH \xrightarrow[0\ °C,\ 15\ min,\ RT\ overnight]{CHCl_3,\ N_2O_4} p\text{-}ClC_6H_4CHO$$
206
98%

The classical oxidants for the conversion of alcohols into aldehydes are compounds of **hexavalent chromium.** The stoichiometry of these oxidations is shown in equation 207, and the mechanism is discussed in the section Oxidation of Secondary Alcohols to Ketones (equations 243–245).

207

$$3\ RCH_2OH + K_2Cr_2O_7 + 4\ H_2SO_4 \longrightarrow 3\ RCHO + Cr_2(SO_4)_3 + K_2SO_4 + 7\ H_2O$$

When **sodium or potassium dichromate** in the presence of dilute sulfuric acid is used, care must be taken to prevent overoxidation of the products to carboxylic acids. Lower boiling aldehydes can be removed by concomitant steam distillation and thus escape from further oxidation. The disadvantage of dichromate oxidations is the need for rather high reaction temperatures, often those of refluxing aqueous solutions [*639, 653*].

"Neutral" sodium dichromate even under these conditions does not cause much overoxidation. Benzyl alcohol is oxidized to benzaldehyde at pH 5.6 at a rate 18 times faster than that of the oxidation of benzaldehyde to benzoic acid [*639*].

A modern version of dichromate oxidation in aqueous media is a "phase-transfer" reaction carried out in a two-phase system. Alkaline dichromate is converted into tetraalkylammonium dichromate, which is soluble in organic solvents such as dichloromethane, chloroform, or benzene ("orange benzene"). The treatment of alcohols with a solution of potassium dichromate in acetic acid in the presence of Adogen 464 (Aldrich's trade name for methyltrialkyl [C_8–C_{10}] ammonium chloride) and benzene gives aldehydes at 55 °C [*651*]. Similar results are obtained with a chloroform solution of tetrabutylammonium chromate at 60 °C [*618*].

The best modifications of the phase-transfer reaction is oxidation with tetrabutylammonium dichromate prepared in situ from sodium dichromate and tetrabutylammonium bisulfate in 3 M sulfuric acid and methylene chloride. Under such conditions, oxidation takes place at room temperature and is finished within minutes [*630*]. The results of oxidations with **pyridinium dichromate, $(C_5H_5NH)_2Cr_2O_7$**, a stable bright-orange solid, mp 144–146 °C, prepared by precipitating with pyridine from a solution of chromium trioxide in a minimum amount of water, depend on the solvent used in the oxidations. In methylene chloride at 25 °C, decanol gives a 98% yield of decanal, and 2-octynol gives a 70% yield of 2-octynal. In dimethylformamide, the oxidation yields carboxylic acids [*603*]. Similar reagents, 3- and 4-carboxypyridinium dichromate prepared by the treatment of nicotinic and isonicotinic acid, respectively, with chromium trioxide in water, oxidize benzylic alcohols to aldehydes in dichloromethane–pyridine solutions at room temperature in 72–90% yields [*663*].

Oxidations with a complex of pyridine with silver dichromate in benzene do not offer any advantages, and refluxing is necessary [*659*].

Instead of dichromates, **chromic acid** obtained by dissolving chromium trioxide in water and sulfuric acid and acetic acid is used [184, 573].

A solution of chromium trioxide in dilute sulfuric acid used in aqueous acetone is called **Jones reagent** [572]. Other solvents of chromium trioxide are ether [538] and hexamethylphosphoric triamide (HMPA) [543]. Oxidations are also carried out with *chromium trioxide adsorbed on Celite (diatomaceous earth)* [538], *silica gel* [537], *or an ion exchanger* such as Amberlyst A26 (a macroreticular quaternary ammonium salt anion exchanger) [571, 617]. Such oxidations often take place at room temperature and can be used not only for saturated alcohols but also for unsaturated and aromatic alcohols (equations 208 and 209).

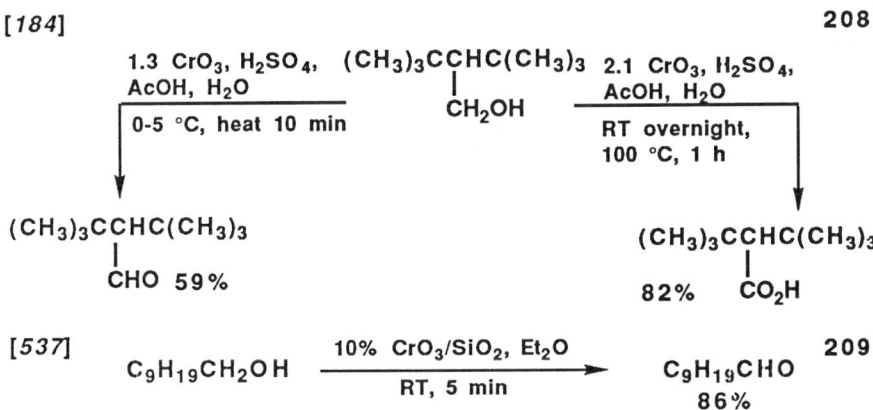

Collins reagent, a complex of **chromium trioxide with two molecules of pyridine,** is prepared by adding, in small portions, chromium trioxide to pyridine. The mixture is stirred and cooled in an ice bath. Reverse addition may cause ignition [599]. The reagent may also be prepared by adding chromium trioxide to a solution of pyridine in methylene chloride [596] or by adding pyridine to a suspension in petroleum ether of chromium trioxide adsorbed on silica gel [597]. The best solvent for oxidations with Collins reagent is dichloromethane, in which the solubility of the reagent is 12.5 g/100 mL [595]. The pyridine–chromium trioxide complex is very hygroscopic and is readily converted into insoluble dipyridinium dichromate, $(C_5H_5NH)_2Cr_2O_7$.

Collins reagent is not acidic and, consequently, is very suitable for oxidations of acid-sensitive substrates, especially because most of the oxidations are carried out at room temperature [595, 596]. The disadvantages are that up to a 1:6 ratio of the substrate to the reagent is sometimes to be used to obtain best yields [594, 595, 596, 600] and that saturated aliphatic alcohols often give low yields [599] (equation 210).

A similar complex obtained by adding one equivalent of *3,5-dimethylpyrazole* to a suspension of *chromium trioxide* in dichloromethane

[595] 210

$$C_6H_{13}CH_2OH \xrightarrow[25\ °C,\ 5\text{-}15\ min]{6\ mol\ 2\ C_5H_5N.CrO_3,\ CH_2Cl_2} C_6H_{13}CHO$$

93%

gives very high yields of allylic and benzylic aldehydes at room temperature after 30 min (the ratio of substrate to oxidant is 1:2.5) [602].

Another pyridine–chromium trioxide complex, **pyridinium chlorochromate, $C_5H_5NHCrO_3Cl$ (PCC)**, is prepared by adding pyridine to a solution of chromium trioxide in 6 M hydrochloric acid [605]. This complex is superior to Collins reagent in that much a smaller excess is needed, with the ratio of the substrate to the oxidant being 1:1.5–2 (equation 211).

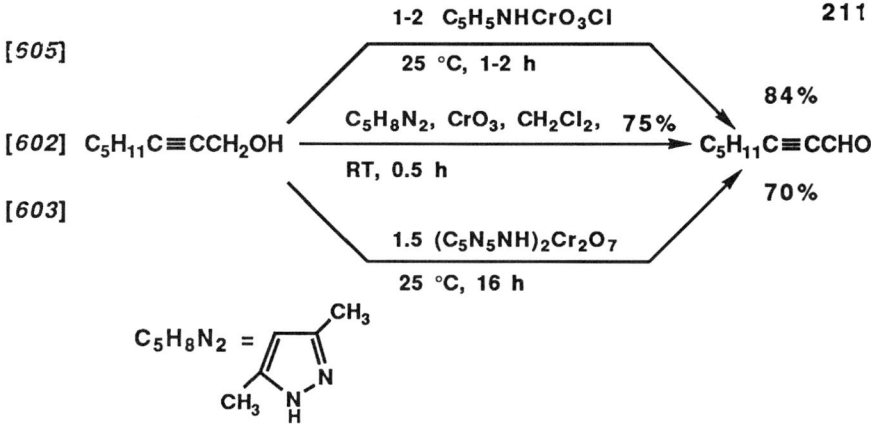

Because the reagent is slightly acidic, it cannot be used to oxidize acid-sensitive compounds [605]. For easier isolation of the products, the complex may be formed *in the presence of alumina*. After the reagent is stirred with the alcohol at room temperature for 2 h, the aldehyde is isolated by filtration and evaporation of the filtrate [604]. Another solid-support oxidant of pyridinium chlorochromate is prepared by treatment of cross-linked *poly(vinylpyridine) with chromium trioxide and hydrochloric acid* (equation 212) [612].

[612] 212

$$C_5H_{11}CH_2OH \xrightarrow[CH_2Cl_2,\ RT,\ 3\ days]{poly(vinylpyridinium)\ chlorochromate} C_5H_{11}CHO$$

91%

Instead of pyridine, 4-dimethylaminopyridine can be converted into a chlorochromate, and the complex can be used to oxidize benzylic alcohols to aldehydes [530]. Also, tetrabutylammonium chlorochromate gives good yields of unsaturated and aromatic aldehydes from the respective alcohols [619].

Other oxidants of hexavalent chromium are **chromyl chloride** and **di-*tert*-butyl chromate.** Chromyl chloride adsorbed on alumina–silica gel from its solution in dichloromethane oxidizes aliphatic and aromatic alcohols at room temperature within hours in 77–100% yields [*675*]. Di-*tert*-butyl chromate, prepared in situ from chromyl chloride in *tert*-butyl alcohol at −70 °C, gives comparable results under similar conditions [*674*]. Di-*tert*-butyl chromate, prepared from chromium trioxide and *tert*-butyl alcohol, oxidizes primary aliphatic and aromatic alcohols to the corresponding aldehydes even at low temperatures (1–2 °C) [*677, 678*].

Benzylic alcohols are successfully transformed into aldehydes by potassium hypochlorite containing 24 or 35% of available chlorine. In aqueous methanolic solution, a spontaneous exothermic reaction takes place and is finished at 36–40 °C overnight to yield 77% of benzaldehyde from benzyl alcohol [*702*].

Both tert-*butyl hypochlorite* and N-*chlorosuccinimide* dehydrogenate primary benzylic alcohols to the aldehydes (equation 213), whereas primary aliphatic alcohols are converted into esters of the corresponding acids [*709*]. Butyl alcohol dissolved in carbon tetrachloride and treated with *tert*-butyl hypochlorite in the presence of pyridine at 40–45 °C gives, after 2 h, a 66% yield of butyl butyrate [*709*].

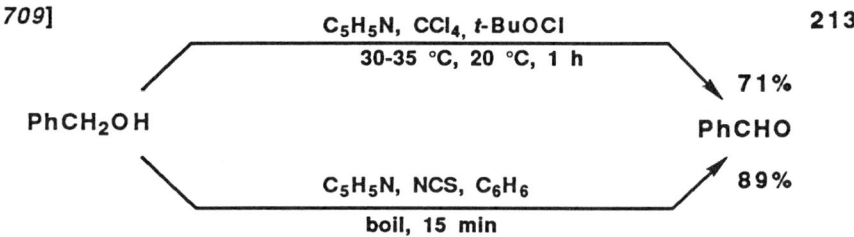

A useful addition to the roster of oxidants converting primary alcohols into aldehydes is **manganese dioxide,** which can be prepared by several methods [*805, 806, 807, 808, 809, 810*]. It is used as a suspension in petroleum ether [*805, 808*], ether [*806, 808, 811*], hexane [*806, 811*], benzene [*808, 813*], chloroform [*808, 811, 813*], and carbon tetrachloride [*808, 813*]. The oxidations are carried out at room temperature and are especially suitable for allylic and benzylic alcohols, which are oxidized more readily than the saturated alcohols [*808, 810*]. Yields vary widely depending on the substrate, on the ratio of the substrate to oxidant (which in turn depends on the particle size of the oxide [*810*]), on the solvent, and on the reaction time [*808, 811*] (equation 214).

Manganates and permanganates are rarely used to convert alcohols into aldehydes because permanganate oxidizes aldehydes to acids. However, when adsorbed on molecular sieves, potassium permanganate in benzene at 70 °C oxidizes benzyl alcohol and cinnamyl alcohol to the aldehydes in 80 and 94% yields, respectively. Saturated aliphatic alcohols give very

[811] $C_6H_5CH_2OH$ $\xrightarrow{MnO_2, RT}$ C_6H_5CHO 214

Reaction time (h)	Yield (%)		
	Hexane	Chloroform	Ether
1	61	61	70
24	78	89	78

low yields of aldehydes [863]. Benzyl alcohol [833] and other aromatic alcohols [832] are also oxidized to aldehydes by barium manganate.

The rarely used potassium ferrate transforms benzylic and allylic alcohols to aldehydes at room temperature in good to high yields. Saturated alcohols give unsatisfactory results [917, 918].

The oxidation of benzyl alcohol and cinnamyl alcohol with high-valency nickel oxide results in the formation of aldehydes or acids depending on the reaction conditions [929, 932].

Nickel peroxide, obtained by oxidation of nickel sulfate hydrate with an alkaline 6% solution of sodium hypochlorite, oxidizes alcohols either to aldehydes, when benzene is used as a solvent, or to acids, when the reaction is done in aqueous alkaline solution (equation 215) [932]. Allenic alcohols are converted into the aldehydes in ether at 20 °C in good yields [1137].

[932]   215

$$PhCH=CHCHO \; 86\% \xleftarrow[50\,°C, 1\,h]{1.2\;NiO_2,\;C_6H_6} PhCH=CHCH_2OH \xrightarrow[50\,°C, 6\,h]{2.0\;NiO_2,\;NaOH,\;H_2O} 90\% \; PhCH=CHCO_2H$$

The oxidation of alcohols with ruthenium tetroxide prepared by oxidation of ruthenium trichloride hydrate with sodium bromate takes place at room temperature. However, aldehydes may undergo further oxidation to carboxylic acids [940].

Very selective oxidizing agents suitable for the conversion of primary alcohols into aldehydes are high-potential quinones such as **tetrachloro-o-benzoquinone, tetrachloro-p-benzoquinone**, and **2,3-dichloro-5,6-dicyano-p-benzoquinone** [973]. Such dehydrogenations are carried out in chloroform, carbon tetrachloride, or ethanol, usually under very mild conditions at room temperature or in refluxing ether, and give fair to good yields (equation 216) [973].

Dehydrogenations of saturated alcohols, as well as benzylic alcohols, to aldehydes are carried out by refluxing with a benzene solution of diethyl azodicarboxylate, but the yields are only fair [977].

Dimethyl sulfide and **chlorine** or **N-chlorosuccinimide (NCS)** oxidize primary alcohols to aldehydes in high yields under very mild conditions.

[973]

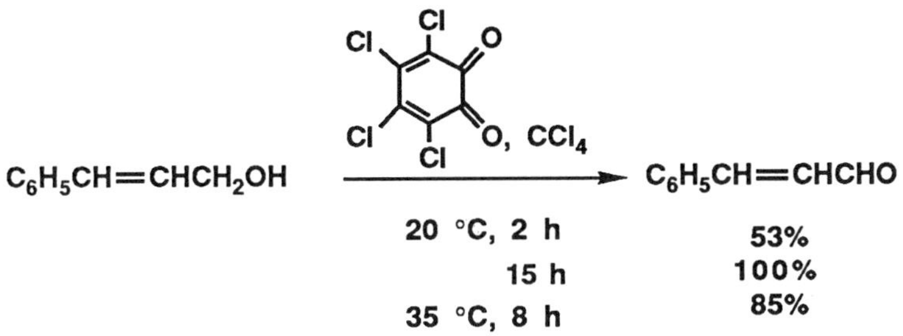

$$C_6H_5CH=CHCH_2OH \xrightarrow[\substack{20\ °C,\ 2\ h \\ 15\ h \\ 35\ °C,\ 8\ h}]{\text{tetrachlorobenzoquinone, CCl}_4} C_6H_5CH=CHCHO \quad \begin{array}{c}53\% \\ 100\% \\ 85\%\end{array}$$ **216**

1-Octanol is converted into octanal in 96% yield on treatment with dimethyl sulfide and *N*-chlorosuccinimide in toluene at –25 °C [*1138*].

A similar oxidation by **dimethyl sulfoxide (DMSO)** has attracted much attention because of the gentle conditions under which it takes place and the high yields of the resulting carbonyl compounds, especially those obtained from sterically hindered alcohols. In addition to dimethyl sulfoxide, an "activator" of DMSO is needed for the formation of a complex with the alcohol, and a base is required for the abstraction of a proton from the complex to initiate its collapse to the aldehyde (or ketone). The course of the reactions is summarized in equation 217.

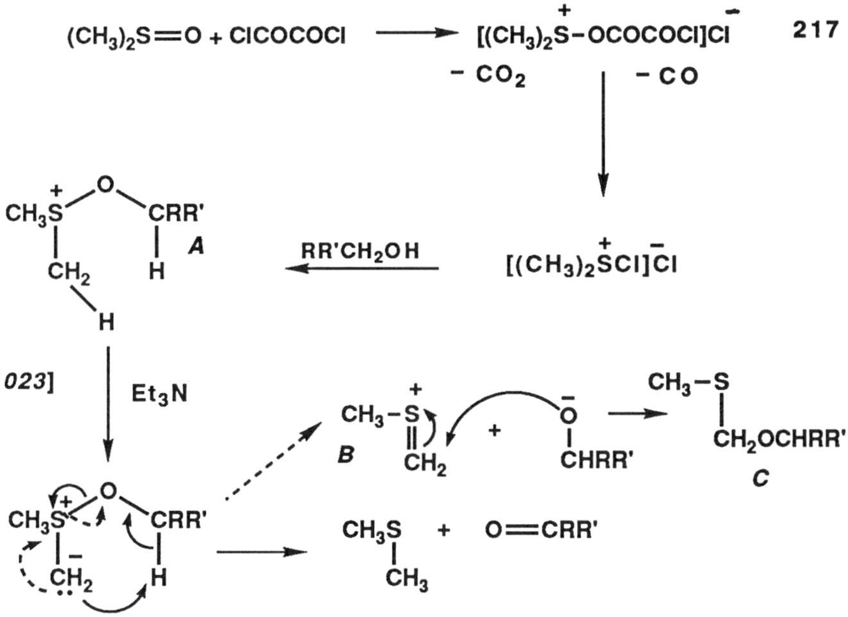

$$(CH_3)_2S=O + ClCOCOCl \longrightarrow [(CH_3)_2\overset{+}{S}-OCOCOCl]\overset{-}{Cl} \quad \textbf{217}$$

[*1023*]

The roles of the various reagents are as follows [*1023*]: The activator converts dimethyl sulfoxide into a reactive sulfonium intermediate, which reacts with the alcohol and forms a complex **A**. A base, usually triethylamine, abstracts a proton from one of the methyl groups linked to sulfur. An intramolecular shift of electrons causes the disintegration of the complex **A** into dimethyl sulfide and the carbonyl compound. A side reaction results from a competitive shift of electrons along the dotted arrows to form a sulfur ylide **B**, which recombines with the alkoxide thus generated to form a methylthiomethyl alkyl ether **C**. The side reaction forming an ether takes place at somewhat higher temperatures (above –60 °C) and is favored by some DMSO activators such as dicyclohexylcarbodiimide and trifluoroacetic anhydride. Formation of the methylthiomethyl alkyl ethers is suppressed and yields of the carbonyl compounds are increased by using bases bulkier than triethylamine, especially triisopropylamine [1012, 1023].

The best activator seems to be oxalyl chloride [1022, 1023], but chlorine [1020], phosgene [1023], thionyl chloride [1023], sulfur trioxide–pyridine [1023], phosphoric acid [1016], dicyclohexylcarbodiimide [1010, 1015], pyridinium trifluoroacetate [1010], trifluoroacetic anhydride [1012, 1023, 1139], and other acyl chlorides and anhydrides give very good yields [1023] (equations 218 and 219). The use of more than one equivalent of oxalyl chloride may lead to α-chloroketones [1024].

[1022]   218

$$CH\equiv C(CH_2)_2CH_2OH \xrightarrow[\text{Et}_3\text{N, 5 min} \longrightarrow \text{RT}]{\substack{(COCl)_2,\ CH_2Cl_2,\ Me_2SO,\\ -60\ \text{to}\ -50\ °C,\ 22\ \text{min}}} CH\equiv C(CH_2)_2CHO$$

99.6%

[1015]   219

$$CH_3C\equiv CC\equiv CCH_2CH_2CH_2OH \xrightarrow[\text{reflux 4 h}]{\substack{Me_2SO,\ H_3PO_4,\\ Et_2O\\ C_6H_{11}NCNC_6H_{11}\\ (DCC)}} CH_3C\equiv CC\equiv CCH_2CH_2CHO$$

73%

A modification of alcohol oxidation with dimethyl sulfoxide is the reaction of DMSO with alkyl chloroformates, which are formed from alcohols and phosgene (equations 220 and 221) [1021].

The oxidation of alcohols to aldehydes can also be accomplished by benzeneseleninic anhydride, $(C_6H_5SeO)_2O$, either as such [525] or prepared in situ from diphenyldiselenide, $(C_6H_5)_2Se_2$, and *tert*-butyl hydroperoxide [1140]. Benzylic alcohols are oxidized more rapidly than allylic

[1021]   220

$$RR'CHOH \xrightarrow{COCl_2} RR'CHOCOCl \xrightarrow[-Cl]{DMSO} RR'CHO\underset{(CH_3)_2S+}{\overset{CO}{\diagup}\negthickspace\diagdown}O \xrightarrow{-CO_2}$$

$$\underset{RR'C\overset{H}{\diagup}O\overset{+}{-}S(CH_3)_2}{} \xrightarrow[-H^+]{Et_3N} RR'C=O + (CH_3)_2S$$

[1021]

$(CH_3)_2CHCH_2OH \xrightarrow[\text{evaporation}]{\begin{array}{c}COCl_2,\\ Et_2O\\ RT,\end{array}} (CH_3)_2CHCH_2OCOCl \xrightarrow{\begin{array}{c}1.\ Me_2SO,\\ 15\ °C,\\ 2\ min,\\ RT,\\ 15\ min\\ 2.\ Et_2O,\\ cooling,\\ RT,\\ 20\ min\end{array}} (CH_3)_2CHCHO$ 221

70%

and saturated alcohols [525]. The fairly selective reaction usually requires refluxing in benzene [525, 1140] (equation 222).

[1140]   222

$CH_3(CH_2)_8CH_2OH \xrightarrow[\text{reflux 5 h}]{Ph_2Se_2,\ t\text{-BuOOH},\ C_6H_6} CH_3(CH_2)_8CHO$

92%

Examples of oxidizing agents and conditions for the conversion of primary alcohols into aldehydes are displayed in equations 223–227.

EXAMPLES OF OXIDATION OF PRIMARY ALCOHOLS TO ALDEHYDES

$C_7H_{15}CH_2OH \longrightarrow C_7H_{15}CHO$ 223

[55]	O_2/Pt, AcOEt, RT, 20 h	21%
[349]	CuO, He, 250-300 °C	98%
[630]	$Na_2Cr_2O_7$, 3 M H_2SO_4, Bu_4NHSO_4, CH_2Cl_2, RT, 1 min	95%
[571]	CrO_3 on Amberlyst A26 in Cl cycle, dried at 50 °C for 5 h, C_6H_6, reflux, 9 h	89%
[617]	CrO_3, Bu_4NCl (catalyst) CH_2Cl_2, 90 min	60%
[596]	$CrO_3 \cdot 2C_5H_5N$, CH_2Cl_2; RT, 15 min	90%
[675]	CrO_2Cl_2 in CH_2Cl_2 adsorbed at Al_2O_3-SiO_2, RT, 5 h	94%
[1138]	$NCS \cdot Me_2S$, PhMe, -25 °C, 1.5 h; Et_3N, 5 min	96%
[801]	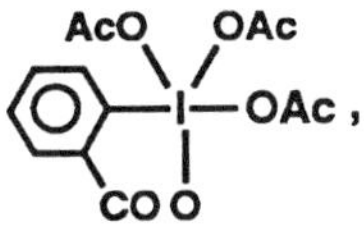 (AcO)(OAc)-I-OAc benziodoxole, AcOH, CH_2Cl_2, 25 °C, 1 h	93%

$CH_2=CHCH_2OH \longrightarrow CH_2=CHCHO$ **224**

Ref	Conditions	Yield
[7]	air/Ag, 230-300 °C	70-75%
[380]	AgO, H_3PO_4, 20 °C, 7 min	70%
[442]	$Pb(OAc)_4$, C_5H_5N, RT, h	35%
[808]	MnO_2, petroleum ether, RT, 19 h	99%

$(CH_3)_2C=CHCH_2CH_2C(CH_3)=CHCH_2OH \longrightarrow$ **225**
$(CH_3)_2C=CHCH_2CH_2C(CH_3)=CHCHO$

Ref	Conditions	Yield
[597]	H_2CrO_4, C_5H_5N, SiO_2, AcOH, CH_2Cl_2, 30-35 °C, 4 h	92%
[599]	$CrO_3 \cdot 2C_5H_5N$, C_5H_5N, cooling with ice, RT, 15-22 h	63%
[806]	MnO_2, C_6H_{14}, RT, 96 h	61-79%
[530]	Me_2N-C$_6$H$_4$-N·HCl, CrO_3, CH_2Cl_2, RT, 15 h	88%
[674]	CrO_2Cl_2, C_5H_5N, t-BuOH, CH_2Cl_2, -78 to -70 °C, 15 min, RT, 2 h	87%
[572]	CrO_3, Me_2CO, 0 °C, ~20 min	91%
[543]	CrO_3, HMPA, RT, 12 h	85%
[1140]	$(PhSe)_2$, t-BuOOH, CCl_4, reflux 1 h	100%

$C_6H_5CH_2OH \longrightarrow C_6H_5CHO$ **226**

Ref	Conditions	Yield
[345]	air/Cu, Ag on pumice, 300 °C	77%
[55]	O_2/Pt, AcOEt, RT, 24 h	72%
[56]	air or O_2, PtO_2, C_7H_{16}, 60 °C, 1 h	78%
[1136]	O_2/Co_2O_3, C_6H_6, 80 °C, 30 min	81%
[380]	AgO, HNO_3, -10 °C, 3 min	71%
[442]	$Pb(OAc)_4$, C_5H_5N, RT, a few hours	85%
[808]	MnO_2, CCl_4, RT, 4.3 h	69%
[349]	CuO, He, 250-300 °C	92%
[420]	$(NH_4)_2Ce(NO_3)_6$, 50% AcOH, 90 °C, 8 h	95%
[599]	$CrO_3 \cdot 2C_5H_5N$, cooling with ice, RT, 15-22 h	63%

$C_6H_5CH_2OH \longrightarrow C_6H_5CHO$ **226**

Ref	Conditions	Yield
[597]	H_2CrO_4, C_5H_5N, SiO_2, AcOH, CH_2Cl_2, 30-35 °C, 4 h	82%
[674]	CrO_2Cl_2, C_5H_5N, t-BuOH, CH_2Cl_2, -78 to -70 °C, 15 min, RT, 2 h	85%
[678]	CrO_3, t-BuOH, cooling with ice, C_6H_6; 1-2 °C, 6 h	90%
[639]	$Na_2Cr_2O_7$, H_2O, 98 °C, 3 h	85%
[630]	$Na_2Cr_2O_7$, Bu_4NHSO_4, 3 M H_2SO_4, CH_2Cl_2, RT, 1 min	100%
[618]	CrO_3, Bu_4NCl, H_2O; CCl_4, 60 °C, 1 h	81%
[572]	CrO_3, Me_2CO, 0 °C, ~20 min	76%
[543]	2 CrO_3, HMPA, RT, 3 h	85%
[538]	1-2 CrO_3, Et_2O, CH_2Cl_2; SiO_2, 35 min	75%
[596]	$CrO_3 \cdot 2C_5H_5N$, CH_2Cl_2, RT, 15 min	89%
[595]	6 $CrO_3 \cdot 2C_5H_5N$, CH_2Cl_2, 25 °C, 5-15 min (Collins reagent)	95%
[603]	1.5 $CrO_3 \cdot 2C_5H_5N$, CH_2Cl_2, 25 °C, 20-24 h	83%
[530]	3 CrO_3, HCl, N(C$_6$H$_4$)—NMe$_2$, CH_2Cl_2, RT, 3 h	64%
[602]	2.5 (2-methyl-1-pyrroline), CrO_3, CH_2Cl_2, RT, 45 min	83%
[675]	CrO_2Cl_2, CH_2Cl_2, Al_2O_3-SiO_2, RT, 5 h	94%
[863]	$KMnO_4$, molecular sieves 4Å, C_6H_6, 70 °C, 2 h	80%
[833]	$BaMnO_4$, C_6H_6, reflux 9 min	90%
[895]	bis(2,2'-bipyridyl)Cu(MnO$_4$)$_2$, CH_2Cl_2, RT, 15 min	95%
[692]	10% NaOCl, Bu_4NHSO_4, H_2O, 24 °C, 75 min	76%
[709]	t-BuOCl, C_5H_5N, CCl_4, 30-35 °C, 20 °C, 1 h	71%
[709]	NCS, C_5H_5N, C_6H_6, 100 °C, 15 min	89%
[917]	K_2FeO_4, H_2O, RT, 4 min	80%
[929]	NiO_2, C_6H_6, reflux 1 h	100%
[940]	RuO_4, CCl_4, 10-15°, RT, overnight	90%
[973]	p-$C_6Cl_4O_2$, Et_2O, reflux 3 h,	66%
[1138]	$NCS \cdot Me_2S$, PhMe, -25 °C, 1.5 h; Et_3N, 5 min	90%
[1011]	Me_2SO (DMSO), $(CF_3CO)_2O$, CH_2Cl_2, -50 °C, 40 min; Et_3N, 10 min, RT, 40 min	80%
[1023]	Me_2SO (DMSO), $(COCl)_2$, CH_2Cl_2, -60 °C, 35 min; Et_3N, 5 min	98%
[236]	$(Me_3SiO)_2/(Ph_3P)_3RuCl_2$, CH_2Cl_2, 25 °C, 2 h	91%
[1140]	$(PhSe)_2$, t-BuOOH, CCl_4, reflux 2 h	100%
[977]	$EtO_2CN=NCO_2Et$, C_6H_6, reflux 4 h	61%

$C_6H_5CH_2OH \longrightarrow C_6H_5CHO$ **226**

[663]	$(3-C_5H_4NCO_2H)_2Cr_2O_7$, C_5H_5N, CH_2Cl_2, RT, 25 min	85%
[619]	Bu_4NCrO_3Cl, $CHCl_3$, RT, 18 h	68%
	reflux 4 h	65%
[702]	KOCl, H_2O, MeOH, 46 °C, 36-40 °C overnight	77%

$PhCH=CHCH_2OH \longrightarrow PhCH=CHCHO$ **227**

[7]	air/Ag, 300 °C	60-70%
[1136]	air, Co_2O_3, C_6H_6, 80 °C, 0.5 h	94%
[380]	AgO, AcOH, 20 °C, C 3 h, 40 °C, 100 min	75%
[442]	$Pb(OAc)_4$, C_5H_5N, RT, h	91%
[805]	MnO_2, CCl_4, RT, 30 min	70%
[806]	MnO_2, Et_2O, RT, 188 h	77%
[599]	$CrO_3 \cdot 2C_5H_5N$, cooling with ice, RT, 15-23 h	81%
[597]	H_2CrO_4, C_5H_5N, SiO_2, AcOH, CH_2Cl_2, 30-35 °C, 4 h	87%
[674]	CrO_2Cl_2, C_5H_5N, t-BuOH, CH_2Cl_2, -78 to -70 °C, 15 min, RT, 2 h	78%
[572]	CrO_3, Me_2CO, 0 °C, ~20 min	84%
[543]	CrO_3, HMPA, RT, 1 h	92%
[596]	$CrO_3 \cdot 2C_5H_5N$, CH_2Cl_2; RT, 15 min	96%
[863]	$KMnO_4$, molecular sieves 4 A, C_6H_6, 70 °C, 2 h	94%
[833]	$BaMnO_4$, C_6H_6, reflux 45 min	100%
[895]	bis(2,2'-bipyridyl)Cu$(MnO_4)_2$, CH_2Cl_2, RT, 1 h	95%
[618]	2 Bu_4NHCrO_4, $CHCl_3$, 60 °C	91%
[604]	2 $CrO_3 \cdot HCl \cdot C_5H_5N$, Al_2O_3, C_6H_{14}, RT, 2 h	84%
[917]	K_2FeO_4, H_2O, RT, 7 min	75%
[918]	K_2FeO_4, 10% KOH, t-BuOH, RT, 1 h	96%
[973]	$o-C_6Cl_4O_2$, CCl_4, RT, 15 h	99%
[1011]	Me_2SO (DMSO), $(CF_3CO)_2O$, CH_2Cl_2, -50 °C, 40 min; Et_3N, 10 min, RT, 40 min	81%
[1023]	Me_2SO (DMSO), $(COCl)_2$, CH_2Cl_2, -60 °C, 35 min; Et_3N, 5 min	97%
[1140]	$(PhSe)_2$, t-BuOOH, CCl_4, reflux 1 h	87%
[639]	$Na_2Cr_2O_7$, H_2O, 98 °C, 3 h	91%
[602]	(3,5-dimethylpyrazole)·CrO_3, CH_2Cl_2, RT, 45 min	90%
[663]	$(3-C_5H_4NCO_2H)_2Cr_2O_7$, C_5H_5N, CH_2Cl_2, RT, 30 min	80%
[619]	Bu_4NCrO_3Cl, $CHCl_3$, RT, 52 h	84%
	reflux 3 h	82%
[932]	1.2 NiO_2, C_6H_6, 50 °C, 1 h	86%

Oxidation of Primary Alcohols to Carboxylic Acids

The conversion of primary alcohols into carboxylic acids is not a difficult task. Often, the same oxidant that has been used to oxidize alcohols to aldehydes is applicable to the oxidation to acids when used in appropriate amounts, in different solvents, at higher temperatures, or at longer reaction times. An example is the oxidation of primary alcohols with **air or oxygen** with platinum-on-charcoal or, better still, platinum dioxide as catalyst (equation 228) [56].

[56]
$$C_{11}H_{23}CHO \xleftarrow[\text{60 °C, 15 min}]{O_2/PtO_2,\ C_7H_{16}} C_{11}H_{23}CH_2OH \xrightarrow[\text{60 °C, 2 h}]{O_2/PtO_2,\ C_7H_{16}} C_{11}H_{23}CO_2H$$

77% 96% 228

Electrolytic oxidation with nickel anodes and stainless steel cathodes and sodium hydroxide or potassium hydroxide as the electrolyte converts saturated, unsaturated, acetylenic, and aromatic alcohols into acids at 25–75 °C in yields ranging from 51 to 92% (equation 229) [116].

[116]   229

$$RCH_2OH \xrightarrow[\text{4 A, 25 °C, 5 h}]{\text{Ni anode, 1 M NaOH}} RCO_2H$$

$R = C_6H_{13}$ 84%
$R = CH \equiv C$ 51%

Potassium hydroxide can act as an oxidant at higher temperatures. Heating heptadecyl alcohol with 3 times its weight of powdered potassium hydroxide for 15 min at 240–250 °C results in a 76% yield of heptadecanoic acid [484]. Under much milder conditions, benzyl alcohol is converted into benzoic acid in a 75% yield on heating for 10–60 min at 25–80 °C with an excess of powdered potassium hydroxide in aqueous *tert*-butyl alcohol and carbon tetrachloride [954].

Vanillyl alcohol yields 93% of vanillic acid on heating with aqueous sodium hydroxide and silver oxide for a few minutes at 75 °C. The oxidation most probably involves a Cannizzaro-type reaction [366].

The oxidation of primary alcohols to acids has been accomplished with **nitric acid.** Concentrated nitric acid (d 1.42) at 25–30 °C oxidizes 3-chloropropanol to 3-chloropropanoic acid in 78–79% yield after several hours [469] and converts 6-bromohexanol into 6-bromohexanoic acid in 91% yield after heating for 45 min at 100 °C [465].

Oxidations with dilute nitric acid (d 1.15) are commonly used to transform sugars into dicarboxylic (glycaric) acids. Evaporation of a solution of

100 g of lactose in 1200 mL of dilute nitric acid (d 1.15) on a steam bath to a volume of 200 mL gives 40 g (64%) of mucic (galactaric) acid [467].

Chromic acid dissolved in water and sulfuric acid, acetic acid, or both can be used to oxidize primary alcohols to carboxylic acids, provided that enough oxidant and more energetic conditions than those for oxidation to aldehydes are applied [184, 574] (equations 208 and 230).

[574] 230

$$FCH_2(CH_2)_8CH_2OH \xrightarrow[\text{5 °C, RT overnight}]{\text{4 CrO}_3, \text{ AcOH, H}_2\text{O}} FCH_2(CH_2)_8CO_2H$$
93%

Not very many other chromium compounds are suitable for the preparation of carboxylic acids from alcohols. One such compound is zinc dichromate, which oxidizes benzyl alcohol to benzoic acid in 90% yield in dichloromethane solution at room temperature [660].

Pyridinium dichromate in dichloromethane solution converts primary alcohols into aldehydes. In dimethylformamide at 25 °C, carboxylic acids are formed. Cyclohexylmethanol thus gives cyclohexanecarboxylic acid in 84% yield [603]. Oxidations of aliphatic alcohols with tert-*butyl chromate* yield mixtures of acids with aldehydes and esters [677].

Manganese dioxide, an oxidant of primary alcohols to aldehydes, can also yield acids under proper conditions (equation 231) [1141].

[1141] 231

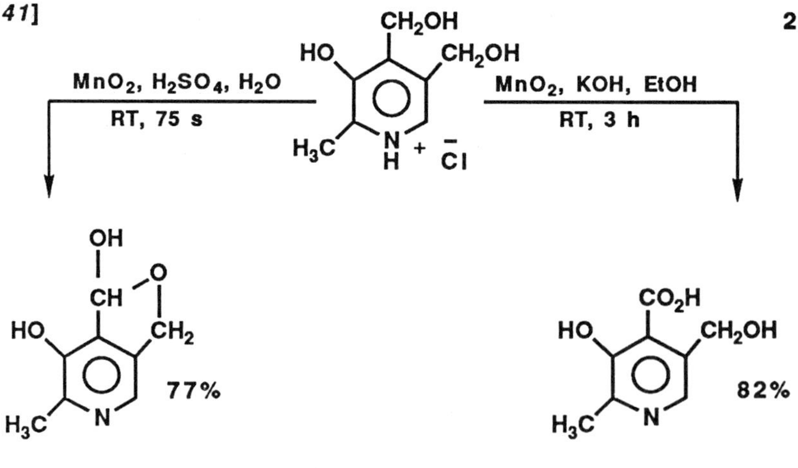

Oxidations with **permanganates** are suitable for the preparation of carboxylic acids from saturated and benzylic alcohols. Unsaturated alcohols may suffer cleavage of double bonds. Conventional oxidations are carried out in aqueous media, usually in the presence of alkali hydroxides. Thus

2-benzoylaminobutanol is converted into α-benzoylaminobutyric acid in 67–72% yield by stirring in aqueous sodium hydroxide and potassium permanganate at autogenous temperature (40 °C) [*884*].

In nonaqueous media, the oxidation of alcohols to acids is accomplished in heterogeneous systems by refluxing primary alcohols with solid **sodium permanganate monohydrate** in hexane (equation 232). The reaction is applicable even to allylic alcohols [*835*].

[*835*]   232

$$CH_3(CH_2)_6CH_2OH \xrightarrow[69\ °C,\ 5\ h]{NaMnO_4 \cdot H_2O,\ C_6H_{14}} CH_3(CH_2)_6CO_2H \quad 67\%$$

Similar results are obtained when saturated and benzylic alcohols are stirred at room temperature or refluxed in dichloromethane with cupric permanganate octahydrate, $Cu(MnO_4)_2 \cdot 8H_2O$. The octahydrate reacts faster than the anhydrous salt. Decanol and benzyl alcohol give 91 and 84% yields of decanoic acid and benzoic acid, respectively, at room temperature after 24 h [*894*].

Homogeneous oxidations are accomplished with *tetrabutylammonium permanganate* prepared by treating an aqueous solution of potassium permanganate with tetrabutylammonium bromide and isolating the crystalline precipitate [*845, 899*]. Tetrabutylammonium permanganate is soluble in dichloromethane, chloroform, acetone, and pyridine and sparingly soluble in benzene ("purple benzene").

Other quaternary ammonium salts can be used to dissolve the permanganate [*845*]. The reaction can also be carried out in a two-phase system by preparing the quaternary ammonium permanganate in situ while using only a catalytic amount of the quaternary salt (equation 233) [*845*].

[*845*]   233

$$CH_3(CH_2)_6CH_2OH \xrightarrow[Bu_4NBr\ (catalyst),\ RT,\ 3\ h]{KMnO_4,\ H_2O,\ C_6H_6} CH_3(CH_2)_6CO_2H \quad 47\%$$

Another method of solubilizing potassium permanganate in organic solvents is the addition of catalytic amounts of crown ethers, for example, dicyclohexano-18-crown-6 ether, to a mixture of potassium permanganate and benzene rolled in a ball mill for 2 h. Heptanol affords a 70% yield of heptanoic acid, and benzyl alcohol gives a 100% yield of benzoic acid [*841*]. The ball-milling step and the rather high price of the crown ether make this method less attractive than the catalytic tetrabutylammonium permanganate process [*845*].

Nickel peroxide, prepared by adding an alkaline solution of 6% sodium hypochlorite to a solution of nickel sulfate hydrate at 20 °C, is used to oxidize alcohols to carbonyl compounds or carboxylic acids. The oxidation to carbonyl compounds is carried out in benzene solutions, and the oxidation to carboxylic acids takes place in aqueous sodium hydroxide. Fair to high yields of saturated and unsaturated acids, as well as aromatic acids, are obtained at 5–50 °C (equations 215 and 234) [932].

[932]

$$CH\equiv CCH_2OH \xrightarrow[\text{5 °C, 0.5 h}]{\text{NiO}_2,\ \text{NaOH},\ \text{H}_2\text{O}} CH\equiv CCO_2H$$

234

50%

Sodium ruthenate [937] and *potassium ruthenate* [196] oxidize allylic and benzylic alcohols to carboxylic acids at room temperature. Cinnamyl alcohol is transformed into cinnamic acid with sodium ruthenate in 1 M sodium hydroxide at 10 °C after 1 h in 70% yield [937]. In oxidations with potassium ruthenate, only catalytic amounts can be used in the presence of a persulfate, which reoxidizes the reduced ruthenium salt [196].

An example of **biooxidation** is the conversion of the benzyl ether of kojic acid into the benzyl ether of comenic acid by *Arthrobacter ureafaciens* in a phosphate buffer at pH 7.2 in up to 97% yield (equation 235) [1044].

[1044]

235

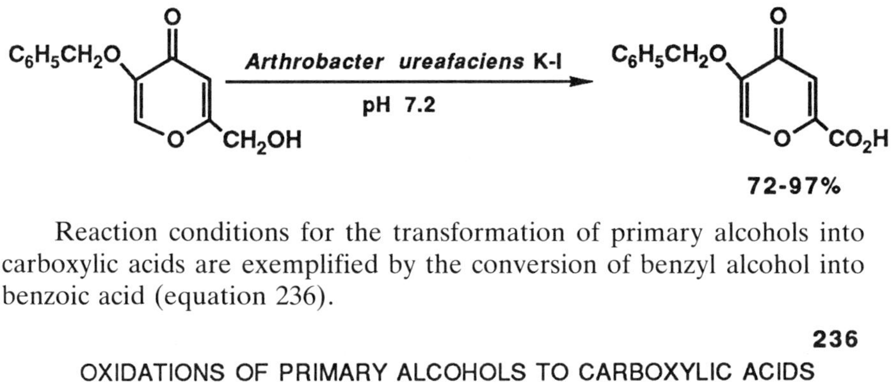

[Reaction scheme: benzyl ether of kojic acid → benzyl ether of comenic acid via *Arthrobacter ureafaciens* K-I, pH 7.2]

72–97%

Reaction conditions for the transformation of primary alcohols into carboxylic acids are exemplified by the conversion of benzyl alcohol into benzoic acid (equation 236).

236

OXIDATIONS OF PRIMARY ALCOHOLS TO CARBOXYLIC ACIDS

$$C_6H_5CH_2OH \longrightarrow C_6H_5CO_2H$$

Ref	Conditions	Yield
[56]	O_2/Pt(C), H_2O, reflux 10 h	97%
[116]	Ni anode, 1 M NaOH, 25 °C, 1.5 h	86%
[954]	KOH, H_2O, *t*-BuOH, CCl_4, 25–80 °C, 10–60 min	75%
[660]	$ZnCr_2O_7 \cdot 3H_2O$, CH_2Cl_2, RT, 6 h	90%
[835]	$NaMnO_4 \cdot H_2O$, hexane, 69 °C, 6 h	81%
[894]	$Cu(MnO_4)_2 \cdot 8H_2O$, CH_2Cl_2, RT, 24 h	84%
[899]	Bu_4NMnO_4, C_5H_5N, RT	98%
[845]	$KMnO_4$, H_2O, C_6H_6; Bu_4NBr (catalyst), RT, 3 h	92%
[841]	$KMnO_4$, dicyclohexano-18-crown-6 ether, RT, 2 h	100%
[932]	1.5 NiO_2, C_6H_6, 30 °C, 3 h	97%
[937]	Na_2RuO_4, NaOH, H_2O, 25 °C, 1 h	97%
[196]	K_2RuO_4, $K_2S_2O_8$, RT, 1.5 h	98%

Oxidation of Primary Alcohols to Esters

In acidic oxidations of primary alcohols, esterification of the alcohol by the acid produced by its oxidation sometimes takes place. Thus the oxidation of ethanol with potassium peroxymonosulfate (**Oxone**) in the presence of concentrated sulfuric acid at 70 °C gives a quantitative yield of ethyl acetate [205].

Butyl alcohol treated with **sodium dichromate and sulfuric acid** at 20–35 °C furnishes a 41–47% yield of butyl butyrate [640]. Butyl butyrate is also obtained in 82% yield by oxidation of 1-butanol by bromine and potassium bromate in 40% hydrobromic acid [742] (equation 237).

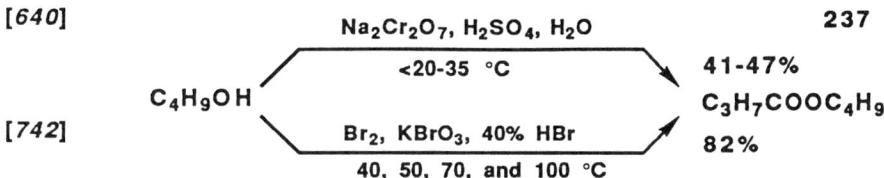

The oxidation of aliphatic alcohols in benzene or petroleum ether with *tert*-butyl chromate at 1–2 °C for 6 h leads to mixtures of aldehydes, acids, and their esters. 1-Butanol gives 30% of butanal, 27% of butyric acid, and 36% of butyl butyrate [677]. Also, electrolysis of aliphatic alcohols on platinum or carbon electrodes in aqueous potassium iodide at room temperature results in 80–83% yields of the corresponding esters [121].

Ester formation is especially easy in oxidations of diols (with at least one primary hydroxyl group) that can cyclize to five- or six-membered lactones after partial oxidation [355, 1035] (*see* equations 292–295).

A different type of ester formation takes place when α,β-unsaturated alcohols are oxidized with **manganese dioxide** and subsequently treated with sodium cyanide, acetic acid, manganese dioxide, and methanol. The final result is shown by the sequence of reactions in equation 238 [382].

[382]   238

$$C_6H_5CH{=}CHCH_2OH \xrightarrow[0\ °C,\ 30\ min]{MnO_2/C_6H_{14}} C_6H_5CH{=}CHCHO$$

$$\downarrow\ NaCN,\ AcOH,\ 20\text{-}25\ °C$$
$$MnO_2,\ MeOH,\ 12\ h$$

$$\left[C_6H_5CH{=}CHCH{<}^{OH}_{CN} \longrightarrow C_6H_5CH{=}CHCOCN \right]$$

$$\downarrow MeOH$$

$$C_6H_5CH{=}CHCO_2Me \qquad 95\%$$

In the presence of ammonia, some primary alcohols are oxidized to amides with **nickel peroxide** (equation 239) [*1137*].

[*1137*] 239

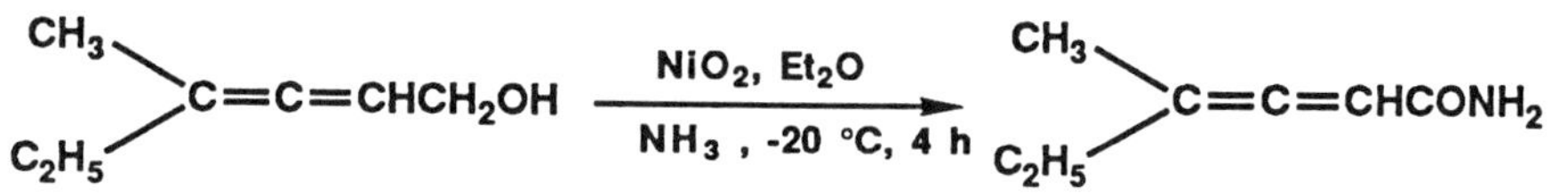

$$\underset{C_2H_5}{\overset{CH_3}{>}}C=C=CHCH_2OH \xrightarrow[NH_3, -20\,°C, 4\,h]{NiO_2,\ Et_2O} \underset{C_2H_5}{\overset{CH_3}{>}}C=C=CHCONH_2$$

Oxidation of Secondary Alcohols to Ketones

All oxidants used for the oxidation of primary alcohols to aldehydes can be applied to secondary alcohols. Because the products, ketones, are much less sensitive to overoxidation than aldehydes, more intensive reaction conditions, such as an excess of the oxidant, higher temperatures, or longer reaction times, can be used. Examples of oxidations of secondary alcohols to ketones are shown in equations 265–268.

Catalytic dehydrogenation of volatile alcohols is carried out by passing their vapors over **copper** at 300 °C [*344*] or over copper chromite at 275–325 °C for 1.7–4 h, with yields ranging from 20 to 80% [*354*]. Dehydrogenation in the liquid phase is accomplished by refluxing the alcohol with **copper chromite** in xylene for 4 h [*356*] or by heating the alcohol in paraffin oil with **Raney nickel** and a catalytic amount of potassium hydroxide at 150–180 °C for 6 h. Thus *endo*-norborneol is transformed into norcamphor in 95% yield [*928*].

Secondary alcohols are converted into ketones in 70–98% isolated yields when refluxed with Raney nickel in benzene for 1–24 h with azeotropic removal of water [*927*]. The addition of 1-octene as a hydrogen acceptor does not affect the yields. Primary alcohols, under such reaction conditions, usually suffer decarbonylation and yield hydrocarbons with one less carbon [*927*] (equation 240).

[*344*] 240

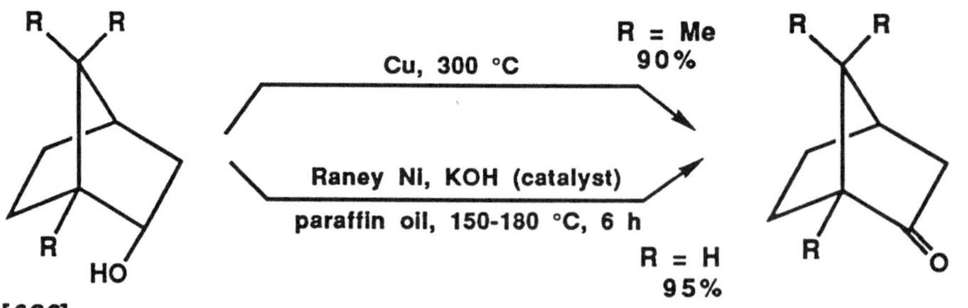

Cu, 300 °C R = Me 90%

Raney Ni, KOH (catalyst)
paraffin oil, 150-180 °C, 6 h R = H 95%

[*928*]

A similar dehydrogenation is accomplished in 90–100% yield by passing vapors of the alcohol in a current of air (less than the stoichiometric amount) over a copper or, better still, silver catalyst at 350–400 °C [*7*].

Photooxidation of alcohols to ketones in high yields is performed at room temperature with oxygen in the presence of at least 1 mol of a base (Na, NaH, NaOH, KOH, or *tert*-BuOK) under irradiation with rose bengal as a sensitizer [*45*].

Catalytic oxidation is effected by passing **oxygen** through solutions of alcohols in the presence of metal catalysts.

Alcohols in ethylene carbonate containing sodium acetate and palladium chloride are oxidized by oxygen at room temperature in 62–98% yields [*70*]. Oxygen passed at room temperature under irradiation through a solution of catalytic amounts of chloroplatinic acid and cuprous chloride in alcohols produces ketones in yields of up to 98% [*57*]. Other catalysts used for this purpose are platinum [*55*], platinum-on-charcoal [*56*], and, better still, platinum oxide [*56*]. Such oxidations are carried out usually at room temperature and give fair to high yields.

Electrolysis of secondary alcohols in a solution of potassium iodide in aqueous *tert*-butyl alcohol results in 57–98% yields of ketones at room temperature when 4–15 F/mol of electricity in cells with platinum or carbon electrodes is used [*121*].

Relatively rare reactions are chemical oxidations of secondary alcohols to ketones by derivatives of hydrogen peroxide: potassium peroxymonosulfate (Oxone) [*205*] and *m*-chloroperoxybenzoic acid [*276*]. These compounds do not offer any advantages over more-common oxidants.

Oxides of copper and **silver** are used to oxidize secondary alcohols in the vapor phase at 250–300 °C [*349*] or in the liquid phase at room temperature [*380*], respectively. A similar effect is achieved with lead tetraacetate in pyridine at room temperature. Benzhydrol is thus converted into benzophenone in 80% yield [*442*]. The oxidation of codeine with silver carbonate requires 1 h of refluxing in benzene to give a 75% yield of codeinone [*376*].

Ceric ammonium nitrate or **ceric sulfate** is used to oxidize saturated and unsaturated secondary alcohols to ketones. The ceric salts are used only in catalytic amounts with sodium bromate as a reoxidant (equation 241) [*741*].

[*741*]   241

$$C_3H_7\underset{\underset{OH}{|}}{C}HC_8H_{17} \xrightarrow[\text{MeCN, H}_2\text{O, 80 °C, 0.8 h}]{0.2\ (NH_4)_2Ce(NO_3)_6,\ 2.0\ NaBrO_3} \underset{94\%}{C_3H_7COC_8H_{17}}$$

Dinitrogen tetroxide passed through an ice-cooled chloroform solution of secondary benzylic alcohols produces aromatic ketones in 88–98% yields [*456*].

The oxidation of secondary alcohols to ketones by compounds of **hexavalent chromium** is not only the most common but also the best known reaction as far as the mechanism is concerned. The stoichiometry is the same as that of primary alcohols (equation 242).

$$3\ RR'CHOH + 2\ CrO_3 + 3\ H_2SO_4 \longrightarrow 3\ RR'CO + Cr_2(SO_4)_3 + 6\ H_2O \qquad 242$$

The initial step in the mechanism is the formation of an ester of the alcohol and chromic acid [561, 577, 1142]. The ester decomposes to the carbonyl compound and chromium(IV) acid (equation 243). The tetrava-

$$\underset{\underset{H}{|}}{\overset{R}{\underset{R'}{>}}}C-OH + CrO_3 \longrightarrow \underset{\underset{H}{|}}{\overset{R}{\underset{R'}{>}}}C-O-Cr\overset{OH}{\underset{O}{\diagup}} \longrightarrow \overset{R}{\underset{R'}{>}}C=O + \underset{\underset{OH}{|}}{\overset{OH}{\underset{|}{Cr}}}=O \qquad 243$$

lent chromium(IV) acid reacts with chromic oxide to form chromium(V) acid (equation 244), and the chromium(V) acid oxidizes the alcohol via its ester (equation 245).

$$\overset{IV}{(HO)_2Cr=O} + \overset{VI}{CrO_3} \longrightarrow 2\ HOCr\overset{V}{\underset{O}{\diagup}}\overset{O}{\underset{}{}} \qquad 244$$

$$\underset{\underset{H}{|}}{\overset{R}{\underset{R'}{>}}}C-OH + HO-\overset{V}{Cr}\overset{O}{\underset{O}{\diagup}} \longrightarrow \underset{\underset{H}{|}}{\overset{R}{\underset{R'}{>}}}C-O-Cr=O \longrightarrow \overset{R}{\underset{R'}{>}}C=O + \overset{III}{HOCrO} \qquad 245$$

The presence of three oxidizing species, chromium(VI), chromium(V), and chromium(IV) acids, accounts for differences in the oxidation products of some alcohols. Thus in the reaction of cyclobutanol with chromic acid, not only cyclobutanone but also 4-hydroxybutanal is obtained, evidently by oxidation with chromium(V) or chromium(IV) acid via a free-radical pathway [577, 621, 654, 655, 1143] (equation 246).

[577]

$$\underset{}{\square}\text{-OH} \xrightarrow{H_2CrO_4} \underset{}{\square}=O \longrightarrow \underset{CH_2-CH_2}{HOCH_2\ \ \ CHO} \qquad 246$$

Reactions with chromium(IV) involve one-electron transfer via free radicals [577].

The rates of oxidation are affected by electronic and steric effects. In oxidations of *p*-substituted 1-phenylethanols, electron-releasing substituents increase the reaction rate, whereas electron-withdrawing substituents decrease the reaction rate (equation 247) [1144].

[1144] **247**

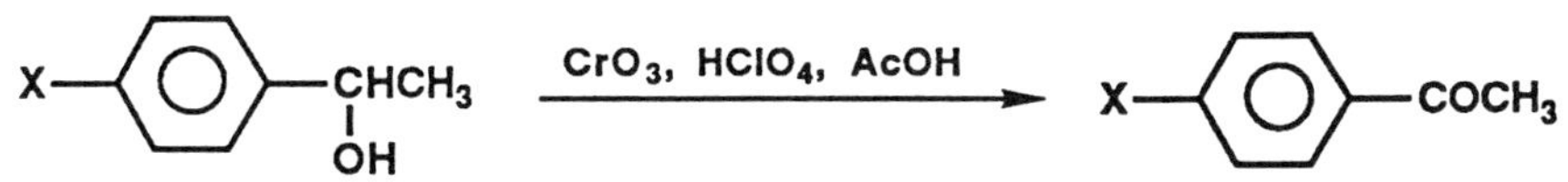

X—⟨C₆H₄⟩—CHCH₃(OH) →[CrO₃, HClO₄, AcOH]→ X—⟨C₆H₄⟩—COCH₃

Relative reaction rates

X = CH₃O   2.7
X = H   1
X = NO₂   0.2

Steric effects are evident in the oxidation of sterically hindered alcohols: *exo*-isoborneol is oxidized 2 times faster than *endo*-borneol, and *endo*-norborneol is oxidized 2.5 times faster than *exo*-norborneol (equation 248) [*1145*].

[1145] **248**

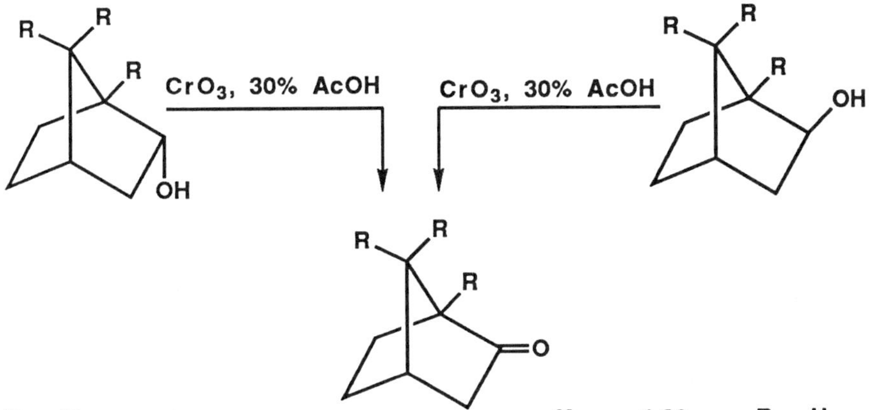

[norbornyl alcohol] →[CrO₃, 30% AcOH]→ [norbornanone] ←[CrO₃, 30% AcOH]← [norbornyl alcohol]

R = H K_2 = 3.23 K_2 = 1.30   R = H
R = CH₃ K_2 = 8.46 K_2 = 16.6   R = CH₃

Oxidations with **chromium trioxide** (chromic oxide or chromic anhydride) and with **chromic acid** are carried out in different solvents, usually by adding solutions of chromic oxide or chromic acid to the solutions of the alcohols. When chromium trioxide dissolved in 80% acetic acid is added to a stirred solution of *cis*-2-phenylcyclohexanol in acetic acid at 50 °C and the mixture is allowed to stand at room temperature for 1 day, an 80% yield of 2-phenylcyclohexanone is obtained [*576*]. Other solvents used are dimethylformamide [*542*], hexamethylphosphoric triamide (HMPA) [*543*], acetone [*578, 581*], ether [*538*], dichloromethane [*538, 617*], and benzene [*571*] (equation 249).

An especially favored oxidizing agent is **Jones reagent,** which is prepared by diluting to 100 mL with water a solution of 26.7 g of chromium

[542]   249

CrO₃, DMF, H₂SO₄ (catalyst), RT → 79%

trioxide in 23 mL of sulfuric acid [579]. The resulting solution is added to a solution of the alcohol in acetone (equation 250) [578].

[578]   250

$$CH_3CHC\equiv CC_4H_9 \xrightarrow[\text{5-10 °C, 2.5 h}]{\text{CrO}_3, \text{H}_2\text{O}, \text{H}_2\text{SO}_4, \text{Me}_2\text{CO}} CH_3COC\equiv CC_4H_9$$
$$\text{OH} \qquad\qquad\qquad\qquad\qquad\qquad\qquad\qquad 74\%$$

The addition of chromium trioxide to solutions of alcohols in ether and dichloromethane in the presence of Celite furnishes ketones in 71–93% yields after 35 min at room temperature [538]. Oxidation can also be performed by refluxing the alcohols in solvents such as chloroform, ether, hexane, or benzene with chromium trioxide on an anion exchanger, Amberlyst A26 [571]. Very good results are obtained when chromium trioxide is converted into **tetrabutylammonium chromate** by addition of catalytic amounts of tetrabutylammonium chloride in dichloromethane [617].

Chromic acid is also the reacting compound in oxidations with **sodium dichromate** or **potassium dichromate** in dilute sulfuric acid [641, 642, 643, 644, 1146], perchloric acid [621, 655], or acetic acid [627, 628]. Modifications lie in the use of excess chromic acid [536] and of additional solvents such as benzene [627], ether [536, 641], or dimethyl sulfoxide [629] and in the control of the reaction temperature at 50–55 °C [644], 30–68 °C [642], 25 °C [536, 641], or 0 °C [536]. Results of some of the different procedures are tabulated in Table II.

Table II. Oxidation of Menthol to Menthone at Various Conditions

Conditions	Yield (%)
H_2CrO_4, water, 50–55 °C	90
H_2CrO_4, 90% acetic acid, 25 °C	71
H_2CrO_4, acetone, 5–10 °C	86
H_2CrO_4, water, diethyl ether, 25–30 °C	97
H_2CrO_4 (20% excess), dilute sulfuric acid, 30–68 °C	94

SOURCE: The data were taken from references 641, 642, and 644.

Oxidations with dichromates can be accomplished in nonaqueous media or in a two-phase system by using phase-transfer reagents. The addition of 1-phenylethanol to a slurry of potassium dichromate and Adogen 464 (methyltrialkylammonium chloride) in benzene heated at 55 °C for 15 h results in an 80% yield of acetophenone. However, 2-octanol gives only a 33% yield of the ketone after being heated at 55 °C for 24 h [*651*]. Other solvents suitable for the reaction are dichloromethane, chloroform, and carbon tetrachloride, but not hexane [*651*]. Better results are obtained when the alcohol is shaken for 1 min at room temperature with a mixture of a catalytic amount (10%) of tetrabutylammonium bisulfate and sodium dichromate in dichloromethane and aqueous 3 M sulfuric acid [*630*].

Hexavalent chromium compounds soluble in organic solvents are pyridinium chromate [*597*] and pyridinium dichromate [*603*]. **Pyridinium chromate** is prepared by adding 2 mol of pyridine to 1 mol of chromium trioxide adsorbed on silica gel.

Pyridinium dichromate, prepared from chromium trioxide in a minimum amount of water and pyridine, forms a bright-orange solid and is soluble in water, dimethylformamide, dimethyl sulfoxide, and dimethylacetamide; sparingly soluble in dichloromethane, chloroform, and acetone; and almost insoluble in hexane, toluene, ether, and ethyl acetate. Allylic secondary alcohols are oxidized more rapidly than their saturated analogues. Oxidations are carried out in dichloromethane solutions at 25 °C, and ketones are obtained in high yields (equation 251) [*603*].

[*603*]   251

$$CH_2=CHCHC_5H_{11} \atop OH \quad \xrightarrow[25\ °C,\ 9\ h]{1.5\ (C_5H_5NH)_2Cr_2O_7,\ CH_2Cl_2} \quad CH_2=CHCOC_5H_{11} \atop 80\%$$

A chromium(VI) oxidant that is applicable to oxidations of acid-sensitive substrates is the **complex of chromium trioxide with two molecules of pyridine (Collins reagent).** As described on pages 22 and 274, its preparation requires the portionwise addition of chromium trioxide to dry pyridine at 15–20 °C (addition of pyridine to chromium oxide could cause ignition) [*592, 595, 599*]. Up to 6 mol of the complex is used to oxidize alcohols in dichloromethane solutions at 25 °C, and the reaction is finished in 5–15 min [*595*]. Alternatively, the oxidation can be carried out in pyridine cooled with an ice bath and is finished at room temperature within 15–22 h [*592, 599*].

Instead of pyridine, **3,5-dimethylpyrazole** may be used for **complexing chromium trioxide.** It forms a 1:1 complex that can be generated in situ in dichloromethane at room temperature. The reaction is finished within 10 min. Addition of the alcohols and stirring of the mixture for 30 min at room temperature result in 78–100% yields of ketones (equation 252) [*602*].

[602]

[Mechanism scheme showing oxidation of RR'CHOH by pyrazole-chromium species yielding O=CRR']

The addition of pyridine to chromium trioxide in 6 M hydrochloric acid produces **pyridinium chlorochromate, $C_5H_5NHCrO_3Cl$ (PCC),** a yellow-orange solid stable in air. If used in a 1:1.5 ratio, it oxidizes secondary alcohols to ketones in solutions in dichloromethane at room temperature within 1–2 h in high yields [605].

The reagent can be also *adsorbed on silica gel* [604] or formed with *cross-linked poly(vinylpyridine)* resin [612]. The yields of carbonyl compounds with the reagent adsorbed on cross-linked resin depend on the relative amounts used, the solvent (dichloromethane, cyclopentane, tetrahydrofuran, benzene, cycloheptane, and, best of all, cyclohexane), the temperature, and the reaction time [612].

Tetrabutylammonium chlorochromate is prepared by adding a solution of chromium trioxide in concentrated hydrochloric acid to a solution of tetrabutylammonium hydrogen sulfate in 5 N hydrochloric acid. Refluxing 3–6 mol of the precipitated product with alcohols in chloroform affords good yields of ketones [619].

Halogens and their oxygen-containing compounds are useful and selective oxidants for secondary alcohols. The addition of a **chlorine** solution in carbon tetrachloride to a chloroform solution of 5α-cholestane-3β,19-diol below 30 °C gives a 77% yield of 5α-cholestan-19-ol-3-one after 15 min [681].

Similar selectivity is found with chlorine (or bromine) in the presence of hexamethylphosphoric triamide (HMPA). The addition of solutions of chlorine in chloroform to stirred solutions of the alcohols in a mixture of HMPA, dichloromethane, and an aqueous solution of sodium dihydrogen phosphate at 0–5 °C results in very good yields of ketones. Competitive experiments with equimolar mixtures of primary and secondary alcohols show a 95–97% predominance of ketones over aldehydes [682].

Hypochlorites are very good oxidizers of alcohols and are frequently selective enough to oxidize secondary alcohols in preference to primary alcohols (*see* equations 288–291). Solutions of *sodium hypochlorite* in acetic acid react exothermically with secondary alcohols within minutes [*693*]. *Calcium hypochlorite* in the presence of an ion exchanger (IRA 900) oxidizes secondary alcohols at room temperature in yields of 60–98% [*705*]. *Tetrabutylammonium hypochlorite,* prepared in situ from 10% aqueous sodium hypochlorite and a 5% dichloromethane solution of tetrabutylammonium bisulfate, oxidizes 9-fluorenol to fluorenone in 92% yield and benzhydrol to benzophenone in 82% yield at room temperature in 35 and 150 min, respectively [*692*]. Cyclohexanol is oxidized to cyclohexanone by tert-*butyl hypochlorite* in carbon tetrachloride in the presence of pyridine. The exothermic reaction must be carried out with due precautions [*709*].

Like chlorine, **bromine** is used to convert secondary alcohols into ketones. A convenient way is to apply the addition product of 2 mol of bromine with 1,4-diazabicyclo[2,2,2]octane (Dabco), a nonhygroscopic yellow solid, prepared by mixing carbon tetrachloride solutions of bromine and Dabco. The compound decomposes at 155–160 °C [*726*]. Oxidation with this addition product is carried out in acetonitrile solution at 50 °C and results in 50 and 71% yields of cyclopentanone and cyclohexanone, respectively, but very low yields of 2-pentanone [*726*].

Better results are obtained on treatment of alcohols with bromine in dichloromethane solution in the presence of hexamethylphosphoric triamide (HMPA) and aqueous sodium bicarbonate. This reagent is highly selective for the oxidation of secondary alcohols in preference to primary alcohols [*682*].

A very selective method of oxidizing secondary alcohols is brominolysis of tributyltin ethers prepared by the treatment of hydroxy compounds with bis(tributyl)tin oxide. The reaction is regio- and stereospecific and is an important means of oxidizing unprotected glycosides [*734*].

Thus, refluxing methyl β-D-glucoside; two equivalents of tributyltin oxide, $[(C_4H_9)_3Sn]_2O$; and molecular sieves (3 Å) in chloroform for 2–3 h followed by treatment with two equivalents of bromine at 0 °C furnishes methyl β-D-3-dehydro-3-ketoglucoside in more than 90% yield. Similarly, methyl β-D-xyloside gives the 3-oxo compound, whereas methyl α-D-glucoside, α-D-galactoside, α-D-xyloside, and β-L-arabinoside yield 70–80% of the corresponding 4-keto derivatives [*734*].

Sodium bromite oxidizes secondary alcohols in preference to primary alcohols. Menthol treated with sodium bromite in aqueous acetic acid at room temperature affords an 86% yield of menthone after 5.5 h [*739*].

Secondary alcohols are oxidized in preference to primary alcohols also by **sodium bromate** in the presence of catalytic amounts of ceric ammonium nitrate or ceric sulfate [*741*].

Reaction conditions for the oxidation of secondary alcohols by halogens and their compounds are shown in equations 265–268.

Like primary alcohols, secondary alcohols are oxidized by **manganese dioxide**. The reaction is of free-radical nature and results in the formation of ketones and manganous(II) oxide [*815*]. Manganese dioxide is used in a large molar excess (14.7:1 [*809*], 16:1 [*824*], or 21:1 [*822*]), but a molar ratio as low as 2.3:1 gives very good yields [*808*]. Oxidations in carbon tetrachloride and petroleum ether usually result in higher yields than those in chloroform or benzene [*808*]. Aryl carbinols react faster than allylic alcohols, and these, in turn, react faster than saturated alcohols [*815*]. The reaction is accelerated by removal of the reaction water by azeotropic distillation with benzene [*815*] (equation 253).

[*822*]   253

$$CH_3O-\underset{CH_3O}{\underset{|}{\bigcirc}}-\underset{\underset{OH}{|}}{CHCH_3} \xrightarrow[\text{RT, 15 h}]{21 \text{ mol } MnO_2, Me_2CO} CH_3O-\underset{CH_3O}{\underset{|}{\bigcirc}}-COCH_3$$

Oxidations with manganese dioxide are well suited for the preparation of unsaturated and acetylenic ketones [*824*]. The yields depend on the method of preparation, dryness and particle size of the oxidant, and, to a smaller extent, on reaction temperatures and times [*808*] (equation 254).

[*824*]   254

$$CH_3CH=CHCHC\equiv COEt \underset{|}{\underset{OH}{}} \xrightarrow[\text{RT, 4.25 h}]{16 \text{ mol } MnO_2, CH_2Cl_2} CH_3CH=CHCOC\equiv COEt$$
$$40\%$$

Barium manganate, prepared from potassium manganate and barium chloride [*833*] or by the reduction of potassium permanganate with potassium iodide in the presence of barium chloride and sodium hydroxide [*832*], is used for the quantitative oxidation of benzhydrol to benzophenone. The reaction mixture is refluxed in benzene for 0.5–2 h [*833*]. The result is comparable with and even better than that of oxidation with manganese dioxide [*180, 813*].

Sodium permanganate, which is used as its monohydrate in solid form in refluxing dichloromethane or hexane, oxidizes allylic alcohols more slowly and in lower yields than the saturated alcohols. 2-Cyclohexen-1-ol, after a 24-h reflux in hexane, furnishes a 47% yield of the ketone, whereas cyclohexanol gives a 100% yield after 1.5 h [*835*].

Aqueous **potassium permanganate** is hardly ever used to oxidize secondary alcohols to ketones, because further oxidation of ketones to carboxylic acids occurs even under mild conditions (15–30 °C) [*1147*]. On the other hand, oxidations of alcohols to ketones are successfully accomplished in organic solvents with **solid or supported permanganates.** As a bonus, isolation of the products obtained by this method is much easier than that from oxidations in aqueous media; the products are isolated just by simple filtration.

Potassium permanganate adsorbed on *molecular sieves* is prepared by adding 20 g of Linde 13X molecular sieves (1/16-in. [0.16-cm] pellets) to 500 mL of 0.06 M potassium permanganate, evaporating the water under reduced pressure, and removing the unadsorbed potassium permanganate by screening through a 20-mesh gauze. In this way, up to 0.27 mmol of potassium permanganate is loaded in 1 g of the reagent [*863*]. Cyclododecanol (1.36 mmol) in 20 mL of benzene refluxed with 15 g of the reagent for 1.5 h gives a 90% yield of cyclododecanone [*863*].

An even better result is obtained with potassium permanganate in toluene in the presence of **neutral alumina.** After a mixture of 5 g of alumina, 7.9 g (0.05 mol) of potassium permanganate, and 0.01 mol of cyclododecanol in 50 mL of toluene has been stirred for 30 h at room temperature, a 95% yield of cyclododecanone is obtained [*865*].

The role of the supports in the reactions just described is not clear. Even less clear is the function of *copper sulfate pentahydrate*, in the presence of which oxidations with solid potassium permanganate result in excellent yields of ketones. Traces of water are essential. Primary alcohols, on the other hand, give poor yields of aldehydes and acids [*844*] (equation 255).

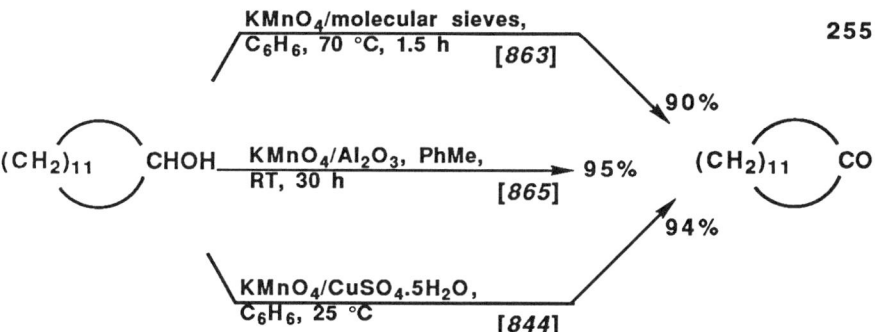

Similar results are obtained with *copper permanganate octahydrate*, $Cu(MnO_4)_2 \cdot 8H_2O$. Addition of this salt to a solution of 2-decanol in dichloromethane results in an exothermic reaction and boiling of the mixture for 5 min. After 10 additional minutes at room temperature, 2-decanone is isolated in 93% yield [*894*]. Allylic alcohols are oxidized in 84–85% yields to α,β-unsaturated ketones after boiling in dichloromethane for 24 h [*894*].

In contrast to the previous examples of oxidations with solid permanganates insoluble in organic solvents, potassium permanganate may be dissolved in organic solvents after forming a *complex with crown ethers*. Equimolar amounts of dicyclohexano-18-crown-6 ether and potassium permanganate stirred in benzene at 25 °C give a purple solution (half-life 48 h at 25 °C), which can be used directly to oxidize secondary alcohols to ketones. Because large amounts of the expensive crown ether are necessary, this method can hardly compete with the ones previously described [*841*].

Another way of solubilizing potassium permanganate is its conversion into *tetrabutylammonium permanganate*. The crystalline purple compound, which is obtained as a precipitate after aqueous solutions of tetrabutylammonium bromide and potassium permanganate have been mixed, is sparingly soluble in benzene but easily soluble in dichloromethane, chloroform, acetone, and especially pyridine. The reagent oxidizes alcohols to carbonyl compounds in pyridine at room temperature [*899*]. The disadvantage of tetrabutylammonium permanganate is the possibility of a spontaneous ignition [*901*].

Potassium ferrate in aqueous solution at room temperature transforms 1-phenylethanol in 18 min into acetophenone in 90% yield. However, under similar conditions, cyclohexanol gives a very low yield of the ketone (20–30%). The reagent does not attack double bonds and does not oxidize aldehydes [*917*].

High yields of ketones result from the gentle oxidation of alcohols with compounds of ruthenium. **Ruthenium tetroxide** oxidizes cyclohexanol to cyclohexanone in carbon tetrachloride at room temperature in 93% yield [*940*]. Instead of the rather expensive ruthenium tetroxide, which is required in stoichiometric amounts, catalytic amounts of ruthenium trichloride may be used in the presence of sodium hypochlorite as a reoxidant with the same results [*701*]. **Sodium ruthenate** [*937*] and **potassium ruthenate** [*196*], which are prepared from ruthenium dioxide and sodium periodate in sodium hydroxide and from ruthenium trichloride and potassium persulfate, respectively, also effect oxidations to ketones at room temperature.

Carbonyl compounds act as hydrogen acceptors in the ***Oppenauer oxidation*** of alcohols to aldehydes or ketones. The reaction is based on hydride transfer from the alkoxide ion of the starting alcohol prepared in situ from anhydrous bases, aluminum isopropoxide, or, better still, *tert*-butoxide (equation 256).

The byproduct $R^3R^4CH-O^-$ is a strong base that perpetuates the reaction so that only catalytic amounts of aluminum *tert*-butoxide are needed to initiate the oxidation. The reaction is reversible and leads to an equilibrium, which must be systematically destroyed by removal of the byproduct to drive the reaction to completion. Carbonyl compounds used as hydrogen acceptors are acetone, cyclohexanone, benzaldehyde, cinna-

$$\underset{R^2}{\overset{R^1}{\diagdown}}\underset{H}{\overset{|}{C}}-OH \xrightarrow[R^3R^4C=O]{t\text{-}BuO^-} \begin{array}{c} R^1 \diagdown \\ R^2 \diagup C - \bar{O} \\ H \\ R^3 \diagdown \\ R^4 \diagup C=O \end{array} \longrightarrow \begin{array}{c} R^1 \diagdown \\ R^2 \diagup C=O \\ + \\ R^3 \diagdown \\ R^4 \diagup CH - \bar{O} \end{array} \quad 256$$

maldehyde, and benzophenone [962, 1148]. The product or byproduct is removed by distillation or azeotropic distillation. The Oppenauer oxidation is eminently suited for high-boiling alcohols such as hydroxy steroids (equation 257) [963, 964].

[964]   257

$$\text{HO-cholesterol} \xrightarrow[75-85\,°C,\,8\,h]{Me_2CO,\,Al(O\text{-}t\text{-}Bu)_3,\,C_6H_6} \text{70-81\%}$$

[963]

$$\xrightarrow[\text{distillation}]{Al(O\text{-}i\text{-}Pr)_3,\,\text{cyclohexanone},\,PhMe} \text{81-93\%}$$

Chloral is used as the hydrogen acceptor in the quantitative oxidation of cyclobutanol to cyclobutanone. The reaction takes place on Woelm alumina at room temperature in carbon tetrachloride after 24 h [958]. Under similar conditions, primary alcohols are oxidized in preference to secondary alcohols [959].

Another catalyst for the dehydrogenation of alcohols by carbonyl compounds is **Raney nickel,** which is used to convert dihydrocholesterol into 3-cholestanone in 80% yield by refluxing for 24 h with cyclohexanone in toluene [961].

Allylic alcohols in steroids are oxidized by **2,3-dichloro-5,6-dicyano-p-benzoquinone (DDQ),** either alone (equation 258) [975] or as a catalyst with periodic acid as a reoxidant [974]. Saturated alcohols and other func-

[975]   258

(steroid structure) —DDQ, C₆H₆, RT, 5-15 h→ (steroid structure) 70%

tional groups present in the molecules remain untouched. DDQ competes successfully with manganese dioxide, because it reacts at room temperature and the oxidation is finished within hours [974, 975].

Both saturated and benzylic alcohols are dehydrogenated by refluxing with *diethyl azodicarboxylate* in benzene for 10 h. The azodicarboxylate is converted into hydrazodicarboxylate, while the alcohols give ketones in 51–87% yields [977].

Cyclohexanol is converted into cyclohexanone by heating for a few minutes on a steam bath with **N-chlorosuccinimide (NCS)** in pyridine and benzene [709].

A mixture of N-*chlorosuccinimide and dimethyl sulfide* oxidizes alcohols to ketones under very mild conditions and in high yields. The treatment of 4-*tert*-butylcyclohexanol with this mixture in toluene at 0–25 °C, followed by the addition of triethylamine in toluene, results in a 90–93% yield of 4-*tert*-butylcyclohexanone [721].

Dimethyl sulfoxide (DMSO), which is successfully used to dehydrogenate primary alcohols to aldehydes, converts secondary alcohols into ketones in very high yields and under very gentle conditions. The mechanism is discussed in a previous section, Dehydrogenation and Oxidation of Primary Alcohols to Aldehydes (equation 217). The first oxidations were carried out in the presence of dicyclohexylcarbodiimide and an acid catalyst such as pyridinium trifluoroacetate [1016], which protonates the diimide and facilitates the attack by dimethyl sulfoxide (equation 259).

[1016]   259

$C_6H_{11}N=C=NC_6H_{11}$

$\xrightarrow[C_5H_5N]{CF_3CO_2H}$ $C_6H_{11}N=\overset{+}{C}-NHC_6H_{11}$ ⟶ $C_6H_{11}N=C-NHC_6H_{11}$

(with DMSO O=S(CH₃)–CH₃ attached, then sulfonium intermediate $^+S-CH_3$ / CH_3)

The intermediate reacts with the alcohol as shown by the arrows in equation 260 and yields dicyclohexylurea and an alkoxysulfonium cation.

$$RR'CHOH \longrightarrow \begin{array}{c} C_6H_{11}N=C-NHC_6H_{11} \\ H-O \\ RR'CHO-\overset{+}{S}-CH_3 \\ CH_3 \end{array} \longrightarrow C_6H_{11}NHCONHC_6H_{11} + RR'CH-O-\overset{+}{S}-CH_3 \quad 260$$
$$CH_3$$

A base (the trifluoroacetate ion) abstracts a proton from one of the methyl groups of the oxysulfonium ion, and a bond shift through a five-membered transition state gives the ketone and dimethyl sulfide [1016] (equation 261).

$$\xrightarrow{-H^+} RR'C\cdots\overset{O}{\underset{H}{|}}\cdots\overset{+}{S}-CH_3 \longrightarrow RR'C=O + S(CH_3)_2 \quad 261$$

In more recent works, the use of dicyclohexylcarbodiimide has been abandoned because the reaction works satisfactorily with acid catalysts alone, followed by bases such as triethylamine or diisopropylethylamine, which gives, sometimes, even better yields than triethylamine [1012, 1023]. Several activators are being used, and the best seems to be oxalyl chloride (the **Swern oxidation**) [1023, 1149]. Other activators are mentioned in the section Oxidation of Primary Alcohols to Aldehydes. The advantages of oxidations with dimethyl sulfoxide lie in the mildness of the reagent and in the low temperatures, sometimes −45 °C [1020] or −60 °C [1023], at which the reactions are run.

In a typical example, 10 mmol of an alcohol, 11 mmol of oxalyl chloride, and 24 mmol of DMSO in 40 mL of dichloromethane react at −60 °C. The mixture is then made alkaline with 50 mmol of triethylamine [1023] (equation 262). In other instances, the molar ratios of the alcohol to DMSO and to the activator (benzoic anhydride) were 1:47 and 1:17, respectively; with phosphorus pentoxide as the activator, the respective molar ratios were 1:47 and 1:1 [1009], and with pyridine–sulfur trioxide, they were 1:70 and 1:3 [1018]. Dichloromethane and toluene [1012] are the best solvents.

[1023]

cyclohex-2-enol $\xrightarrow[\text{Et}_3\text{N, -60 °C}]{2.4^* \text{ DMSO, } 1.1^* \text{ (COCl)}_2, \text{ CH}_2\text{Cl}_2, -60 \text{ °C}}$ cyclohex-2-enone 262

80%

*Values are moles per mole of alcohol.

Dimethyl sulfoxide oxidizes primary allylic alcohols in preference to cyclic secondary alcohols [*1018*] and is suitable for oxidations of sterically hindered alcohols in high yields [*1009, 1139*] (equation 263).

[*1139*]   263

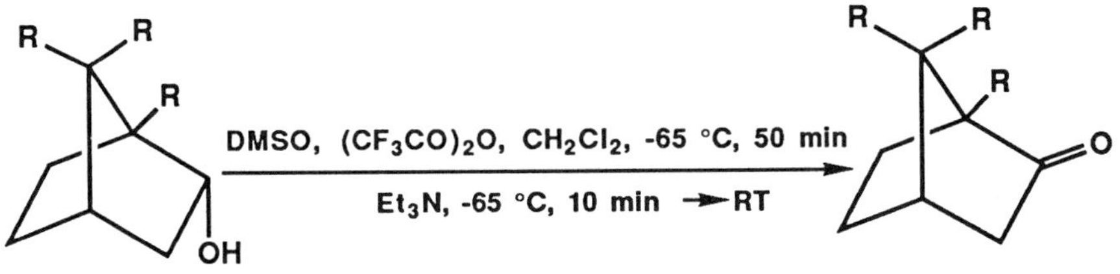

DMSO, $(CF_3CO)_2O$, CH_2Cl_2, -65 °C, 50 min
Et_3N, -65 °C, 10 min → RT

R = H   96% (85%)*
R = CH_3 98% (81%)*

*Values in parentheses are isolated yields.

The disadvantages of oxidations with dimethyl sulfoxide are the large excess of the oxidant needed and the obnoxious smell of the dimethyl sulfide formed. Therefore, oxidations with this reagent should be carried out in hoods.

A very peculiar oxidation of secondary alcohols and primary–secondary diols to ketones in 53–84% yields is a reaction with *triphenylmethylfluoroborate*, which reacts according to equation 264 and which affects exclusively secondary alcohols [*981*].

[*981*]   264

$RRCHOH \xrightarrow[-HBF_4]{Ph_3CBF_4} RR'CHOCPh_3 \xrightarrow[catalyst]{Ph_3CBF_4} RR'CO + Ph_3CH$

Several biochemical oxidations can be applied to the conversions of secondary alcohols into ketones. In a complex system containing **horse liver alcohol dehydrogenase**, $(\pm)$-*trans*-3-methylcyclohexanol and $(\pm)$-*cis*-2-methylcyclopentanol are dehydrogenated to $(-)$-(S)-3-methylcyclohexanone (yield 50%; ee 100%) and to $(+)$-(S)-2-methylcyclopentanone (yield 55%; ee 96%), respectively [*1036*].

The oxidation of a secondary alcoholic group in the presence of primary alcoholic groups by **Acetobacter suboxydans** converts adonit into adonose [*1041*]. The treatment of androstenediol and dehydroandrosterone with **yeast** yields Δ^4-androstenedione [*1088*]. Steroidal hydroxy ketones are dehydrogenated to dicarbonyl compound with **Corynebacterium simplex** [*1056*] (*see* equations 446 and 447). Examples of oxidations of secondary alcohols to ketones are shown in equations 265–268.

Oxidations 147

EXAMPLES OF OXIDATION OF SECONDARY ALCOHOLS TO KETONES

CH$_3$CHC$_6$H$_{13}$ ⟶ CH$_3$COC$_6$H$_{13}$ **265**
   |
  OH

Ref	Conditions	Yield
[815]	MnO$_2$, C$_6$H$_6$, distillation	42-65%
[835]	NaMnO$_4$·H$_2$O, C$_6$H$_{14}$, 69 °C, 2.5 h	95%
[863]	KMnO$_4$/molecular sieves, C$_6$H$_6$, 70 °C, 7 h	92%
[844]	KMnO$_4$/CuSO$_4$·5H$_2$O, C$_6$H$_6$, 25 °C	94%
[927]	Raney Ni, C$_6$H$_6$, 125-135 °C, 4 h	93%
[1023]	DMSO, (COCl)$_2$, CH$_2$Cl$_2$, -60 °C, 35 min, Et$_3$N, -60 °C, 20 min	98%
[981]	2 Ph$_3$CBF$_4$, CH$_2$Cl$_2$, 25 °C, 10 h	84%
[276]	AcO$_2$H/catalyst*, AcOEt, CH$_2$Cl$_2$, 0 °C, 0.35 h	99%
[70]	O$_2$/PdCl$_2$, AcONa, CO(OCH$_2$)$_2$, RT, 36 h	76%
	133 h	86%
[55]	O$_2$, AcOEt, RT, 20 h	21%
[56]	O$_2$/PtO$_2$, C$_7$H$_{16}$, 20 °C, 96 h	80%
[326]	m-ClC$_6$H$_4$CO$_3$H, THF, HCl, RT, 1 h	75%
[675]	CrO$_2$Cl$_2$/SiO$_2$-Al$_2$O$_3$, CH$_2$Cl$_2$, 25 °C, 24 h	94%
[682]	Cl$_2$, CHCl$_3$, HMPA, CH$_2$Cl$_2$, H$_2$O, NaH$_2$PO$_4$, 0-5 °C, 2 min	95%
[682]	Br$_2$, CH$_2$Cl$_2$, HMPA, H$_2$O, NaHCO$_3$, 0-5 °C	95%
[693]	NaOCl, H$_2$O, AcOH, 15-25 °C	96%
[121]	electrolytic oxidation, KI, H$_2$O, t-BuOH, 6-10 F/mol	92-99%
[543]	CrO$_3$, HMPA, RT, 15 H	40%
[538]	CrO$_3$, CH$_2$Cl$_2$, Et$_2$O, Celite, 35 min	80%
[571]	CrO$_3$, Amberlyst A26, C$_6$H$_6$, reflux, 9 h	97%
[599]	3 (2 C$_5$H$_5$N·CrO$_3$), C$_5$H$_5$N, ice bath, 30 min, RT,15-22 h	18%
[595]	6 (2 C$_2$H$_5$N·CrO$_3$), CH$_2$Cl$_2$, 25 °C, 5-15 min	97%
[602]	(pyrrole)N·CrO$_3$, CH$_2$Cl$_2$, RT, 30 min	93%
[612]	12 mol of poly(vinylpyridine)·HCl·CrO$_3$, CH$_2$Cl$_2$, RT, 4 days	76%
[617]	CrO$_3$, Bu$_4$NO$_3$SCF$_3$, CH$_2$Cl$_2$, RT, 1 h	80%

*The catalyst is Me$_2$C—CH$_2$—CMe$_2$.
           | |
           O—CrO$_2$—O

cyclohexanol ⟶ cyclohexanone   **266**

Ref	Conditions	Yield
[45]	O$_2$, t-BuOK, hν, rose bengal, 25-27 °C	97%
[70]	O$_2$/PdCl$_2$, AcONa, CO(OCH$_2$)$_2$, RT, 56 h	88%
[56]	O$_2$/PtO$_2$, C$_7$H$_{16}$, 20 °C, 1.5 h	92%
[55]	O$_2$/Pt, AcOEt, RT, 18 h	68%
[55]	air/Pt, AcOEt, RT, 24 h	74%
[121]	electrolytic oxidation, KI, H$_2$O, t-BuOH, 8 F/mol, RT	74%
[326]	m-ClC$_6$H$_4$CO$_3$H, THF, HCl, RT, 1 h	85%
[380]	2 AgO, H$_3$PO$_4$, 20 °C, 8 h	55%

cyclohexanol → cyclohexanone   **266**

Ref	Conditions	Yield
[349]	CuO, He, 250-300 °C	90%
[917]	K_2FeO_4, H_2O, RT, 1.5 h	20-30%
[682]	Cl_2, $CHCl_3$, HMPA, CH_2Cl_2, H_2O, NaH_2PO_4, 0-5 °C, 2 min	81%
[682]	Br_2, CH_2Cl_2, HMPA, H_2O, $NaHCO_3$, 0-5 °C, 5 min	87%
[693]	NaOCl, H_2O, AcOH, 15-25 °C, 2 h	96%
[705]	$Ca(OCl)_2$, IRA 900 (OCl), <40 °C, RT, 3 h	80%
[709]	t-BuOCl, C_5H_5N, CCl_4, -5 °C	90.5%
[726]	$2Br_2 \cdot DABCO$, MeCN, 50 °C, 3 h	71%
[739]	$NaBrO_2$, H_2O, AcOH, RT, 5.5 h	100%
[709]	NCS, C_5H_5N, C_6H_6, 100 °C, <5 min	50%
[940]	RuO_4, CCl_4 0-15 °C, RT, overnight	93%
[701]	$RuCl_3$, H_2O, NaOCl, 0 °C, 0.5 h	90-95%
[937]	RuO_2, $NaIO_4$, NaOH, 25 °C, 1 h	85%
[196]	$RuCl_3$, $K_2S_2O_8$, RT, 2 h	64%
[543]	CrO_3, HMPA, RT, 15 h	49%
[571]	CrO_3, Amberlyst A26, C_6H_{14}, 3 h	77%
[536]	$Na_2Cr_2O_7 \cdot 2H_2O$ (equivalent amount), H_2SO_4, Et_2O, 25 °C	92%
[536]	$Na_2Cr_2O_7 \cdot 2H_2O$ (100% excess), H_2SO_4, Et_2O, 0 °C	98%
[629]	$Na_2Cr_2O_7 \cdot 2H_2O$, H_2SO_4, DMSO, 70 °C, >30 min	89%
[641]	$Na_2Cr_2O_7 \cdot 2H_2O$, H_2SO_4, Et_2O, 25 °C, 2.25 h	92%
[617]	CrO_3, Bu_4NCl, CH_2Cl_2, RT, 2 h	70%
[630]	$Na_2Cr_2O_7$, Bu_4NHSO_4 (10%), 3 M H_2SO_4, RT, 1 min	88%
[597]	2 $(C_5H_5N)_2CrO_4$ on SiO_2, AcOH, 30-35 °C, 9 h	64%
[597]	2 $(C_5H_5N)_2CrO_4$ on SiO_2, AcOH, C_6H_6, reflux 3 h	76%
[599]	3 (2 $C_5H_5N \cdot CrO_3$), C_5H_5N, cooling with ice, 30 min, RT, 15-22 h	45%
[595]	6 (2 $C_5H_5N \cdot CrO_3$), CH_2Cl_2, 25 °C, 5-15 min	98%
[612]	poly(vinylpyridine)·HCl·CrO_3, C_6H_{12}, 75 °C, 24 h	94%
[815]	MnO_2, C_6H_6, reflux	49%
[835]	$NaMnO_4 \cdot H_2O$, C_6H_{14}, 69 °C, 1.5 h	100%
[1023]	DMSO, $(COCl)_2$, CH_2Cl_2, -60 °C, 35 min; Et_3N, -60 °C, 5 min	94%
[981]	Ph_3CBF_4, CH_2Cl_2, 25 °C, 13.5 h	66%
[801]	(o-carboxyphenyl)-I(OAc)$_3$-type reagent, CH_2Cl_2, RT, 30 min, NaOH, 10 min	90%

$C_6H_5CHCH_3$(OH) → $C_6H_5COCH_3$ **267**

Ref	Conditions	Yield
[456]	N_2O_4, $CHCl_3$, 0 °C, 1 h, RT, 6 h	98%
[602]	pyrazole·CrO_3, CH_2Cl_2, RT, 30 min	100%

Oxidations 149

$C_6H_5CHCH_3 \longrightarrow C_6H_5COCH_3$ **267**
|
OH

[619]	Bu$_4$NCrO$_3$Cl, CHCl$_3$, reflux 2 h	80%
[651]	K$_2$Cr$_2$O$_7$, Adogen 464*, 3 M H$_2$SO$_4$, C$_6$H$_6$, 55 °C, 15 h	80%
[675]	CrO$_2$Cl$_2$/SiO$_2$-Al$_2$O$_3$, CH$_2$Cl$_2$, 25 °C, 5 h	100%
[682]	Cl$_2$, CHCl$_3$, HMPA, CH$_2$Cl$_2$, H$_2$O, NaH$_2$PO$_4$, 0-5 °C, 15 min	98%
[682]	Br$_2$, CH$_2$Cl$_2$, HMPA, H$_2$O, NaHCO$_3$, 0-5 °C, 15 min	98%
[815]	MnO$_2$, C$_6$H$_6$, distillation	72%
[809]	MnO$_2$ (15 equivalents), C$_6$H$_6$, RT, 0.5 h	70%
[917]	K$_2$FeO$_4$, H$_2$O, RT, 18 min	90%
[959]	CCl$_3$CHO, Al$_2$O$_3$ (Woelm), CCl$_4$, 25 °C, 24 h	90%
[977]	EtO$_2$CN=NCO$_2$Et, C$_6$H$_6$, reflux, 10 h	87%

*Adogen 464 is methyltrialkyl (C$_8$-C$_{10}$) ammonium chloride.

$C_6H_5CHC_6H_5 \longrightarrow C_6H_5COC_6H_5$ **268**
|
OH

[45]	O$_2$, KOH, t-BuOH, hv, rose bengal, 25-26 °C	91%
[442]	Pb(OAc)$_4$, C$_5$H$_5$N, RT, a few hours	80%
[456]	N$_2$O$_4$, CHCl$_3$, 0 °C, 1 h, RT, 6 h	89%
[543]	CrO$_3$, HMPA, RT, 2 h	100%
[538]	CrO$_3$, CH$_2$Cl$_2$, Et$_2$O, Celite, RT, 35 min	93%
[571]	CrO$_3$, Amberlyst A26, C$_6$H$_6$, 1 h	77%
[599]	3 (2C$_5$H$_5$N.CrO$_3$), C$_5$H$_5$N, cooling with ice, 30 min, RT, 15-22 h	71%
[595]	6 (2C$_5$H$_5$N.CrO$_3$), CH$_2$Cl$_2$, 25 °C, 5-15 min	96%
[602]	(pyrrole).CrO$_3$, CH$_2$Cl$_2$, RT, 30 min	98%
[605]	C$_5$H$_5$N.CrO$_3$Cl, CH$_2$Cl$_2$, 25 °C, 1-2 h	100%
[619]	Bu$_4$NCrO$_3$Cl, CHCl$_3$, reflux 1 h	82%
[660]	ZnCr$_2$O$_7$.3H$_2$O, CH$_2$Cl$_2$, RT, 6 min	95%
[815]	MnO$_2$, C$_6$H$_6$, distillation	92%
[692]	NaOCl, H$_2$O, Bu$_4$NHSO$_4$, CH$_2$Cl$_2$, RT, 2.5 h	82%
[833]	2 BaMnO$_4$, C$_6$H$_6$, reflux, 0.5-2 h	100%
[863]	KMnO$_4$/molecular sieves, C$_6$H$_6$, 70 °C, 7 h	100%
[844]	KMnO$_4$/CuSO$_4$.5H$_2$O, C$_6$H$_6$, 70 °C, 4 h	100%
[841]	KMnO$_4$, dicyclohexyl-18-crown-6 ether, C$_6$H$_6$, 25 °C	100%
[899]	Bu$_4$MnO$_4$, C$_5$H$_5$N, RT	97%
[977]	EtO$_2$CN=NCO$_2$Et, C$_6$H$_6$ reflux, 10 h	71%
[1023]	DMSO (COCl)$_2$, CH$_2$Cl$_2$, -60 °C, 35 min; Et$_3$N, 20 min	98%
[525]	(PhSeO)$_2$O, THF, RT, 3 h	85%

Oxidation of Secondary Alcohols to α-Hydroxy Hydroperoxides

A unique reaction leading to unique compounds takes place when secondary alcohols are treated with oxygen in the presence of benzophenone as a sensitizer for photooxidation. The products, which are isolated by distillation at room temperature at 0.1–0.8 mm of Hg in 16–25% yields, show the presence of "active" oxygen and react with 2,4-dinitrophenylhydrazine to form 2,4-dinitrophenylhydrazones of the corresponding ketones.

The compounds must be handled with utmost care because they **tend to explode** (equation 269) [*44*].

[*44*] $\underset{\underset{OH}{|}}{RCHR'} \xrightarrow[12\ °C,\ 17\text{-}57\ h]{O_2,\ h\nu,\ Ph_2CO,} \underset{\underset{OH}{|}}{\overset{\overset{OOH}{|}}{RCR'}}$ $\begin{array}{l}R\ =\ CH_3\\R'\ =\ CH_3,\ C_2H_5,\ C_3H_7\end{array}$ 269

16-25%

Oxidation of Secondary Alcohols to Carboxylic Acids

The treatment of secondary alcohols with powerful oxidants such as **nitric acid** [*466, 468*], **chromium trioxide** [*580*], or **potassium permanganate** [*1147*] results not only in oxidation to ketones but also in subsequent oxidation of the ketones to carboxylic acids. Thus cyclohexanol gives adipic acid [*468, 1147*], 4-isopropylcyclohexanol gives β-isopropyladipic acid [*466*], and 2-methylcyclohexanol gives 6-ketoenanthoic (δ-acetylvaleric or heptanon-6-oic) acid [*580*] (equations 270 and 271).

[*468*]   50% HNO₃/NH₄VO₃   270
             55-60 °C, >1 h   58-60%
  cyclohexyl-OH   HO₂C(CH₂)₄CO₂H
             KMnO₄, Na₂CO₃, H₂O   70%
[*1147*]   15-30 °C

[*580*] 2-methylcyclohexanol $\xrightarrow[30\ °C,\ 1\ h,\ RT]{CrO_3,\ H_2SO_4,\ H_2O}$ CH₃CO(CH₂)₄CO₂H   271
                                                                                      46-55%

Oxidation of Tertiary Alcohols

Tertiary alcohols are resistant to oxidation. *tert*-Butyl alcohol is frequently used as a solvent in oxidations. However, some tertiary alcohols are converted into tertiary hydroperoxides on treatment with *hydrogen peroxide in sulfuric acid* [*177, 179*]. Dimethylphenylcarbinol added to a mixture of 87% hydrogen peroxide and sulfuric acid at a temperature below 0 °C gives a 94% yield of cumyl hydroperoxide after 3.5 h [*177*]. Similarly, acetylenic alcohols with the tertiary hydroxyl group adjacent to the triple bonds are converted into the corresponding hydroperoxides in high yields [*179*] (equation 272).

[179]

$$Me_2CC\equiv CH-OH \xrightarrow[-4 \text{ to } -0\,°C,\ 5.5\ h]{50\%\ H_2O_2,\ H_2SO_4} Me_2CC\equiv CH-OOH \quad 272$$

98% (73% pure)

In acid media, for example, in **chromic acid in sulfuric acid**, tertiary alcohols that can suffer dehydration form alkenes, which are degraded to ketones and carboxylic acids (***Barbier–Loquin*** and ***Wieland*** **degradation** of carboxylic acids) [*1150*] (equation 273).

[*1150*]

$$Me_2CHCH_2CO_2Et \xrightarrow[Et_2O]{MeMgI} Me_2CHCH_2CMe_2OH \xrightarrow[\text{reflux}]{CrO_3,\ 10\%\ H_2SO_4} Me_2CHCO_2H \quad 273$$

Rather rare examples of the oxidation of tertiary alcohols are the degradation of 3-ethyl-3-pentanol to 3-pentanone and iodoethane in respective yields of 90 and 84% by acetyl hypoiodite in acetic acid and chlorobenzene under irradiation for 1 h at 20–25 °C [*780*] and the conversion of *trans*-9,10-dihydroxy-1,4,5,8,9,10-hexahydronaphthalene into 3,8-cyclodecadiene-1,6-dione and its tetramethylacetal by lead tetraacetate (equation 274) [*447*].

[*447*]   274

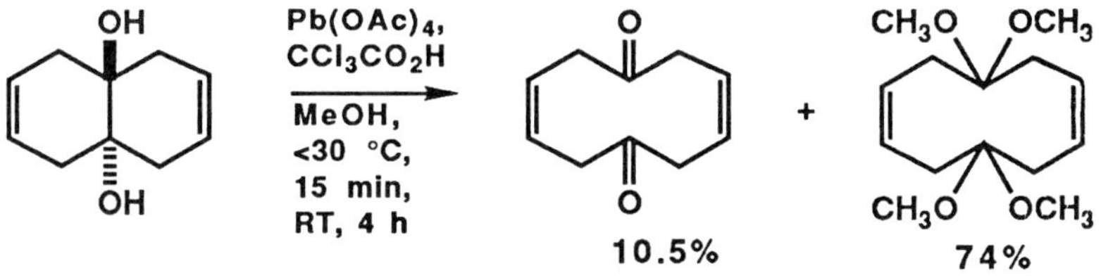

Reagents: Pb(OAc)₄, CCl₃CO₂H, MeOH, <30 °C, 15 min, RT, 4 h

Products: dione (10.5%) + tetramethylacetal (74%)

Oxidation of Unsaturated Alcohols at Multiple Bonds

The enol form of mesitylphenylacetaldehyde, a **vinylic alcohol**, when dissolved in acetic acid and treated with lead tetraacetate at 40 °C, gives 2-acetoxy-2-mesitylphenylacetaldehyde in 86% yield [*1151*]. The result can be interpreted as the addition of acetoxyls across the enol double bond followed by elimination of one molecule of acetic acid (equation 275).

The oxidation of the alcoholic groups of ***allylic alcohols*** and ***phenylallylic alcohols*** was thoroughly discussed in previous sections on the oxidation of primary and secondary alcohols. In this section, only such reactions in which the double bond is oxidized while the alcoholic group is preserved will be discussed.

[1151]   275

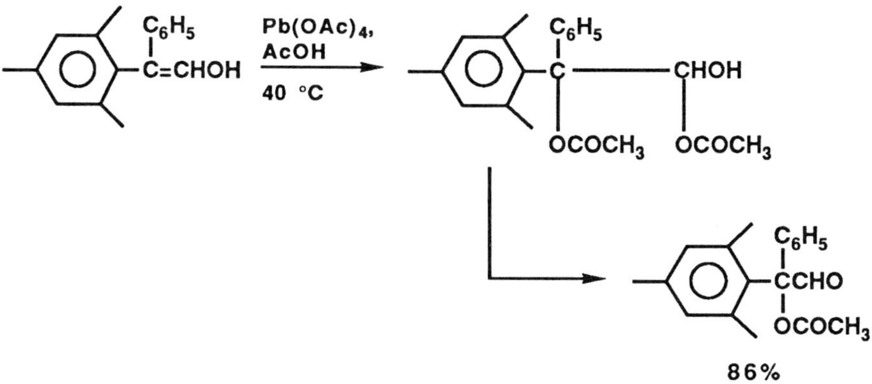

[reaction scheme: aryl-C(C6H5)=CHOH + Pb(OAc)4, AcOH, 40 °C → aryl-C(C6H5)(OCOCH3)-CHOH(OCOCH3) → aryl-C(C6H5)(OCOCH3)-CHO, 86%]

The regio- and stereospecificity of *epoxidation* depend on the structure of the alcohols and on the oxidants. Most of the examples of epoxidation include allylic alcohols. In geraniol, where both allylic and a nonallylic double bonds are present, epoxidation occurs almost exclusively at the allylic double bond [215].

Stereoselectivity in cyclic allylic alcohols depends on the ring size and on the reagent used. 2-Cyclohexen-1-ol, on treatment with **peroxybenzoic acid,** gives a *cis* epoxide, probably because of the hydrogen bond of the hydroxyl hydrogen in the transition state. If the acetate of the alcohol is treated with peroxybenzoic acid, the *trans* epoxide predominates over the *cis* epoxide in a ratio of 4:1 (equation 276) [298].

[298]   276

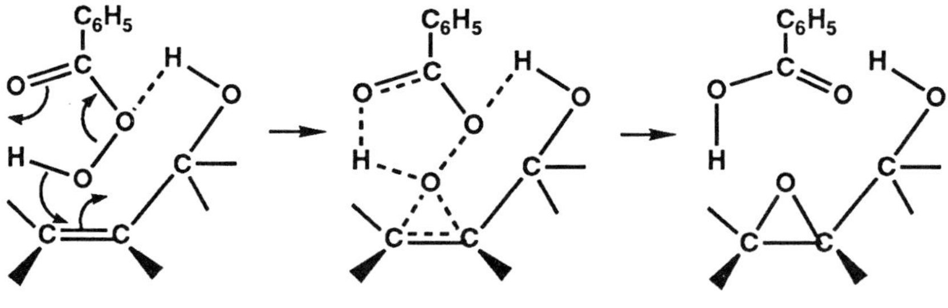

[transition state diagrams showing epoxidation mechanism]

In *cis*-2-cycloocten-1-ol, oxidation with ***tert*-butyl hydroperoxide** catalyzed by vanadium acetylacetonate yields almost exclusively the *cis* epoxide, whereas ***m*-chloroperoxybenzoic acid** produces exclusively the *trans* epoxide. In *trans*-2-cycloocten-1-ol, both oxidants furnish predominantly the *cis* epoxide (equation 277) [216].

In the epoxidations of acyclic allylic alcohols by *m*-chloroperoxyben-

[216]   277

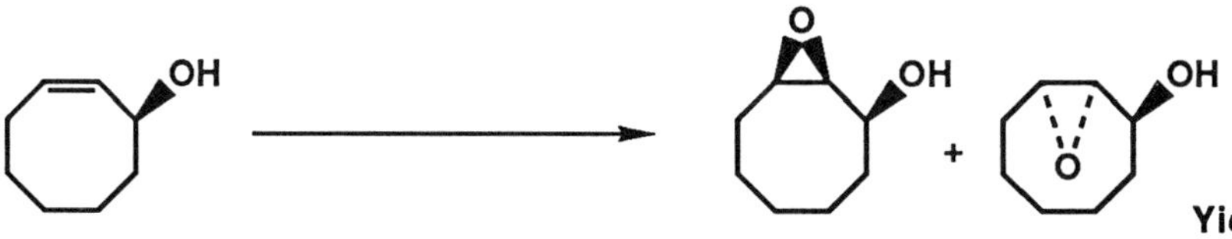

cis-2-cycloocten-1-ol (above)

			Yield
t-BuOOH/VO(acac)$_2$, C$_6$H$_6$, 40 °C, 24 h	97%	3%	83%
t-BuOOH/MoO$_2$(acac)$_2$, C$_6$H$_6$, 80 °C, 5 h	42%	58%	78%
m-ClC$_6$H$_4$CO$_3$H, CH$_2$Cl$_2$, 0 °C, 24 h	0.2%	99.8%	81%

trans-2-cycloocten-1-ol (not shown)

t-BuOOH/VO(acac)$_2$, C$_6$H$_6$, 40 °C, 24 h	93%	7%	62%
m-ClC$_6$H$_4$CO$_3$H, CH$_2$Cl$_2$, 0 °C, 24 h	84%	16%	58%

zoic acid, *p*-nitroperoxybenzoic acid, and *tert*-butyl hydroperoxide with vanadyl acetylacetonate and molybdenum hexacarbonyl, fairly large differences occur in the ratios of the diastereomers formed. The main factor determining the ratios of stereoisomers seems to be the structure of the allylic alcohol (equations 278 and 279) [218, 328].

For the synthesis of natural products containing oxirane rings, it is desirable to prepare not only the correct diastereomers but also the proper enantiomers. To this effect, different chiral oxidants were tested: *tert*-butyl hydroperoxide and *chiral hydroxamic acids* as ligands to *vanadyl acetylacetonate* [1031] and especially *tert*-butyl hydroperoxide, titanium tetra-

Compound	Epoxide	Oxidant			
		p-O$_2$NC$_6$H$_4$CO$_3$H*	m-ClC$_6$H$_4$CO$_3$H	t-BuOOH + VO(acac)$_2$	t-BuOOH + Mo(CO)$_6$
[218]	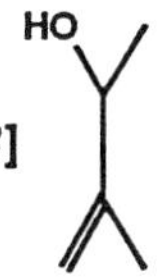 threo/erythro	32/68	45/55	5/95	16/84 (278)
[218]	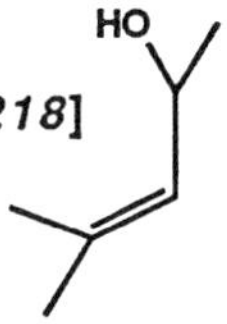 threo/erythro	96/4	95/5	86/14	95/5 (279)

*Data on the oxidation with peroxy-*p*-nitrobenzoic acid are from reference 328.

isopropoxide, and **(+)** or **(–) dialkyl tartrate (Sharpless reagent)** [*221, 222, 1026, 1027*], with which 90–95% enantiomeric excesses at 70–82% yields are obtained (equation 280) [*221*].

[*221*] **280**

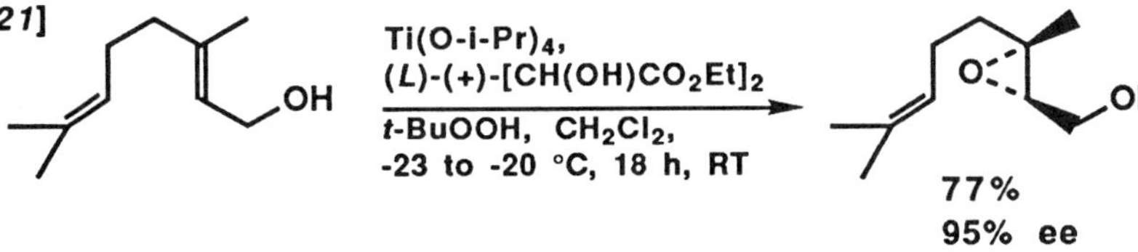
OH
Ti(O-i-Pr)4,
(L)-(+)-[CH(OH)CO2Et]2
t-BuOOH, CH2Cl2,
-23 to -20 °C, 18 h, RT
O
OH
77%
95% ee

Not only allylic alcohols but also homoallylic alcohols can be epoxidized. A peculiar transannular oxidation takes place when 1-methyl-4-cycloocten-1-ol is treated with the **Fieser reagent, [CrO_3–$(CH_3CO)_2O$–CH_3COOH],** or with **pyridinium chlorochromate.** Two diastereomeric keto oxides are formed in different ratios (equation 281) [*535*].

[*535*] **281**

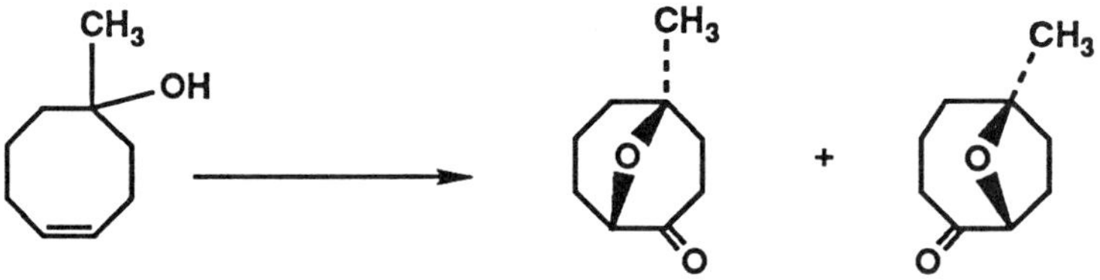
CH3
OH
CH3
O
O
+
CH3
O
O

CrO_3, AcOH, Ac_2O, RT, 12 h	33%	:	67%	
Na_2CrO_4, AcOH, 90 °C, 1 h	0	:	100%	(yield 53%)
$C_5H_5NHCrO_3Cl$, CH_2Cl_2, reflux 40 h	0	:	100%	

When 1-(buten-3-yl)cyclohexanol and 1-(penten-4-yl)cyclohexanol are treated with **chromium trioxide** in acetic acid and acetic anhydride, oxidation at the double bonds results in the formation of carboxylic acids, which cyclize to form γ- and δ-lactones, respectively [*568*]. The same reaction occurs with the cycloheptanol analogues in better yields (equation 282) [*568*].

[*568*] **282**

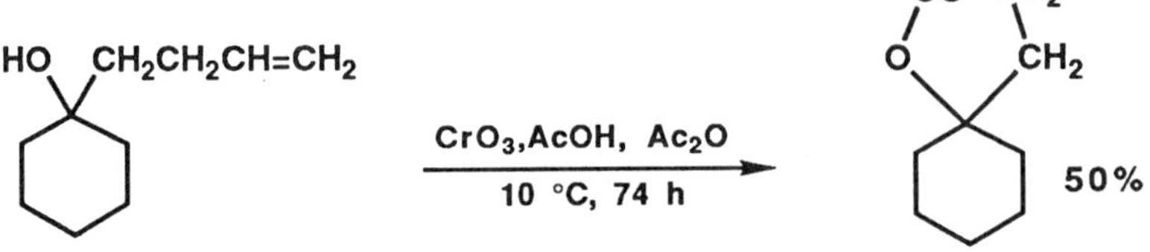
HO CH2CH2CH=CH2
CrO3,AcOH, Ac2O
10 °C, 74 h
CO–CH2
O
CH2
50%

Oxidation of the alcoholic group in acetylenic alcohols is discussed in previous sections (equations 218, 219, 250, 254, and 272). Oxidations affecting the rest of the molecule, that is, acetylenic hydrogen, are shown in equation 283. Such oxidations are carried out analogously to those of simple terminal acetylenes and lead to diacetylenic diols [2, 58].

[58]

$(CH_3)_2CC\equiv CH$ with OH group:
- $O_2/Cu(OAc)_2$, C_5H_5N-MeOH-Et_2O, reflux 7 h → 80%
- $CuSO_4 \cdot 5H_2O$, AcONa, H_2O, AcOH, CuCl, H_2O, 70 °C, 0.5 h → 58%
- $(AcO)_2Cu$, C_5H_5-MeOH-Et_2O, reflux 20 min → 88%

Product: $(CH_3)_2CC\equiv CC\equiv CC(CH_3)_2$ with OH, OH groups

283

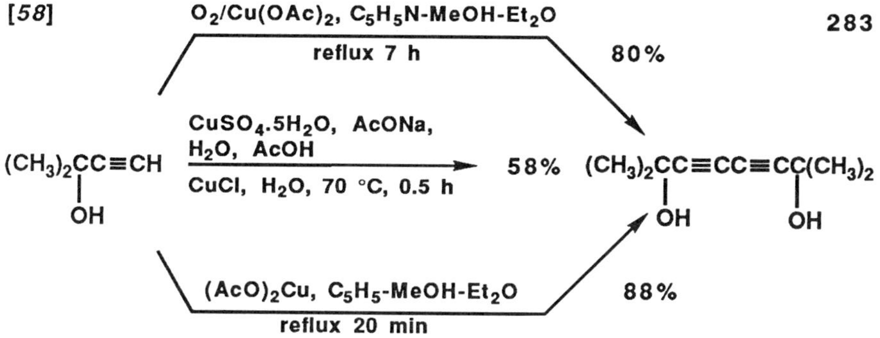

Oxidation of Diols

The oxidation of diols having alcoholic groups of the same nature, for example, both alcoholic groups are primary, secondary, allylic, or benzylic, is usually carried out at both groups to yield **dialdehydes** [832] or **diketones** [582]. Such reactions are achieved by chromium trioxide [582], barium manganate [832], dimethyl sulfoxide activated with acetic anhydride [1013], and others (equations 284 and 285).

[832]   284

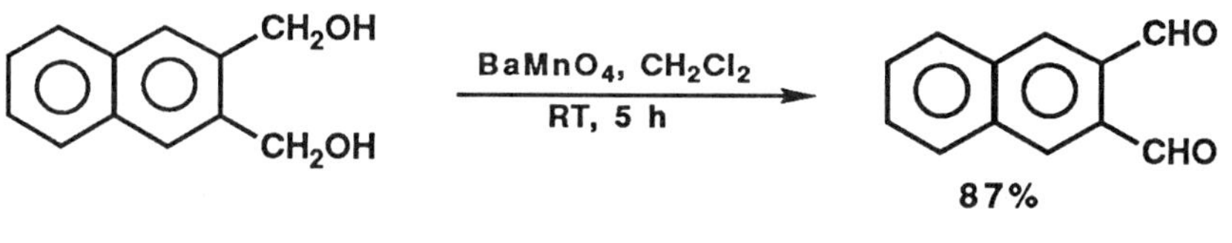

naphthalene-2,3-diyl-bis(CH$_2$OH) — BaMnO$_4$, CH$_2$Cl$_2$, RT, 5 h → naphthalene-2,3-dicarbaldehyde (CHO, CHO), 87%

[582]   285

HO-cyclopentenyl-OH — CrO_3, H_2SO_4, H_2O, CH_2Cl_2, −5 to 0 °C, 2 h → cyclopentenedione, 65–80%

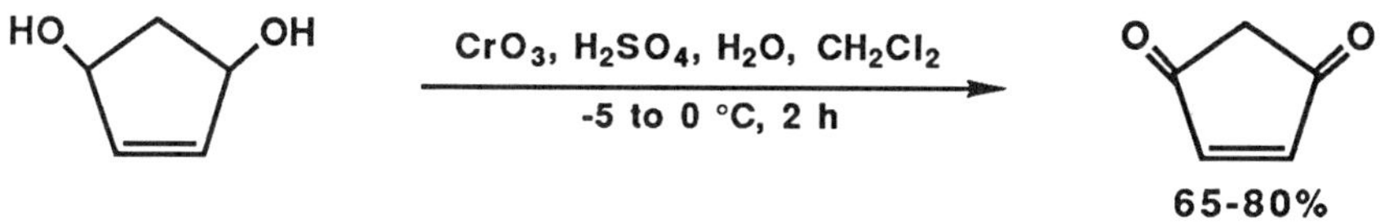

If the hydroxyl groups in a diol are tertiary and secondary, only the secondary alcoholic group is oxidized (equation 286) [1019].

Selective oxidations can be accomplished when the molecule contains hydroxyls of different nature. Many examples of selective oxidations have been carried out with steroids, where, in addition to the chemical nature

[1019]   286

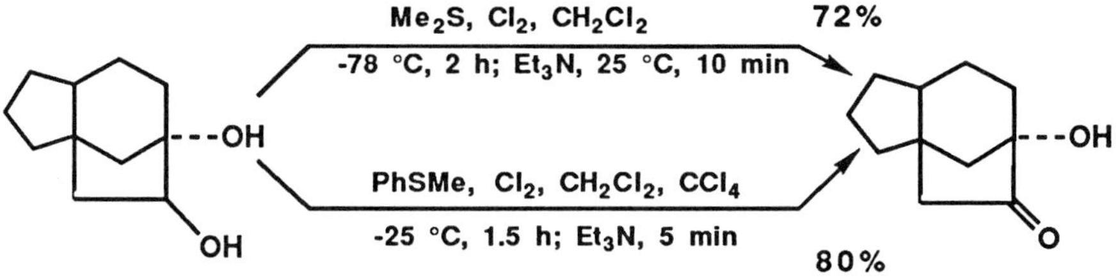

Me₂S, Cl₂, CH₂Cl₂, 72%
-78 °C, 2 h; Et₃N, 25 °C, 10 min

PhSMe, Cl₂, CH₂Cl₂, CCl₄
-25 °C, 1.5 h; Et₃N, 5 min
                                                                                            80%

of the alcoholic group, its accessibility and stereochemistry may play a significant role [681, 743, 975, 1059].

A general rule is that **allylic alcohols** are more readily oxidized than **saturated secondary alcohols** [975], and these, in turn, are **more readily oxidized than saturated primary alcohols** [681, 741, 1041, 1152]. Ceric sulfate [741], ceric ammonium nitrate [741], chlorine [681], sodium hypochlorite [1152], and 2,3-dichloro-5,6-dicyano-p-benzoquinone [975] are successfully used for this purpose (equations 287–289).

[975]   287

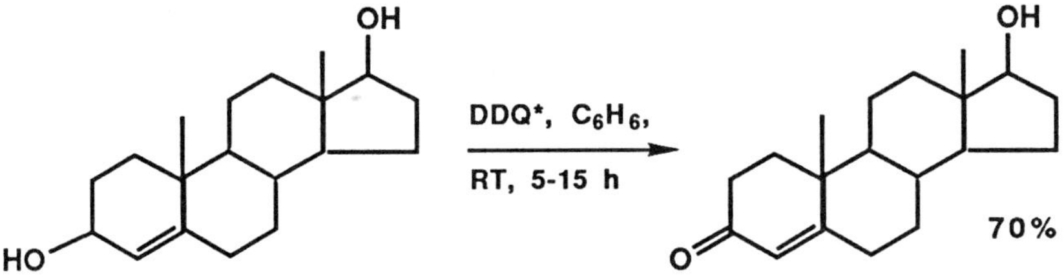

DDQ*, C₆H₆,
RT, 5-15 h
                                                                                            70%

*DDQ is 2,3-dichloro-5,6-dicyano-p-benzoquinone.

[741]   288

$CH_3CH(CH_2)_8CH_2OH$ $\xrightarrow{\begin{array}{c}Ce(SO_4)_2/NaBrO_3,\\ \text{aqueous MeCN}\\ \text{reflux 2 h}\end{array}}$ $CH_3C(CH_2)_8CH_2OH$
         | ||
         OH   O   86%

[681]   289

Cl₂, CCl₄,
CHCl₃, C₅H₅N
RT, 20 min
                                                                                            77%

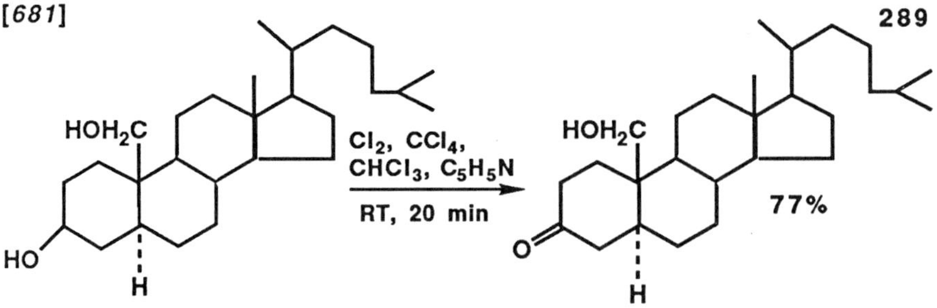

However, exceptions do occur, especially when the allylic grouping contains electron-withdrawing groups in the vicinity of the hydroxylic function (equation 290) [743].

[743] 290

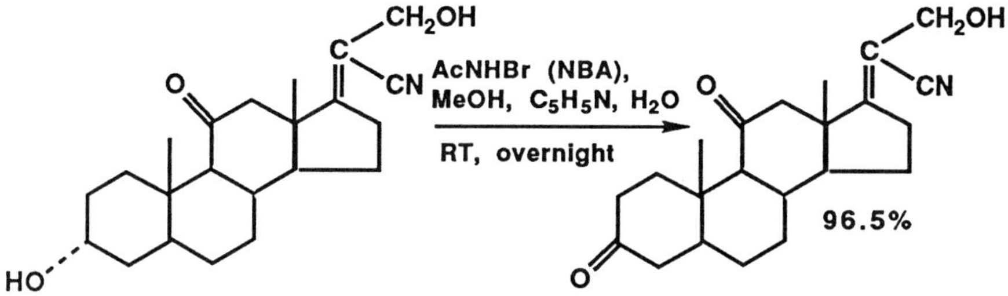

$$\text{Steroid-CH}_2\text{OH, C=CN} \xrightarrow[\text{RT, overnight}]{\text{AcNHBr (NBA), MeOH, C}_5\text{H}_5\text{N, H}_2\text{O}} \text{Steroid-CH}_2\text{OH, C=CN} \quad 96.5\%$$

Secondary steroidal alcohols react in preference to primary allylic alcohols in biochemical oxidations using **Rhizopus arrhizus** and **Helicostylum piriforme** or **Cunninghamella blakesleeana**. However, the reaction is complicated by hydroxylations in positions 6 (with *Rhizopus arrhizus*) and 9 (with *Helicostylum piriforme* or *Cunninghamella blakesleeana*) [1059].

Another example of the preferential oxidation of secondary hydroxyls versus primary hydroxyls is the formation of L-adonose from L-adonit by bakers' yeast (*Acetobacter suboxydans*) (equation 291) [1041].

[1041] 291

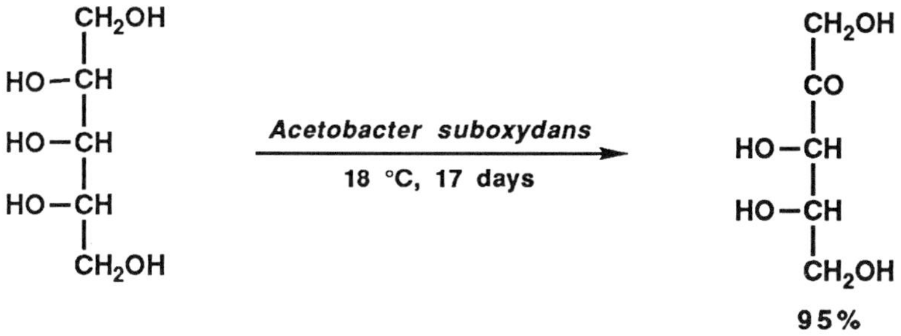

$$\begin{array}{c} \text{CH}_2\text{OH} \\ | \\ \text{HO-CH} \\ | \\ \text{HO-CH} \\ | \\ \text{HO-CH} \\ | \\ \text{CH}_2\text{OH} \end{array} \xrightarrow[\text{18 °C, 17 days}]{\textit{Acetobacter suboxydans}} \begin{array}{c} \text{CH}_2\text{OH} \\ | \\ \text{CO} \\ | \\ \text{HO-CH} \\ | \\ \text{HO-CH} \\ | \\ \text{CH}_2\text{OH} \end{array} \quad 95\%$$

Oxidation of the *primary alcoholic group to a carboxyl group* in diols with primary and secondary hydroxyls is accomplished by **silver carbonate** [377]. Unfortunately, an extremely large excess of the reagent is needed. Similar results are obtained with a rather exotic oxidant, 4-methoxy-2,2,6,6-tetramethyl-1-oxopiperidinium chloride, which is prepared by treatment with chlorine of a stable radical, 4-methoxy-2,2,6,6-tetramethylpiperidin-1-oxyl. The compound oxidizes 1,4-butanediol to γ-butyrolactone in 100% yield (isolated yield 81%) and 1,5-pentanediol to δ-valerolactone in 61% yield (isolated yield 40%) [995] (equation 292).

$$\text{HOCH}_2(\text{CH}_2)_n\text{CH}_2\text{OH} \longrightarrow \underset{292}{\text{CH}_2\text{—O—CO—(CH}_2)_n}$$

[377] Ag$_2$CO$_3$, Celite, C$_6$H$_6$, reflux

    n = 2, 24 equivalents of Ag$_2$CO$_3$, 6 h   90%
    n = 3, 25 equivalents of Ag$_2$CO$_3$, 3 h   94%
    n = 4, 23 equivalents of Ag$_2$CO$_3$, 2.5 h   96%

[995] 4-MeO-2,2,6,6-tetramethylpiperidine-1-oxoammonium Cl$^-$, CH$_2$Cl$_2$

    n = 2, 10 min   81%
    n = 3, 30 min   40%

Diols with primary and tertiary hydroxyls also give lactones on treatment with silver carbonate, but diols with primary and secondary hydroxyls yield, besides lactones, hydroxy ketones, resulting from preferential oxidation at the secondary center (equation 293) [377].

[377]   CH$_3$CH(OH)(CH$_2$)$_2$CH$_2$OH $\xrightarrow[\text{reflux}]{\text{Ag}_2\text{CO}_3 \text{ (10 equivalents)}}$ CH$_3$CH—CH$_2$—CH$_2$—O—CO (lactone) + CH$_3$CO(CH$_2$)$_2$CH$_2$OH   293

    C$_6$H$_6$, 1 h   40.5%   49.5%
    CHCl$_3$, 5 h   72%   8%

Because the conversion of primary diols into lactones is accomplished not only by oxidizing agents but also by *catalytic dehydrogenation,* it is likely that the reaction takes place stepwise via hemiacetal intermediates (equation 294) [355].

[355]   3-methyl-1,5-pentanediol $\xrightarrow[\text{200 °C, 1.5–3 h}]{\text{CuCr}_2\text{O}_4}$ 4-methyl-2-hydroxytetrahydropyran $\longrightarrow$ 4-methyl-δ-valerolactone   294

     90–95%

A similar conversion of a bis-primary diol into a lactone is achieved by ***biochemical dehydrogenation*** (equation 295) [*1035*].

[*1035*]   295

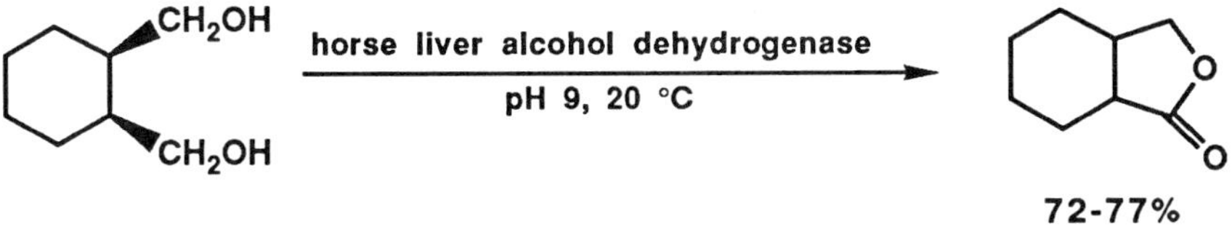

$$\text{cyclohexane-1,2-diyl-bis(CH}_2\text{OH)} \xrightarrow[\text{pH 9, 20 °C}]{\text{horse liver alcohol dehydrogenase}} \text{bicyclic lactone}$$

72-77%

Cleavage of Vicinal Diols to Carbonyl Compounds

The carbon–carbon bond between two ***vicinal free hydroxyl groups*** can be cleaved to give ***two carbonyl compounds.*** This reaction is invaluable in the structure determination and transformation of sugars. In connection with hydroxylation of double bonds, it can be used as an alternative to the cleavage of double bonds by ozone [*953*] (*see* equation 305).

Two oxidants essentially dominate these oxidations: **lead tetraacetate** in organic solvents and **periodic acid** in aqueous media. On occasion, other oxidation reagents cause the cleavage of vicinal diols: **ceric ammonium nitrate** [*424*], *sodium bismuthate* [*482, 483*], *chromium trioxide* [*482, 588*], *potassium dichromate* with perchloric acid [*949*], *manganese dioxide* [*817*], and *trivalent* [*779, 789*] or *pentavalent* [*798*] *iodine compounds*.

The mechanism of the cleavage of vicinal diols can be represented by the reaction of diols with **periodic acid, HIO$_4$ or H$_5$IO$_6$** (equation 296).

296

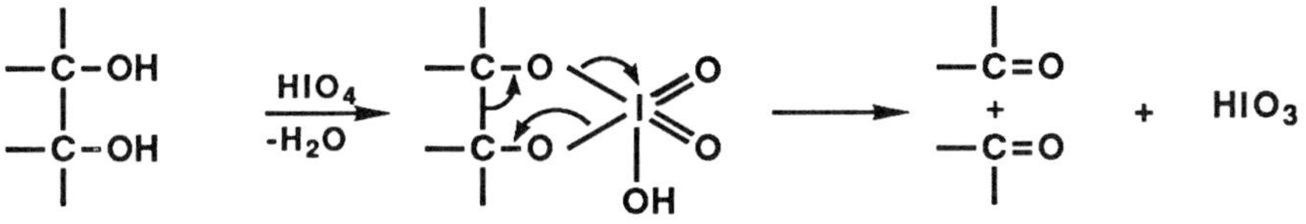

$$\begin{array}{c} -\overset{|}{\text{C}}\text{-OH} \\ -\overset{|}{\text{C}}\text{-OH} \end{array} \xrightarrow[-\text{H}_2\text{O}]{\text{HIO}_4} \begin{array}{c} -\overset{|}{\text{C}}-\text{O} \\ -\overset{|}{\text{C}}-\text{O} \end{array}\hspace{-2pt}\overset{\text{O}}{\underset{\text{OH}}{\overset{\|}{\text{I}}}}\hspace{-2pt}\overset{\text{O}}{\|} \longrightarrow \begin{array}{c} -\overset{|}{\text{C}}=\text{O} \\ + \\ -\overset{|}{\text{C}}=\text{O} \end{array} + \text{HIO}_3$$

The cleavage of secondary straight-chain diols is a good source of aldehydes [*433, 448, 775, 789, 798*] (equations 297–299).

[*775*]   297

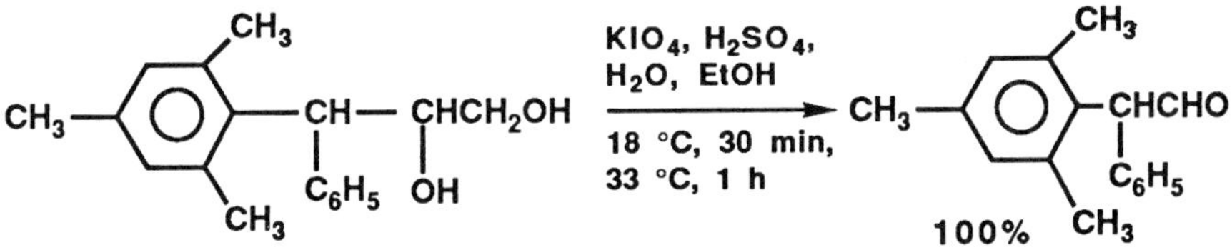

$$\text{2,4,6-trimethylphenyl-CH(C}_6\text{H}_5\text{)-CHCH}_2\text{OH(OH)} \xrightarrow[\substack{18\,°\text{C, 30 min,}\\33\,°\text{C, 1 h}}]{\text{KIO}_4,\ \text{H}_2\text{SO}_4,\ \text{H}_2\text{O, EtOH}} \text{2,4,6-trimethylphenyl-CH(C}_6\text{H}_5\text{)-CHO}$$

100%

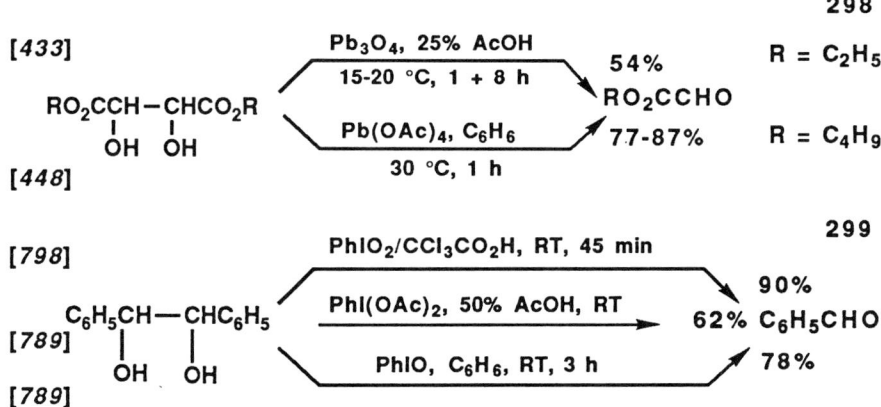

In saccharides, cleavage by **periodic acid** or **sodium periodate** occurs between any carbons possessing *free* hydroxylic groups. Thus glucosazone treated with periodic acid in aqueous ethanol at room temperature is degraded to the 1,2-bis(phenylhydrazone) of mesoxalaldehyde (propanone-dial) in 85% yield [759].

With proper protection of certain hydroxyl groups, a desired degradation can take place. 3,4:5,6-Diisopropylidene-D-mannitol treated with sodium periodate in water at 0–5 °C for 30 min furnishes an 89% yield of 2,3:4,5-diisopropylidene-D-arabinose [1153]. Ethyl 2-acetamido-2-deoxy-1-thio-α-D-galactofuranoside yields 80% of a 4-formylthioacetalfuranoside on oxidation with an aqueous solution of metaperiodate at 0–3 °C (equation 300) [762].

[762]   300

2,4-Benzal-D-sorbitol is degraded by periodic acid in aqueous dioxane at 35–45 °C to 2,4-benzal-L-xylose [756] (equation 301).

The rate of oxidation of diols with lead tetraacetate depends strongly on their configurations: *cis* diols react 200–3000 times faster than *trans* diols [1154], and the racemates of certain diols react about 15 times faster than the *meso* forms [1154]. The rates of oxidation of pinacols prepared from cyclopentanone, cyclohexanone, and cycloheptanone are in the ratio

[756]

301

CH₂OH—[O>CHC₆H₅ / O]—OH, HO—, —CH₂OH

1. HIO₄, dioxane, H₂O, 35-45 °C, 30 min
2. H₂O, 100 °C, 1 h

→ HO—[CH₂OH / O>CHC₆H₅ / O]—CHO   80%

1:314:62,000, evidently because of the increasing accessibility of the hydroxyls [*1154*]. Yields of oxidation with ceric ammonium nitrate are comparable with those of oxidation with lead tetraacetate (equation 302) [*424*].

[*424*]

302

Pb(OAc)₄, 75% AcOH, 25 °C → 94%

(NH₄)₂Ce(NO₃)₆, 75% AcOH, 25 °C → 94%

bicyclohexyl-diol → cyclohexanone=O

Cyclic vicinal diols are split to dialdehydes or diketones, depending on the atoms or groups bonded to the carbons carrying the hydroxyls. *cis*-1,2-Cyclohexanediol and **lead tetraacetate** yield adipic aldehyde [*446, 1155*].

The stereochemistry of the diols often affects the yields of carbonyl compounds. Thus, the oxidation of *cis*-1,2-cyclohexanediol with **sodium bismuthate** in aqueous phosphoric acid and ether at 30 °C gives only 23% of adipic aldehyde, whereas the *trans* isomer gives a 49% yield [*483*] (equation 303).

303

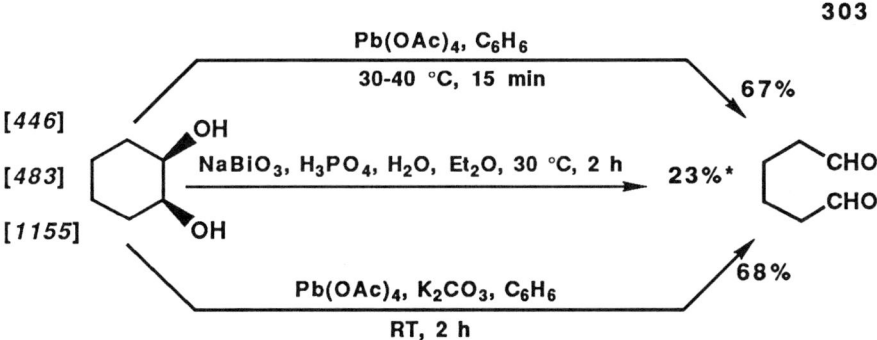

[*446*]
[*483*]
[*1155*]

Pb(OAc)₄, C₆H₆, 30-40 °C, 15 min → 67%

NaBiO₃, H₃PO₄, H₂O, Et₂O, 30 °C, 2 h → 23%*

Pb(OAc)₄, K₂CO₃, C₆H₆, RT, 2 h → 68%

*The yield of adipic aldehyde from the *trans* isomer was 49%.

1,2-Dimethyl-1,2-cyclopentanediol, *potassium dichromate*, and aqueous perchloric acid yield, after 1.5–2 h at room temperature, up to 96% of 2,6-heptanedione [*949*]. *cis*-Decalin-9,10-diol produces cyclodecane-1,6-dione on treatment with *manganese dioxide* in 90% yield (equation 304) [*817*].

[*817*] **304**

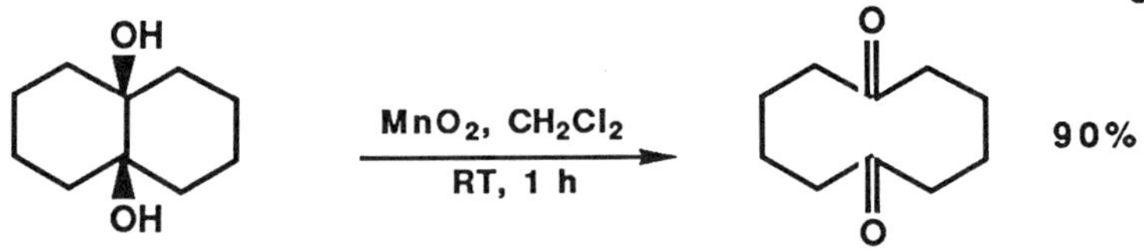
OH
OH
MnO2, CH2Cl2
RT, 1 h
O
O
90%

A mixture of *cis*- and *trans*-1,2-dimethylacenaphthene-1,2-diol dissolved in acetic acid is oxidized with *chromium trioxide* to 1,8-diacetylnaphthalene in 89% yield [*588*]. The treatment of 17β,17aβ-dimethyl-D-homoandrostane-3β,17α,17aα-triol 3β-acetate with *chromium trioxide* or with *sodium bismuthate* yields 3β-acetoxy-17,17a-dimethyl-17,17a-seco-homoandrostane-17,17a-dione (equation 305) [*482*].

[*482*] **305**

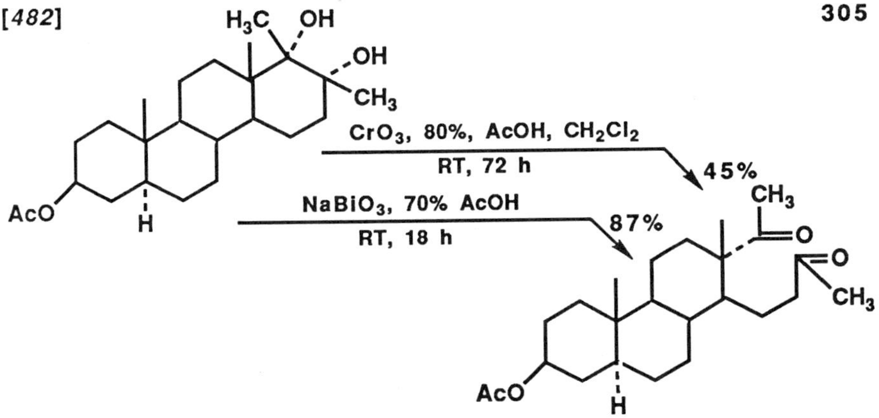
H3C OH
OH
CH3
AcO
H
CrO3, 80%, AcOH, CH2Cl2
RT, 72 h
45%
NaBiO3, 70% AcOH
RT, 18 h
87%
CH3
O
O
CH3
AcO
H

An example of a ***cleavage of a double bond*** by using a sequence of hydroxylation by **osmic acid** and the subsequent fission of the diol by **sodium periodate** is the conversion of 1-allyl-1-ethyl-7-methoxy-2-tetralone into 1-ethyl-1-formylmethyl-7-methoxy-2-tetralone (equation 306) [*953*].

[*953*] **306**

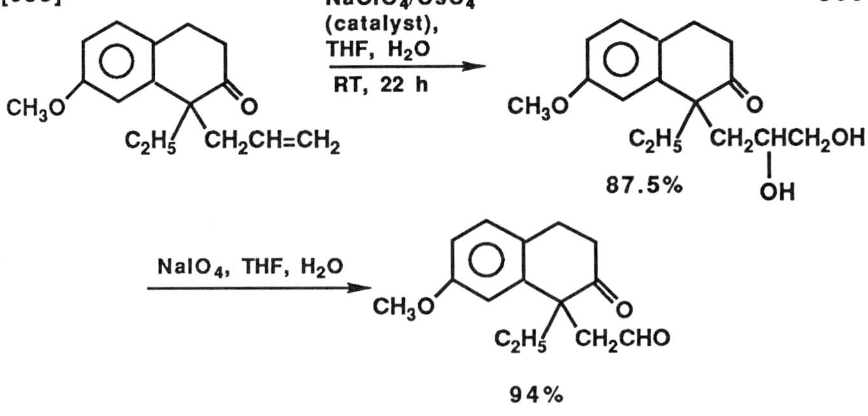
CH3O
O
C2H5 CH2CH=CH2
NaClO4/OsO4
(catalyst),
THF, H2O
RT, 22 h
CH3O
O
C2H5 CH2CHCH2OH
OH
87.5%
NaIO4, THF, H2O
CH3O
O
C2H5 CH2CHO
94%

Vicinal diols are also oxidized to **carboxylic acids**. 1,2-Dihydroxydecane in benzonitrile shaken with *oxygen* at atmospheric pressure at 100 °C in the presence of cobalt laurate as a catalyst furnishes, after 2 h, a 66% yield of pelargonic (nonanoic) acid [*69*]. Electrolysis at 20–30 °C in aqueous potassium carbonate with nickel electrodes in an undivided cell and a current of 0.3 A yields 74–84% of dicarboxylic acids from both *cis* and *trans* cyclic diols having six to eight carbons in the rings (equation 307) [*122*].

[*122*]   307

$$(CH_2)_3\begin{matrix}CH_2-CHOH\\ \\ CH_2-CHOH\end{matrix} \xrightarrow[0.3\text{ A, RT, 2.7 h}]{Ni,\ K_2CO_3,\ H_2O} (CH_2)_3\begin{matrix}CH_2CO_2H\\ \\ CH_2CO_2H\end{matrix}$$

79%

PHENOLS

Apart from **dehydrogenative coupling** in *para* and *ortho* positions [*1095*] (equation 44), phenols undergo **oxidations to free radicals** [*922, 923*], **hydroxylation** in the aromatic rings [*197, 1039*], and especially **oxidation to quinones**. Peroxy acids cleave the aromatic rings and give **carboxylic acids** [*249, 273, 274*]. Phenols containing side chains with double bonds conjugated with the aromatic rings are oxidized to phenol aldehydes [*86, 989*].

Potassium ferricyanide effects the one-electron oxidation of sterically hindered phenols to **phenoxyl radicals** (equation 308) [*922, 923*].

[*922*]   308

R = C(CH$_3$)$_3$ K$_3$Fe(CN)$_6$, KOH, H$_2$O, C$_6$H$_6$, RT, 1–2 h   99–100%

$$R\text{-}\underset{R}{\overset{R}{\bigcirc}}\text{-OH} \longrightarrow R\text{-}\underset{R}{\overset{R}{\bigcirc}}\text{-O}\cdot$$

R = C$_6$H$_5$ K$_3$Fe(CN)$_6$, NaOH, H$_2$O, Et$_2$O, RT, 20 min   81–91%
[*923*]

Hydroxylations in the aromatic rings take place in the *para* or *ortho* positions with respect to the hydroxylic group and are achieved by alkaline **persulfate** *(Elbs hydroxylation)* (equation 309) [*197*].

Another route to hydroxylated phenols is the application of mushroom **polyphenol oxidase** in chloroform. The primary products of oxidation,

[197]

$$\text{2-chlorophenol} \xrightarrow[\text{20 °C, 3-4 h, overnight}]{K_2S_2O_8,\ 10\%\ NaOH} \text{3-chlorocatechol (62%)} \quad 309$$

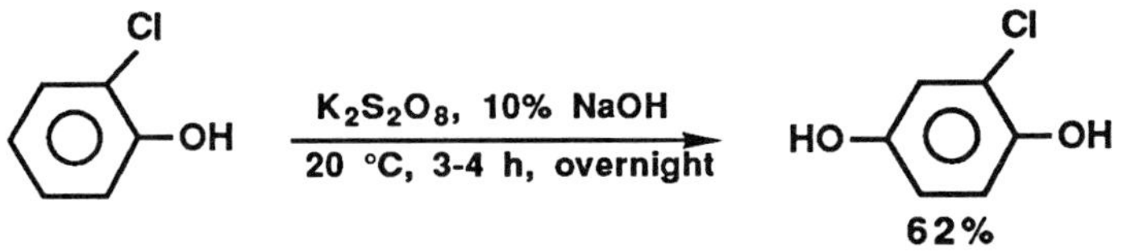

hydroxy phenols, are further oxidized to quinones, which can be easily reconverted into dihydroxy compounds by ascorbic acid [1039].

Conversion to ortho- and para-quinones is by far the most common oxidation of phenols. **Mercuric oxide** or **mercuric trifluoroacetate** [383], **lead dioxide** [430], **chromium trioxide** [559], **bromine** [732], **2,3-dichloro-5,6-dicyano-p-benzoquinone (DDQ)** [971], **Fremy salt** [487, 488, 489], and **hydrogen peroxide** in the presence of **horseradish peroxidase** [1038] are the most widely used oxidants (equations 310 and 311).

[488]

$$\text{3,4-dimethylphenol} \xrightarrow[\text{25 °C, 20 min}]{\cdot ON(SO_3K)_2,\ H_2O,\ Et_2O} \text{3,4-dimethyl-p-benzoquinone (49-50%)} \quad 310$$

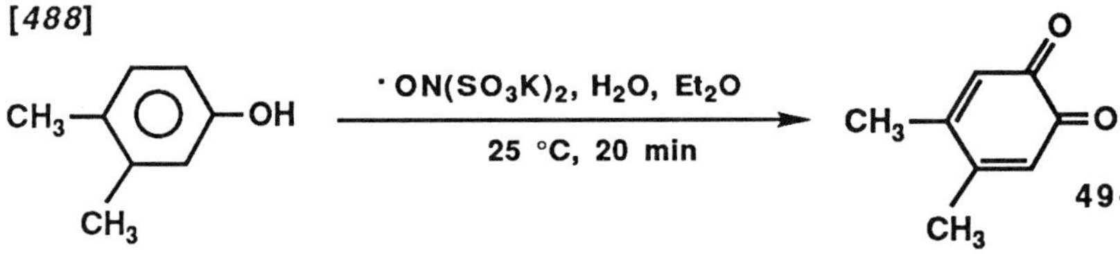

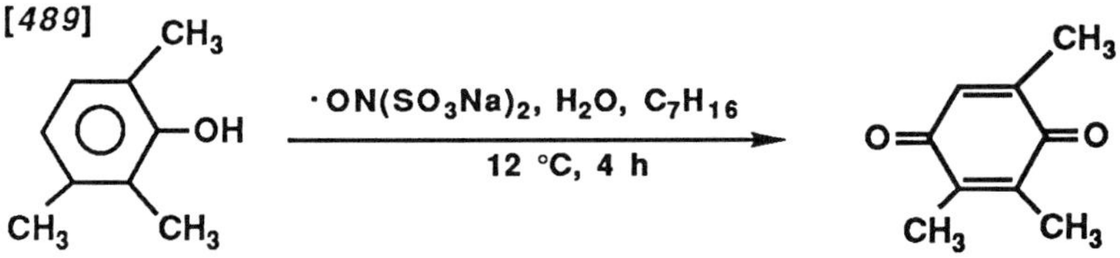

[489]

$$\text{2,3,6-trimethylphenol} \xrightarrow[\text{12 °C, 4 h}]{\cdot ON(SO_3Na)_2,\ H_2O,\ C_7H_{16}} \text{2,3,5-trimethyl-p-benzoquinone (77-79%)} \quad 311$$

Oxidations to p-benzoquinones occur even if the *para* position is occupied by a substituent [383, 559, 732] and even if an *ortho* position is free [383]. The substituent in the *para* position is either eliminated or modified. 2,4,6-Tribromo-m-cresol is oxidized by **chromium trioxide** in 70% acetic acid at 70–75 °C within 10 min to dibromo-m-toluquinone (3,5-dibromo-2-methyl-p-benzophenone) in 77% yield [559]. 2-Bromo-4-hydroxy-5-methoxybenzyl alcohol treated with Fremy salt at pH 6 at room temperature for 1 h gives an 84% yield of 2-bromo-5-methoxy-p-benzoquinone [487]. 4-Benzyl-2,6-dibromophenol, on oxidation with **bromine** in a sealed tube, yields 2,6-dibromophenylchinomethide [732] (equation 312). The oxidation of 2,5-di-*tert*-butyl-4-methoxyphenol with mercuric oxide or

[732]

$$C_6H_5CH_2\text{-(2,6-dibromophenol)} \xrightarrow[\text{100 °C, 4 h}]{Br_2,\ CCl_4} C_6H_5CH\text{=(2,6-dibromoquinomethide)} \quad (80\text{-}85\%) \quad 312$$

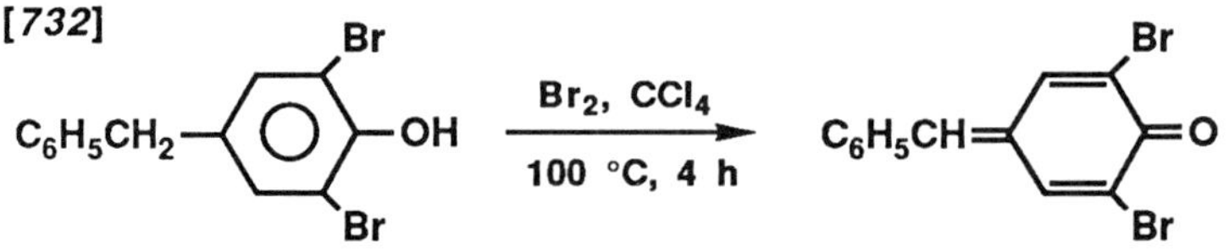

mercuric trifluoroacetate results in elimination of the methoxy group and formation of 2,5-di-*tert*-butyl-*p*-benzoquinone [*383*] (equation 313).

[*383*]   313

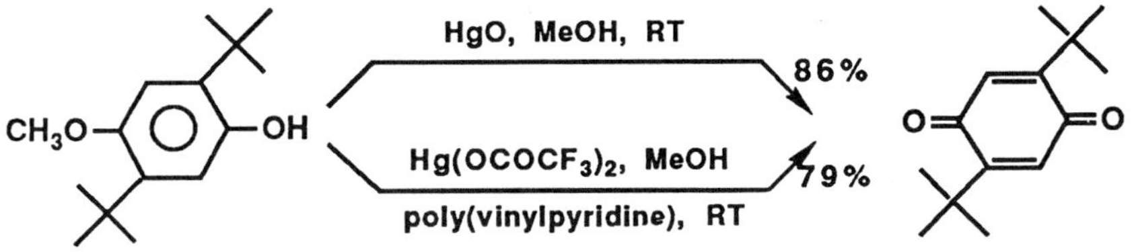

Sometimes the oxidation to quinones is accompanied by doubling of the molecule (equation 314) [*430, 971, 1038*].

314

[*430*]

[*971*]

[*1038*]

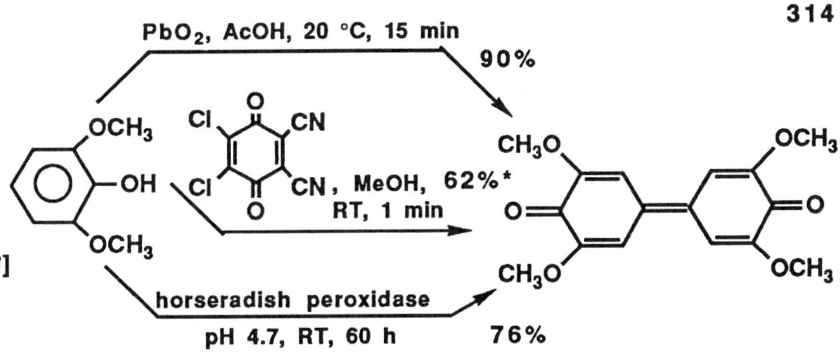

*In addition, a 13% yield of 2,6-dimethoxy-*p*-benzoquinone was obtained.

The oxidation of naphthols to 1,4-naphthoquinones is accomplished by ceric ammonium nitrate (equation 315) [*423*].

[*423*]   315

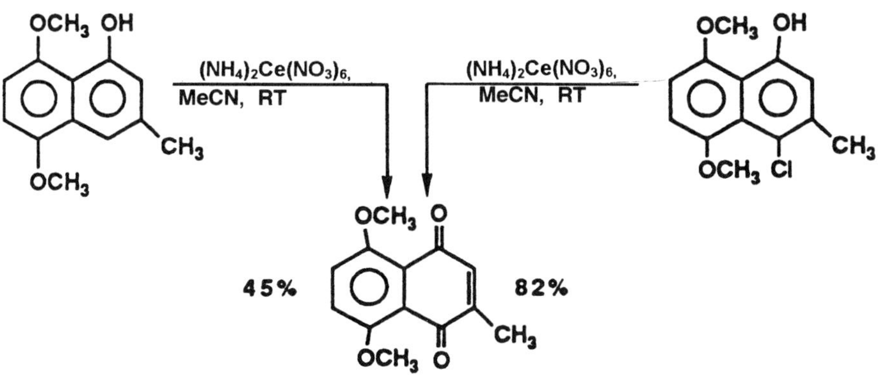

Oxidations to quinones are easier if two hydroxylic groups in *para* or *ortho* positions are present in the aromatic rings. A classical example is the oxidation of hydroquinone and its substitution derivatives to *p*-benzoquinones by **mercuric oxide** [*384*], **ceric ammonium nitrate** [*422*], **lead dioxide** [*431*], *dinitrogen tetroxide* [*457*], *potassium chromate* [*615*], **sodium dichromate** [*636*], **sodium chlorate** [*719*], *barium manganate* [*833*], **ferric chloride** [*907*], **ferric sulfate** [*914*], and others (equation 316). These oxidations are usually very fast.

316

HO—⟨benzene⟩—OH ⟶ O=⟨cyclohexadiene⟩=O

[*422*]	$(NH_4)_2Ce(NO_3)_6$, 75% aqueous MeCN, RT, 2 min	83%
[*636*]	$Na_2Cr_2O_7$, H_2SO_4, H_2O, 20-30 °C, 30-45 min	76-81%
[*719*]	$NaClO_3/V_2O_5$, H_2SO_4, H_2O, 40 °C, 4 h	92-96%
[*833*]	$BaMnO_4$, C_6H_6, reflux 20 min	75%

Oxides of silver [*171, 368*], **mercury** [*384*], **lead** [*431*], and nitrogen [*457*] react at room temperature. 2,5-Hydroquinone diacetic acid and mercuric oxide in ether, after several hours at room temperature, give *p*-benzoquinone-2,5-diacetic acid in 90% yield [*384*].

The dehydrogenating agent 2,3-dichloro-5,6-dicyano-*p*-benzoquinone **(DDQ)** is prepared from 2,3-dichloro-5,6-dicyanohydroquinone by oxidation with lead dioxide [*431*] or dinitrogen tetroxide [*457*] (equation 317).

[*431*] PbO_2, 5% HCl, EtOH, C_6H_6 RT, 5 min   83%   317

2,3-dichloro-5,6-dicyanohydroquinone ⟶ DDQ

[*457*] N_2O_4 (HNO_3 + As_2O_3), CCl_4 RT, 10 min   93%

Oxidants especially suitable for the preparation of quinones are salts of trivalent iron. Bromo-*m*-xyloquinone (2-bromo-3,5-dimethyl-*p*-benzoquinone) is obtained in 84% yield by steam distillation of 2-bromo-3,5-dimethylhydroquinone with **ferric sulfate** and dilute hydrochloric acid [*914*]. The treatment of 2-mercaptohydroquinone with 2 N ferric chloride in ethanol at room temperature gives a quantitative yield of *p*-benzoquinone disulfide [*907*]. Just shaking halogenated hydroquinones with an aqueous solution of **ferric chloride** and ether results in up to quantitative yields of the halogenated *p*-benzoquinones (equation 318) [*1156*].

[*1156*]

$$\text{HO-C}_6\text{F}_4\text{-OH} \xrightarrow[\text{RT}]{\text{FeCl}_3, \text{H}_2\text{O}, \text{Et}_2\text{O}} \text{O=C}_6\text{F}_4\text{=O} \quad \mathbf{318}$$

100%

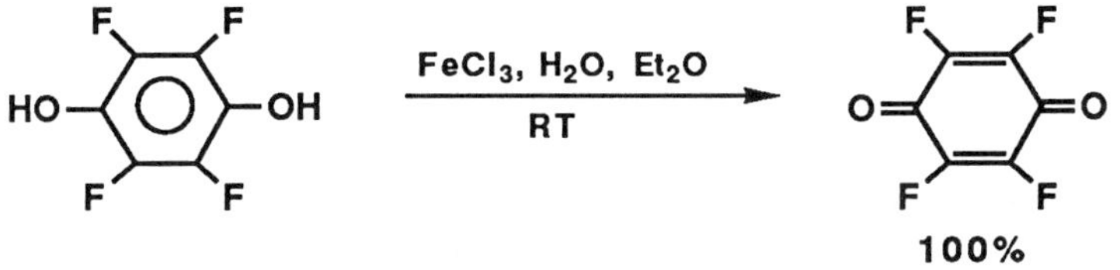

p,p'-Dihydroxybiphenyls are converted with **lead tetraacetate** into diphenoquinones (equation 319) [*369, 440*]. *p,p'*-Dihydroxystilbene is oxidized to stilbene quinone with **silver oxide** [*369*].

[*440*] X = Y = H

Pb(OAc)$_4$, AcOH, 30 °C, RT overnight, 75% → X = Y = H **319**

[*369*] X = Cl, Y = H

Pb(OAc)$_4$, AcOH, dioxane, 30 °C, 7 min, 71% → X = Cl, Y = H

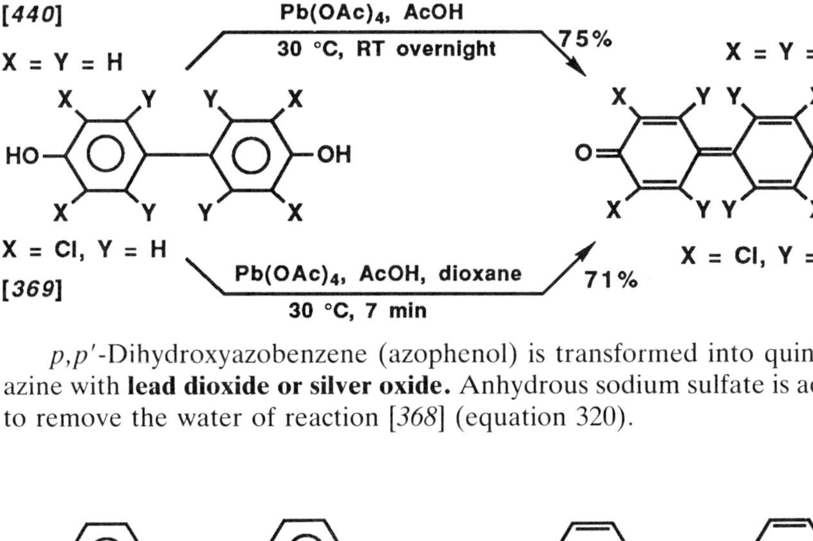

p,p'-Dihydroxyazobenzene (azophenol) is transformed into quinoneazine with **lead dioxide or silver oxide.** Anhydrous sodium sulfate is added to remove the water of reaction [*368*] (equation 320).

320

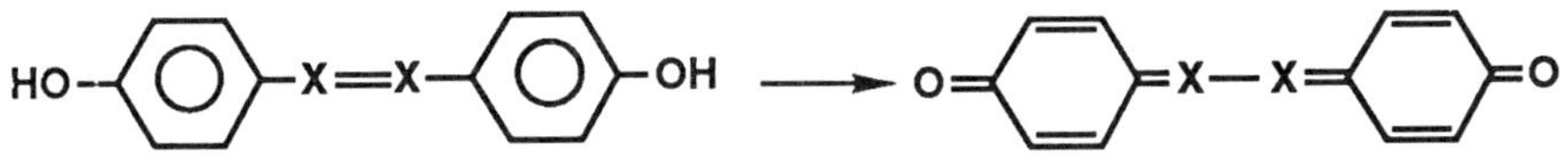

$$\text{HO-C}_6\text{H}_4\text{-X=X-C}_6\text{H}_4\text{-OH} \longrightarrow \text{O=C}_6\text{H}_4\text{=X-X=C}_6\text{H}_4\text{=O}$$

Ref	X	Reagents	Yield
[*368*]	X = N:	Ag$_2$O, Na$_2$SO$_4$, Et$_2$O, RT, 1 h	90%
[*369*]	X = CH:	Ag$_2$O, Me$_2$CO, reflux 1 h	40–45%
[*368*]	X = N:	PbO$_2$, Na$_2$SO$_4$, Et$_2$O, RT	70%

Silver oxide and sodium sulfate are frequently used to oxidize 1,2-dihydroxy aromatic compounds to *ortho*-quinones. Phenanthrene furnishes, after being shaken for 15 s with the mixture in ether at room temperature, a 65% yield of 3,4-phenanthrenequinone [*171*]. Another oxidant used to prepare *ortho*-quinones is **sodium iodate** (equation 321) [*754*].

[754] 321

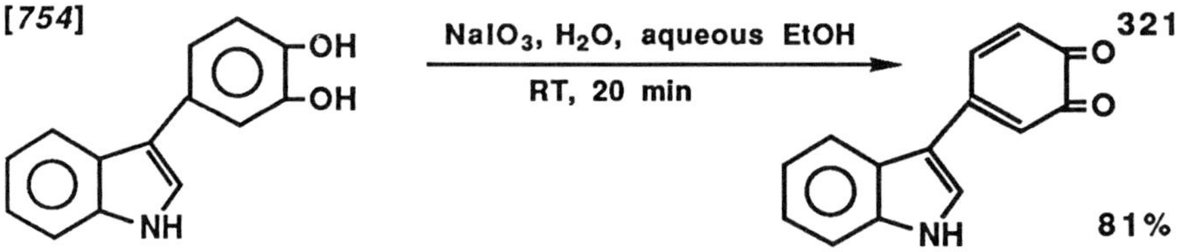

Organic peroxy acids oxidize phenols and *ortho*-dihydroxy aromatic compounds with the concomitant opening of the rings to dicarboxylic acids (equations 322 and 323) [249, 273, 274]. The reaction is very exothermic and requires external cooling.

[273]   322

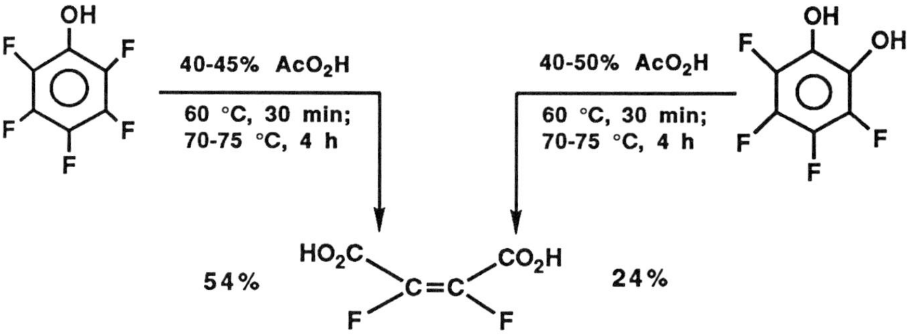

[249]   323

[274]

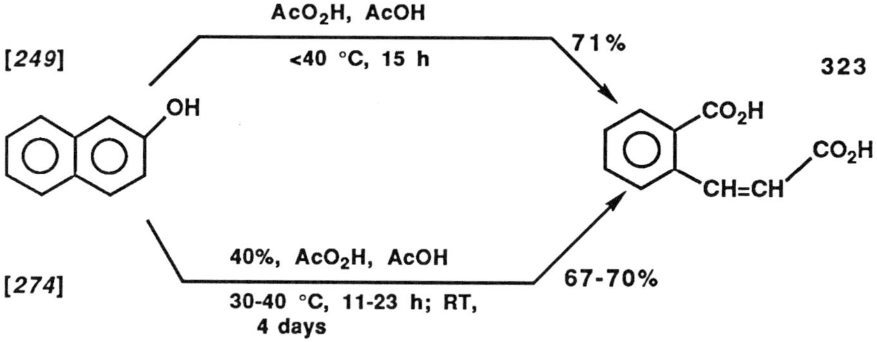

ETHERS

Ethers having α-hydrogens form *hydroperoxides*, which are highly explosive in the dry state. Hydroperoxide formation is facilitated by light. Although no absorption of oxygen was noticed when tetrahydrofuran and dioxane were exposed to oxygen at 12 and 15 °C for 30 min under irradiation [44], it is strongly recommended to store diethyl ether, other dialkyl ethers, tetrahydrofuran, and dioxane in metal containers or dark-glass bottles and away from direct light.

In the presence of benzophenone as a sensitizer and under irradiation with ultraviolet light, a 46% yield of tetrahydrofuran α-hydroperoxide is obtained when **oxygen** is bubbled through tetrahydrofuran at 12 °C for 28.5 h. Dioxane, methyl butyl ether, diisopropyl ether, dibutyl ether, and diisoamyl ether, under similar conditions, give α-hydroperoxides in yields of 10, 15, 12, 6, and 8%, respectively, based on the reacted ethers [*44*].

By far the most important oxidation of ethers is their *conversion into esters and lactones*. Only a few reagents are capable of such a transformation: **ozone** [*110*], **chromium trioxide** [*539, 586*], **zinc dichromate** [*660*], **potassium permanganate** [*855*], **zinc permanganate** [*898*], **benzyltriethylammonium permanganate** [*902*], and **ruthenium tetroxide** (which gives the best yields) [*774, 940*] (equations 324 and 325).

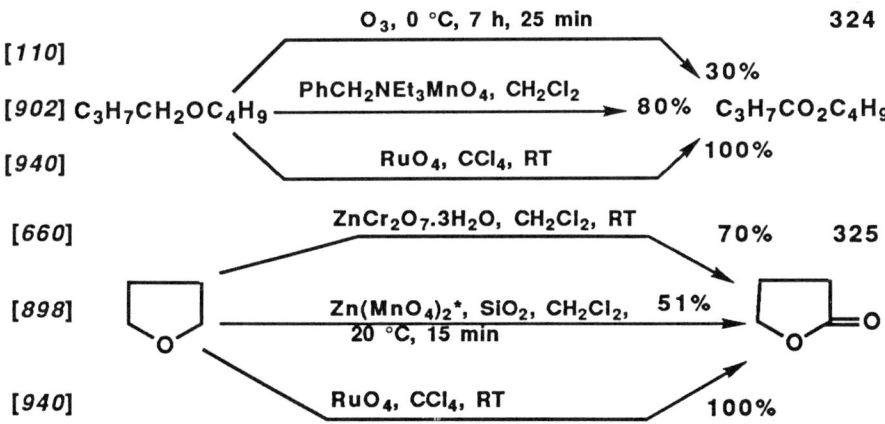

*Support on silica gel is essential for such work, because zinc permanganate is a potentially dangerous reagent [*898*].

In nonsymmetrical ethers, the oxidation of various groups takes place preferentially in the following sequence: benzyl > phenylalkyl > primary alkyl > secondary alkyl > methyl [*902*] (equation 326).

[*774*] 326

R = CH$_3$

$\mathrm{RuCl_3 \cdot H_2O, NaIO_4, MeCN, CCl_4, H_2O}$

RT, 2 h 89%

C$_6$H$_5$CH$_2$OR C$_6$H$_5$CO$_2$R

R = C$_4$H$_9$ C$_6$H$_5$CH$_2$NEt$_3$MnO$_4$, CH$_2$Cl$_2$ 99%

[*902*]

The oxidation products of mixed ethers may differ when different oxidants are used. Whereas **ruthenium tetroxide** oxidizes benzyl methyl ether to methyl benzoate in 89% yield [774], **chromium trioxide** in acetic acid converts hexadecyl methyl ether into palmitic acid and hexadecyl formate and hexadecyl ethyl ether into palmitic acid and hexadecyl acetate (equation 327) [539].

[539]   327

$$C_{16}H_{33}OC_2H_5 \xrightarrow[20\ °C,\ 1\ h]{CrO_3,\ AcOH,\ CH_2Cl_2} C_{15}H_{33}CO_2H + C_{16}H_{33}OCOCH_3$$
$$\phantom{C_{16}H_{33}OC_2H_5 \xrightarrow[20\ °C,\ 1\ h]{CrO_3,\ AcOH,\ CH_2Cl_2}}\ \ \ 55\%\ \ \ \ \ \ \ \ \ \ \ 9\%$$

In secondary benzyl ethers, cleavage to the ketone usually predominates over the formation of alkyl benzoates (equation 328) [586].

[586]   328

$$\underset{\underset{CH_3}{|}}{C_6H_{13}CH}OCH_2C_6H_5 \xrightarrow[0\ °C,\ 12\ h]{\substack{4\ CrO_3,\ H_2SO_4,\\ Me_2CO}} C_6H_{13}COCH_3 + C_6H_5CO_2\underset{\underset{CH_3}{|}}{CH}C_6H_{13}$$
$$79\%\ \ \ \ \ \ \ \ \ \ \ 21\%$$

A similar cleavage of benzyl methyl ethers takes place when a solution of equimolar amounts of **bromine in carbon tetrachloride** is added dropwise to a refluxing solution of the ether in carbon tetrachloride under irradiation. Benzyl methyl ether gives a 77% yield of benzaldehyde, and benzhydryl methyl ether gives a 36% yield of benzophenone [731].

Oxidation of Unsaturated Ethers at Multiple Bonds

Vinyl ethers and *peroxy acids* give primarily *epoxides*, which can be isolated [297] or which can react further with the reaction medium [314] (equations 329 and 330).

[297]   329

$$\underset{\underset{OC_2H_5}{|}}{C_6H_5C}=C(CH_3)_2 \xrightarrow[0\ °C,\ 30\ s]{BzO_2H,\ Et_2O} \underset{\underset{OC_2H_5}{|}}{C_6H_5C}\overset{\overset{O}{\triangle}}{\text{—}}C(CH_3)_2$$
$$90\%$$

[314]   330

dihydropyran $\xrightarrow[\substack{\rightarrow\ RT,\ 30\ min}]{\substack{m\text{-}ClC_6H_4CO_3H,\\ MeOH\\ 8\text{-}10\ °C,\ 40\ min,}}$ [epoxide intermediate] → tetrahydropyran-OH with OCH$_3$ (67%) + tetrahydropyran-OH with OCH$_3$ (8%)

Stronger oxidants convert ***enol ethers*** into ***esters or lactones.*** Ethyl vinyl ether added to a suspension of **pyridinium chlorochromate (PCC)** in dichloromethane at room temperature furnishes ethyl acetate in 75% yield after 1 h (equation 331) [*608*]. Glucals produce lactones (equation 332) [*609*].

[*608*] **331**

$$CH_2{=}CHOC_2H_5 \xrightarrow[\text{RT, 1 h}]{C_5H_5NHCrO_3Cl,\ CH_2Cl_2} \underset{75\%}{CH_3COOC_2H_5}$$

[*609*] **332**

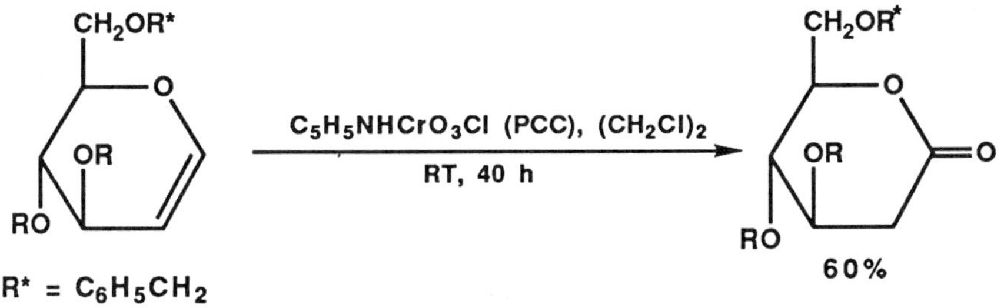
CH2OR*
O
OR
RO
C5H5NHCrO3Cl (PCC), (CH2Cl)2
RT, 40 h
CH2OR*
O
OR
O
RO
60%

$R^* = C_6H_5CH_2$

Even stronger oxidative agents such as dilute **nitric acid** transform 3,4-dihydro-2*H*-pyran at 100 °C into glutaric acid in 70–75% yields [*1158*].

Allylic ethers are ***epoxidized*** stereoselectively to predominantly *syn* products with **peroxytrifluoroacetic acid** in dichloromethane. In tetrahydrofuran, *anti* epoxidation prevails. Also, ***m*-chloroperoxybenzoic acid** in dichloromethane gives a higher proportion of *anti*-addition products (equation 333) [*287*].

[*287*] **333**

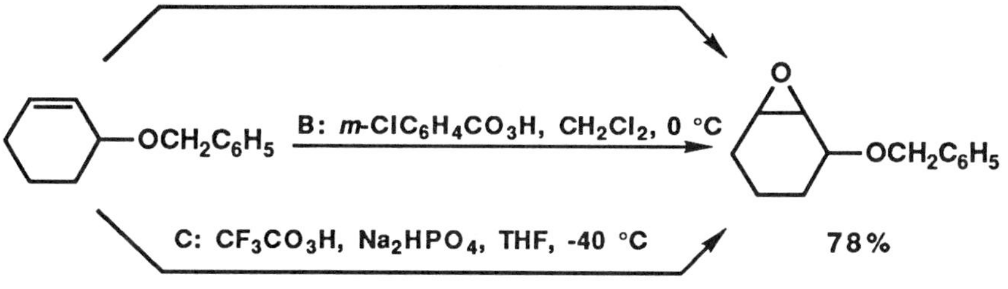
A: CF3CO3H, Na2HPO4, CH2Cl2, -40 °C
OCH2C6H5
B: m-ClC6H4CO3H, CH2Cl2, 0 °C
O
OCH2C6H5
C: CF3CO3H, Na2HPO4, THF, -40 °C
78%

	syn/anti
A	**1.4:1**
B	**1:2.7**
C	**1:5**

Selenium dioxide converts ***aryl and aralkyl allyl ethers*** into aldehydes and the corresponding alcohols. Allyl benzyl ether refluxed with selenium dioxide in acetic acid and dioxane for 1 h gives a 50% yield of benzyl alcohol and acrolein. Cinnamyl methyl ether is converted into methanol and cinnamaldehyde in 66% yield (equation 334) [*511*].

[*511*]   334

$$C_6H_5CH=CHCH_2OCH_3 \xrightarrow[\text{reflux 1 h}]{\text{SeO}_2,\ \text{AcOH, dioxane}} CH_3OH + C_6H_5CH=CHCHO \quad 66\%$$

Alkyl acetylenyl ethers treated with iodosobenzene in dichloromethane in the presence of 1% of tris(triphenylphosphine)ruthenium dichloride for 15 min at room temperature furnish keto esters in 59–70% yields (equation 335) [*786*].

[*786*]   335

$$C_6H_5C\equiv C\text{-}OC_2H_5 \xrightarrow[\text{RT, 15 min}]{\text{PhIO/(Ph}_3\text{P)}_3\text{RuCl}_2,\ \text{CH}_2\text{Cl}_2} C_6H_5COCO_2C_2H_5 \quad 70\%$$

A special category of ethers are *trimethylsilyl ethers*. Trimethylsilyl ethers of primary alcohols, on treatment with Jones reagent, give acids [*590*]. On treatment with *N*-bromosuccinimide under irradiation, trimethylsilyl ethers yield esters [*744*]. Secondary alkyl trimethylsilyl ethers are converted into ketones by oxidation with both reagents [*590, 744, 981*]. Oxidation with Jones reagent is regiospecific: the 2-*tert*-butyldimethylsilyl 11-*tert*-butyldiphenylsilyl ether of 2,11-dodecanediol is oxidized only in the sterically less hindered position [*590*]. Trimethylsilyl ethers of tertiary alcohols are degraded by periodic acid to carboxylic acids with shorter chains [*755*] (equations 336–339).

Trimethylsilyl enol ethers treated with chromyl chloride in dichloromethane at –78 °C furnish α-hydroxy ketones in 62–82% yields [*676*]. The same products are obtained by oxidation with *m*-chloroperoxybenzoic acid

[*744*]   336

$$C_5H_{11}CH_2OSi(CH_3)_3 \xrightarrow[0\ °C,\ 5\ h]{\text{NBS, CCl}_4,\ h\nu} C_5H_{11}CO_2CH_2C_5H_{11} \quad 80\%$$

[*744*]   337

$$\underset{\underset{CH_3}{|}}{C_6H_5CHOSi(CH_3)_3} \xrightarrow[\text{RT, 3.5 h}]{\text{NBS, C}_5\text{H}_5\text{N, } h\nu} C_6H_5COCH_3 \quad 76\%$$

[*590*]   338

$$t\text{-BuMe}_2\text{SiO}\underset{\underset{CH_3}{|}}{C}H(CH_2)_8\underset{\underset{CH_3}{|}}{C}HOSiPh_2Bu\text{-}t \xrightarrow[\substack{\text{acetone,}\\0\ °C,\ 1\ h}]{\substack{\text{Jones}\\ \text{reagent,}\\ \text{KF}}} \underset{\underset{CH_3}{|}}{C}O(CH_2)_8\underset{\underset{CH_3}{|}}{C}HOSiPh_2Bu\text{-}t \quad 90\%$$

Oxidations

[755]   339
$$(CH_3)_2CHCHCHCOC(CH_3)_2OSi(CH_3)_3 \xrightarrow[\substack{0-5\ °C,\\ 15\ min,\\ 25\ °C,\\ 1.5\ h}]{H_5IO_6,\ THF} (CH_3)_2CHCHCHCO_2H$$
         | | | |
         HO CH$_3$ HO CH$_3$
                                                                    77-83%

[1159] or with triphenyl phosphite ozonide followed by triphenylphosphine [996] (equations 340 and 341).

[676]   340
$$(CH_3)_3CCOCH_3 \xrightarrow[\text{reflux}]{\substack{Me_3SiCl,\\ Et_3N,\\ MeCN}} (CH_3)_3CC{=}CH_2 \xrightarrow[\substack{-78\ °C,\\ 30\ min}]{CrO_2Cl_2,\ CH_2Cl_2} (CH_3)_3CCOCH_2OH$$
                                         |
                                       OSi(CH$_3$)$_3$ 82%
                                        62%

[1159]   341

1. *m*-ClC$_6$H$_4$CO$_3$H, C$_6$H$_{14}$, -15 °C, 20 min, 30 °C, 2 h   70-73%
2. Et$_3$N·HF, CH$_2$Cl$_2$, RT, 2 h

OSi(CH$_3$)$_3$ HO

1. (PhO)$_3$PO$_3$, CH$_2$Cl$_2$, -78 °C, 3 h   79%
[996]   2. Et$_3$N, Ph$_3$P, Et$_2$O, RT, 1 h

The bis(trimethylsilyl) ether of hydroquinone is oxidized by pyridinium chlorochromate to *p*-benzoquinone (equation 342) [610].

[610]   342
$$(CH_3)_3SiO{-}\langle{\bigcirc}\rangle{-}OSi(CH_3)_3 \xrightarrow[25\ °C,\ 2\ h]{C_5H_5NHCrO_3Cl,\ CH_2Cl_2} O{=}\langle{\bigcirc}\rangle{=}O$$
                                                                          99%

Oxidation of Epoxides (Oxiranes)

The oxidative cleavage of epoxides with **hydrogen peroxide** gives vicinal hydroxy hydroperoxides [178]. With **dimethyl sulfoxide** in the presence of trifluoromethanesulfonic acid and diisopropylethylamine, epoxides are converted into α-hydroxy ketones [1014], and with **periodic acid,** dicarbonyl compounds are formed [761] (equations 343 and 344).

[178]   343

$(C_6H_5)_2C$—CH_2 $\xrightarrow{\text{98% } H_2O_2,\ Et_2O,\ RT,\ 14\ days}$ $(C_6H_5)_2C(OOH)$—CH_2OH 92%

[1014]   DMSO, CF_3SO_3H, 23 °C, 30-45 min / CH_2Cl_2, -78 °C, i-Pr_2NEt   (cyclohexene oxide) $\xrightarrow{HIO_4,\ H_2O,\ 45\ °C}$ 344 [761]

2-hydroxycyclohexanone 61%   82%   OHC–(CH$_2$)$_4$–CHO

ALDEHYDES

Aldehydes are easily oxidized to carboxylic acids; the reaction sometimes occurs spontaneously on contact of the aldehyde with air. Benzaldehyde, for example, when left in an open bottle, is slowly converted into benzoic acid at room temperature. Such oxidation is accelerated if the aldehyde is exposed to **oxygen,** especially under irradiation with ultraviolet light. The same effect is achieved by using catalysts, such as salts or oxides of metals. Oxygen in the presence of copper acetate and nickel acetate converts methacrolein into methacrylic acid in 99% yield at 62% conversion at room temperature and a pressure of 14 atm [4]. Platinum catalyzes the oxidation of sugars to aldonic acids [5, 6] (*see* equation 366). Furfural, when treated with oxygen in aqueous alkaline solutions in the presence of cuprous oxide and silver oxide at 50–55 °C, is converted into furoic acid in high yields [63]. The same acid is obtained as a product of the Cannizzaro reaction [656] and by heating with potassium dichromate and sulfuric acid [656] (equation 345).

345

[63] O_2, NaOH, H_2O, Cu_2O, Ag_2O, 50-55 °C, 20-25 min, 40 °C, 15-30 min   86-90%

[656]   furfural–CHO $\xrightarrow{\text{NaOH, } H_2O}$ 73-76%   furan–CO_2H

[656] $K_2Cr_2O_7$, H_2O, H_2SO_4, 100 °C   75%

Aldehydes are transformed into carboxylic acids by **hydrogen peroxide** and its derivatives. In the presence of selenium dioxide as a catalyst, acrolein is oxidized to acrylic acid in 90% yield on treatment with a 12% solution of hydrogen peroxide in *tert*-amyl alcohol at 40 °C for 3 h [*125*]. Methacrolein is converted into methacrylic acid by 90% hydrogen peroxide in the presence of selenium dioxide at 60 °C [*125*] or by peroxysulfuric acid [*200*] (equation 346).

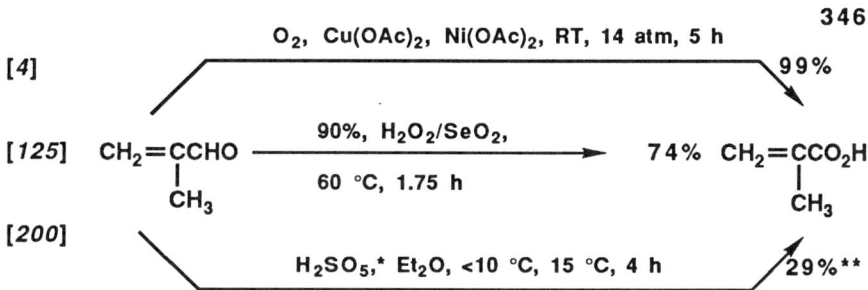

346

*H_2SO_5 is prepared either by adding ammonium persulfate to 85% sulfuric acid below 15 °C, or by adding 90% hydrogen peroxide to concentrated sulfuric acid below 15 °C, and keeping the mixture for 2 h at room temperature.

**When methanol was used as the solvent, a 91-97% yield of methyl methacrylate was obtained [*602*].

Potassium peroxymonosulfate ($2KHSO_5 \cdot KHSO_4 \cdot K_2SO_4$) oxidizes benzaldehyde in chloroform and dilute sulfuric acid to benzoic acid at room temperature within 2 days in a 70% yield [*205*].

The reagent of choice for the conversion of aldehydes into carboxylic acids is **silver oxide**, which is often prepared in situ from silver nitrate in alkaline medium [*365*]. Enanthol heated with silver oxide affords enanthoic acid [*77*]. 3-Cyclohexene-1-carboxaldehyde (1,2,3,6-tetrahydrobenzaldehyde) gives a 62.5% yield of the carboxylic acid on treatment with silver nitrate and ethanolic potassium hydroxide at room temperature [*365*] (equation 347).

Other examples of oxidations using silver oxide in alkaline media are variations of the preceding reaction [*206, 362, 366, 367*] (equation 348).

Aldehydes are also oxidized to carboxylic acids in high yields by *argentic oxide* at room temperature in aqueous tetrahydrofuran. However,

[367]

$\underset{S}{\text{CHO}}$ $\xrightarrow[\text{cooling with ice, 5 min}]{\text{AgNO}_3,\ \text{NaOH, H}_2\text{O}}$ $\underset{S}{\text{CO}_2\text{H}}$

348

95-97%

this method has the disadvantage of requiring large excesses of the fairly expensive oxidant (4–10 equivalents) [382] (equation 347).

Some aldehydes are oxidized to acids by very strong oxidizing agents. β-Chloropropionaldehyde gives a 60–65% yield of β-chloropropionic acid on treatment with fuming **nitric acid** for 25 min at 30–35 °C [470]. Vanillin is converted into vanillic acid in 89–95% yield by heating for 5 min with a mixture of *sodium hydroxide and potassium hydroxide* at 180–195 °C [1160]. Furfural furnishes a 75% yield of furoic acid by heating for 30–45 min with **potassium dichromate** and aqueous sulfuric acid at 100 °C [656] (equation 345). Heating enanthal with sulfur and ammonium sulfide affords only a 46% yield of enanthoic acid [1161].

An oxidant very suitable for the conversion of aldehydes into acids is **sodium chlorite** [712]. Because the byproducts of the reaction, NaOCl and ClO$_2$, are themselves oxidants and could decrease yields, scavengers such as hydrogen peroxide [712], sulfamic acid [1162], or resorcinol [1162] are added to the solution of the substrate and sodium chlorite in water, aqueous acetonitrile, or *tert*-butyl alcohol. Under such conditions, unsaturated aldehydes can be oxidized to unsaturated acids without destruction of the double bond (equation 349).

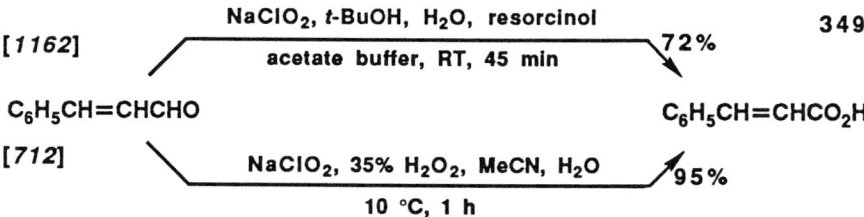

The oxidation of aromatic aldehydes to carboxylic acids with **manganese dioxide** is somewhat surprising because manganese dioxide is used to oxidize toluenes to benzaldehydes. The transformation of benzaldehyde into benzoic acid is achieved in 75% yield by refluxing with manganese dioxide in petroleum ether for 24 h [813].

On the other hand, **potassium permanganate** is frequently used to prepare carboxylic acids from both aliphatic [868] and aromatic aldehydes [885]. High yields of carboxylic acids are also obtained by oxidation with **tetrabutylammonium permanganate** [899] or **benzyltriethylammonium permanganate** [903] (equation 350). Recently, doubts about the safety of work with quaternary ammonium permanganates have been raised [901].

Oxidations

[899]

[934] Reaction scheme: Piperonal (3,4-methylenedioxybenzaldehyde, —CHO) is converted to piperonylic acid (—CO$_2$H) by three methods:
- Bu$_4$NMnO$_4$, C$_5$H$_5$N, RT — 99% (**350**)
- NiO$_2$, NaOH, H$_2$O, 60 °C, 3 h — 100%
- PhCH$_2$NEt$_3$MnO$_4$, AcOH, CH$_2$Cl$_2$, 1.25 h — 80%

[903]

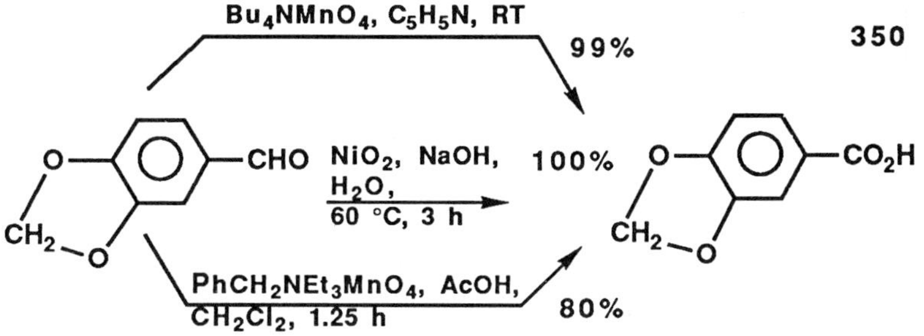

The unusual oxidant **nickel peroxide** converts aromatic aldehydes into carboxylic acids at 30–60 °C after 1.5–3 h in 58–100% yields [934]. The oxidation of aldehydes to acids by pure ruthenium tetroxide results in very low yields [940]. On the contrary, **potassium ruthenate,** prepared in situ from ruthenium trichloride and potassium persulfate in water and used in catalytic amounts, leads to a 99% yield of *m*-nitrobenzoic acid at room temperature after 2 h. Another oxidant, iodosobenzene in the presence of tris(triphenylphosphine)ruthenium dichloride, converts benzaldehyde into benzoic acid in 96% yield at room temperature [788]. The same reaction with a 91% yield is accomplished by treatment of benzaldehyde with osmium tetroxide as a catalyst and cumene hydroperoxide as a reoxidant [1163].

The biochemical oxidation of phenylacetaldehyde with **cyclohexanone oxygenase** produces phenylacetic acid, in addition to smaller amounts of benzyl formate and benzyl alcohol (equation 351) [1034].

[1034]

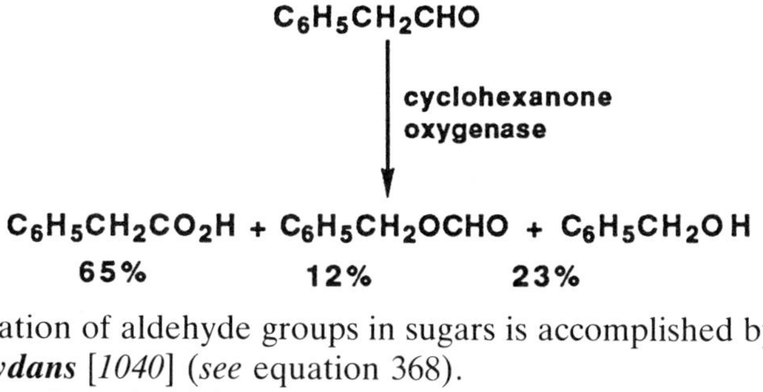

$$C_6H_5CH_2CHO \xrightarrow{\text{cyclohexanone oxygenase}} C_6H_5CH_2CO_2H + C_6H_5CH_2OCHO + C_6H_5CH_2OH$$

65%   12%   23%   **351**

The oxidation of aldehyde groups in sugars is accomplished by *Acetobacter suboxydans* [1040] (see equation 368).

The conversion of aldehydes into carboxylic acids as a result of disproportionation occurs in the Cannizzaro and Guerbet reactions. The **Cannizzaro reaction** takes place when aromatic aldehydes are treated with alkali hydroxides and results in the formation of carboxylic acids and the corresponding alcohols. Yields of the acids can therefore be 50% at a maximum. The reaction is catalyzed by silver [363] and silver oxide [362], which is sometimes generated in situ from silver nitrate [364] (see equation 357).

The **Guerbet reaction** is a complex combination of oxidation of a primary alcohol to an aldehyde, aldol condensation followed by dehydration, hydrogenation of the α,β-unsaturated aldehyde to a saturated alcohol, and oxidation of the aldehyde to an acid. The reaction is catalyzed by sodium and copper bronze and is carried out by heating the mixture in an autoclave at 270–300 °C and 50–60 atm for 6–10 h (equation 352).

[346]   352

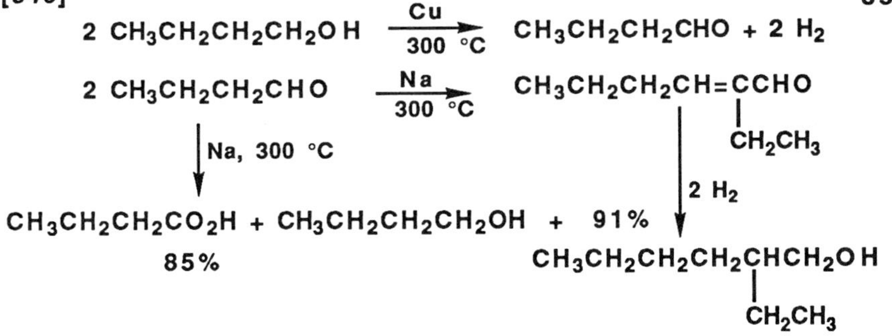

$$2\ CH_3CH_2CH_2CH_2OH \xrightarrow[300\ °C]{Cu} CH_3CH_2CH_2CHO + 2\ H_2$$

$$2\ CH_3CH_2CH_2CHO \xrightarrow[300\ °C]{Na} CH_3CH_2CH_2CH{=}CCHO$$
$$\qquad\qquad\qquad\qquad\qquad\qquad\qquad\quad |$$
$$\qquad\qquad\qquad\qquad\qquad\qquad\qquad\quad CH_2CH_3$$

Na, 300 °C ↓   ↓ 2 H$_2$

$CH_3CH_2CH_2CO_2H + CH_3CH_2CH_2CH_2OH\ +\ 91\%$
85% $CH_3CH_2CH_2CH_2CHCH_2OH$
                                                                         $|$
                                                                        CH_2CH_3

Because the purpose of the reaction is to prepare a branched saturated alcohol, the Guerbet reaction cannot be considered a method of oxidation [346].

An elegant method for the gentle and selective oxidation of aldehydes to carboxylic acids is the oxidation of the **aldehyde–bisulfite** addition compound by **dimethyl sulfoxide.** Depending on the workup of the reaction mixture, the acid, its ester, or its amide can be prepared at room temperature (equation 353) [713].

[713]   353

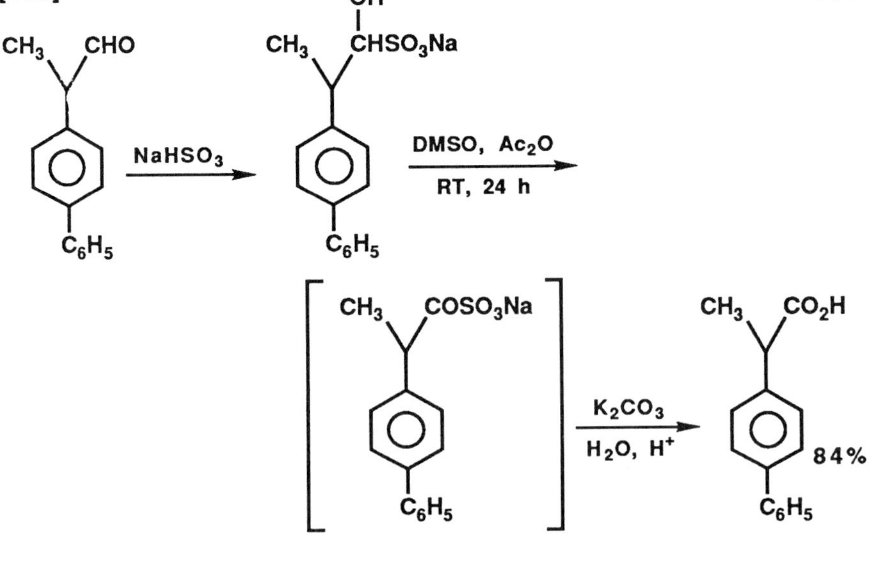

4-phenyl-α-methylbenzaldehyde (CH$_3$, CHO on benzene ring with C$_6$H$_5$ para substituent) $\xrightarrow{NaHSO_3}$ CH$_3$, CH(OH)SO$_3$Na analogue $\xrightarrow[RT,\ 24\ h]{DMSO,\ Ac_2O}$ [CH$_3$, COSO$_3$Na analogue] $\xrightarrow[H_2O,\ H^+]{K_2CO_3}$ CH$_3$, CO$_2$H analogue, 84%

A somewhat similar oxidation takes place when an aldehyde is treated with **sodium cyanide** and **manganese dioxide** or **argentic oxide** [382]. The reaction is assumed to proceed through a cyanohydrin, which is oxidized to an α-keto nitrile. The α-keto nitrile in turn is converted into an acid (by argentic oxide in methanol) or an ester (by manganese dioxide in methanol). The method is especially suited for α,β-unsaturated aldehydes [382] (equation 354).

[382]   354

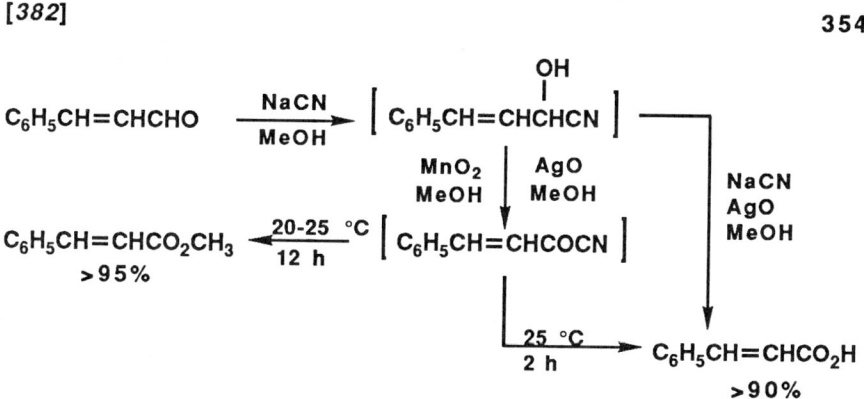

Examples of other oxidations of aldehydes to carboxylic acids are given in equations 355–357.

EXAMPLES OF OXIDATION OF ALDEHYDES TO CARBOXYLIC ACID

	$C_6H_{13}CHO$	→	$C_6H_{13}CO_2H$	355
[77]	Ag_2O, 10% NaOH, 95 °C, 7 h; HNO_3, 70 °C, → RT, 1 h		97.5%	
[1161]	$(NH_4)_2S$, S, C_5H_5N, 140 °C, 5 h		46%	
[868]	$KMnO_4$, H_2SO_4, H_2O, 15-20 °C		76-78%	
[940]	RuO_4, CCl_4, RT		30%	

	C_6H_5CHO	→	$C_6H_5CO_2H$	356
[1163]	$PhCMe_2OOH$, OsO_4 catalyst, 0 °C, 1.5 h		91%	
[205]	2 $KHSO_5 \cdot KHSO_4 \cdot K_2SO_4$, H_2SO_4, H_2O, $CHCl_3$, RT, 2 days		70%	
[1162]	$NaClO_2$, t-BuOH, resorcinol, pH 4, RT, 18 h		88%	
[788]	2 PhIO, $(Ph_3P)_3RuCl_2$, CH_2Cl_2, RT, 24 h		96%	
[813]	MnO_2, petroleum ether, reflux 24 h		75%	
[903]	$PhCH_2NEt_3MnO_4$, AcOH, RT, 1.25 h		84%	
[934]	NiO_2, NaOH, H_2O, 30 °C, 1.5 h		94%	
[940]	RuO_4, CCl_4, RT		33%	

357

$$\text{2-hydroxy-3-methoxybenzaldehyde} \longrightarrow \text{2-hydroxy-3-methoxybenzoic acid}$$

[366]	Ag$_2$O, NaOH, H$_2$O, reflux 2 h	50%
[206]	Ag$_2$O, NaOH, H$_2$O, 55-60 °C, 10 min	83-95%
[364]	AgNO$_3$, NaOH, H$_2$O, 55-85 °C	100%
[363]	NaOH/Ag, reflux 1 h	50%
[1160]	NaOH, KOH, 180-195 °C, 5 min	89-95%
[885]*	KMnO$_4$, 10% KOH, 70-80 °C, 40-45 min + 1 h	90-96%* (78-84% pure)
[1162]	NaClO$_2$, H$_2$NSO$_3$H, H$_2$O, RT, 1 h	84%

*The reference shows the reaction conditions for the oxidation of piperonal [3,4-(methylenedioxy)benzaldehyde].

Oxidation of aldehydes to peroxy acids is accomplished by passing **oxygen** under irradiation through a solution of benzaldehyde in acetone [*1164*] or by passing a current of ozone-containing oxygen through a solution of benzaldehyde in ethyl acetate at 20–22 °C for 1–2 h [*112*] (equation 358).

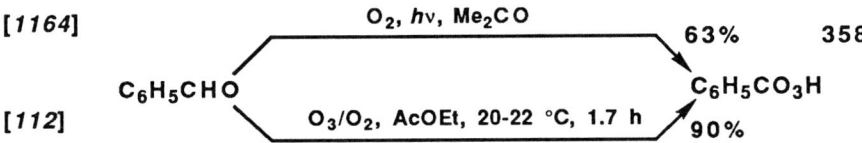

[*1164*] O$_2$, hv, Me$_2$CO — 63%
C$_6$H$_5$CHO → C$_6$H$_5$CO$_3$H   358
[*112*] O$_3$/O$_2$, AcOEt, 20-22 °C, 1.7 h — 90%

Aromatic aldehydes, especially those containing hydroxyl or alkoxy groups, undergo ***oxidation to aryl formates (Dakin reaction)*** on treatment with hydrogen peroxide or peroxy acids [*302*]. Because the formyl esters of phenols are easily hydrolyzed, the formates may not even be intercepted, and phenols are isolated as the oxidation products [*302*]. The reaction may be interpreted as a nucleophilic attack by the oxygen of the peroxy compound at the carbon of the carbonyl group, followed by migration of one of the substituents of the carbonyl group.

Hydride migration results in carboxylic acids (the usual reaction). Migration of the aryl group gives aryl formates but practically only when electron-releasing groups are present in the benzene ring in the *ortho* or *para* positions with respect to the aldehyde group (equation 359) [*302*].

Oxidations of aromatic aldehydes to formyl esters of phenols and to phenols are accomplished by **hydrogen peroxide** [*171, 172, 173*] or by **organic peroxy acids** such as **peroxyacetic acid** [*259*], **peroxybenzoic acid** [*302*], and ***m*-chloroperoxybenzoic acid** [*315, 318*] (equations 360–363).

Oxidations

[302]

359
ArC=O
 |
 O—H

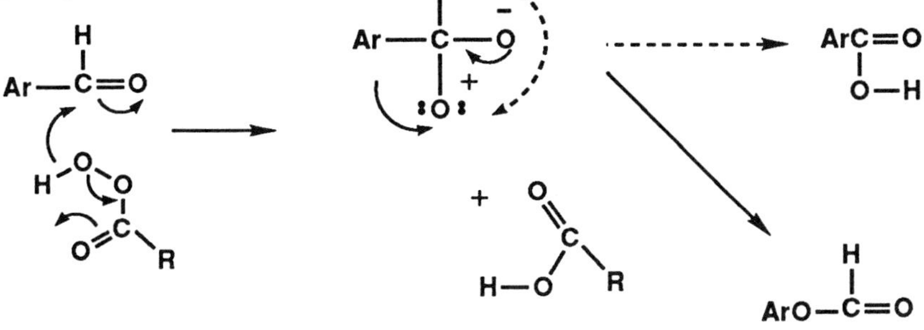

360
     H
     |
ArO—C=O

+
   O
   ‖
H—O—C—R

[172] X = H
3% H$_2$O$_2$, NaOH, 45-50 °C, 15-20 h
69-73%

6% H$_2$O$_2$, NaOH, 40-50 °C, 1.5 h

[173] X = MeO
68-80%

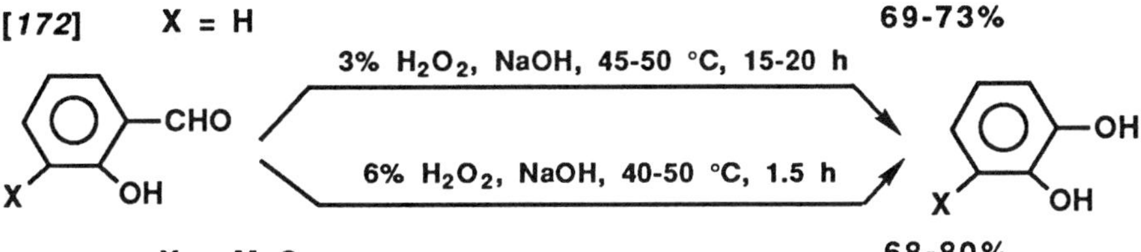

[171]
30.8% H$_2$O$_2$, 12.5 N KOH, C$_5$H$_5$N
boil for a few seconds
361
80%

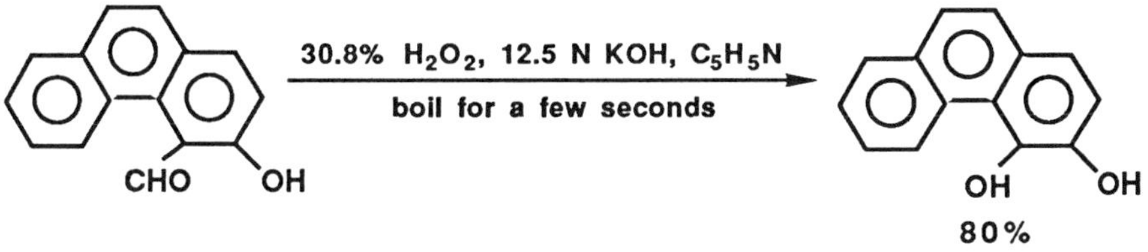

[259]
20%, AcO$_2$H, AcOH, TosOH
30 °C, RT overnight
362
60%

[318] X = H, Y = OCH$_3$
m-ClC$_6$H$_4$CO$_3$H, CH$_2$Cl$_2$
reflux, 5 h

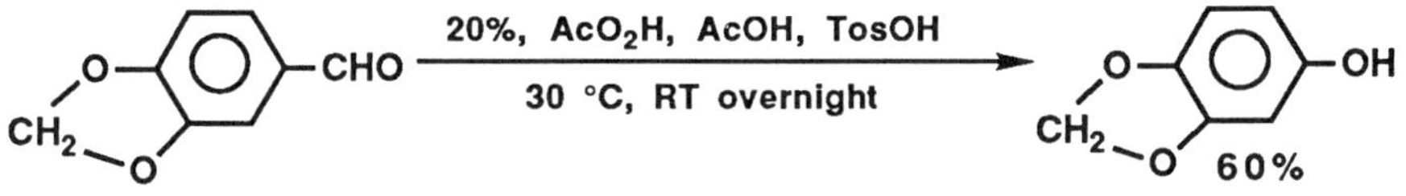

[315] X = OCH$_3$, Y = OH
m-C$_6$H$_4$CO$_3$H, KF, CH$_2$Cl$_2$
RT

77%

10% KOH, RT
363
92%

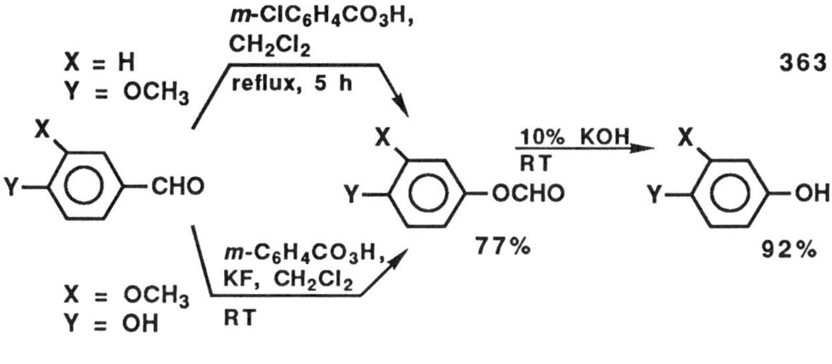

Oxidation of Aldehydes Having Other Functionalities

Oxidations of *unsaturated aldehydes,* as long as the oxidation affected only the aldehyde group, have been discussed with the oxidations of saturated and aromatic aldehydes. In this section, only such oxidations that affect the double bonds are described.

Because of the easy oxidizability of the aldehydes to carboxylic acids, any oxidation at the double bond is a delicate operation. Only a few oxidants are suitable to produce aldehyde epoxides, if used under controlled conditions.

Whereas hydrogen peroxide in acids [125] and peroxysulfuric acid [200] convert methacrolein into methacrylic acid, **hydrogen peroxide** in alkaline medium (best at pH 8–8.5) forms the epoxy aldehydes (equation 364) [146].

[146]   364

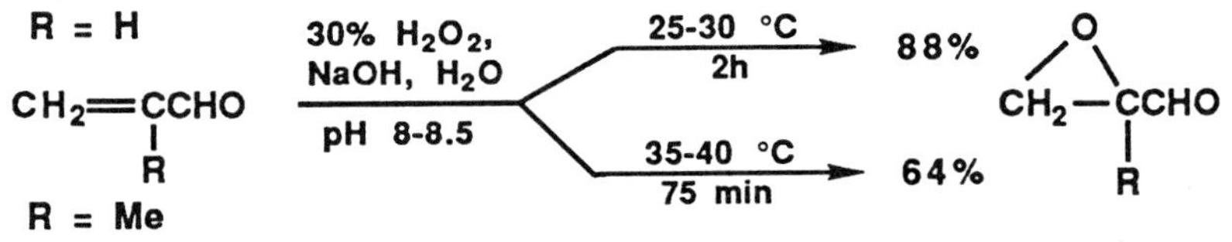

R = H, CH$_2$=CCHO, R; R = Me   30% H$_2$O$_2$, NaOH, H$_2$O, pH 8.5   25–30 °C, 2h → 88%   35–40 °C, 75 min → 64%   CH$_2$—CCHO (epoxide), R

The procedure just mentioned fails with cinnamaldehyde. The epoxidation of cinnamaldehyde is accomplished by tert-*butyl hydroperoxide* in aqueous methanolic sodium hydroxide (equation 365) [219].

[219]   365

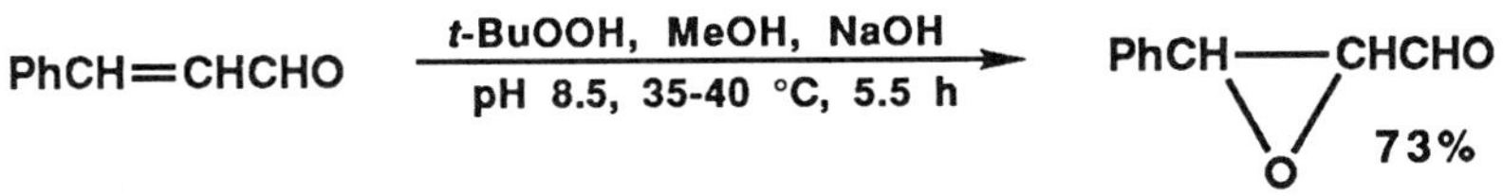

PhCH=CHCHO $\xrightarrow{\text{t-BuOOH, MeOH, NaOH}}_{\text{pH 8.5, 35–40 °C, 5.5 h}}$ PhCH—CHCHO (epoxide) 73%

Oxidation of hydroxy aldehydes may affect only the aldehyde group or both the aldehyde and alcoholic groups. Both types are especially common in sugar chemistry; for example, aldoses may be converted into aldonic acids or aldaric acids.

Arabinose, xylose, glucose, galactose, and mannose are transformed into the corresponding *aldonic acids* by air or oxygen passed through their solutions in dilute sodium hydroxide in the presence of 5% platinum on carbon as a catalyst [5, 6] (equation 366). The acids are usually isolated as calcium salts after calcium chloride has been added to the reaction mixture. The classical oxidant for this purpose is **bromine water** [1165] (*see* equation 368).

Both the aldehyde and the alcoholic groups are affected in δ-phenyl-δ-hydroxyvaleraldehyde, which is oxidized by **potassium permanganate** to γ-benzoylbutyric acid (equation 367) [749].

[*5*] **366**

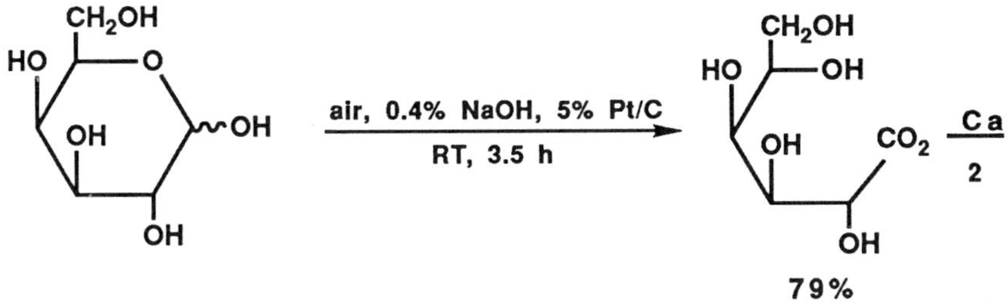
CH2OH
HO
O
OH
OH
OH
air, 0.4% NaOH, 5% Pt/C
RT, 3.5 h
CH2OH
HO
OH
OH
CO2
Ca
2
OH
79%

[*749*] **367**

$$C_6H_5CH(OH)(CH_2)_3CHO \xrightarrow[\text{RT}]{KMnO_4,\ NaOH,\ H_2O} C_6H_5CO(CH_2)_3CO_2H \quad 90\%$$

Biochemical oxidations of sugars are very common and versatile. Depending on the microorganism used, hydroxylic groups in various positions can be oxidized to keto groups, and the aldehyde group is converted into a carboxyl group.

Glucose fermentation at 25–30 °C with ***Acetobacter suboxydans*** first gives gluconic acid and then oxidizes the alcoholic group on C-5 to give 5-ketogluconic acid in 90% yield. An unnamed bacterium converts glucose into 2-ketogluconic acid in 82% yield (equation 368) [*1040*].

The oxidation of both the aldehyde group and the terminal primary alcoholic group in sugars is traditionally carried out with **dilute nitric acid**

[*1040*] **368**

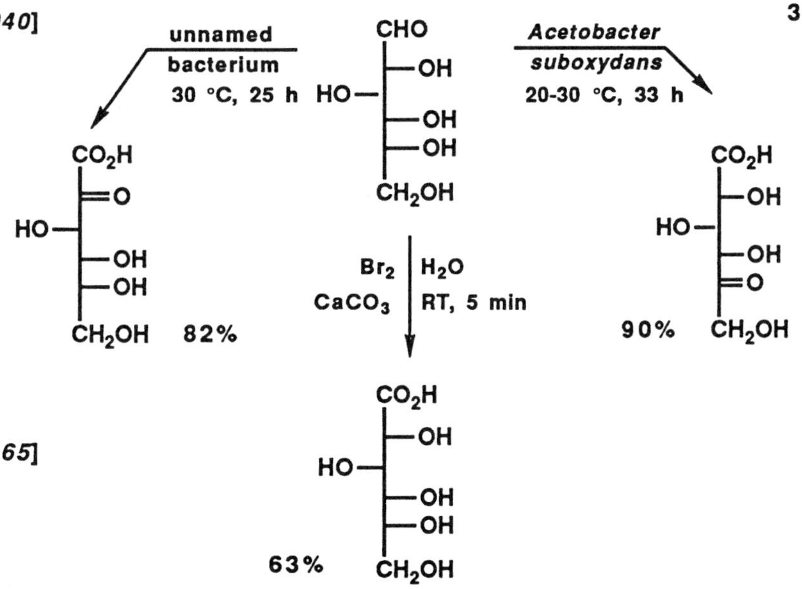
unnamed
bacterium
30 °C, 25 h
CHO
OH
HO
OH
OH
CH2OH
Acetobacter
suboxydans
20-30 °C, 33 h
CO2H
O
HO
OH
OH
CH2OH
82%
Br2
CaCO3
H2O
RT, 5 min
CO2H
OH
HO
OH
O
90%
CH2OH
CO2H
OH
HO
OH
OH
63%
CH2OH

[*1165*]

at 100 °C and leads to *aldaric acids* (dicarboxylic acids). Thus a solution of lactose (100 g) in 1200 mL of dilute nitric acid (25%; d 1.15) evaporated on a steam bath to a volume of 200 mL gives 40 g (67% yield) of mucic acid (*galacto*-tetrahydroxyadipic acid) [467].

Oxidation of Aldehyde Derivatives

Acetals of aldehydes are converted into esters by the action of **ozone** [111] or **peroxy acids** [277, 327]. Butyraldehyde diethylacetal, when treated for 30 min with peroxyacetic acid and sulfuric acid as a catalyst, furnishes ethyl butyrate in a 69% yield after 11 h at 40 °C [277]. Enanthaldehyde (heptanal) dimethyl acetal reacts with ozone in dichloromethane at –78 °C for 15 h or at room temperature for 1.5 h to produce methyl heptanoate in 90% yield (equation 369) [111].

[111]   369

$$C_6H_{13}CH(OCH_3)_2 \xrightarrow[\text{-78 °C, 15 h}]{O_3, CH_2Cl_2} \left[C_6H_{13}C(OCH_3)(O-CH_3)(H)(O-O) \right] \longrightarrow C_6H_{13}C(OCH_3)(=O) \quad 90\%$$

Cyclic acetals react faster; their reactions require less than 2 h at –78 °C [111]. Tetrahydropyranyl ethers are cleaved to hydroxy esters (equations 370 and 371) [111].

[111]   370

$$C_6H_{13}CH\langle O-CH_2/O-CH_2 \rangle \xrightarrow[\text{-78 °C, 10 min}]{O_3, CH_2Cl_2} C_6H_{13}CO_2CH_2CH_2OH \quad 100\%$$

[111]   371

(tetrahydropyran-2-yl methyl ether) $\xrightarrow[\text{-78 °C, 1.5 h}]{O_3, AcOEt}$ (5-hydroxypentyl acetate)   97%

The oxidation of glycosides to esters by ozone is stereoselective. It affects only the β anomers; α anomers do not react [111]. Ozone can also be used to deprotect ethylidene and benzylidene glycosides, because it cleaves the acetal bonds (equation 372) [111].

m-**Chloroperoxybenzoic acid** converts γ-lactols into lactones. δ-Lactols give low yields (equation 373) [327].

Unsaturated aldehyde acetals may be transformed either into *unsaturated esters* or into *epoxy acetals*. **Peroxyacetic acid** in ethyl acetate epox-

[111]   372

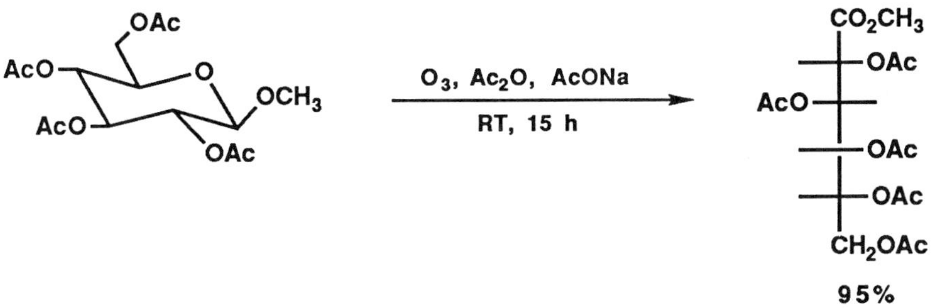

[Structure: tetra-acetylated methyl glycoside] → O₃, Ac₂O, AcONa / RT, 15 h →

CO₂CH₃
—OAc
AcO—
—OAc
—OAc
CH₂OAc

95%

[327]   373

[C₄H₉CH-CH₂-CH₂-O-CH(OCH₃)- cyclic acetal] → m-ClC₆H₄CO₃H, BF₃·Et₂O / CH₂Cl₂, RT, 3 h →

C₄H₉CH with CH₂-CH₂ ring, O—C=O

94%

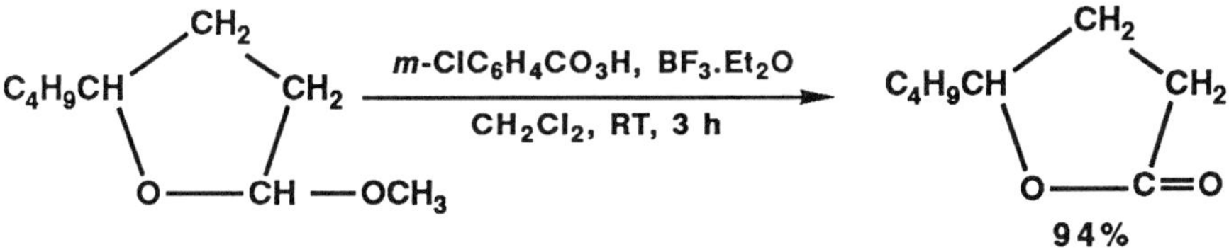

idizes 2-(1-propenyl)-4-methyl-1,3-dioxolane after 11 h in 46% yield [277]. On the other hand, crotonaldehyde dibutylacetal, with the same reagent and solvent at 60 °C, gives butyl crotonate in 73% yield (equations 374 and 375) [277].

[277]   374

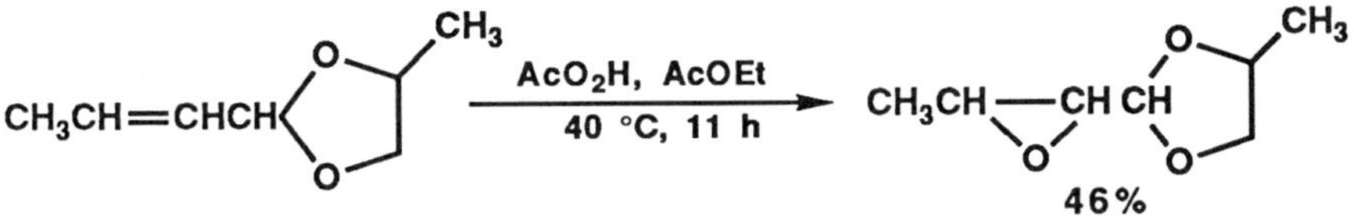

CH₃CH=CHCH(dioxolane with CH₃) → AcO₂H, AcOEt / 40 °C, 11 h → CH₃CH—CHCH(dioxolane with CH₃) with epoxide

46%

[277]   375

CH₃CH=CHCH(OC₄H₉)₂ → AcO₂H, AcOEt / 60 °C, 1 h → CH₃CH=CHCO₂C₄H₉

73%

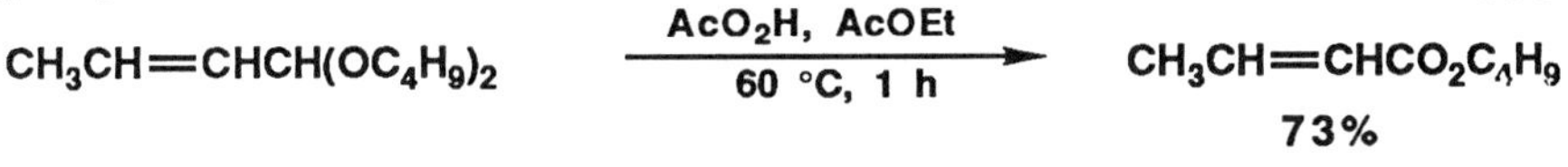

The treatment of **acrolein diethyl acetal** with **potassium permanganate** results in the **hydroxylation** of the double bond (equation 376) [871].

[871]   376

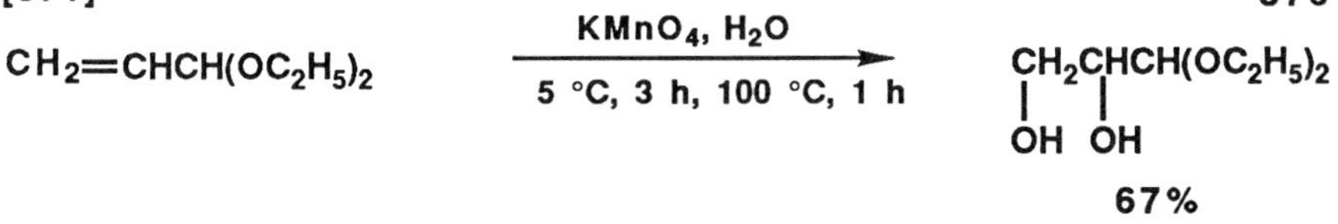

CH₂=CHCH(OC₂H₅)₂ → KMnO₄, H₂O / 5 °C, 3 h, 100 °C, 1 h → CH₂CHCH(OC₂H₅)₂
                                                                  | |
                                                                  OH OH

67%

KETONES

The results of the oxidation of ketones depend on the oxidants used. **Aqueous hydrogen peroxide,** in the absence of mineral acid, converts cyclohexanone into 1,1'-dihydroxydicyclohexyl *peroxide,* whereas with hydrochloric acid as a catalyst, 1-hydroxy-1'-hydroperoxydicyclohexyl peroxide is formed (equation 377) [*181*].

[*181*]   377

Dicycloalkyl-2,2'-diones prepared from cyclopentanone, cyclohexanone, 4-methylcyclohexanone, and cycloheptanone undergo peroxidation with hydrogen peroxide to yield homologues of 3,6-dihydroxy-1,2-dioxanes (equation 378) [*180*].

[*180*]   378

Oxidation of Ketones to Esters (Baeyer–Villiger Reaction)

The oxidation of ketones to esters is known as the ***Baeyer–Villiger reaction*** and is effected mainly by hydrogen peroxide and peroxy acids, especially the organic peroxy acids.

The mechanism is the same as that of the ***Dakin reaction*** of aromatic aldehydes (equation 359). The reaction is acid-catalyzed. The peroxy acid transfers oxygen onto the carbon of the carbonyl group and generates an unstable intermediate. A rearrangement of the groups bonded to the original carbonyl carbon results in the formation of *esters* or, with cyclic ketones, *lactones* [*262, 303*] (equation 379).

Because migration of the groups attached to the carbonyl carbon takes place with the bonding electron pair, the migratory aptitudes of the two

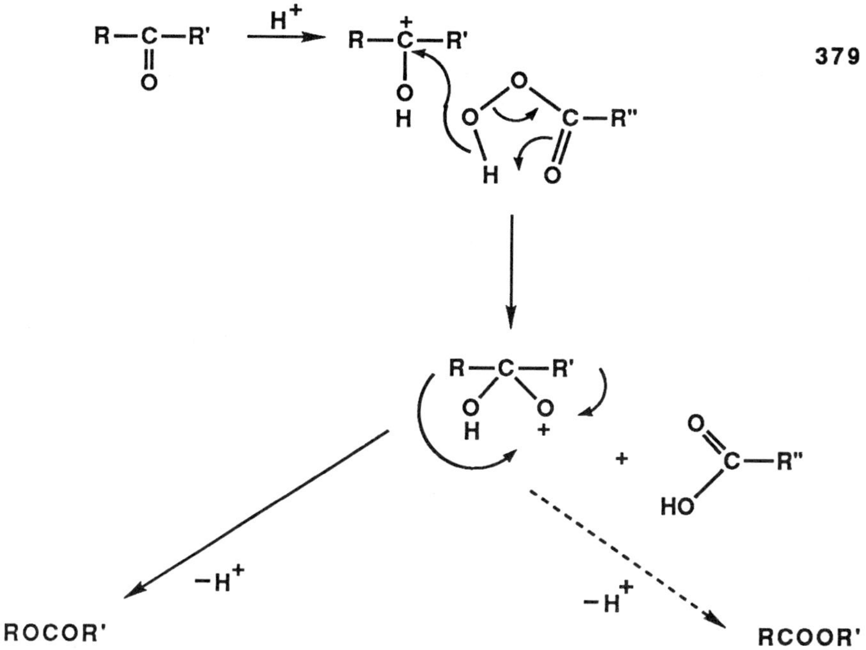

$$\text{R-C(=O)-R'} \xrightarrow{H^+} \text{R-}\overset{+}{\text{C}}\text{-R'} \quad \textbf{379}$$

ROCOR'   RCOOR'

groups are determined by their relative nucleophilicities. Such group will migrate preferentially that has more nucleophilic character, that is, higher electron density owing to electrons or electron-releasing substituents.

Although there are deviations and differences due to the type of ketones, the migratory aptitudes can be arranged in the following sequence [282, 284]: tert-butyl > cyclohexyl ~ secondary alkyl ~ benzyl ~ phenyl > primary alkyl > cyclopropyl >> methyl [282, 284, 286, 304]. In substituted benzophenones, the benzene ring with electron-releasing groups rearranges preferentially [1166]. The migration occurs intramolecularly. As a consequence, the rearrangement takes place with complete retention of configuration and chirality [306, 1034].

Hydrogen peroxide alone rarely converts ketones into esters [148, 1167]. It is used usually in the presence of an acid [128] or as a reoxidant [343] (equations 380 and 381).

Its derivative, bis(trimethylsilyl) peroxide, effects the **Baeyer–Villiger reaction** in aprotic solvents in the presence of trimethylsilyl trifluoromethanesulfonate [237] (see equation 388).

[128] cyclobutanone $\xrightarrow[\text{RT, 24 h}]{\text{30\% H}_2\text{O}_2,\ \text{CF}_3\text{CH}_2\text{OH}}$ γ-butyrolactone **380**

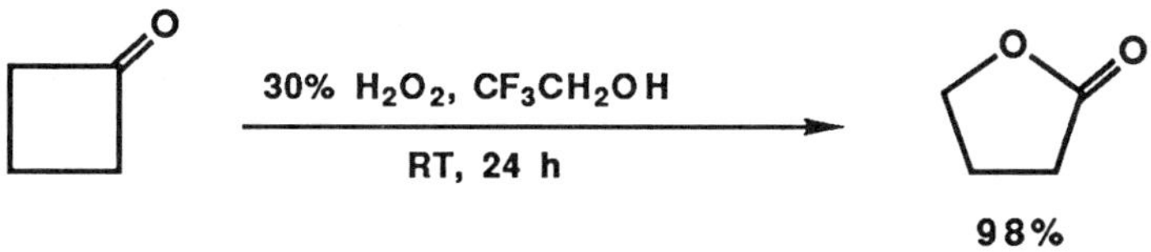

98%

[343]

2-methylcyclohexanone → (1.25 PhSe(O)OH, 10 mol of 30% H$_2$O$_2$, CH$_2$Cl$_2$, RT, 1 h) → ε-methyl-ε-caprolactone **381**

83%

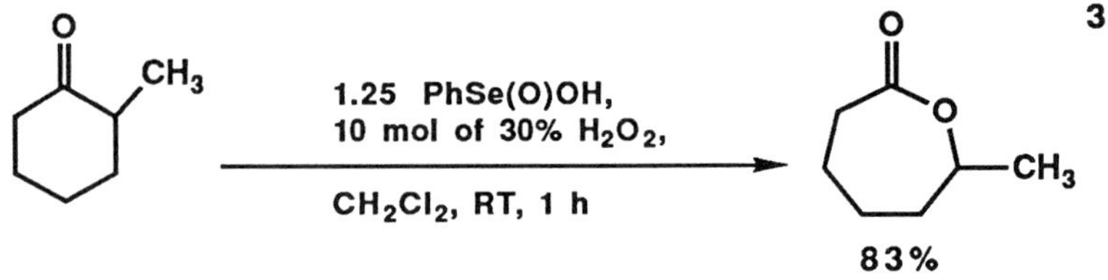

Potassium peroxymonosulfate oxidizes benzophenone to phenyl benzoate (yield 77%), cyclopentanone to δ-valerolactone (yield 35%), and cyclohexanone to ε-caprolactone (yield 46%) [205]. In an excess of sulfuric acid and in the presence of alcohol, the δ-valerolactone is immediately transesterified to ethyl δ-hydroxyvalerate [201] (equation 382).

382

[205] 2 KHSO$_5$·KHSO$_4$·K$_2$SO$_4$, H$_2$O, 0–5 °C, 30 min ← cyclopentanone → K$_2$S$_2$O$_8$, H$_2$SO$_4$, EtOH, −26 °C, 26 h [201]

δ-valerolactone 35%

HO(CH$_2$)$_4$CO$_2$C$_2$H$_5$

quantitative yield

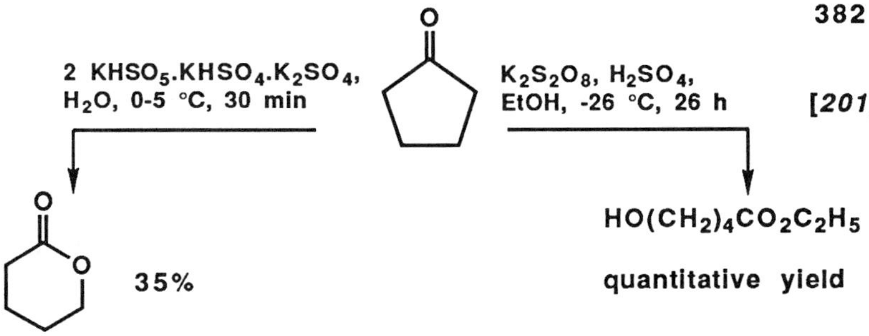

Peroxyacetic acid is used frequently to oxidize all kinds of ketones to esters [261] and lactones [262]. The usual concentration of commercial peroxyacetic acid is 40%, but acids of concentrations of 20–28% can also be used. The reactions are exothermic and are run at temperatures ranging from room temperature to 40 °C, sometimes in the presence of sulfuric acid [261]. In oxidations of aromatic ketones, the rings containing activating substituents, such as methyl and methoxy groups, migrate preferentially. In benzophenones containing a phenyl group and chlorophenyl, bromophenyl, and nitrophenyl groups, the phenyl group migrates preferentially [261]. Cyclopentanone and cyclohexanone are oxidized to δ-valerolactone and ε-caprolactone in 84 and 70–94% yields, respectively. Little polymerization of the lactone to polyesters was observed [262]. Cyclohexanones with alkyl groups in the α position with respect to the carbonyl give consistently ε-alkyl-ε-caprolactones, in agreement with the results obtained with other reagents [262, 343] (equation 383).

An important addition to the arsenal of oxidants for the Baeyer–Villiger reaction is **peroxytrifluoroacetic acid** [282, 283, 284]. Although this reagent is less easily accessible than peroxyacetic acid and aromatic peroxy acids, it is more reactive. The yields of esters obtained by oxidation of ketones with peroxytrifluoroacetic acid are high enough to justify the use of this oxidant for the quantitative determination of aldehydes and

[262]

25% AcO₂H, AcOEt → 383

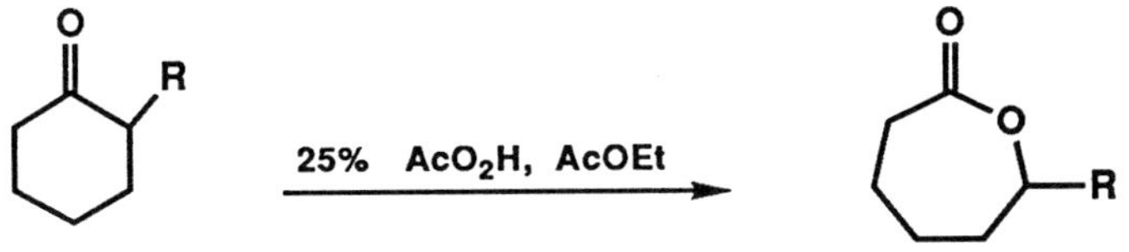

R	Time (h)	Temperature (°C)	Yield (%)
H	6.5	40	90
Me	8.5	40	92
i-Pr	9	50	84.5
s-Bu	13	50	92
cyclo-C_6H_{11}	10	50	82

ketones with an accuracy within 3–10% [280]. The disadvantage is the necessity of 90% hydrogen peroxide for the preparation of peroxytrifluoroacetic acid from trifluoroacetic anhydride [283, 284] (equation 384).

[284]

▷—COCH₃ $\xrightarrow{\text{90\% H}_2\text{O}_2,\ (CF_3CO)_2O,\ CH_2Cl_2,\ Na_2HPO_4}_{\text{RT, 25 min, reflux 1 h}}$ ▷—OCOCH₃   384

53%*

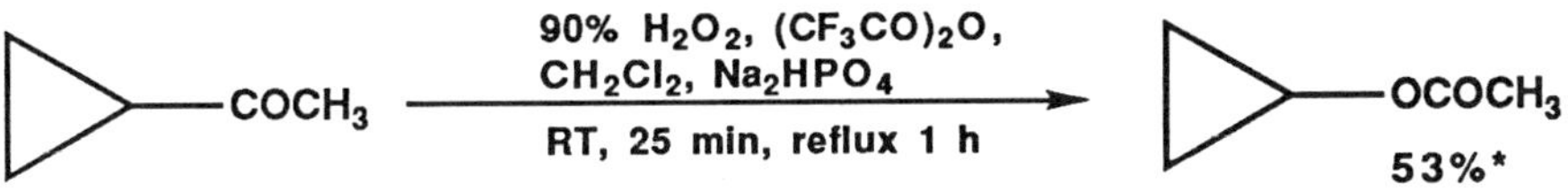

*The purity of the product is 96%, 4% being methyl cyclopropanecarboxylate.

Peroxybenzoic acid is used to oxidize cyclanones [303], as well as alkyl and phenyl cycloalkyl ketones [304, 305]. Yields of cycloalkyl acetates obtained on treatment of cycloalkyl methyl ketones with a 10–15% stoichiometric excess of peroxybenzoic acid in chloroform at room temperature range from 58 to 72% (equation 385) [305].

[305]

$(CH_2)_n$CHCOCH₃ $\xrightarrow[\text{23-25 °C}]{\text{BzO}_2\text{H, CHCl}_3}$ $(CH_2)_n$CHOCOCH₃   385

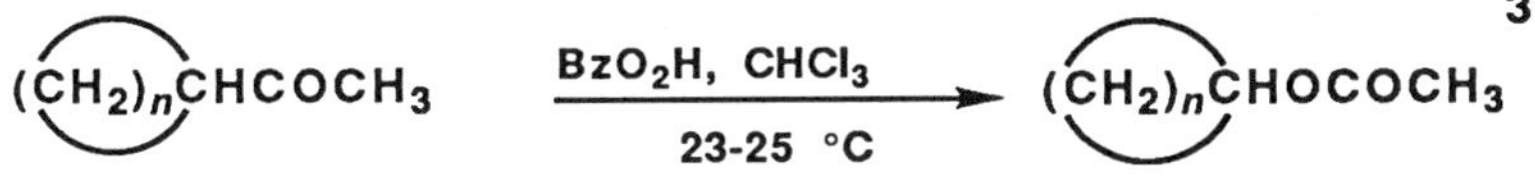

n	Time (h)	Yield (%)
3	45	58
4	73	61
5	77	72
6	49	69

Cyclohexyl phenyl ketone oxidized by peroxybenzoic acid in chloroform at 24–26 °C for 239 h furnishes an 85% yield of esters, of which 71% is cyclohexyl benzoate and 14% is phenyl cyclohexanecarboxylate. This result shows that the migratory aptitude of the cyclohexyl group is 5 times greater than that of the phenyl group [304].

***m*-Chloroperoxybenzoic acid** gives almost identical results as peroxytrifluoroacetic acid in the oxidation of alkyl cyclopropyl ketones [*286*]. On treatment of methyl cyclopropyl ketone with a dichloromethane solution of *m*-chloroperoxybenzoic acid at 25 °C for 35 days or with a dichloromethane solution of peroxytrifluoroacetic acid at room temperature for 25 min and at reflux for 1 h, the ratio of cyclopropyl acetate to methyl cyclopropanecarboxylate is 95:5 or 96:4, respectively. In contrast, the corresponding ratios of cyclopropyl alkanoate to alkyl cyclopropanecarboxylate are 1:99 and 6:94, respectively, for the two acid treatments of isopropyl cyclopropyl ketone and 3:97 for phenyl cyclopropyl ketone (the product ratios are the same with both acids) [*286*].

Peroxymaleic acid, prepared from 90% hydrogen peroxide and maleic anhydride in dichloromethane with cooling with ice, oxidizes ketones to esters or lactones in 40–83% yields on refluxing in dichloromethane for 2–12 h [*338*]. Because the peroxymaleic acid solution in dichloromethane decomposes to the extent of 5% in 6 h at room temperature and because the oxidation of ketones with this reagent requires refluxing in dichloromethane, peroxymaleic acid does not offer any advantages over the more readily accessible peroxy acids (equation 386).

[*338*]   386

$$CH_3COC_6H_{13} \xrightarrow[\text{cooling with ice; reflux 2 h}]{90\% \ H_2O_2, \ \underset{CH-CO}{\overset{CH-CO}{\|}}\!\!>\!\!O, \ CH_2Cl_2} CH_3CO_2C_6H_{13}$$

71%

Another dicarboxylic peroxy acid, **peroxyphthalic acid,** is used for the Baeyer–Villiger reaction of β-diketones and α-keto esters [*334*]. Besides peroxy acids, *ceric ammonium nitrate* oxidizes 2-adamantanone to 2-oxahomoadamantanone in aqueous acetonitrile in 73% yield at 60 °C after 3 h [*422*]. Tetracyclone (2,3,4,5-tetraphenyl-2,4-cyclopentadien-1-one) is oxidized by ceric ammonium nitrate in aqueous acetonitrile to tetraphenyl-2-pyrone in 77% yield [*422*].

The Baeyer–Villiger reaction is also effected by ***biochemical oxidation*** using the enzyme **cyclohexanone oxygenase** from *Acinetobacter* strain NCIB 9871. Cyclohexanone is thus converted into ε-caprolactone [*1043*], and phenylacetone (1-phenyl-2-propanone) is transformed into benzyl acetate. The formation of benzyl acetate from phenylacetone involves the same migration as that in oxidation with peroxytrifluoroacetic acid (equation 387) [*1034*]. More examples of biochemical Baeyer–Villiger reactions occur in diketones and steroids (*see* equation 397).

[*1034*]   387

$$C_6H_5CH_2COCH_3 \xrightarrow[\text{Enz-FAD, NADPH, H}^+]{\text{cyclohexanone oxygenase, } O_2} C_6H_5CH_2OCOCH_3$$

Baeyer–Villiger Oxidation of Functionalized Ketones

Because of the common mechanisms and the same specific oxidants, the *Baeyer–Villiger oxidation* of ketones possessing other functional groups will be mentioned in this section rather than in the places where it should be discussed according to the system of the book.

The results of oxidation of *unsaturated ketones* depend on the position of the double bond with respect to the carbonyl group and on the oxidant.

An example showing the behavior of a ketone containing a nonconjugated double bond is the oxidation of 2-allylcyclohexanone. Hydrogen peroxide in benzonitrile produces exclusively an epoxide [*148*], whereas peroxyacetic acid [*148*] or bis(trimethylsilyl) peroxide [*237*] yields the lactone of 6-hydroxy-8-nonenoic acid (equation 388).

Most of the discussion of the Baeyer–Villiger reaction of unsaturated ketones is devoted to α,β-unsaturated ketones, such as mesityl oxide [*254*], benzalacetophenone [*307*], and, especially, benzalacetone [*258*]. The oxidation of ionones does not involve the Baeyer–Villiger reaction and is, therefore, discussed elsewhere (*see* equations 440 and 441).

The treatment of mesityl oxide with 45% peroxyacetic acid in chloroform at 20–25 °C for 4.5 h gives a 53% yield of a mixture containing 22% of 4-methyl-3,4-epoxypentan-2-one and 78% of 2-acetoxy-3,3-dimethyloxirane (the combined products of epoxidation and the Baeyer–Villiger reaction) (equation 389) [*254*].

The predominance of oxidation at the keto group over epoxidation in α,β-unsaturated compounds carrying phenyl groups in β positions with respect to carbonyls is mechanistically explained for benzalacetophenone [*258*] and proven by the oxidation of 3,4-diphenyl-3-buten-2-one [*307*] (equations 390 and 391).

The same reaction that converts ketones into lactones transforms α-

192   Oxidations in Organic Chemistry

[*254*]   (CH$_3$)$_2$C=CHCOCH$_3$ 389

           | 45% AcO$_2$H,
           | CHCl$_3$,
           | 20-25 °C,
           | 4.5 h
           ↓

(CH$_3$)$_2$C——CHCOCH$_3$ +   (CH$_3$)$_2$C——CHOCOCH$_3$
        \O/   \O/
        12%   41%

[*258*]   390

C$_6$H$_5$CH=CH—C(=O)—CH$_3$ →$^{H^+}$ C$_6$H$_5$CH=CH—C(=$^+$O—H)—CH$_3$ →

                                                    with H—O—O—C(=O)—R

C$_6$H$_5$CH=CH—C(O—H)(O$^+$)—CH$_3$ →$^{-H^+}$ C$_6$H$_5$CH—CH—C(O—H)—CH$_3$
                                                      \O/
                                                              ↓
                                                   C$_6$H$_5$CH=CHOCOCH$_3$

[*307*]   C$_6$H$_5$CH=C—COCH$_3$ 391
                              |
                              C$_6$H$_5$
                    BzO$_2$H | CHCl$_3$, 27°
                              ↓

C$_6$H$_5$CH——C—OCOCH$_3$ C$_6$H$_5$CH=C—OCOCH$_3$ + C$_6$H$_5$CH——CCOCH$_3$
       \O/ | | \O/ |
            C$_6$H$_5$ C$_6$H$_5$ C$_6$H$_5$

      ↓ H$^+$ ↓ H$^+$ ↓ H$^+$

C$_6$H$_5$CHCOC$_6$H$_5$ C$_6$H$_5$CH$_2$COC$_6$H$_5$ C$_6$H$_5$CHO → C$_6$H$_5$CO$_2$H
      |
      OH   14%   70%   15%

diketo compounds into acid anhydrides, which are sometimes isolable and sometimes hydrolyzed immediately to the corresponding acids.

The treatment of diphenylcyclobutene-1,2-dione with hydrogen peroxide in carbon tetrachloride affords a 79% yield of diphenylmaleic anhydride (equation 392) [*1167*].

[*1167*]   392

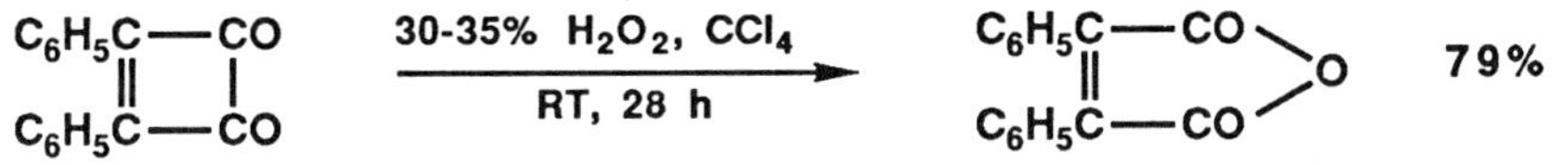

$$C_6H_5C\text{—}CO \atop C_6H_5C\text{—}CO \quad \xrightarrow[\text{RT, 28 h}]{30\text{-}35\% \ H_2O_2, \ CCl_4} \quad \begin{matrix} C_6H_5C\text{—}CO \\ \parallel \\ C_6H_5C\text{—}CO \end{matrix}\!\!>\!\!O \quad 79\%$$

β-Naphthoquinone reacts exothermically with 9% peroxyacetic acid in acetic acid and gives an 83% yield of the anhydride of *o*-carboxyallocinnamic acid [*271*] (equation 393). A similar oxidation of 9,10-diketostearic acid gives a 90% yield of pelargonic acid and a 95% yield of azelaic acid [*271*].

[*271*]   393

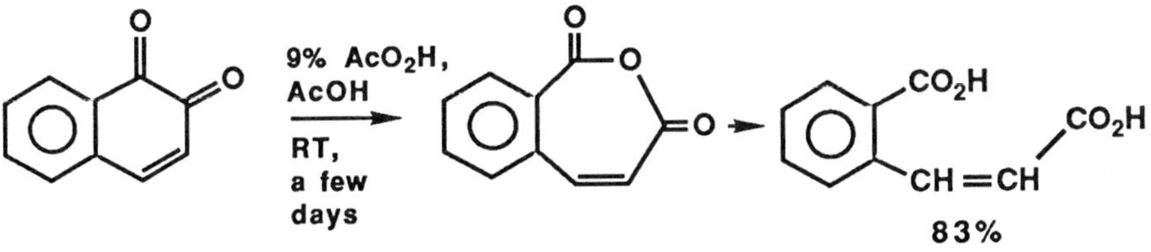

The oxidation of ethyl pyruvate and ethyl phenylglyoxylate with peroxyphthalic acid at 20–22 °C produces, after 48 and 90 h, mixed anhydrides of ethylcarbonic acid and acetic and benzoic acids, respectively [*334*].

Isatin, which can be considered a cyclic α-keto amide, is oxidized by hydrogen peroxide exclusively to isatoic anhydride, whereas with potassium peroxydisulfate, 2,3-dioxo-1,4-benzoxazine is obtained (equation 394) [*260*].

[*260*]   394

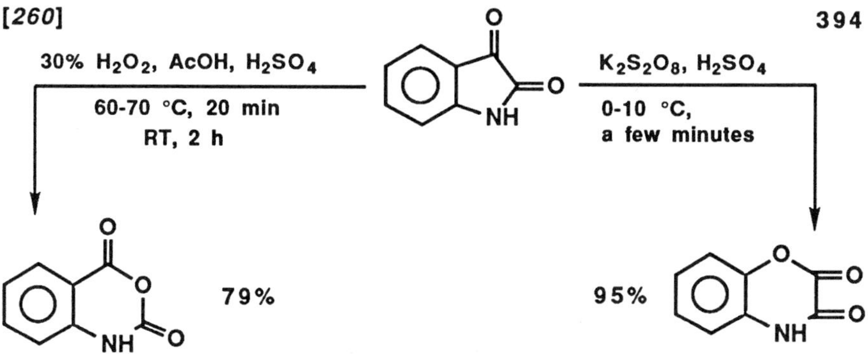

Some isatoic acid anhydrides are readily hydrolyzed during oxidation so that the corresponding *N*-(2-hydroxyphenyl)oxamidic acids are isolated as products [260]. With 7-chloroisatin, treatment with 30% hydrogen peroxide in 5% aqueous sodium hydroxide for 40 min at 65 °C produces chloroanthranilic acid in 78% yield (equation 395) [1168].

[1168]   395

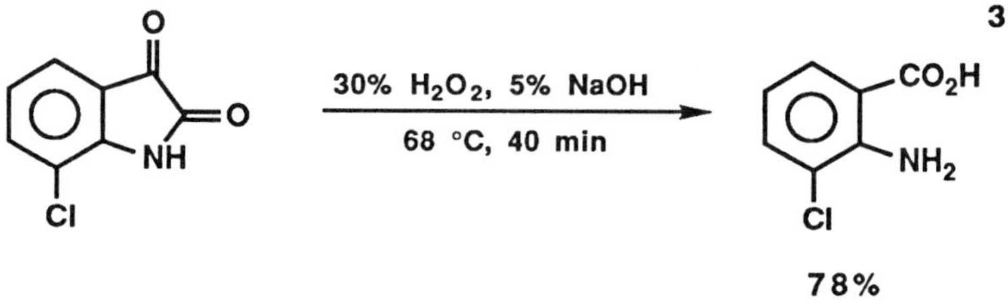

$$\underset{\text{Cl}}{\text{7-chloroisatin}} \xrightarrow[\text{68 °C, 40 min}]{\text{30\% H}_2\text{O}_2,\ 5\% \text{ NaOH}} \underset{\text{Cl}}{\text{chloroanthranilic acid}}$$

78%

The Baeyer–Villiger reaction of **hydroxy ketones** and **unsaturated diketones** is very common in steroids, where it occurs preferentially at the keto group on C-17 and leaves the carbonyls in position 3 intact.

Chemical oxidation is achieved with bis(trimethylsilyl) peroxide in the presence of trimethylsilyl trifluoroacetate (triflate). Androsterone acetate is converted into androstenolactone acetate in 86% yield (equation 396) [237].

[237]   396

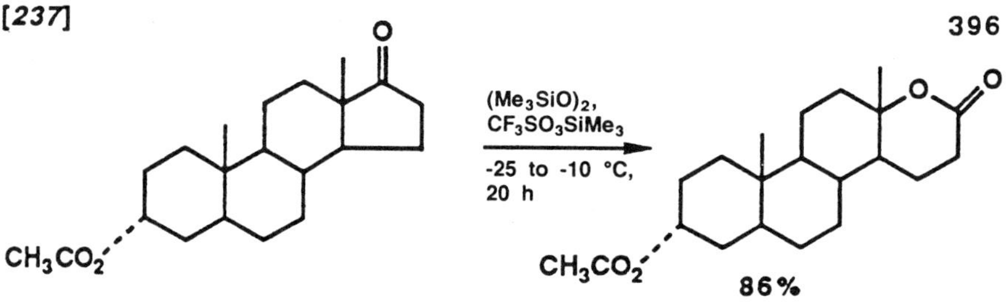

$$\text{androsterone acetate} \xrightarrow[\substack{-25 \text{ to } -10 \text{ °C,}\\ 20 \text{ h}}]{(\text{Me}_3\text{SiO})_2,\ \text{CF}_3\text{SO}_3\text{SiMe}_3} \text{androstenolactone acetate}$$

86%

Most of the **Baeyer–Villiger oxidations** of steroids are accomplished **biochemically**. 19-Nortestosterone treated with ***Aspergillus tamarii*** furnishes a 70% yield of 19-nortestololactone [1049]. Progesterone and testosterone are converted into Δ^1-dehydrotestololactone by fermentation with ***Cylindrocarpon radicicola*** (in ref. 1065, *radicola*) [1065]. Testololactone is obtained from progesterone by oxidation with ***Penicillium chrysogenum*** [1065] and from 4-androstene-3,17-dione by treatment with ***Penicillium lilacinum*** [1072] (equation 397).

Other examples of the **Baeyer–Villiger reaction** are given in equations 398–399.

397

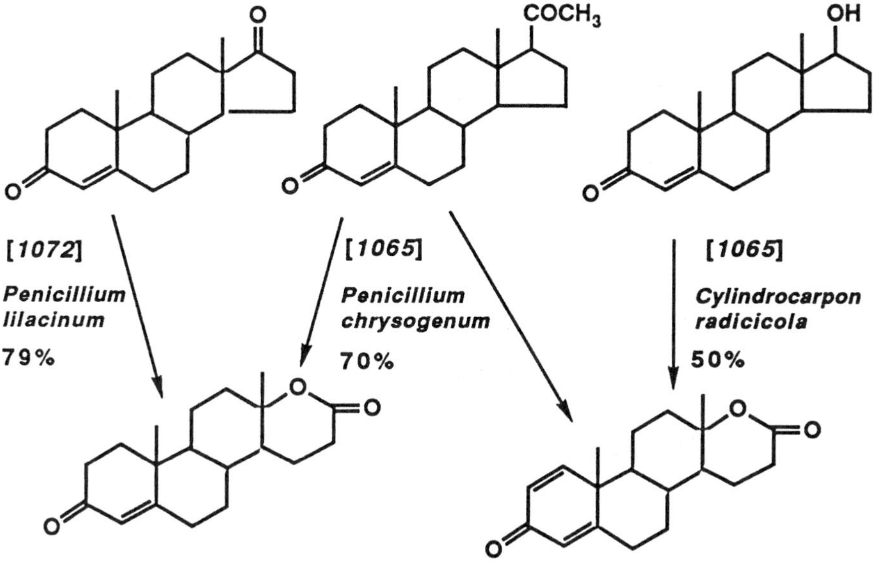

EXAMPLES OF THE BAEYER-VILLIGER REACTION

[1072] *Penicillium lilacinum* 79%

[1065] *Penicillium chrysogenum* 70%

[1065] *Cylindrocarpon radicicola* 50%

398

Cyclohexanone → ε-caprolactone

Ref	Conditions	Yield
[205]	2 KHSO$_5$·KHSO$_4$·K$_2$SO$_4$; H$_2$O, 0-5 °C, 30 min	46%
[262]	20-28% AcO$_2$H, 40 °C, 6.25 h	85%
[303]	BzO$_2$H, CHCl$_3$, 22-25 °C, 6.5 h	71%
[237]	(Me$_3$SiO)$_2$, CF$_3$SO$_3$SiMe$_3$ (catalyst), CH$_2$Cl$_2$, -78 °C, 6 h; -40 °C, 1 h	76%
[1043]	Cyclohexanone oxygenase from *Acinetobacter* strain NCIB 9871	28%

C$_6$H$_5$COC$_6$H$_5$ ⟶ C$_6$H$_5$COOC$_6$H$_5$ **399**

Ref	Conditions	Yield
[205]	2KHSO$_5$·KHSO$_4$·K$_2$SO$_4$, H$_2$SO$_4$, AcOH, <35 °C, 30 min	77%
[261]	40% AcO$_2$H, AcOH, H$_2$SO$_4$, RT, 0.5 h	82%
[284]	90% H$_2$O$_2$, (CF$_3$CO)$_2$O, CH$_2$Cl$_2$, Na$_2$HPO$_4$, RT, 25 min reflux 1 h	88%
[338]	HO$_2$CCH=CHCO$_3$H, CH$_2$Cl$_2$, reflux, 2 h	70%

Hydroxylation of Ketones

Ketones having at least one hydrogen next to the carbonyl group can be hydroxylated to form α-hydroxy ketones (acyloins). Acetophenone is oxidized to α-hydroxyacetophenone by **o-iodosobenzoic acid.** In the presence of potassium hydroxide and methanol, dimethyl acetal is obtained, which on subsequent hydrolysis gives the hydroxy ketone in 83% yield (equation 400) [*790*].

[*790*] 400

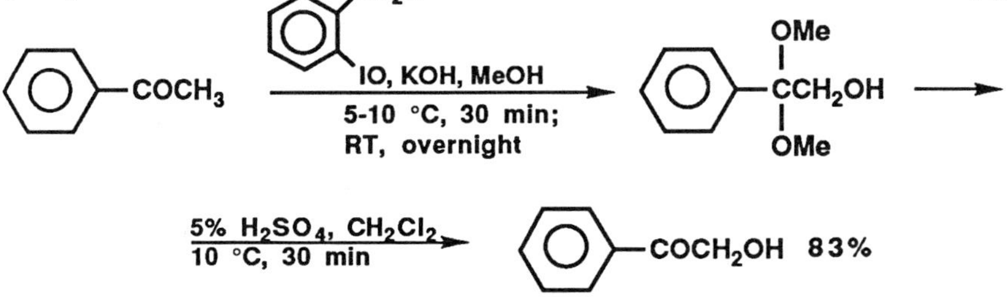

$$C_6H_5\text{-}COCH_3 \xrightarrow[\text{5-10 °C, 30 min;}]{\text{IO, C}_6H_4\text{CO}_2H,\text{ KOH, MeOH}} C_6H_5\text{-}C(OMe)_2CH_2OH \xrightarrow[\text{10 °C, 30 min}]{\text{5\% H}_2SO_4,\text{ CH}_2Cl_2} C_6H_5\text{-}COCH_2OH \quad 83\%$$

The conversion of ketones into α-hydroxy ketones can be achieved by the oxidation of enolates or enol ethers. A special reagent for enolates is the *oxodiperoxy molybdenum complex with pyridine and hexamethylphosphoramide*. The reaction is applied to aromatic aliphatic ketones and cyclic ketones and furnishes 34–81% yields of α-hydroxy ketones with up to 26% of α-diketones (equation 401) [*531*].

[*531*] 401

$$C_6H_5COCH_2C_3H_8 \xrightarrow[\text{2. MoO}_5 \cdot C_5H_5N \cdot \text{HMPA, -44 °C}]{\text{1. LDA, THF, -44 °C}} \underset{\underset{62\%}{OH}}{C_6H_5COCHC_3H_8}$$

Trimethylsilyl enol ethers prepared in 60–91% yields from ketone enolates and trimethylsilyl chloride are converted into α-hydroxy ketones by **chromyl chloride** in 62–82% yields (equation 402) [*676*] (equation 340).

[*676*] 402

$$RCOCH_3 \xrightarrow[\substack{-78\text{ °C,} \\ -30\text{ °C to RT}}]{\text{LDA, THF, Me}_3SiCl} RC(OSi(CH_3)_3)=CH_2 \xrightarrow[\substack{CH_2Cl_2, \\ -78\text{ °C,} \\ 30\text{ min}}]{\text{CrO}_2Cl_2,} RCOCH_2OH$$

R	Yield (%)	
$C(CH_3)_3$	62	82
C_6H_5	85	62

Hydroxylations of steroidal and unsaturated steroidal ketones in different positions are accomplished by **microorganisms,** a fact that shows how much ahead microorganisms are in synthetic chemistry as far as regio-

and stereoselectivity are concerned. The most favored position of hydroxylation is 11α [*1045, 1060, 1061, 1075, 1076, 1077, 1078, 1079*] followed by 11β [*1062, 1067*], 6β [*1062, 1066*], 7α [*1064*], 10β [*1062*], 12α [*1062*], 12β [*575*], 13β [*1062*], 14α [*1062*], 15α [*1064, 1066*], 15β [*575, 1051, 1064*], 17α [*1060*], and 17β [*1062*]. Hydroxylations are applied to progesterone [*575, 1075, 1076*], deoxycorticosterone [*1045, 1064*], pregnanedione [*1078*], androstenedione [*1066*], and their derivatives.

The biochemical hydroxylations are carried out under the conditions required for the cultivation of individual microorganisms. The common denominator in the cultivation of microorganisms is an aerated aqueous solution containing nutrient material, buffered to the required pH, and kept at a temperature of 26 to 28 °C for a few days or weeks. The products are isolated by extraction with dichloromethane, chloroform, ethyl acetate, and the like. The yields are usually low, and chromatography is often used in isolations.

The oxidation of progesterone with **Rhizopus arrhizus** yields 11α-hydroxyprogesterone [*1075*], together with 6β,11α-dihydroxyprogesterone [*1076*]. In contrast, **Calonectria decora** converts progesterone into 12β,15β-dihydroxyprogesterone [*575*] (equation 403).

403

6-Dehydroprogesterone [*1077*] and 3,17-pregnanedione [*1078*] are hydroxylated in the 11α position by **Rhizopus nigricans** in 50–60 and 39% yields, respectively. The same microorganism gives a 73.5% yield of 11α-hydroxy-16α,17α-epoxyprogesterone from 16α,17α-epoxyprogesterone [*1079*].

Desoxycorticosterone (4-pregnen-21-ol-3,20-dione) is converted into Δ^4-pregnene-11α,17α-diol-3,20-dione by **Aspergillus niger** in 67% yield [*1045*] and to 7α-hydroxydesoxycorticosterone in 36% yield by **Peziza sp.** ETH,M.23 [*1064*]. 17α-Hydroxydesoxycorticosterone is hydroxylated in the 11α position by **Cunninghamella blakesleeana** H-334 [*1060, 1061*] and in the 15β position by **Bacillus megaterium** in 50% yield [*1051*] (equations 404 and 405).

Δ^4-Androstene-3,17-dione gives a 24% yield of 15α-hydroxy-Δ^4-androstene-3,17-dione on treatment with **Fusarium lini** at 25–27 °C [1066] and a mixture of the same compound (yield 33%) together with 6β-hydroxy-Δ^4-androstene-3,17 dione (yield 20%) on incubation with **Gibberella saubinetii** (in ref. 1066, *saubinetti*) [1066, 1067] (equation 406).

Esters of hydroxy ketones are prepared by acetoxylation with lead tetraacetate in benzene. Cyclohexanone furnishes a 75% yield of 2-acetoxycyclohexanone and a 3% yield of *cis*-2,6-diacetoxycyclohexanone after heating at 80 °C for 8 h [438]. Similarly, 3α-hydroxy-5β-pregnan-20-one yields 64% of 21-acetoxy-3α-hydroxy-5β-pregnan-20-one [439] after 4 h at room temperature.

Oxidation of Ketones to α-Dicarbonyl Compounds

Methyl and methylene groups adjacent to carbonyl groups are easily oxidized to carbonyls to yield α-keto aldehydes or α-diketones. The reagent of choice is **selenium dioxide** or *selenious acid*. The reaction is catalyzed by acids and by acetate ion and proceeds through transition states involving enols of the carbonyl compounds [518]. The oxidation is carried out by refluxing the ketone with about 1.1 mol of selenium dioxide in water, dilute acetic acid, dioxane, or aqueous dioxane [517]. The byproduct, black selenium, is filtered off, but small amounts of red selenium sometimes remain in a colloidal form and cannot be removed even by distillation of the product. Shaking the product with mercury [523] or Raney nickel [524] takes care of the residual selenium. The α-dicarbonyl compounds are yellow oils that avidly react with water to form white crystalline hydrates (equations 407 and 408).

[517] p-CH$_3$OC$_6$H$_4$COCH$_2$C$_6$H$_5$ $\xrightarrow[\text{reflux 5-8 h}]{\text{1.1 SeO}_2\text{, dioxane, H}_2\text{O}}$ 83%   p-CH$_3$O$_6$H$_4$COCOC$_6$H$_5$ 407

[518] $\xrightarrow[\text{89 °C, 12 h}]{\text{1.1 SeO}_2\text{, 70% AcOH}}$ 100%

[508] cyclohexanone $\xrightarrow[\text{cooling, 3 h, RT, 11 h}]{\text{H}_2\text{SeO}_3\text{, H}_2\text{O, dioxane}}$ 1,2-cyclohexanedione 60%   408

Refluxing 2-acetylmesitylene with selenium dioxide in aqueous dioxane for 5 h yields 82.5% of mesitylglyoxal as a yellow oil [516]. The reaction is applicable to acetophenones [520, 678, 1001]; to deoxybenzoin, its homologues, and their derivatives [517, 518]; and to cyclic ketones [508]. α-Dicarbonyl compounds are produced in good yields (equation 409).

[520] X=H X–C$_6$H$_4$–COCH$_3$ $\xrightarrow[\text{reflux 5 h}]{\text{SeO}_2\text{, H}_2\text{O, dioxane}}$ 69-72% X–C$_6$H$_4$–COCHO   409

X=Br $\xrightarrow[\text{55 °C, 24 h}]{\text{Me}_2\text{SO (DMSO), 48% HBr}}$ 86%
[1001]

The oxidation of methylene groups to carbonyls is especially easy if the methylenes are flanked by two carbonyl groups. The methylene group in barbituric acid is oxidized to the keto group by *chromium trioxide* either directly [555] or after reaction with benzaldehyde [554] (equation 410).

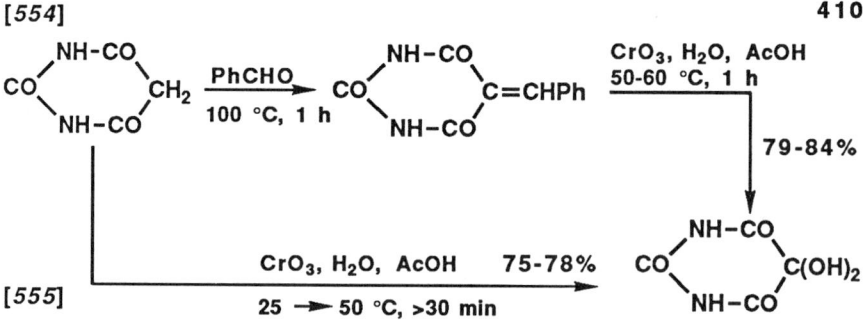

[554]   410

[555]

Oxidations of α-methyl and α-methylene groups are also effected by **dimethyl sulfoxide** in the presence of aqueous hydrobromic acid [1001] or hydrogen bromide [1002] (equation 409).

Heating 1,3-indanedione with dimethyl sulfoxide and anhydrous hydrogen bromide at 70–90 °C, removing the resulting dimethyl sulfide by distillation, and heating the residue for 1 h at 100 °C with 0.3 N hydrochloric acid furnish ninhydrin in 80–82% yield [1002].

Before the discovery of the applications of selenium dioxide and dimethyl sulfoxide, oxidations of methylene groups in β-diketones or β-keto esters were achieved with nitroso compounds. Thus acetylacetone boiled with *p*-**nitrosodimethylaniline** in alcoholic sodium hydroxide gives triketopentane in 55% yield [986].

The oxidants dimethyl sulfoxide and nitroso compounds react easily with **α-bromo ketones** and convert them into **α-dicarbonyl compounds.** The reaction with nitroso compounds is usually carried out in the presence of pyridine and proceeds through a nitrone stage. Phenacyl bromide (α-bromoacetophenone) is thus transformed first into phenacylpyridinium bromide and further, with **nitrosobenzene,** into α-ketoaldonitrone, which is subsequently treated with hydroxylamine to give phenylglyoxal monoxime or with phenylhydrazine to give phenylglyoxal osazone [985] (equation 411).

[985] 411

$C_6H_5COCH_2Br$

[1003]

$\xrightarrow{C_5H_5N}$ $C_6H_5COCH_2N^+\langle C_6H_5\rangle \; Br^-$

$\xrightarrow{Me_2SO \; (DMSO), \; RT, \; 9\,h}$ C_6H_5COCHO \; 71%

PhNO, EtOH
NaOH,
-10 to -5 °C

$C_6H_5COCH=NOH$ \; 75% $\xleftarrow{H_2NOH}$ $C_6H_5COCH=NPh$ \; 76% \; $\overset{|}{O}$

$C_6H_5\underset{\|}{C}CH=NNHC_6H_5$
$\; NNHC_6H_5$
76% $\xleftarrow{2\;PhNHNH_2}$

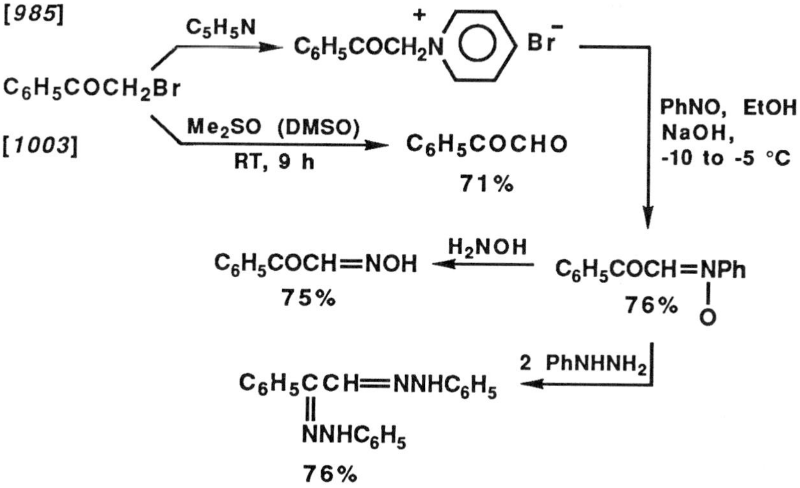

A direct and simpler conversion of ω-bromoacetophenone into phenylglyoxal is achieved by treatment with **dimethyl sulfoxide** at room temperature [1003]. The reaction of α-bromo ketones with dimethyl sulfoxide can be carried out in anhydrous medium [1003], as well as in the presence of water [1001]. The mechanism of the reaction in aqueous medium is more complex, but the results are the same (α-dicarbonyl compounds or their hydrates). Finally, α-bromo ketones may be converted first by silver nitrate into nitrates of α-hydroxy ketones, which are oxidized to α-dicarbonyl compounds by dimethyl sulfoxide [*1004*]. This two-step oxidation is especially useful with sterically hindered α-bromo ketones [*1005*] (equation 412).

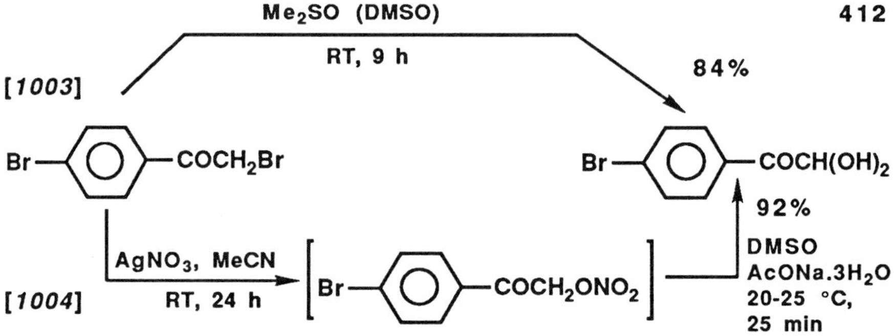

$$\text{[1003]} \quad Br\text{-}C_6H_4\text{-}COCH_2Br \xrightarrow[\text{RT, 9 h}]{Me_2SO\ (DMSO)} Br\text{-}C_6H_4\text{-}COCH(OH)_2 \quad 84\%\ \ 412$$

$$\text{[1004]} \xrightarrow[\text{RT, 24 h}]{AgNO_3,\ MeCN} [Br\text{-}C_6H_4\text{-}COCH_2ONO_2] \xrightarrow[\substack{AcONa\cdot 3H_2O \\ 20\text{-}25\ ^\circ C,\ 25\ min}]{DMSO} \quad 92\%$$

Another variation on the same theme is the oxidation of α-bromo ketones by dimethyl sulfoxide in the presence of potassium iodide and sodium carbonate. In situ conversion of the bromo ketone into iodo ketone evidently facilitates the reaction, because it is successful even with secondary α-bromo ketones, with which simple oxidation fails. Depending on the steric environment, the reaction may occur at room temperature, but it may require a temperature of 120 °C. Yields range from 71 to 95% (equation 413) [*1005*].

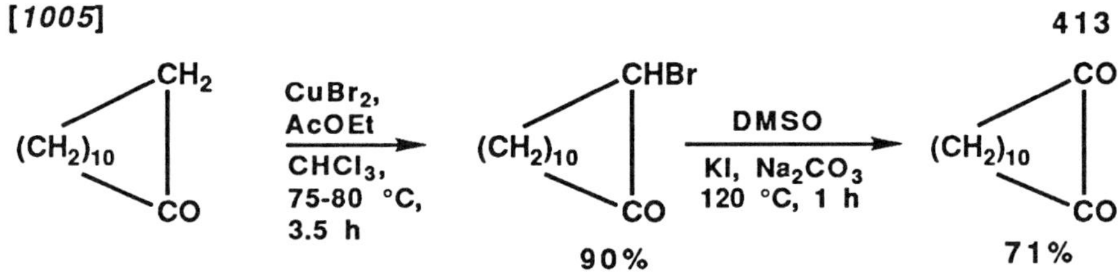

[*1005*] 413

Cyclic: (CH$_2$)$_{10}$ with CH$_2$ and CO $\xrightarrow[\substack{CHCl_3,\\ 75\text{-}80\ ^\circ C,\\ 3.5\ h}]{CuBr_2,\ AcOEt}$ (CH$_2$)$_{10}$ with CHBr and CO (90%) $\xrightarrow[\substack{KI,\ Na_2CO_3 \\ 120\ ^\circ C,\ 1\ h}]{DMSO}$ (CH$_2$)$_{10}$ with CO and CO (71%)

Oxidation of Ketones to Carboxylic Acids

Oxidations of ketones give carboxylic acids with the same or smaller number of carbon atoms.

THE WILLGERODT REACTION

The **Willgerodt reaction,** which involves heating ketones with ammonium polysulfide, is the most important route to carboxylic acids without degradation of the carbon chain. It applies especially to methyl and alkyl ketones of the aromatic series but has been carried out even with aliphatic ketones, although usually with much lower yields. The primary products are thioamides, which are hydrolyzed to amides and carboxylic acids [*501, 1169*]. In the original version of this reaction, a ketone was treated with a mixture of sulfur and ammonium sulfide prepared by saturating a concentrated aqueous solution of ammonia with hydrogen sulfide [*499, 501*]. The temperatures required for the reaction ranged from 160 to 220 °C, and the reaction had to be carried out in sealed tubes or autoclaves. The reaction products, mixtures of amides, thioamides, and sometimes acids, were hydrolyzed with dilute aqueous sodium or potassium hydroxide to give the desired carboxylic acids [*499*].

The **Kindler modification** dodges heating in closed vessels under pressure by using, instead of aqueous ammonium polysulfide, a mixture of sulfur and morpholine. After being refluxed for 6–12 h, the ketone is converted into the thiomorpholide of the carboxylic acid. Subsequent alkaline hydrolysis yields the carboxylic acid as the final product [*501, 503, 504, 1170*].

The uniqueness of the Willgerodt reaction lies in the ultimate formation of acids from ketones regardless of the position of the carbonyl group in the chain.

Despite numerous studies of the mechanism of the Willgerodt–Kindler reaction, the interpretation of the experimental results still leaves something to be desired. It has been demonstrated that the same amides or thiomorpholides that result from the Willgerodt–Kindler reaction of ketones are formed under the same conditions from terminal mercaptans [*1169*], terminal alkenes [*502, 1169*], and terminal alkynes [*502*]. However, the yields of the products from ketones are consistently higher than those obtained from the other starting materials. In addition, some discrepancies have been found in explaining changes in the carbon skeleton that occur sometimes, especially with branched ketones. Generally, the mechanism may be depicted as a series of addition and elimination reactions (equation 414).

$$RCOCH_3 \xrightarrow{R_2NH} R\underset{NR_2}{\overset{OH}{\underset{|}{\overset{|}{C}}}}-CH_3 \xrightarrow{-H_2O} R-\underset{NR_2}{\overset{|}{C}}=CH_2 \xrightarrow{-R_2NH} \quad\quad 414$$

$$RC\equiv CH \xrightarrow{HNR_2} RCH=\underset{NR_2}{\overset{|}{C}}H \xrightarrow{H_2S} RCH_2-CH\overset{SH}{\underset{NR_2}{\diagdown}} \xrightarrow{S}$$

$$RCH_2C\overset{S}{\underset{NR_2}{\diagdown\!\!\!\!\!=}} + H_2S$$

The Willgerodt reaction of isobutyl methyl ketone with ammonium polysulfide at 200 °C gives an 88% yield of isocapronamide after 4 h of heating [*1169*]. Yields of other aliphatic ketones are not nearly as high [*1161, 1169*] (equation 415).

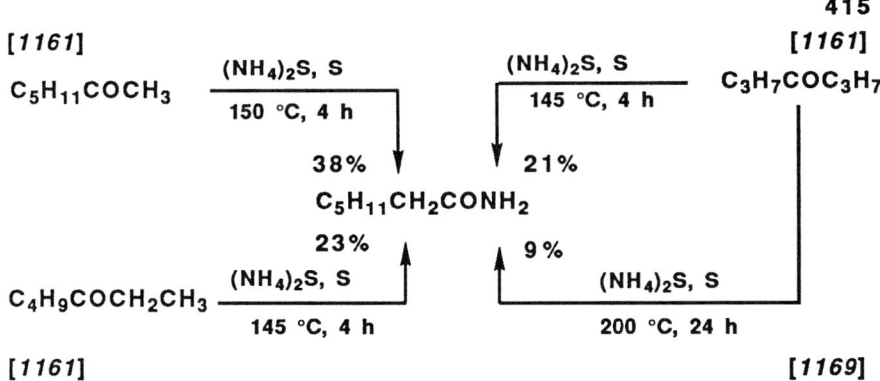

The amides and thioamides obtained by the Willgerodt–Kindler reaction are, sometimes without isolation, converted into acids by refluxing with 15–20% sodium or potassium hydroxide or converted into esters by refluxing for 3 h with alcohols in the presence of gaseous hydrogen chloride [*501*].

The Willgerodt reaction is frequently used to convert alkyl aryl ketones, which are synthesized easily by the Friedel–Crafts reaction, into aryl alkanoic acids with or without isolation of the intermediate thiomorpholides [*499, 504, 1170*] (equation 416).

[*1170*]

$$C_6H_5COCH_2CH_3 \xrightarrow[\text{reflux 6 h}]{S, \text{ HN} \diagup\diagdown O} \left[C_6H_5CH_2CH_2CSN \diagup\diagdown O \right] \quad 416$$

$$PhCH_2CH_2CO_2H \xleftarrow[\text{reflux 6-8 h}]{10\% \text{ NaOH/EtOH}}$$
65%

The reaction with sulfur and morpholine can be applied to the preparation of aromatic acids containing polynuclear residues. Methyl 2-naphthyl ketone refluxed for 16 h with a 50% excess of sulfur and morpholine furnishes a 90% yield of the thiomorpholide of 2-naphthylacetic acid, from which 2-naphthylacetic acid is obtained by refluxing for 5 h with aqueous sulfuric and acetic acid (yield 90%). 1-Acenaphthylacetic acid is prepared by heating 1-acetylacenaphthene with sulfur and ammonium sulfide at

160 °C in a sealed tube for 12 h and by hydrolyzing the product amide by refluxing for 4 h with 15% sodium hydroxide (yield 57%) [*499*].

Heterocyclic ketones, too, undergo the Willgerodt reaction with good yields. 3-Pyridylacetic acid (homonicotinic acid) is obtained in 70% yields from methyl 3-pyridyl ketone on heating with ammonium polysulfide [*501*]. 3-Pyridylacetic acid and 4-pyridylacetic acid are also prepared from methyl pyridyl ketones by refluxing for 12 h with sulfur and morpholine, followed by hydrolysis of the isolated thiomorpholide (yield 80%) by refluxing for 72 h with ethanolic potassium hydroxide (yield 74%) [*503*] (equation 417).

[*501*]   417

[*503*]

OXIDATION OF METHYL KETONES

An interesting reaction, giving an acid with the same number of carbon atoms, takes place when some ketones (having α and α' hydrogens) are stirred for 10–60 min at 25–80 °C with powdered potassium hydroxide, **carbon tetrachloride,** water, and *tert*-butyl alcohol [*954*] (equation 418). Ketones with α hydrogens but not α' hydrogens give the same product as that obtained by oxidation with hypohalites [*954*].

Another way of oxidizing aryl methyl ketones without degradation is by their reaction with **thallium trinitrate** in methanolic solution and in the

[954]   418

(C₆H₅)₂CHCOCH₃ $\xrightarrow[\text{25-80 °C, 10 min to 1 h}]{\text{CCl}_4,\ \text{KOH},\ t\text{-BuOH},\ \text{H}_2\text{O}}$ Ph₂CHCH₂CO₂H
                                                                                                                      70%

presence of perchloric acid. Acetophenone and its substituted derivatives are thus converted into methyl arylacetates in 61–94% yields by treatment with thallium triacetate at room temperature for 2–18 h (equation 419) [*414*].

[*414*]   419

The oxidation of ***methyl ketones to α-keto carboxylic acids*** is rare and is accomplished by treatment with a cold solution of *potassium permanganate*. However, the reaction is not general; acetophenone, *p*-methylacetophenone, and 3,4-dimethylacetophenone are oxidized all the way to the corresponding benzoic acids. On the other hand, 2,4-dimethylacetophenone, when shaken with approximately 1% aqueous potassium permanganate at room temperature, gives 66–72% yields of 2,4-dimethylphenylglyoxylic acid [*888, 889*].

The oxidation of ethyl *p*-xylyl ketone (2,5-dimethylpropiophenone) with cold dilute aqueous potassium permanganate furnishes 2,5-dimethylbenzoylacetic acid (a β-keto acid) in an unstated yield [*890*].

DEGRADATIVE OXIDATIONS OF METHYL KETONES TO ACIDS WITH ONE LESS CARBON ATOM

The conversion of methyl ketones into carboxylic acids having one carbon less is an old method and is popular for its simplicity and high yields. The oxidants are ***hypohalites in alkaline media.*** The reaction consists of two stages, which are carried out in one reaction vessel without the isolation of the intermediates. The methyl group is halogenated to a trichloro-, tribromo-, or, less often, triiodomethyl group, and the trihalomethyl ketone is subsequently hydrolyzed with alkalies to a haloform and a salt of the acid. Instead of hypohalites, halogens in alkaline media can be used [*688, 698, 737*]. Because of the strongly electronegative trihalomethyl group, the hydrolysis is facile and takes place at moderate temperatures.

The hypohalite oxidations are easy to carry out. The methyl ketones, pure or dissolved in dioxane, are added to cooled or warm solutions of hypohalites in water. The reverse order of addition has also been used [736]. An exothermic reaction ensues, and the haloform starts forming a heavy organic layer. The excess hypohalite is destroyed by sodium bisulfite, the heavy layer is separated, the residual haloform is removed by steam distillation, and the aqueous solution of the alkaline salt of the carboxylic acid is treated with sulfuric or hydrochloric acid (equation 420).

$$RCOCH_3 \xrightarrow{3\ NaOX} RCOCX_3 \xrightarrow{NaOH} RCO_2Na + CHX_3 \qquad 420$$

$$X = Cl,\ Br,\ or\ I$$

The reaction is applicable to saturated and unsaturated methyl ketones of the aliphatic and alicyclic series, to aryl methyl ketones, and to methyl ketones of heterocyclic aromatic compounds.

Pinacolone (*tert*-butyl methyl ketone) is transformed into pivalic acid (trimethylacetic acid) in 71–74% yield [737]. Mesityl oxide is converted into β,β-dimethylacrylic acid in 49–53% yield [703]. Isopropenyl methyl ketone gives methacrylic acid in 41% yield [697], and pregnenolone acetate furnishes 3-acetoxyetienic acid in 91–95% yield [1171] (equations 421–424).

Aryl methyl ketones give generally high yields of acids on treatment with hypohalites [688, 696, 698, 736]. The reaction of 1,3,5-triacetylbenzene and sodium hypochlorite results in a 94% yield of trimesic acid [688], and that of methyl β-naphthyl ketone and sodium hypochlorite gives an 87–88% yield of β-naphthoic acid [698]. 4-Acetyl-4′-methoxybiphenyl is converted

[737] 1. Br_2, NaOH, 0–10 °C, 15–20 min   421
       2. 0–10 °C, 1 h, RT, 3 h   71–74%
       3. steam distillation
$(CH_3)_3CCOCH_3$ 4. H_2SO_4 $(CH_3)_3CCO_2H$

[954]   KOH, CCl_4, *t*-BuOH, H_2O 80%
        25–80 °C, 10–60 min

[703]   422
$(CH_3)_2C=CHCOCH_3$ $\xrightarrow{KOCl,\ H_2O,\ dioxane}$ $(CH_3)_2C=CHCO_2H$
                      reflux 3–4 h; H_2SO_4
                                              49–53%

[697]   423
$CH_2=CCOCH_3$ $\xrightarrow{3\ N\ NaOCl,\ 6\ N\ NaOH}$ $CH_2=CCO_2H$
   | <20 °C, >1 h; H_2SO_4 |
   CH_3 CH_3
                                              41%

[1171] Steroid with COCH₃ group and AcO, via NaOBr (Br₂, NaOH), H₂O, dioxane, −5 to 0 °C, 2.5 h; 90 °C, HCl → Steroid with CO₂H and HO, 424, 91-95%

by sodium hypobromite into 4-(4'-methoxyphenyl)benzoic acid (equation 425) [736].

[736] CH_3O–C₆H₄–C₆H₄–COCH₃ $\xrightarrow[\text{dioxane, 35-40 °C, 45 min; H}^+]{\text{Br}_2, \text{NaOH, 0 °C}}$ CH_3O–C₆H₄–C₆H₄–CO_2H 425, 91%

Side reactions are observed in some hypohalite oxidations. Thus 2-acetyl-9,10-dihydrophenanthrene furnishes, depending on the reaction conditions used, either the expected 9,10-dihydrophenanthrene-2-carboxylic acid, or biphenyl-2,2',4-tricarboxylic acid (equation 426) [696].

[696] 2-acetyl-9,10-dihydrophenanthrene:
- 5% NaOCl (pH 10), reflux 9 h → 9,10-dihydrophenanthrene-2-carboxylic acid, 426, 43%
- NaOCl, NaOH, 60-80 °C, 19 h → biphenyl-2,2',4-tricarboxylic acid, 49%

Refluxing *p*-ethylacetophenone with sodium hypochlorite results not only in the oxidation of the methyl ketone but also in the conversion of the ethyl group into carboxyl so that the product is terephthalic acid in 95% yield [696].

In a few instances, not only methyl ketones but other alkyl ketones are degraded to shorter carboxylic acids resulting from halogenation of the **methylene** group adjacent to the carbonyl and subsequent hydrolysis [*103, 160, 1172*]. The reaction between propiophenone and sodium hypobromite at 22–25 °C gives a 96% yield of benzoic acid [*303*]. Under similar conditions, 5-butyl-2-butyrylpyridine is converted into 5-butyl-α-picolinic acid in 79% yield [*160*]. Methyl 3-(α-pyridyl)propyl ketone yields not only the expected 4-(α-pyridyl)butyric acid but also 3-(α-pyridyl)propionic acid [*103*] (equations 427 and 428).

[*160*]   **427**

C_4H_9-pyridyl-COC_3H_7 $\xrightarrow[\text{2. NaHSO}_3, \text{H}_2\text{SO}_4]{\text{1. Br}_2, \text{NaOH, H}_2\text{O, RT, 15 h}}$ C_4H_9-pyridyl-CO_2H 79%

[*103*]   **428**

pyridyl-$CH_2CH_2CH_2COCH_3$ $\xrightarrow[\text{2. SO}_2 \text{ 3. HCl}]{\text{1. Br}_2, \text{NaOH, } -10 \text{ to } -5 \text{ °C, } 45 \text{ °C, 1 h}}$ pyridyl-$CH_2CH_2CH_2CO_2H$ 38%*

                                                                      + pyridyl-$CH_2CH_2CO_2H$ 37%*

*The yield is based on the reacted ketone.

Dimethyldihydroresorcinol (dimedone) is converted into β,β-dimethylglutaric acid on treatment with sodium hypochlorite or sodium hypobromite. As a β-diketone, it suffers alkaline hydrolysis to 3,3-dimethyl-5-ketohexanoic acid, which undergoes regular hypohalite degradation of the methyl ketone end (equation 429) [*735*]. Both the hydrolytic opening of the ring and the hypohalite reaction are accomplished in one step [*699*].

[*735*]   **429**

$(CH_3)_2C(CH_2CO)_2CH_2$ $\xrightarrow[\text{boil 5 min}]{\text{KOH, H}_2\text{O}}$ $(CH_3)_2C(CH_2CO_2K)(CH_2COCH_3)$ $\xrightarrow[\text{HCl}]{\text{Br}_2, \text{NaOH, 1 h, Na}_2\text{SO}_3}$ $(CH_3)_2C(CH_2CO_2H)_2$ ~100%

$(CH_3)_2C(CH_2CO)_2CH_2$ $\xrightarrow[\text{Na}_2\text{SO}_3; \text{HCl}]{\text{Cl}_2, \text{NaOH, RT; KOH, 35-40 °C, 6-8 h, RT}}$ $(CH_3)_2C(CH_2CO_2H)_2$ 91-96%

The vast majority of the preparative oxidations of methyl ketones are carried out with alkaline hypochlorites or hypobromites. The same effect can be achieved by alkaline hypoiodites, but examples are scarce. One example is the opening of the D ring of benzyl estrone to benzyl marrianolic acid by treatment with potassium hypoiodite prepared in situ from iodine and potassium hydroxide (equation 430) [*752*].

[*752*] 430

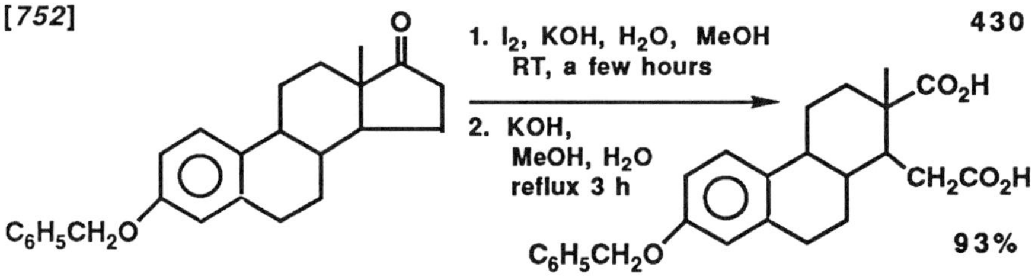
O
1. I_2, KOH, H_2O, MeOH
RT, a few hours
2. KOH,
MeOH, H_2O
reflux 3 h
$C_6H_5CH_2O$
CO_2H
CH_2CO_2H
$C_6H_5CH_2O$
93%

Hypoiodites are used for qualitative tests for methyl ketones ***(Lieben test)***. For this purpose, a compound to be tested is stirred with an aqueous solution of sodium hydroxide (80 mol/mol of methyl ketone). Iodine (4.5 mol of I_2) is added portionwise with stirring, and the mixture is set aside for 20 min at 25 °C before acidification. In the presence of a methyl keto group, a yellow heavy precipitate of iodoform settles at the bottom of the test tube. Iodoform can be identified easily not only by its characteristic smell but also by its melting point (120–123 °C) [*1173*]. This test applies not only to methyl ketones but to any compound that can be converted in the reaction medium into a species containing the $COCH_3$ group, for example, isopropyl or ethyl alcohol.

Hypohalites rarely affect alkyl groups bound to aromatic rings. However, *p*-methyl- and *p*-ethylacetophenone give terephthalic acid when refluxed with sodium hypochlorite [*696*]. The same product is obtained by the oxidation of *p*-methylacetophenone with *nitric acid* and *potassium permanganate* (equation 431) [*891*].

[*696*] 431

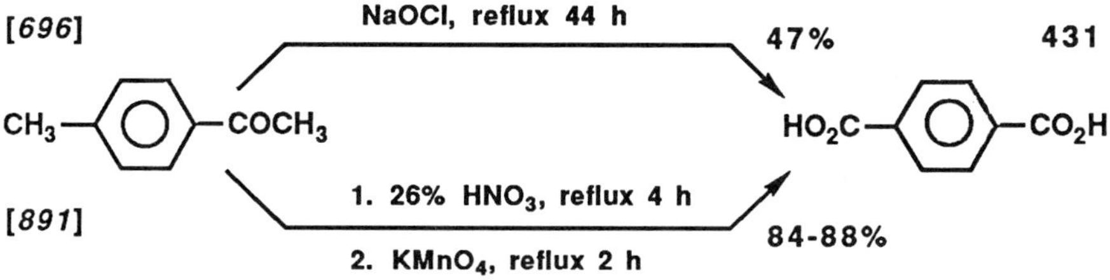
NaOCl, reflux 44 h
47%
CH_3
$COCH_3$
HO_2C
CO_2H
1. 26% HNO_3, reflux 4 h
2. $KMnO_4$, reflux 2 h
84-88%

[*891*]

The oxidation of a methyl ketone to a carboxylic acid with one less carbon by oxidants other than hypohalites is exemplified by the oxidation of 2-acetylfluorene with **sodium dichromate.** In addition to the methyl keto group, the methylene group is also oxidized (equation 432) [*626*].

[*626*] 432

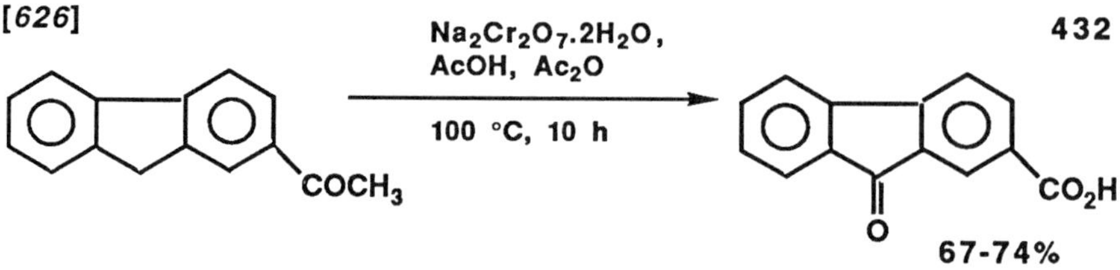
$Na_2Cr_2O_7.2H_2O$,
AcOH, Ac_2O
100 °C, 10 h
$COCH_3$
CO_2H
O
67-74%

OXIDATIVE CLEAVAGE OF KETONES

Strong oxidation agents such as **chromic acid** and **potassium permanganate** oxidize ketones with the simultaneous cleavage of the carbon skeleton next to the carbonyl group. With nonsymmetrical ketones, as many as four products may result. The practical applications of such oxidations are therefore limited to cases where, for structural reasons, the cleavage of the carbon skeleton takes place preferentially at only one side of the carbonyl or where identical products result from the cleavage on either side of the keto group, as occurs with symmetrical cyclic ketones.

The oxidation of cyclopentanone with oxygen [52] or with **nitric acid** [471] gives glutaric acid in 100 or 82% yields, respectively (equation 433).

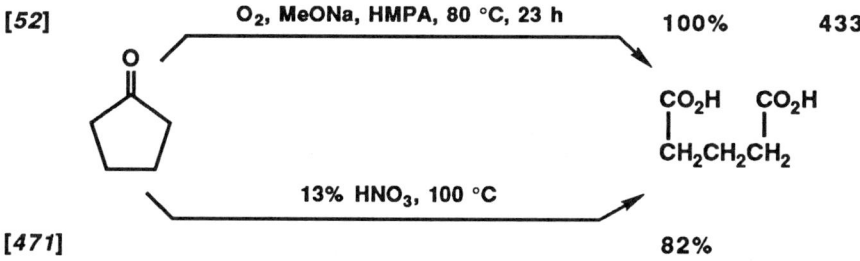

The reaction with nitric acid is exothermic, and careful control of temperature is needed to prevent further oxidation and degradation of the product.

Analogously, cyclohexanone can be oxidized to adipic acid with several oxidants (equation 434) [52, 192, 1174]. As in the case of cyclopentanone, too energetic oxidation gives lower homologous acids, which have to be separated from the main product.

2-Methylcyclohexanone is converted regiospecifically into 6-ketoheptanoic acid (or its esters in the presence of alcohols) by **oxygen** in the presence of ferric salts as catalysts (equation 435) [67].

The oxidation of ketones with **chromic acid** proceeds through enol intermediates, as proven by kinetic measurements [1175]. 2-Chlorocyclohexanone is oxidized predominantly to adipic acid, with 2-chloroadipic acid, glutaric acid, and succinic acid as byproducts (equation 436) [1175].

[67]

$CH_2(CO)(CH_2)(CH_2)(CH_2)(CHCH_3)(CH_2)$ $\xrightarrow{O_2/FeCl_3, MeOH, C_6H_6}{60\ °C,\ 20\ h}$ $CH_2(CO_2CH_3)(CH_2)(CH_2)(CH_2)(COCH_3)(CH_2)$ 435

93%

[1175]   436

2-chlorocyclohexanone $\xrightarrow{CrO_3,\ H_2O,\ HClO_4}{30\ °C}$

$HO_2CCH_2CH_2CH_2CH_2CO_2H$	77%
$HO_2CCH_2CH_2CH_2CHClCO_2H$	14%
$HO_2CCH_2CH_2CH_2CO_2H$	5%
$HO_2CCH_2CH_2CO_2H$	4%

Oxidation of Unsaturated Ketones

Oxidations of unsaturated ketones affecting solely the carbonyl group were discussed in the section Baeyer–Villiger Reaction of Functionalized Ketones (equations 379–399). In this section, only such oxidations that add oxygen to the double bond will be described.

The most dependable reagent for the *epoxidation* of unsaturated ketones is **hydrogen peroxide**, especially in alkaline media [*142, 143, 149, 151*]. Because the Baeyer–Villiger reaction is acid-catalyzed, it does not take place during epoxidations with alkaline hydrogen peroxide or its neutral derivatives, such as *tert*-butyl hydroperoxide [*220*]. Most examples of epoxidation involve unsaturated ketones with conjugated double bonds.

Peroxy acids may epoxidize unsaturated ketones [*299, 332*], but a concomitant Baeyer–Villiger reaction is possible [*254*] (equations 389 and 391). Other ways of forming epoxy ketones are reactions with salts of hypochloric acid [*691, 704*] and with *N*-bromosuccinimide [*746*]. Mesityl oxide is converted into its epoxide, as shown in equation 437 [*142, 220, 254, 746*].

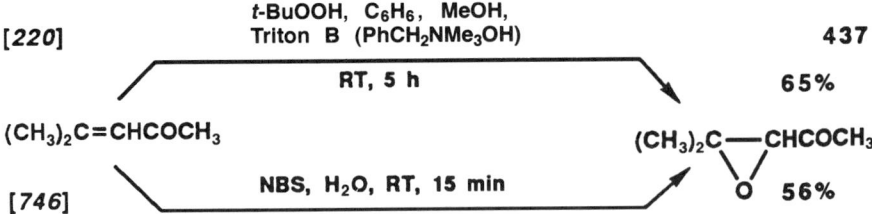

2,3-Cyclohexen-1-one oxide is obtained from 2-cyclohexen-1-one on treatment with hydrogen peroxide [*151*] or *tert*-butyl hydroperoxide [*220*] (equation 438). 1,4-Naphthoquinone is oxidized with sodium [*691*] or cal-

[151] 438

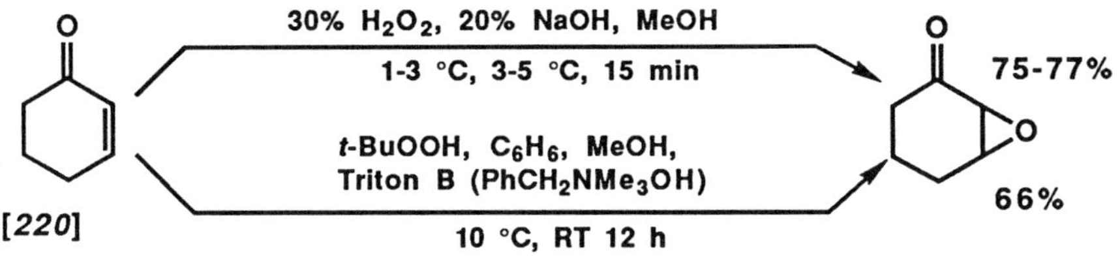

[220]

30% H$_2$O$_2$, 20% NaOH, MeOH
1-3 °C, 3-5 °C, 15 min   75-77%

t-BuOOH, C$_6$H$_6$, MeOH,
Triton B (PhCH$_2$NMe$_3$OH)
10 °C, RT 12 h   66%

cium hypochlorite [704] to 2,3-epoxy-2,3-dihydro-1,4-naphthoquinone [691] (equation 439).

[691] 439

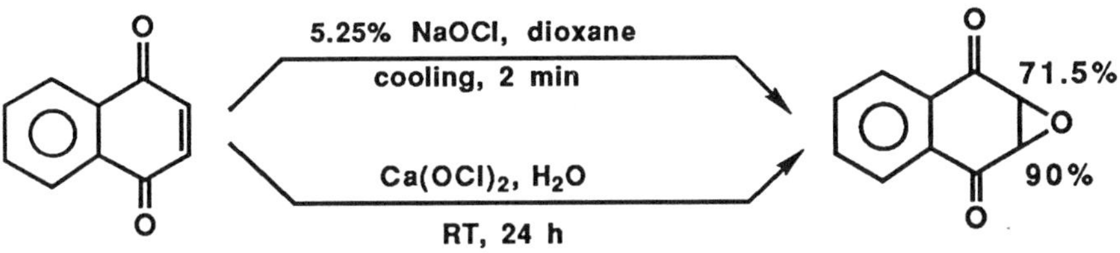

5.25% NaOCl, dioxane
cooling, 2 min   71.5%

Ca(OCl)$_2$, H$_2$O
RT, 24 h   90%

[704]

In α-ionone, peroxybenzoic acid and peroxyphthalic acid epoxidize solely the endocyclic double bond to give 3,4-epoxy-α-ionone in 96.5 and 66% yields, respectively [299]. Alkaline hydrogen peroxide, on the other hand, epoxidizes only the double bond conjugated with carbonyl to give a 28–31% yield of α-ionone-α′,β′-epoxide [332] (equation 440).

[299]

BzO$_2$H, CHCl$_3$, 4 °C, 72 h   96%   440

α-ionone

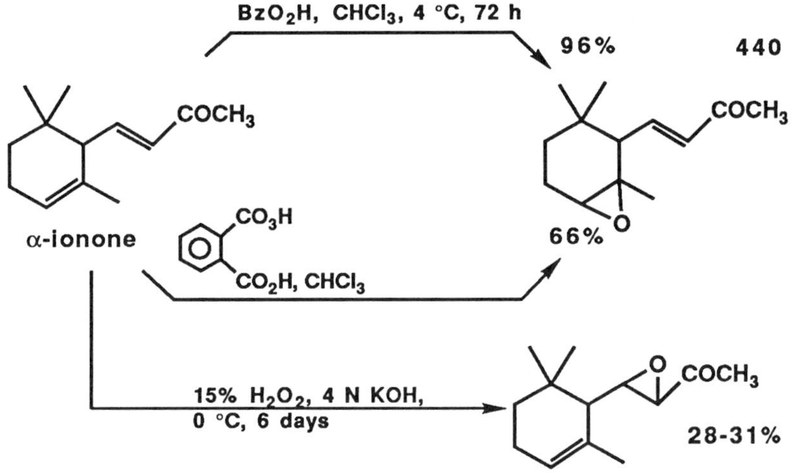

[299]

CO$_3$H
CO$_2$H, CHCl$_3$ 66%

[332]

15% H$_2$O$_2$, 4 N KOH,
0 °C, 6 days   28-31%

β-Ionone, when treated with peroxybenzoic acid [299] or with peroxyphthalic acid [332], is converted into the 2,3-epoxide in respective yields of 86 or 60–70% [332]. The 2,3-epoxide product is further epoxidized at the double bond conjugated with the keto group to yield β-ionone-3,4,α′,β′-diepoxide in a 40% yield (equation 441) [*332*].

[*332*]   441

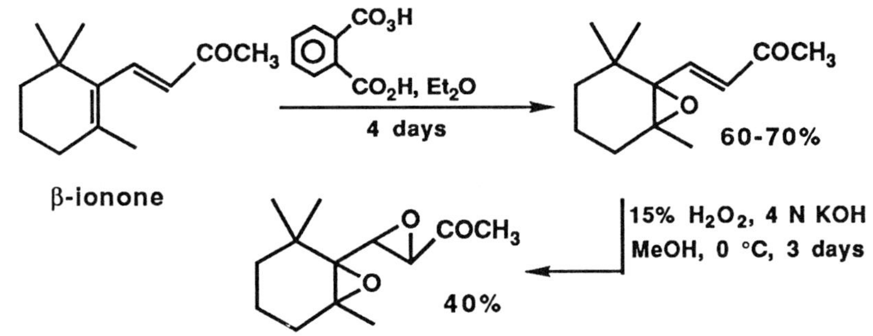

β-ionone

HYDROXYLATION OF UNSATURATED KETONES

The addition of two hydroxylic groups to the double bonds of unsaturated ketones is carried out by the same methods used for hydroxylations of alkenes (equations 71–83). As an example, hydroxylation of the double bond in 3β-hydroxyandrost-5-en-17-one is accomplished by treatment with one equivalent of **osmium tetroxide** in pyridine and subsequent reductive cleavage of the osmate ester with sodium bisulfite in aqueous pyridine (equation 442) [*950*].

[*950*]   442

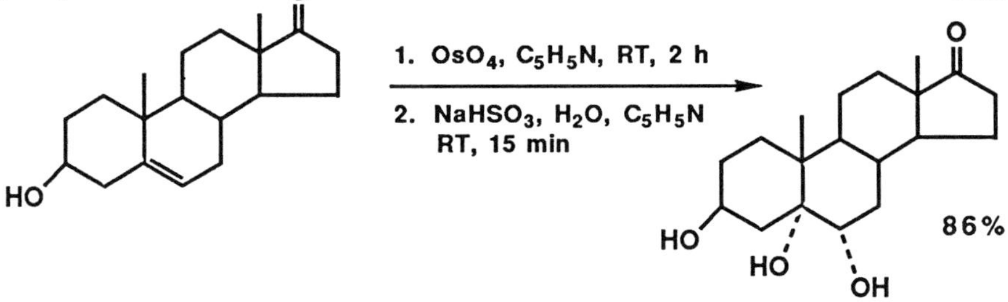

1. OsO_4, C_5H_5N, RT, 2 h
2. $NaHSO_3$, H_2O, C_5H_5N RT, 15 min

86%

Hydroxylation in the γ position of an α,β-unsaturated ketone is exemplified by the biological transformation of cinerone into cinerolone by *Aspergillus niger* or *Streptomyces aureofaciens* (equation 443) [*1050*].

[*1050*]   443

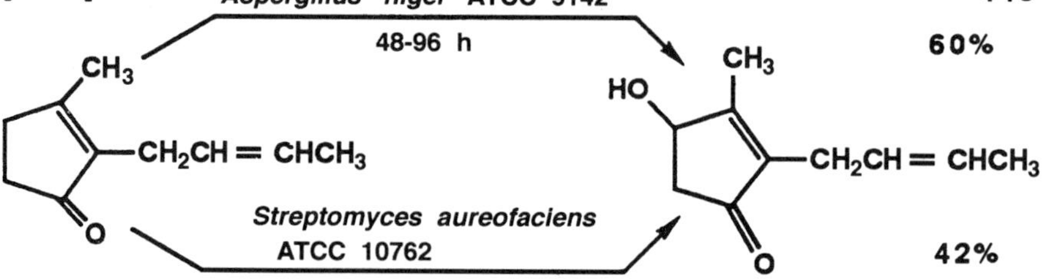

Aspergillus niger ATCC 9142
48–96 h   60%

Streptomyces aureofaciens ATCC 10762   42%

CLEAVAGE OF UNSATURATED KETONES

The oxidative cleavage of unsaturated ketones takes place under the same conditions as that of alkenes or other unsaturated derivatives. The fate of the primary fission product depends on the position of the double bond with respect to the carbonyl group and on the subsequent reactions. *Ozonization* of Δ^4-cholestenone in acetic acid and ethyl acetate, followed by treatment with 30% hydrogen peroxide, gives a keto acid, evidently resulting from the decomposition of the primarily formed diketo acid (equation 444) [*1176*].

Oxidation of Hydroxy Ketones to Diketones

Oxidations of hydroxy ketones to diketones occur frequently in steroidal alcohols. If the alcoholic group, usually secondary, is remote enough from the keto group, its oxidation takes place independently and is achieved by the same reagents that are used for the oxidation of alcohols. A solution of **chromium trioxide** in aqueous sulfuric acid oxidizes 5-pregnen-3β-ol-20-one in acetone solution at room temperature within 2–5 min to 5-pregnen-3,20-dione in 90% yield [*579*]. Similarly, 11β-hydroxytestosterone 17-acetate is transformed by chromium trioxide in 80% acetic acid at room temperature in 30 min into 11-ketotestosterone 17-acetate in 92% yield [*807*].

A reagent suitable for oxidations of steroidal hydroxy ketones to diketones is **dimethyl sulfoxide** in the presence of various activators. It converts testosterone into 4-androstene-3,17-dione in 95–100% yields (equation 445) [*1016, 1018*].

Oxidations of steroidal keto alcohols to diketones are frequently accomplished *biochemically.* In addition, isomerization of double bonds, dehydrogenations, and hydroxylations often take place.

Dehydroandrosterone [*1088*] and dehydroepiandrosterone [*1056*] are converted by **bacterium-infected yeast** and *Corynebacterium simplex* ATCC 6946 into 4-androstene-3,17-dione in 87 and 55% yields, respectively (equation 446) [*1056*].

Oxidations of 11β-hydroxysteroids to 11-ketosteroids are accompanied

[1016] 0.5* H₃PO₄, 5* DCC, 35* DMSO, C₆H₆ → 99% **445**

[1018] 3-3.3* C₅H₅N·SO₃, 70* DMSO, 6.5-12* Et₃N, 25 °C → ~100%

*Values are moles per mole of alcohol.

[1056] *Corynebacterium simplex* ATCC 6946 → 55% **446**

by β-hydroxylation in position 6 with **Rhizopus arrhizus** or in position 9α with **Helicostylum piriforme** or **Cunninghamella blakesleeana** [1059]. Only the secondary alcoholic group is affected. The primary hydroxyl group remains intact (equation 447) [1059].

[1059] *Rhizopus arrhizus* → 34%; *Helicostylum piriforme* → 59% **447**

When the hydroxylic group is in an α position with respect to the keto group, as in *acyloins,* oxidation to α-diketones is easily accomplished by many oxidants.

Air in the presence of cupric sulfate as a catalyst oxidizes benzoin to benzil in 86% yield [62]. **Cupric sulfate** in stoichiometric amounts converts furoin into furil in 63% yield in aqueous pyridine at 100 °C after 2 h [351]. α-Hydroxybenzyl 2-pyrryl ketone is oxidized by stoichiometric amounts of cupric sulfate to phenyl-2-pyrrylglyoxal in 95% yield in aqueous pyridine at 85 °C after 3.5 h (equation 448) [352].

[352]   **448**

(pyrrole)-COCHC$_6$H$_5$ $\xrightarrow{\text{CuSO}_4 \cdot 5\text{H}_2\text{O, C}_5\text{H}_5\text{N, H}_2\text{O}}_{80\ °C,\ 3.5\ h}$ (pyrrole)-COCOC$_6$H$_5$
      |
     OH   95%

Cupric acetate is also used for oxidations of acyloins. Diphenylacetoin is converted almost quantitatively into dibenzylglyoxal when refluxed with cupric acetate in 70% acetic acid for 7 min [358]. Sebacoin (2-hydroxycyclodecanone) gives sebacil in 88–89% (equation 449) yield [359].

[359]   **449**

$(CH_2)_8\begin{smallmatrix}\diagup CO\\ \diagdown CHOH\end{smallmatrix}$ $\xrightarrow{\text{(AcO)}_2\text{Cu, 50% AcOH, MeOH}}_{\text{reflux, 1 min}}$ $(CH_2)_8\begin{smallmatrix}\diagup CO\\ \diagdown CO\end{smallmatrix}$

                                                                                                                    88-89%

A very suitable oxidant for the conversion of acyloins into α-diketones is **ammonium nitrate** in the presence of catalytic amounts of cupric acetate. This reagent converts benzoin into benzil in 90% yield [476]. The same result is obtained with bismuth sesquioxide [481] and **sodium bromate** [740] (equation 450). On the other hand, **ceric ammonium nitrate** does not give benzil but cleaves the bond between the alcoholic and the keto groups and cleaves benzoin into benzaldehyde and benzoic acid [425].

                                                                                                                 450

C$_6$H$_5$COCHC$_6$H$_5$ ⟶   C$_6$H$_5$COCOC$_6$H$_5$
         |
        OH

[62]	O$_2$/CuSO$_4$, C$_5$H$_5$N, H$_2$O, 100 °C, 2 h	86%
[476]	NH$_4$NO$_3$/(AcO)$_2$Cu, 80% AcOH, reflux 1.5 h	90%
[481]	Bi$_2$O$_3$, AcOH, EtOCH$_2$CH$_2$OH, 104 °C, 1 h	95%
[740]	NaBrO$_3$, NaOH, H$_2$O, 100 °C, 5-6 h	84-90%*

*The product is benzilic acid resulting from the rearrangement of benzil, formed as the primary product.

The conversion of pivaloin into pivalil (2,2,5,5-tetramethyl-3,4-hexanedione) is best accomplished with chromium trioxide in acetic acid and dilute sulfuric acid (yield 87%) [406].

Acyloins are tautomeric with 1,2-enediols, and in some cases, enediols predominate if they are stabilized by resonance and hydrogen bonds. Such is the case with quinaldoin, which is easily oxidized to quinaldil by passing air through its boiling solution in dioxane (equation 451) [1177].

[1177]   451

Oxidation of Dicarbonyl Compounds to Carboxylic Acids

The conversion of α-dicarbonyl compounds into carboxylic acids is the monopoly of **hydrogen peroxide and peroxy acids.** The reaction is actually a *Baeyer–Villiger oxidation* producing acid anhydrides, which, in some cases, were isolated. Benzil, when refluxed for 15 min with 95% hydrogen peroxide and 70% perchloric acid, gives benzoic acid. Anisil, under the same conditions, is converted into anisic acid [269]. 9,10-Diketostearic acid, on treatment with 8.7% peroxyacetic acid at room temperature for 1 day, furnishes 90 and 95% yields, respectively, of pelargonic acid and azelaic acid [271].

Cyclic α-diketones and o-quinones give, under comparable conditions, dicarboxylic acids. β-Naphthoquinone is converted in 83% yield into o-carboxyallocinnamic acid by 9% peroxyacetic acid in an exothermic reaction [271] (equation 393).

The treatment of tetrafluoro-o-benzoquinone with 40% peroxyacetic acid at 70–75 °C for 4.75 h degrades the quinone to difluoromaleic acid in 55% yield. The same acid is obtained in 59% yield under similar conditions from fluoranil (tetrafluoro-p-benzoquinone) [273].

The oxidation of β-diketones and β-keto esters with peroxyacetic acid [270] and peroxyphthalic acid [333] leads to complex products that ultimately undergo hydrolysis. In the case of acetylacetone, the products are acetic acid and pyruvic acid [270] (equation 452).

[270]   452

$$CH_3COCH_2COCH_3 \xrightarrow[<30\ °C]{17\%\ AcO_2H} CH_3CO_2H + CH_3COCO_2H$$
$$\phantom{CH_3COCH_2COCH_3 \xrightarrow[<30\ °C]{17\%\ AcO_2H} CH_3CO_2H + }38\%$$

Oxidation of Ketone Derivatives

The treatment of **ketimines** (Schiff bases) with peroxyacetic acid gives *oxaziridines* (equation 453) [*1178*].

[*1178*]   453

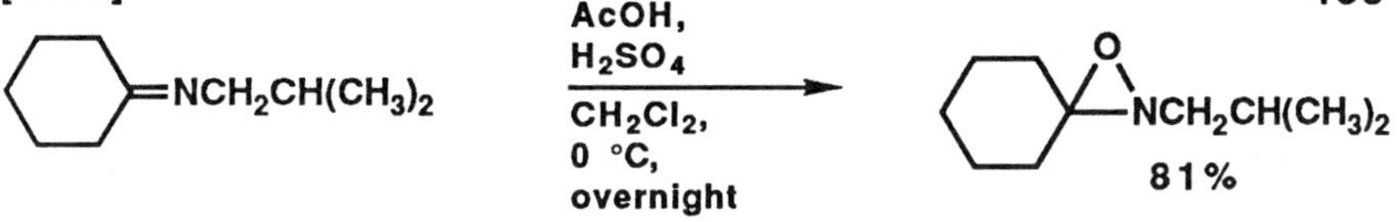

Cyclohexylidene=NCH$_2$CH(CH$_3$)$_2$ $\xrightarrow[\substack{CH_2Cl_2,\\ 0\ °C,\\ \text{overnight}}]{\substack{90\%\ H_2O_2,\\ AcOH,\\ H_2SO_4}}$ [oxaziridine]–NCH$_2$CH(CH$_3$)$_2$ 81%

Oximes are either oxidized to *nitro compounds* [*823*] or converted into their respective *parent ketones* [*422, 427, 527, 660*]. Oxidation to the parent ketones applies also to *semicarbazones* [*427, 527*], dimethylhydrazones [*528, 529*], tosylhydrazones [*527, 528, 529*], *phenylhydrazones* [*527*], *p*-nitrophenylhydrazones [*527*], and 2,4-dinitrophenylhydrazones [*527*] (equations 454–456).

Because such exotic reagents as those shown in equations 455 and 456 are not readily available, cleavage by ozone may be preferred [*1179*].

Ketone derivatives whose oxidations have wide applications in synthesis are **hydrazones** and **vicinal dihydrazones.** Hydrazones are transformed into *diazo compounds*, and vicinal dihydrazones are converted into *acetylenes.* By far the most widely used oxidant is yellow **mercuric oxide**

[*823*]   454

$$(EtO_2C)_2C{=}NOH \rightleftarrows (EtO_2C)_2CHNO \xrightarrow[30\ °C,\ 24\ h]{MnO_2,\ AcOH} (EtO_2C)_2CHNO_2$$
$$88\%$$

                                                                              455

[*427*] — (NH$_4$)$_2$Ce(NO$_3$)$_6$, HNO$_3$, EtOH, −4 to −20 °C, 5 min → 79%

Ph–C(=NOH)–Ph $\xrightarrow[RT]{ZnCr_2O_7 \cdot 3H_2O,\ CH_2Cl_2}$ 70–90% Ph–CO–Ph

[*660*] — (PhSeO)$_2$O, THF, 50–60 °C, 3 h → 89%*

[*527*]

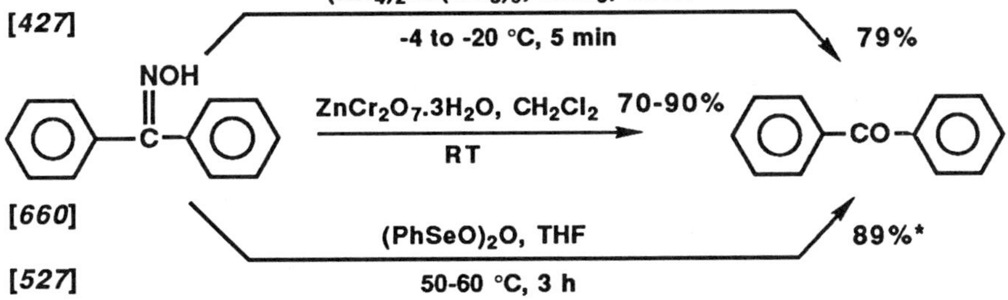

*The same yield was obtained from the semicarbazone after 2 h, and a 90% yield was obtained from the phenylhydrazone after 3 h [*527*].

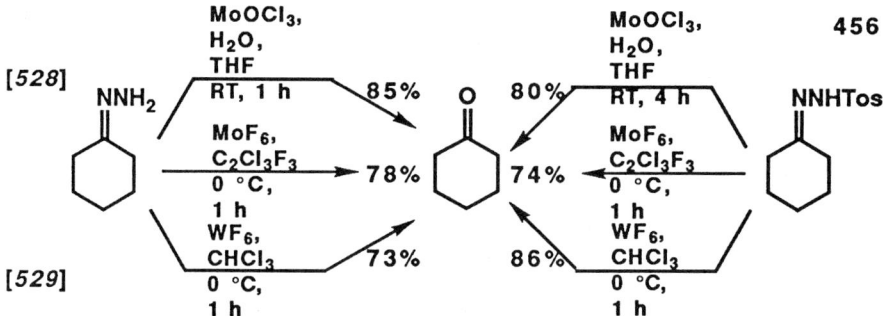

[387, 388, 390, 391, 392]. Red mercuric oxide is less reactive [387]. Less often, oxidations are carried out with silver oxide [370], mercurous trifluoroacetate [405], mercuric trifluoroacetate [406], lead tetraacetate [445], manganese dioxide [370], and nickel peroxide [935]. The most common solvents are pentane, petroleum ether, benzene, toluene, ether, and benzonitrile. Basic catalysts such as alcoholic potassium hydroxide are sometimes used. Anhydrous calcium oxide, sodium sulfate, and magnesium sulfate are frequently added to remove the water of reaction. The reaction temperatures range from room temperature to the reflux temperature of the solvents.

Examples of the oxidation of aliphatic hydrazones are the conversion of acetone hydrazone into 2-diazopropane [386, 392], of hexafluoroacetone hydrazone into 2-diazohexafluoropropane [445], of 4-octanone hydrazone into 4-diazooctane [386], and of the hydrazone of diethyl mesoxalate into diethyl diazomalonate [935] (equations 457 and 458).

[392] $R_2C=NNH_2$ $\xrightarrow{\text{HgO, KOH, Et}_2\text{O, EtOH}}$ $R_2C=N=N$ 70-90% R = CH_3 457

[445] $\xrightarrow{\text{Pb(OAc)}_4\text{, PhCN, 0-25 °C}}$ 76-79% R = CF_3

[935] $(C_2H_5O_2C)_2C=NNH_2$ $\xrightarrow{\text{NiO}_2\text{, Et}_2\text{O, 20 °C, 1 h}}$ $(C_2H_5O_2C)_2C=N=N$ 89% 458

Of aromatic ketone hydrazones, benzophenone hydrazone is oxidized with yellow mercuric oxide [387, 390] or red mercuric oxide [387], fluorenone hydrazone is oxidized with mercuric oxide [388] or nickel peroxide [935], and di-α-thienyl ketone hydrazone and phenyl naphthyl ketone hydrazones are oxidized with silver oxide or manganese dioxide [370] (equations 459–461).

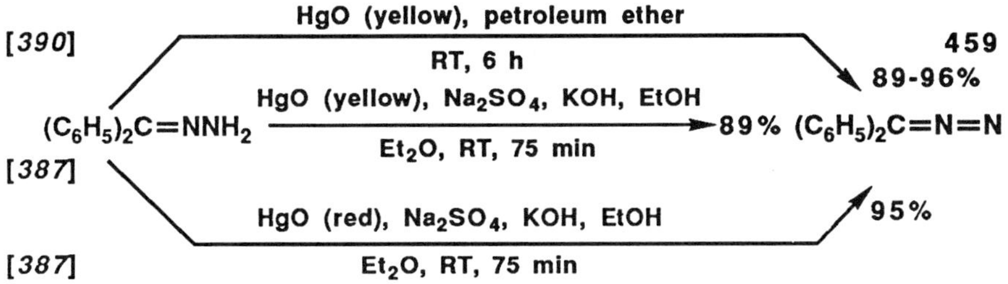

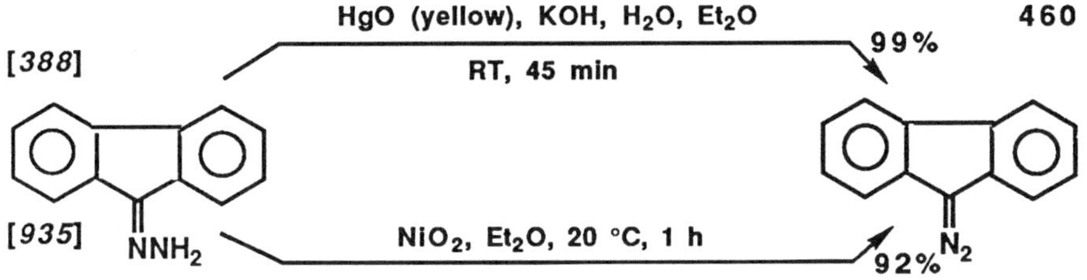

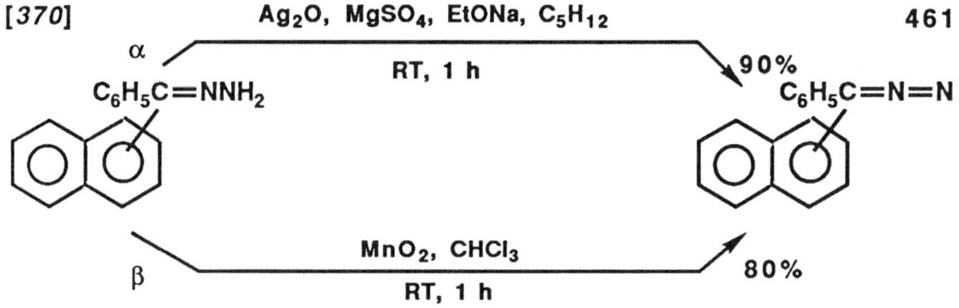

[390] (C₆H₅)₂C=NNH₂ —HgO (yellow), petroleum ether, RT, 6 h→ 459 89-96% (C₆H₅)₂C=N=N

[387] —HgO (yellow), Na₂SO₄, KOH, EtOH, Et₂O, RT, 75 min→ 89%

[387] —HgO (red), Na₂SO₄, KOH, EtOH, Et₂O, RT, 75 min→ 95%

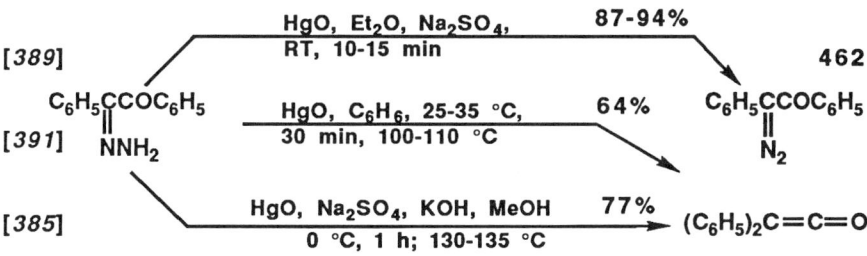

[388] fluorenone hydrazone —HgO (yellow), KOH, H₂O, Et₂O, RT, 45 min→ 460 99% fluorenone diazo

[935] —NiO₂, Et₂O, 20 °C, 1 h→ 92%

[370] α-naphthyl-C(=NNH₂) —Ag₂O, MgSO₄, EtONa, C₅H₁₂, RT, 1 h→ 461 90% α-naphthyl-C=N=N

β —MnO₂, CHCl₃, RT, 1 h→ 80%

Monohydrazones of α-diketones are converted into α-diazoketones [389, 391, 406], which, at higher temperature, give ketenes [385, 406] (equation 462).

[389] C₆H₅C(=NNH₂)COC₆H₅ —HgO, Et₂O, Na₂SO₄, RT, 10-15 min→ 87-94% 462 C₆H₅C(=N₂)COC₆H₅

[391] —HgO, C₆H₆, 25-35 °C, 30 min, 100-110 °C→ 64%

[385] —HgO, Na₂SO₄, KOH, MeOH, 0 °C, 1 h; 130-135 °C→ 77% (C₆H₅)₂C=C=O

Dihydrazones of α-diketones, when treated with mercuric oxide at higher temperatures, yield **acetylenes** [393, 394, 395, 396]. Diphenylacetylene is obtained by refluxing benzil dihydrazone with mercuric oxide [396] in benzene or by refluxing benzil monohydrazone with mercurous trifluoroacetate in ether for 2 h (yield 43%) [405]. The oxidation of dihydrazones of α-diketones is suitable for the synthesis of macrocyclic acetylenes [*393, 394*] (equations 463 and 464).

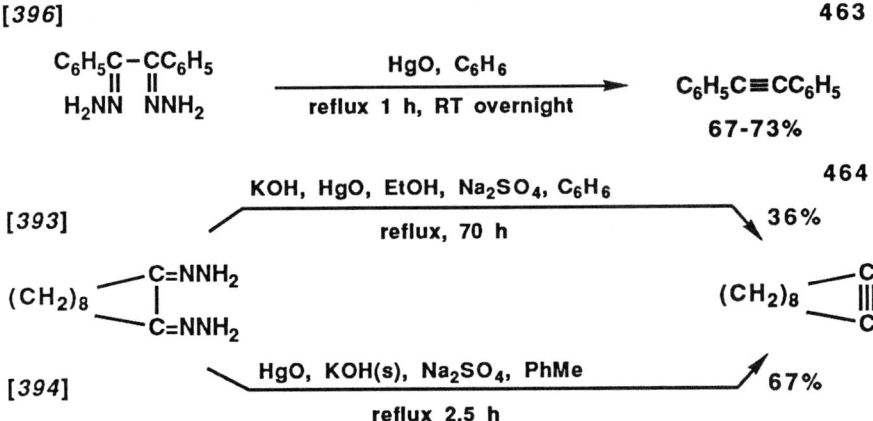

CARBOXYLIC ACIDS AND THEIR DERIVATIVES

Oxidation of Carboxylic Acids and Their Esters

The oxidation of *carboxylic acids to peroxy acids* is carried out by hydrogen peroxide. Lauric acid heated with 90% hydrogen peroxide in the presence of *p*-toluenesulfonic acid and a detergent for 2.25 h furnishes 95% of peroxylauric acid [*174*]. Benzoic acid is converted into peroxybenzoic acid in 85–90% yield on treatment with 70% hydrogen peroxide in the presence of methanesulfonic acid at 25–30 °C for 2.5 h [*176*].

m-Chloroperoxybenzoic acid is obtained in an 80–85% yield by the reaction at 15–25 °C of *m*-chlorobenzoyl chloride with 30% hydrogen peroxide in a solution of sodium hydroxide in aqueous dioxane [*1180*]. Peroxyphthalic acid is similarly prepared from phthalic anhydride (equation 465) [*175, 331*].

Esters of carboxylic acids undergo oxidative **coupling in the α positions** with respect to ester groups when treated in the enolate form with ferric chloride. Thus ethyl acetate stirred with lithium diisopropylamide in tetrahydrofuran at –78 °C for 10 min and subsequently with anhydrous ferric chloride in dimethylformamide yields 69% of diethyl succinate [911].

Under similar conditions but with different oxidants, **esters** are **hydroxylated** in the α positions with respect to the ester groups. Ethyl phenylacetate in tetrahydrofuran treated with lithium diisopropyl amide at –78 °C followed by oxidation with molybdenum oxide complex gives ethyl mandelate in 58% isolated yield [531]. Another way to obtain α-hydroxylated (or methoxylated) esters is by oxidation with iodobenzene diacetate in the presence of potassium hydroxide (or sodium methoxide) (equation 466) [794].

[794] PhI(OAc)$_2$, 3 KOH ← $C_6H_5CH_2CO_2Me$ → PhI(OAc)$_2$, 3 MeONa   466
RT, 3 days

$C_6H_5CHCO_2CH_3$ 50%   70% $C_6H_5CHCO_2CH_3$
  | |
  OH   OCH$_3$

When chiral oxidants such as (+)-2R,8aS and (–)-2S,8aR camphorylsulfonyloxaziridines are used, the hydroxylation of enolates in the α positions with respect to the ester groups occurs enantioselectively, with enantiomeric excesses ranging from 12 to 85.5% and yields ranging from 35 to 88.5% (equation 467) [1032].

[1032]   1. LDA, THF, -78 °C, 25-30 min   467

(+)-2R, 8aS   (-)-R   84%   54% ee

Methylene groups in the α positions with respect to carboxylic groups are changed to carbonyl groups; thus **α-keto esters** are formed. Methyl phenylacetate is oxidized with air in the presence of cobalt benzoate at 110–115 °C to methyl phenylglyoxylate in 86% yield [8]. Diethyl malonate is converted by nitrogen sesquioxide (N_2O_3) into diethyl mesoxalate in 74–76% yield [450].

Esters and ***lactones*** can be oxidized at the alcoholic parts to give carboxylic acids with the same number of carbon atoms. 2-Fluoroheptyl acetate is transformed in 78–84% yield into 2-fluoroheptanoic acid by heating for 25 h at 48 °C with **nitric acid** and acetic acid [*472*]. 3,5-Dimethoxyphthalide is converted into 3,5-dimethoxyphthalic acid in 90% yield by treatment with alkaline **potassium permanganate** at 0 °C for 12 h and at room temperature for 24 h (equation 468) [*879*].

[*879*]   468

3,5-dimethoxyphthalide $\xrightarrow{\text{KMnO}_4,\ \text{NaOH}}_{0\ °C,\ 12\ h,\ RT,\ 24\ h}$ 3,5-dimethoxyphthalic acid   90%

Degradative oxidation of acids to aldehydes with one less carbon is achieved by **tetrabutylammonium periodate.** Phenylacetic acids give benzaldehydes in 50–85% yields when refluxed with the reagent in dioxane [*777*].

Electrolytic decarboxylative coupling of sodium salts of carboxylic acids takes place during their electrolysis. Carbon dioxide is eliminated, and the free radicals thus generated couple to form ***hydrocarbons*** or their derivatives. The reaction is referred to as the ***Kolbe electrosynthesis*** and is exemplified by the synthesis of 1,8-difluorooctane from 5-fluorovaleric acid (equation 469) [*574*]. Yields of homologous halogenated acids range from 31% to 82% [*574*].

[*574*]   469

$$2\ F(CH_2)_4CO_2H \xrightarrow[\text{2. HCl, reflux 1 h}]{\text{1. Na, MeOH; 1.5 A, 3 h}} \left[2\ F(CH_2)_4CO_2Na\right] \xrightarrow{-2\ CO_2} \left[2\ F(CH_2)_4^{\cdot}\right] \longrightarrow F(CH_2)_8F$$

                                                                                                                                                    45%

Destructive oxidation of any methyl-group-containing compound to acetic acid is accomplished by heating the compound with chromic acid and sulfuric acid at 165–170 °C for 30–60 min. The oxidation is known as the ***Kuhn–Roth determination of methyl groups*** [*1181*]. Its significance waned with the advent of nuclear magnetic resonance techniques.

Unsaturated acids undergo many oxidative reactions. ***Epoxidation*** of unsaturated acids is carried out with the same oxidants as those used with simple alkenes and unsaturated carbonyl compounds: **hydrogen peroxide** [*138*], **peroxybenzoic acid** [*295*], and *N***-bromosuccinimide** [*746*] (equations 470 and 471).

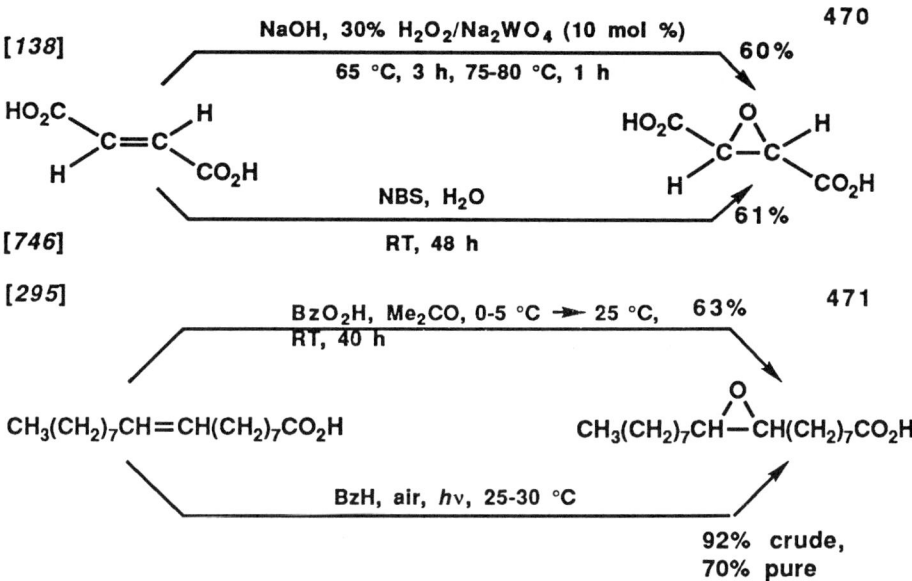

Similar oxidants are used for *epoxidation of esters* of unsaturated carboxylic acids. Methyl oleate is oxidized with peroxybenzoic acid [*295*] or peroxylauric acid [*174*] to methyl 9,10-epoxystearate acid in respective yields of 67 and 76%. Alkaline 50% hydrogen peroxide in methanolic solution transforms diethyl ethylidenemalonate at pH 8.5–9.0 and at 35–40 °C over a period of 1 h into ethyl 2-ethoxycarbonyl-2,3-epoxybutyrate in 82% yield [*145*]. A somewhat exotic oxidizing agent, *dimethyldioxirane*, converts ethyl *trans*-cinnamate into ethyl 2,3-epoxyhydrocinnamate in 63% isolated yield [*210*].

Hydroxylation at double bonds of unsaturated carboxylic acids is accomplished stereoselectively by the same reagents as those used to hydroxylate alkenes. *syn* Hydroxylation is carried out with potassium permanganate [*101*] or **osmium tetroxide** with hydrogen peroxide [*130*], sodium chlorate [*310, 715*], potassium chlorate [*715*], or silver chlorate [*310*] as reoxidant. *anti* Hydroxylation is achieved with peroxyacids, such as peroxybenzoic acid [*310*] or peroxyformic acid, prepared in situ from hydrogen peroxide and formic acid [*101*] (equation 472).

A rare *substitution hydroxylation* is effected by **Fusarium solani**, which converts *o*-hydroxy-*cis*-cinnamic acid into 4-hydroxycoumarin (equation 473) [*1089*].

[101]   472

$$\underset{\underset{5\ °C,\ 10\ min}{KMnO_4,\ KOH}}{\xleftarrow{}} HO_2C(CH_2)_7CH=CH(CH_2)_7CO_2H \xrightarrow[40\ °C,\ 1\ h]{H_2O_2,\ HCO_2H}$$

HO$_2$C(CH$_2$)$_7$CH—CH(CH$_2$)$_7$CO$_2$H
             | |
           OH   OH
             55%

                                      OH
                                       |
                    HO$_2$C(CH$_2$)$_7$CH—CH(CH$_2$)$_7$CO$_2$H
                                  |
                                OH 93%

[1089]   473

o-HOC$_6$H$_4$-CH=CH-CO$_2$H $\xrightarrow[RT]{Fusarium\ solani}$ 4-hydroxycoumarin

52-53%

In the presence of acetic anhydride, **potassium permanganate** oxidizes oleic acid at −5 to 10 °C over a period of 2.5 h to **9,10-diketostearic acid** in 46% isolated yield [*861*].

The oxidation of unsaturated esters to **unsaturated keto esters** is accomplished with **chromium trioxide** in acetic acid and acetic anhydride. Methyl 2-nonenoate is thus converted in benzene solution at 0–20 °C into methyl 4-keto-2-nonenoate in 86% yield [*553*].

Degradative oxidation of unsaturated carboxylic acids leads either to ***aldehydes*** or to ***dicarboxylic acids.*** In the presence of sodium carbonate and benzene, *o*-nitrocinnamic acid is oxidized with **potassium permanganate** at 10 °C to *o*-nitrobenzaldehyde in 64% isolated yield [*842*].

Undecylenic acid (10-undecenoic acid), when treated with **potassium permanganate and sodium periodate** in an aqueous solution of potassium carbonate at 20 °C for 20 h, gives sebacic acid (decanedioic acid) in 73% isolated yield [*763*]. A similar oxidation of oleic or elaidic acid yields a mixture of pelargonic acid and azelaic acid [*763*] (equation 474).

Acetylenic acids are oxidized to ***α-diketo carboxylic acids*** by **potassium permanganate** in aqueous solutions buffered to pH 7.0–7.5 by carbon dioxide (equation 475) [*864*].

Ozone, when combined with hydrogen peroxide [*101*] or **peroxyacetic acid** [*268*], cleaves the triple bond to yield ***carboxylic acids.*** Stearolic acid is converted into a mixture of pelargonic acid and azelaic acid [*268*] (equation 474). 9-Octadecynedioic acid, on treatment with ozone in acetic acid at room temperature followed by oxidation with hydrogen peroxide, gives azelaic acid in 81% yield [*101*].

The triple bond in ***esters of acetylenic alcohols*** behaves similarly. 2,5-Diacetoxy-2,5-dimethyl-3-hexyne is converted by a 2% aqueous solution

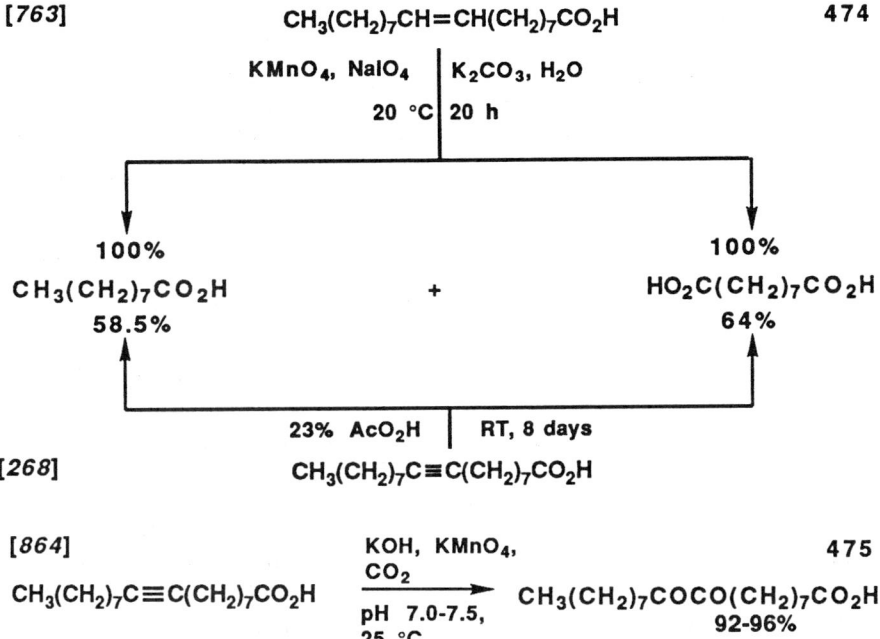

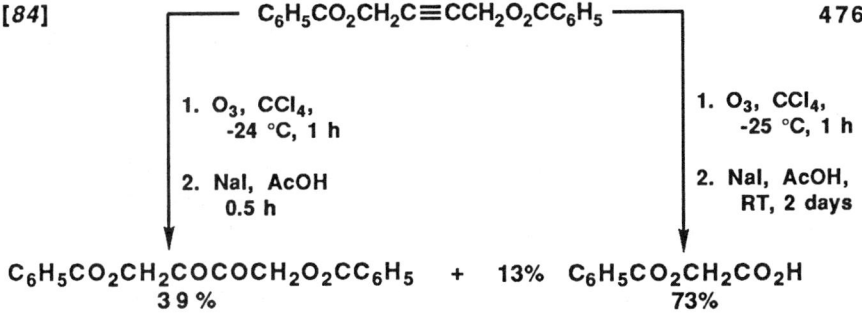

of **potassium permanganate,** while cooled with ice, into 2,5-diacetoxy-2,5-dimethyl-3,4-hexanedione in 67% yield [1117]. 1,4-Dibenzoxy-2-butyne yields, on **ozonization** at −25 °C and subsequent treatment with a solution of sodium iodide in acetic acid, 39% of 1,4-dibenzoxy-2,3-butanedione and 13% of benzoylglycolic acid. After 2 days at room temperature, the yield of the benzoylglycolic acid rises to 73% (equation 476) [84].

Hydroxy carboxylic acids and their esters are oxidized to ***keto acids and their esters,*** respectively. *o*-Nitromandelic acid is oxidized to *o*-nitrophenylglyoxylic acid with a dilute solution of **potassium permanganate** at room temperature in 54% yield [886]. The oxidation of ethyl 3-hydroxycyclobutanecarboxylate with **ruthenium tetroxide and sodium periodate** in a mixture of water and carbon tetrachloride at room temperature overnight gives ethyl 3-cyclobutanonecarboxylate in 78% yield [*941*]. 4-Hydroxyundecanoic acid (in the form of its lactone) is oxidized with **bromine** in aqueous sodium hydroxide at pH 6–6.5 at 0 °C to give a 75–85% yield of 4-ketoundecenoic acid [*729*]. Mandelic acid [*887*] and ethyl mandelate [*839*] are oxidized with potassium permanganate to phenylglyoxylic acid and its ethyl ester, respectively (equation 477).

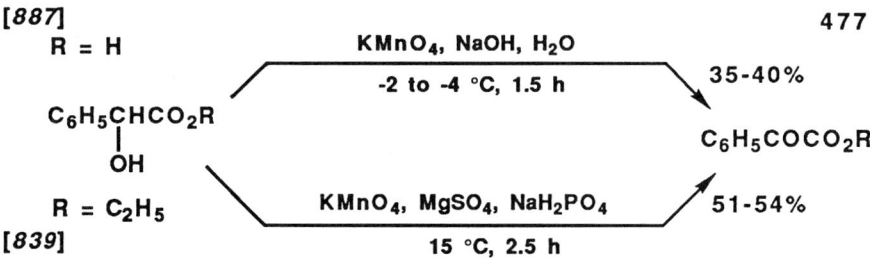

Degradative oxidation of α-hydroxy carboxylic acids furnishes ***aldehydes*** or ***ketones***. Glycolic acid, lactic acid, and mandelic acid are converted into formaldehyde, acetaldehyde, and benzaldehyde, respectively, in 50% yields on refluxing with **N-bromosuccinimide** in water [*745*]. **Periodates**, with phase-transfer reagents, convert α-hydroxyhexanoic acid and mandelic acid into valeraldehyde and benzaldehyde in respective yields of 90 and 88% (equation 478) [*776*, *778*].

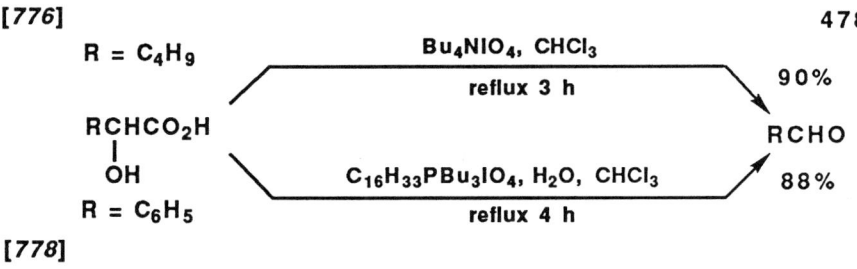

The conversion of α-hydroxy acids into aldehydes with one less carbon has great importance in the chemistry of sugars. Oxidation with **bromine water** transforms aldoses into the corresponding aldonic acids, which, in the form of their calcium salts, are treated with aqueous **hydrogen peroxide** in the presence of ferrous or ferric sulfate **(*Fenton reagent*)** and are degraded to aldoses with one less carbon (***Ruff degradation***) (equation 479) [*87*].

If the hydroxylic group next to the carboxyl is tertiary, decarboxylative oxidation gives ketones. Benzilic acid is degraded to benzophenone in 85–90% yield on refluxing for 20 min with water and **N-bromosuccinimide** [*745*]. 2-(*p*-Methoxybenzyl)-*p*-methoxymandelic acid is converted into des-

[87]   479

$$\begin{array}{c} CO_2 \frac{Ca}{2} \\ HO \!-\!\!\!\!\!\!-\!\!\!\!\!\!-\! H \\ H \!-\!\!\!\!\!\!-\!\!\!\!\!\!-\! OH \\ H \!-\!\!\!\!\!\!-\!\!\!\!\!\!-\! OH \\ CH_2OH \end{array} \xrightarrow[\text{2. } H_2O_2, \text{ 40-60 °C, RT, overnight}]{\text{1. } Ba(OAc)_2 \cdot H_2O, \text{ } Fe_2(SO_4)_3, \text{ } H_2O, \text{ reflux}} \begin{array}{c} CHO \\ H \!-\!\!\!\!\!\!-\!\!\!\!\!\!-\! OH \\ H \!-\!\!\!\!\!\!-\!\!\!\!\!\!-\! OH \\ CH_2OH \\ 46\% \end{array}$$

oxyanisoin (*p*-methoxybenzyl *p*-methoxyphenyl ketone) in 98% yield on heating with **red lead oxide** (Pb_3O_4) in acetic acid at 65–70 °C [*143*].

The oxidation of **α-*amino acids*** gives either ***aldehydes or acids*** with one less carbon. ***Aldehydes*** result from the treatment of the amino acids with a slightly alkaline solution of **sodium hypochlorite.** Valine, sodium hypochlorite, and 10 mol % of sodium hydroxide, on steam distillation, furnish isobutyraldehyde in 79% yield [*700*]. Oxidation to carboxylic acids is achieved on heating of the amino acid with 2.5% **hydrogen peroxide** to 70 °C. Glutamic acid is transformed into succinic acid in 47% yield [*188*]. A more general method seems to be oxidation with *argentic oxide* at 25–70 °C to give 49–100% yields of ***carboxylic acids*** (equation 480) [*381*].

[*381*] $RCHCO_2H$ $\xrightarrow[\text{25-70° C}]{AgO, H_2O}$ RCO_2H 480
                |
               NH_2

   $R = (CH_3)_2CH$ 100%

   $R = (CH_3)_2CHCH_2$ 61%

α-*Keto acids* are oxidized to carboxylic acids having one less carbon with hydrogen peroxide. For example, 3,4-dimethoxyphenylpyruvic acid is degraded to homoveratric acid with 30% **hydrogen peroxide** at room temperature [*189*]. **α-*Keto esters*** undergo the ***Baeyer–Villiger reaction*** on treatment with **peroxyphthalic acid** and give mixed anhydrides (equation 481) [*334*].

[*334*]   481

$C_6H_5COCO_2C_2H_5$ $\xrightarrow[\text{22 °C, 90 h}]{\text{o-}C_6H_4(CO_3H)(CO_2H)}$ $C_6H_5COOCO_2C_2H_5$

Oxidation of Amides, Hydrazides, and Nitriles

The oxidation of amides, when it takes place at carbon atoms not adjoining the nitrogen atom, is discussed in the section Oxidation of the Carbon Chain of Amines (*see* equations 516–518).

The oxidation of *primary amides* with **lead tetraacetate** gives *isocyanates,* when carried out in dimethylformamide; *carbonates,* when carried out in the presence of alcohols; and *ureas,* when carried out in the presence of amines (equation 482) [*1182*].

[*1182*]   482

$(CH_3)_3CCONH_2 \xrightarrow[50\text{-}60\ ^\circ C]{Pb(OAc)_4,\ DMF,\ Et_3N} (CH_3)_3CNCO$

44-96%

Hydrazides are converted into *aldehydes* by **potassium ferricyanide** in the presence of aqueous ammonia in 23–64% yields [*921*]. On treatment with **ceric ammonium nitrate,** the corresponding *carboxylic acids* are formed in 65–90% yields [*428*] (equation 483).

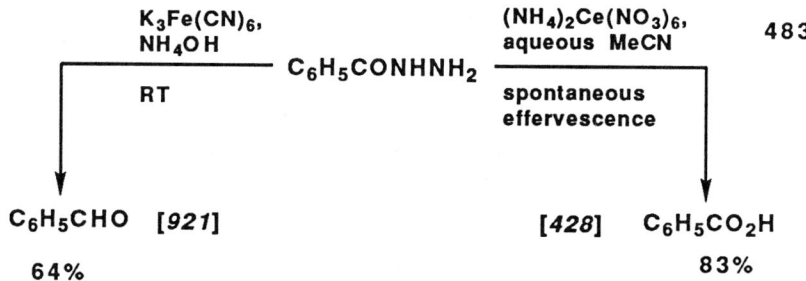

Unsaturated nitriles are *epoxidized* with **hydrogen peroxide.** Tetracyanoethylene, after treatment with 30% hydrogen peroxide in acetonitrile at 10–12 °C, furnishes tetracyanoepoxyethane in 59–68% yield [*150*]. A similar treatment of acrylonitrile gives a 62% yield of glycidamide (epoxypropionamide) [*147*].

NITROGEN COMPOUNDS

Oxidations of nitrogen compounds include oxidation at nitrogen as well as oxidations at carbons carrying the nitrogen functional groups.

Oxidation of Nitro Compounds

Because the nitrogen in nitro compounds is at the highest oxidation state, nitro compounds can be oxidized only in the carbon chain to which the nitro group is attached. The oxidation of *primary and secondary nitro compounds* in the form of their nitronic acids in alkaline medium by **potassium permanganate** represents a variation of the *Nef reaction* and gives the same products: *aldehydes* from primary nitro compounds and *ketones* from secondary nitro compounds (equation 484) [*867*].

[867]   484

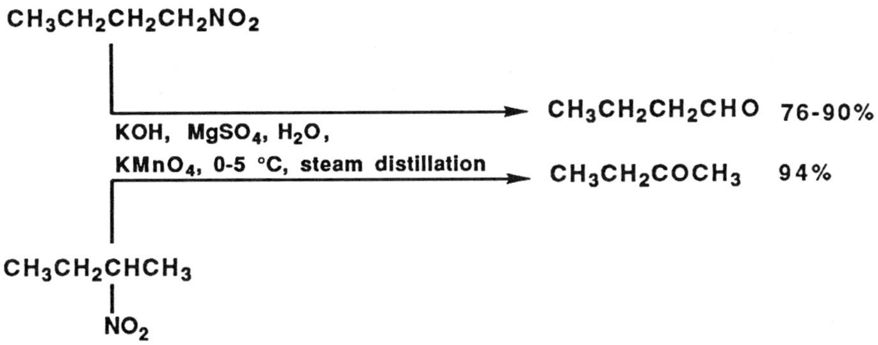

$CH_3CH_2CH_2CH_2NO_2$

$\quad\quad\quad\quad\quad\quad\quad\quad\quad\quad\quad\quad$ → $CH_3CH_2CH_2CHO$ 76–90%

KOH, $MgSO_4$, H_2O,
$KMnO_4$, 0–5 °C, steam distillation → $CH_3CH_2COCH_3$ 94%

$CH_3CH_2CHCH_3$
       |
     NO_2

Oxidation of Nitroso Compounds

Nitroso compounds are oxidized to *nitro compounds* by **hydrogen peroxide** or its derivatives and also by **potassium dichromate** (equation 485) [*658*].

[*658*]   485

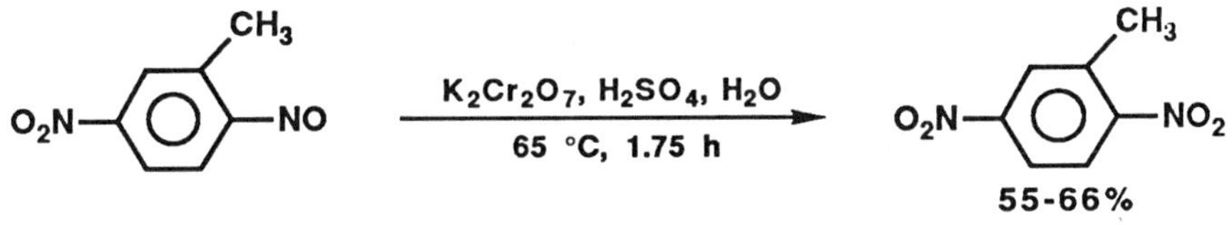

O_2N–⟨Ar(CH_3)⟩–NO $\xrightarrow{K_2Cr_2O_7,\ H_2SO_4,\ H_2O}_{65\ °C,\ 1.75\ h}$ O_2N–⟨Ar(CH_3)⟩–NO_2

                                                                             55–66%

The oxidation of 3-nitroso-2,6-diaminopyridine with 30% hydrogen peroxide at 35 °C for 1.5 h furnishes 3-nitro-2,6-diaminopyridine in 96% yield [*278*]. Under these conditions, the primary amino groups are not affected. Nitrosoamines are oxidized with **peroxytrifluoroacetic acid** to nitroamines in 73–95% yields (equation 486) [*291*].

[*291*]   486

$(C_2H_5)_2NNO$ $\xrightarrow[\text{RT, 30 min; reflux 1 h}]{90\%\ H_2O_2,\ (CF_3CO)_2O,\ CH_2Cl_2}$ $(C_2H_5)_2NNO_2$

                                                                             76%

Oxidation of Hydroxylamines

Because nitroso compounds cannot be prepared by the reduction of nitro compounds, the oxidation of *hydroxylamines* is often the best way for their preparation. Nitroso compounds are obtained by treatment of hydroxylamines with **silver oxide** [*373*], **silver carbonate** [*378*], **sodium dichromate** [*645*], **potassium dichromate** [*657*], **manganese dioxide** [*816*], and diethyl azodicarboxylate [*978*] (equation 487).

$C_6H_5NHOH \longrightarrow C_6H_5NO$ 487

[378]	Ag_2CO_3, CH_2Cl_2, RT, 2 min	85%
[645]	$Na_2Cr_2O_7$, H_2SO_4, H_2O, 0 °C, 3 min	49-53%
[657]	$K_2Cr_2O_7$, H_2SO_4, H_2O, 0 °C	90%
[816]	MnO_2, H_2O, 0 °C, 3 h	40%
[978]	$EtO_2CN=NCO_2Et$, Et_2O, 0 °C, a few hours	89%

N,N-disubstituted hydroxylamines are converted into free-radical *nitroxyls* by **silver oxide** in ether in the presence of anhydrous sodium sulfate (equation 488) [*374*].

[*374*]   488

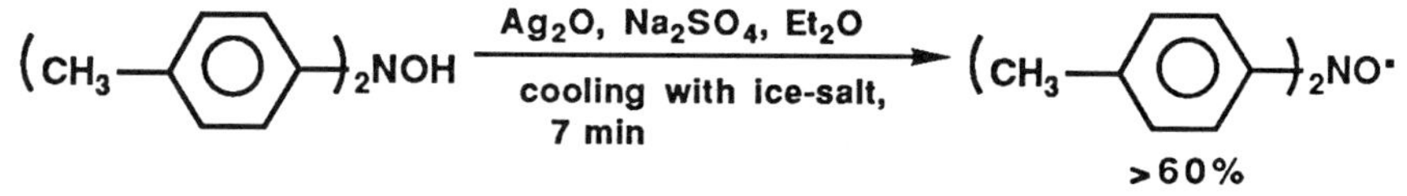

$(CH_3-C_6H_4-)_2NOH \xrightarrow[\text{cooling with ice-salt, 7 min}]{Ag_2O,\ Na_2SO_4,\ Et_2O} (CH_3-C_6H_4-)_2NO^{\bullet}$

                                                                           >60%

Oxidation of Azo Compounds and Azides

Azo compounds are oxidized with **peroxybenzoic acid** to mixtures of *azoxy compounds* (equation 489) [*312*].

[*312*]   489

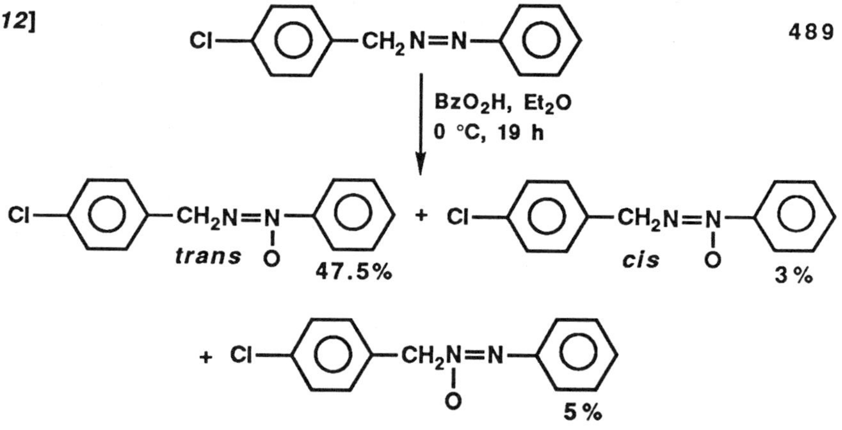

$Cl-C_6H_4-CH_2N=N-C_6H_5 \xrightarrow[0\ °C,\ 19\ h]{BzO_2H,\ Et_2O}$

$Cl-C_6H_4-CH_2N=N(\to O)-C_6H_5$ (trans) 47.5% + $Cl-C_6H_4-CH_2N=N(\to O)-C_6H_5$ (cis) 3%

+ $Cl-C_6H_4-CH_2N=N-C_6H_5$ with O on CH₂N nitrogen   5%

Alkyl *azides* are converted into *nitro compounds* by the sequence of reactions shown in equation 490 [*1183*].

[*1183*]   490

$C_8H_{17}N_3 \xrightarrow[\text{RT, 5 h}]{Ph_3P,\ CH_2Cl_2} [C_8H_{17}N=PPh_3] \xrightarrow[-78\ °C,\ 10\ min]{O_3,\ CH_2Cl_2} \begin{array}{l} C_8H_{17}NO_2 \ \ 70\% \\ + \\ C_7H_{15}CHO \ \ 13\% \end{array}$

Oxidation of Hydrazo Compounds

Hydrazo compounds are dehydrogenated to *azo compounds* by many oxidants. 2,2'-Bisazopropane is prepared by the oxidation of *sym*-diisopropylhydrazine with **cupric oxide** (equation 491) [*350*].

[*350*]   CuO   **491**
$(CH_3)_2CHNHNHCH(CH_3)_2 \cdot HCl \xrightarrow{\text{7 days, heat}} (CH_3)_2CHN=NCH(CH_3)_2$
                                                                        95%

Hydrazobenzene is transformed into azobenzene by **air**, by **manganese dioxide** [*825*], by diethyl azodicarboxylate [*977*], and by diphenylpicrylhydrazyl [*983*] (equation 492).

[*825*] MnO_2, C_6H_6, 81 °C, 24 h   **492**
                                                   97%

[*977*] $C_6H_5NHNHC_6H_5 \xrightarrow[\text{reflux 30 min}]{EtO_2CN=NCO_2Et, C_6H_6} 100\% \; C_6H_5N=NC_6H_5$

[*983*]   **diphenylpicrylhydrazyl**, $CHCl_3$ 100%

Diethyl and di-*tert*-butyl hydrazodicarboxylate are dehydrogenated to diethyl- and di-*tert*-butylazodicarboxylate, respectively, by **nitric acid** [*1184*] and by **N-bromosuccinimide** [*1185*] (equation 493).

[*1184*]   **493**
R = C_2H_5 yellow fuming HNO_3
                           5 °C, 2 h   70-80%

$RO_2CNHNHCO_2R$ $RO_2CN=NCO_2R$

                     NBS, C_5H_5N, CH_2Cl_2 90-95%
R = $(CH_3)_3C$ RT, 11-12 min
[*1185*]

The dehydrogenation of 1,1'-dicyanohydrazocyclohexane by **bromine** in ethanolic hydrochloric acid at 15 °C furnishes 1,1'-dicyanoazocyclohexane in 84–90% yield [*733*]. The treatment of 3,3-pentamethylenediaziridine with **silver oxide** gives the corresponding diazirine (equation 494) [*372*].

[*372*]   **494**

cyclohexane-spiro-diaziridine (NH–NH) $\xrightarrow[\text{RT, 35-65 min}]{Ag_2O, Et_2O}$ cyclohexane-spiro-diazirine (N=N)

                                                       65-75%

Oxidation of Primary Amines at Nitrogen

Primary amines are oxidized to azo, azoxy, nitroso, or nitro compounds, depending on the oxidants used. Most of the examples include aromatic amines, because aliphatic amines that have hydrogens on the carbons carrying the amino groups may undergo dehydrogenations to carbonyl compounds, imines, and nitriles (*see* equations 508–514).

CONVERSION OF PRIMARY AMINES INTO AZO COMPOUNDS

The dehydrogenation of *primary amines into azo compounds* is achieved by sodium perborate, which converts N-acetyl-p-phenylenediamine into p,p'-diacetamidoazobenzene in 58% yield on heating at 50–60 °C with boric acid and acetic acid for 6 h [193].

More common oxidation reagents are **oxygen** in the presence of cobalt peroxide [1136], **manganese dioxide** [813], barium manganate [833], silver permanganate [897], and nickel peroxide [936] (equation 495).

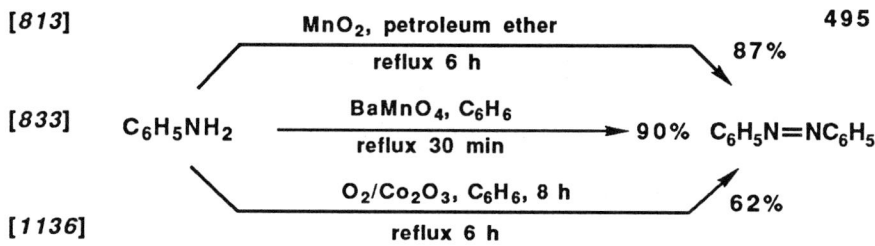

p-Chloroaniline is oxidized to p,p'-dichloroazobenzene on refluxing for 15 min with a silver permanganate–pyridine complex in benzene (yield 85%) [897]. p-Nitroaniline is converted into p,p'-dinitroazobenzene on refluxing with nickel peroxide in benzene for 6 h (yield 44%) [936].

Hydrogen peroxide [129, 249] and **peroxy acids** [1186] oxidize primary aromatic amines to *azoxy compounds*. Peroxyacetic acid oxidizes 2,6-difluoroaniline to 2,2',6,6'-tetrafluoroazoxybenzene in 48% yield [1186]. The oxidation of aniline with hydrogen peroxide or with peroxyacetic acid yields azoxybenzene (equation 496) [129, 249]. Azoxy compounds are usually accompanied by nitroso and nitro compounds.

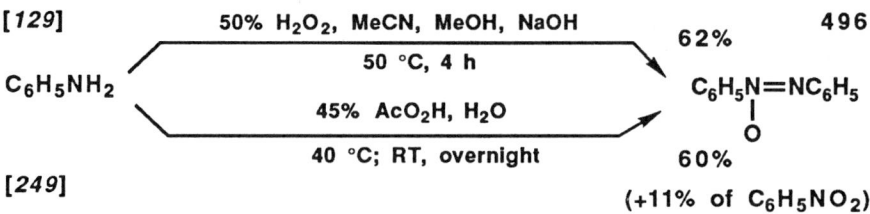

OXIDATION OF PRIMARY AMINES TO NITROSO AND NITRO COMPOUNDS

Oxidants suitable for the partial oxidation of amines to *nitroso compounds* are peroxy acids: **Caro acid,** which is prepared in situ from potassium persulfate and sulfuric acid [*195, 199*]; **potassium peroxysulfate (Oxone)** [*295*]; **peroxyacetic acid** [*153*], and **peroxybenzoic acid** [*1186*].

3-Nitro-*o*-toluidine [*195*] and 5-nitro-*o*-toluidine [*199*] in aqueous–alcoholic solutions, when treated with a mixture of potassium persulfate and concentrated sulfuric acid, give 3-nitro-2-nitrosotoluene and 5-nitro-2-nitrosotoluene in respective yields of 60 and 55–66%. Organic peroxy acids convert 2,6-dihaloanilines into 2,6-dihalonitrosobenzenes (equation 497) [*153, 1186*]. *p*-Phenylenediamine (1,4-diaminobenzene) is oxidized by Oxone ($2KHSO_5 \cdot KHSO_4 \cdot K_2SO_4$) in an aqueous suspension at room temperature to *p*-dinitrosobenzene in a quantitative yield [*205*].

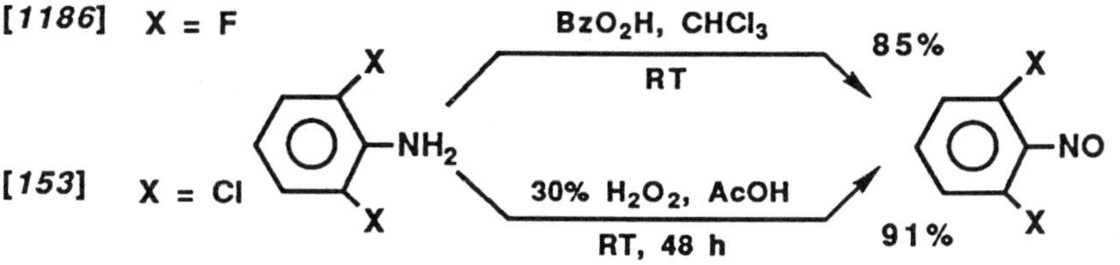

[*1186*] X = F
[*153*] X = Cl

BzO_2H, $CHCl_3$, RT → 85%
30% H_2O_2, AcOH, RT, 48 h → 91%

(equation 497)

The oxidation of some *aliphatic amines* is a good route to *aliphatic nitro compounds.* *tert*-Butylamine is oxidized in 83% yield to 2-methyl-2-nitropropane by **potassium permanganate** in water at 45 °C for 8 h and at 55 °C for 8 h [*738, 892*]. Under similar conditions, 4-amino-2,2,4-trimethylpentane is converted into 4-nitro-2,2,4-trimethylpentane in 69–82% yields [*859*]. Refluxing 3α-acetoxy-20α-amino-5β-pregnane with a chloroform solution of **m-chloroperoxybenzoic acid** for 40 min furnishes 3α-acetoxy-20α-nitro-5β-pregnane in 66% yield [*319*]. 2-Aminobutane is converted into 2-nitrobutane by **peroxyacetic acid** [*253*] or **dimethyldioxirane** [*211*] (equation 498).

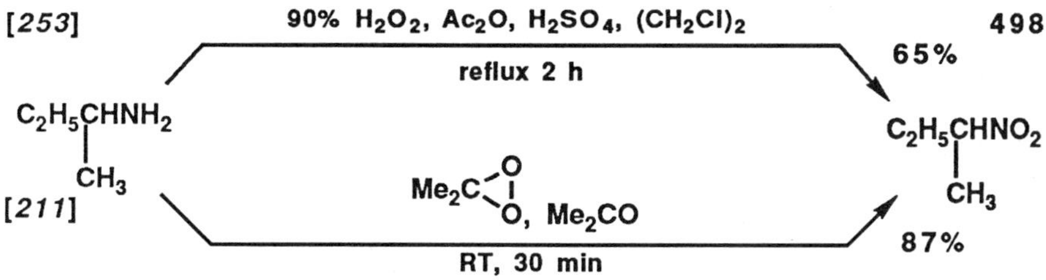

[*253*] 90% H_2O_2, Ac_2O, H_2SO_4, $(CH_2Cl)_2$, reflux 2 h → 65%

$C_2H_5CHNH_2$ | CH_3

[*211*] $Me_2C{<}^O_O$, Me_2CO, RT, 30 min → 87%

$C_2H_5CHNO_2$ | CH_3

(equation 498)

Primary aromatic amines are oxidized to *nitro compounds* by **peroxy acids:** perboric acid [*194*], peroxysulfuric acid [*203*], peroxyacetic acid [*253*], peroxytrifluoroacetic acid [*281, 285, 290*], and peroxymaleic acid [*338*] (equations 499 and 500).

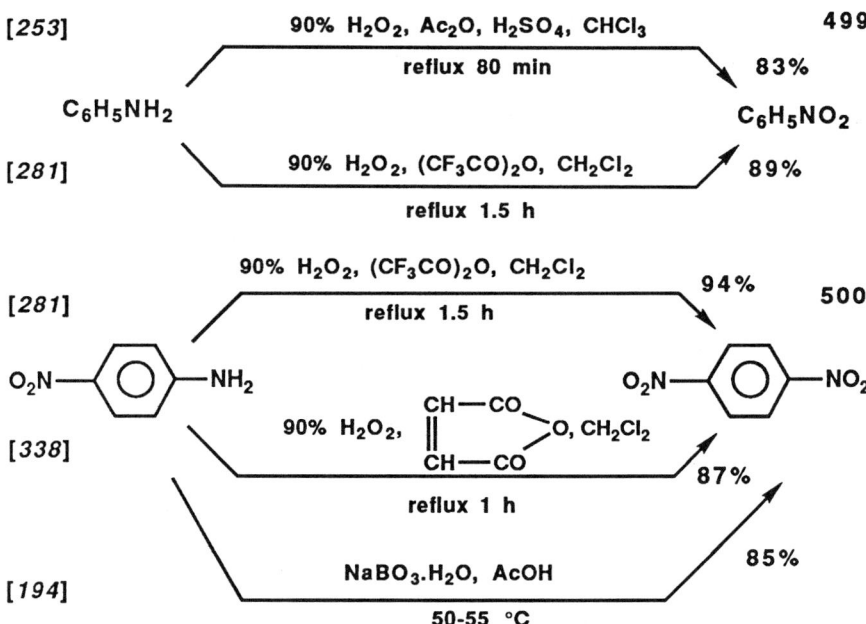

Whereas peroxyacetic acid oxidizes 2,6-dichloroaniline to 2,6-dichloronitrosobenzene [*153*], peroxytrifluoroacetic acid carries the oxidation to the nitro compound. Refluxing 2,6-dichloroaniline with peroxytrifluoroacetic acid, which is prepared in situ from 90% hydrogen peroxide and trifluoroacetic anhydride in dichloromethane, gives 2,6-dichloronitrobenzene in 89–92% crude yields and 59–73% pure yields [*290*]. 2-Amino-4-methylpyridine treated with 30% hydrogen peroxide in fuming sulfuric acid at 10–25 °C for 50 h yields 68% of 4-methyl-2-nitropyridine [*203*].

The **biochemical oxidation** of an amino group to a nitro group is exemplified by the treatment of *p*-aminobenzoic acid with ***Streptomyces thioluteus***. At pH 7 and 30 °C, 90% of the amino acid is converted into *p*-nitrobenzoic acid in 2 h [*1085*].

Oxidation of Secondary and Tertiary Amines at Nitrogen

The oxidation of *secondary amines* with dilute (6%) hydrogen peroxide at room temperature furnishes *dialkylhydroxylamines*. N,N-Diethylhydroxylamine is prepared with a yield not higher than 50%. With dipropylamine, the yield is not stated [*154*].

The oxidation of *tertiary amines* to *amine oxides* is carried out by peroxy compounds, most often **hydrogen peroxide** and **peroxyacetic acid**.

Other peroxides such as **tert-butyl hydroperoxide** [*226, 227*], performic acid [*247, 248*], peroxytrifluoroacetic acid [*158*], peroxydichloromaleic acid [*339*], and dimethyldioxirane [*210*] are used less frequently.

Trialkylamines such as tributylamine are oxidized either with aqueous–methanolic 30–35% hydrogen peroxide at 0 °C to room temperature over 26.5 h or with 40% peroxyacetic acid at 0 °C to room temperature for 4.5 h. The yields of tributylamine oxide are comparable: 98 and 95%, respectively [*156*]. Dimethyldodecylamine oxide is prepared by heating the tertiary amine with *tert*-butyl hydroperoxide in the presence of vanadium catalysts (equation 501) [*226, 227*].

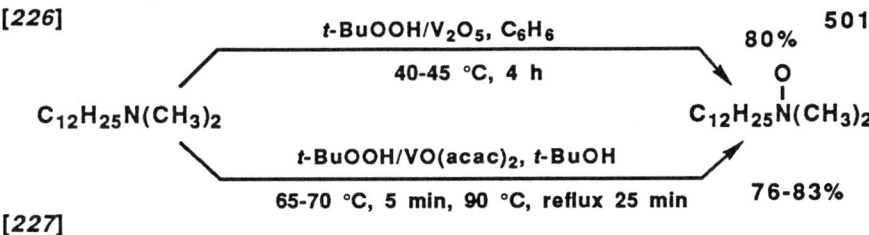

The purpose of preparing aliphatic amine oxides is usually their thermal decomposition to *cis* alkenes and *N,N*-dialkylhydroxylamines (Cope rearrangement) [*156, 161, 1187*]. Thus *N,N*-dimethylcyclohexylmethylamine is oxidized with 30% hydrogen peroxide in methanol to its oxide, whose decomposition at 90–100 °C at 10 mm of Hg and at 160 °C for 2 h furnishes 79–88% of methylenecyclohexane and 78–90% of *N,N*-dimethylhydroxylamine [*161*]. Another example is the preparation of *cis*-cyclooctene from dimethylcyclooctylamine (equation 502) [*1187*].

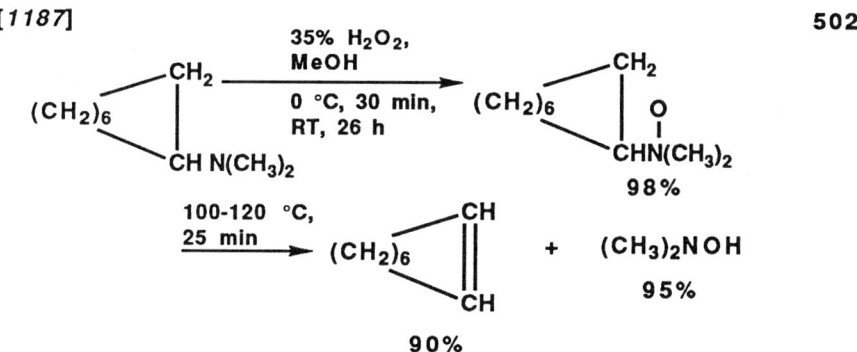

An easy route to amine oxides is the oxidation of tertiary amines with **m-chloroperoxybenzoic acid** in chloroform solution at 0–5 °C for 3 h. Trimethylamine oxide, tribenzylamine oxide, and dimethylaniline oxide are obtained in 96, 96, and 94% yields, respectively [*320*].

Pyridine is frequently oxidized to pyridine oxide (equation 503) [129, 210, 263]. Pyridine oxide is an oxidant capable of hydroxylating aromatic rings [994]. But more important, the presence of oxygen on the nitrogen of the pyridine ring reverses the direction of electrophilic substitutions in the pyridine ring. Whereas electrophilic attacks on pyridine occur in β positions, attacks on pyridine oxide occur in α and γ positions. After the introduction of the electrophiles, the pyridine oxide is converted into pyridine by mild reductions, such as treatment with salts of iron or titanium.

[129] 50% H_2O_2, PhCN, MeOH, NaOH, pH 8, 25-30 °C, 3 h → 79% 503

[263] 40% AcO_2H, 85 °C, 50-60 min → 78-83%

[210] $Me_2C\overset{O}{\underset{O}{<}}$, H_2O, KOH (pH 7.5-8.0), RT, 2 h → 93%

Like pyridine, pyridine homologues and derivatives are converted into amine oxides. α-Picoline oxide is prepared by oxidation with hydrogen peroxide in acetic acid in 83% yield [1188]. Under identical conditions, 2,5-dimethylpyridine oxide and nicotinamide oxide are obtained in 84% yield [160] and 73–82% yield [162], respectively. Pentachloropyridine is oxidized to its oxide by peroxyacetic or peroxytrifluoroacetic acid (equation 504) [158].

[158] 90% H_2O_2, AcOH, H_2SO_4, 20 °C, 72 h → 95% 504

90% H_2O_2, CF_3CO_2H, H_2SO_4, 60 °C 18 h → 95%

Interesting oxidations by peroxyformic acid take place with α- or γ-dimethylaminopolychloro- and polyfluoropyridines. Normal N-oxides are obtained only when a hydroxyl or a dimethylamino group is present in the γ position of the polyhalogenated pyridine ring (equation 505) [247, 248].

Without the presence of the electron-releasing hydroxyl or dimethylamino group, the primary products of oxidation, amine oxides, rearrange

[248] R = OH X = Cl **505** 65%

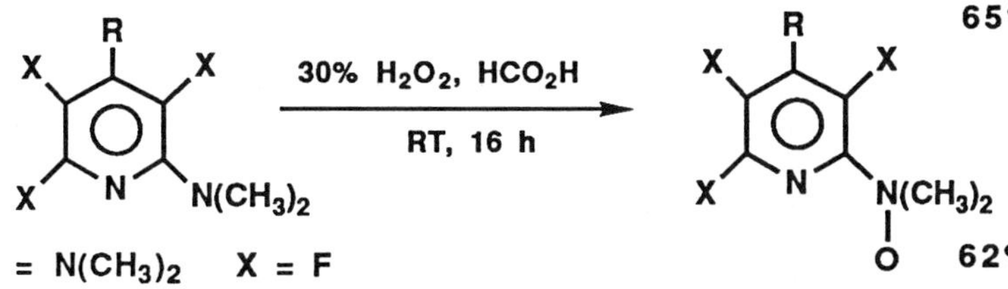
R
X
X
X
N
N(CH3)2
30% H2O2, HCO2H
RT, 16 h
R
X
X
X
N
N(CH3)2
O

[247] R = N(CH3)2 X = F 62%

intramolecularly to *N,N*-dimethylhydroxylamines [*247, 248*] (equation 506).

[247] **506**

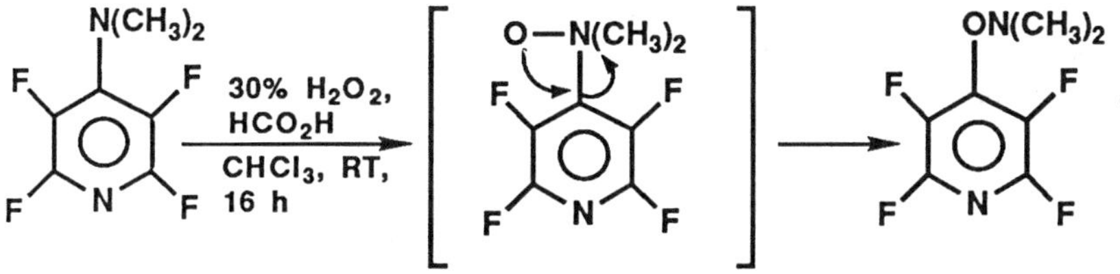
N(CH3)2
F
F
F
N
F
30% H2O2,
HCO2H
CHCl3, RT,
16 h
O—N(CH3)2
F
F
F
N
F
ON(CH3)2
F
F
F
N
F

74%

Pyridazines are oxidized to mono-*N*-oxides by peroxydichloromaleic acid prepared in situ from 90% hydrogen peroxide and dichloromaleic anhydride (equation 507) [*339*].

[339] **507**

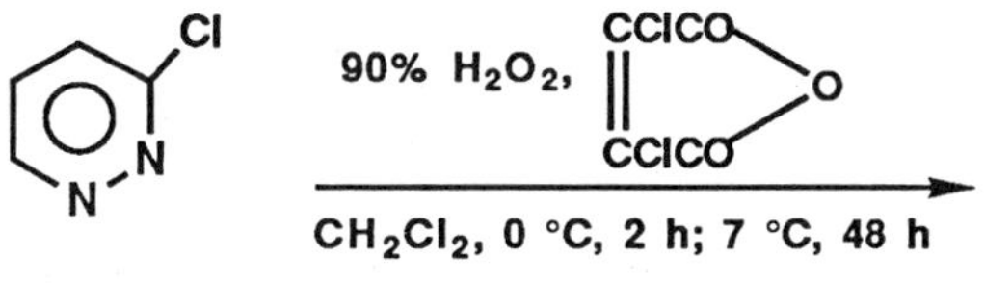
Cl
N
N
90% H2O2,
CClCO
O
CClCO
CH2Cl2, 0 °C, 2 h; 7 °C, 48 h
Cl
N
N
O

79%

m-Phenanthroline and hydrogen peroxide in acetic acid give *m*-phenanthroline di-*N*-oxide in 71% yield after a 2-h reflux [*155*].

Dehydrogenation of Primary Amines to Imines and Aldehydes or Ketones

The elimination of hydrogen from nitrogen of the amino group and from the carbon of the adjacent methylene or methine group converts ***primary amines into imino*** compounds and, in the presence of water, ***into aldehydes or ketones.*** Imines are sometimes intercepted [*198, 833, 850, 917, 918, 957, 1189*] (equation 508).

In amines with the primary amino group bonded to a secondary carbon, dehydrogenation results in the formation of ***ketones*** [*850, 851, 692*] (equation 509).

$C_6H_5CH_2NH_2 \longrightarrow C_6H_5CHO$ **508**

[198]	$Na_2S_2O_8/AgNO_3$, NaOH, H_2O, 0 → 20 °C, 1-3 h	96%*
[850]	$KMnO_4$, $ZnSO_4$, t-BuOH, H_2O, 50-75 °C, steam distillation	61%
[833]	$BaMnO_4$, C_6H_6, reflux 1 h	95%
[918]	K_2FeO_4, 10% KOH, t-BuOH, RT, 30 min	44%
[917]	K_2FeO_4, H_2O, RT, 1 min	70%
[957]	40% CH_2O, HCl, reflux; $C_6H_{12}N_4$, H_2O, reflux 15 min; HCl, reflux 5 min	66%
[1189]	$NaNO_2$, CF_3CO_2H, Me_2SO (DMSO) (mole ratio 1:2:3), 100 °C, 20 h	82%

*$C_6H_5CH=NCH_2C_6H_5$, product of a reaction of benzaldimine and benzylamine, was isolated and hydrolyzed to benzaldehyde and benzylamine.

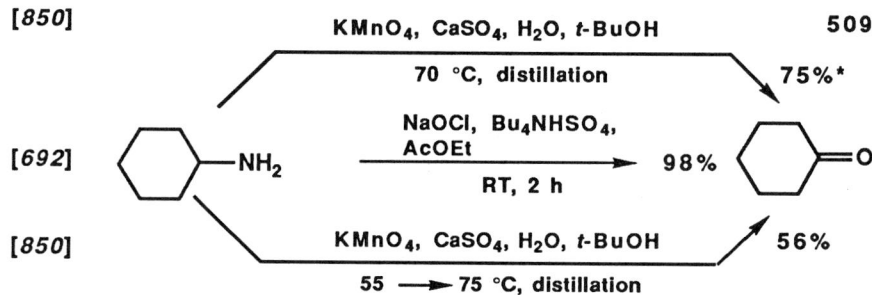

*The product was isolated as the 2,4-dinitrophenylhydrazone.

The oxidation of benzhydrylamine with **potassium permanganate** gives different products depending on the cation of the buffer used in the reaction (equation 510) [851].

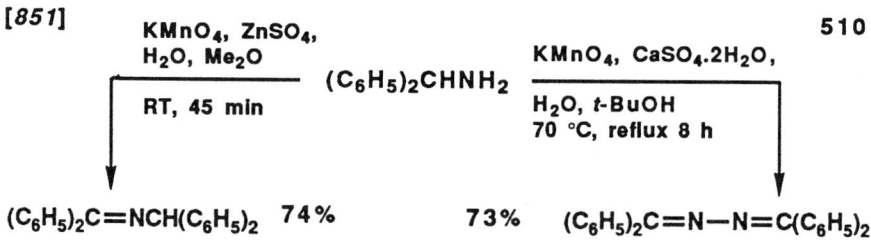

Dehydrogenation of Secondary Amines to Imines

Secondary amines having hydrogens on the α carbon are dehydrogenated to imines by **mercuric acetate** or **manganese dioxide** [825, 1091]. Piperidines give Δ^1-piperideines [1091] (equations 511 and 512).

[825] $C_6H_5CH_2NHC_6H_5$ $\xrightarrow[81\ °C]{MnO_2,\ C_6H_6}$ $C_6H_5CH=NC_6H_5$ 84%   511

[1091]

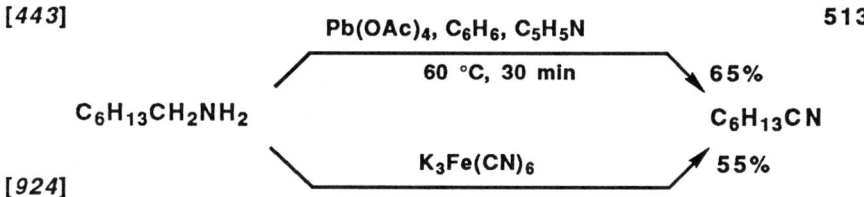

R	Temp. (°C)	Yield (%)
Me	64	31
Pr	86	42
Bu	86	32
t-Bu	~100	75

Dihydropyridine derivatives resulting from the Hantzsch synthesis are dehydrogenated to pyridines by heating with a dilute mixture of nitric acid and sulfuric acid (yields 58–65%) [458] (*see* equation 41).

Dehydrogenation of Primary Amines to Nitriles

More intensive dehydrogenation of primary amines having two hydrogen atoms on the adjacent carbons leads to nitriles. Such dehydrogenations are accomplished by argentic oxide [381], cobalt peroxide [1136], nickel peroxide [936], **lead tetraacetate** [443, 444], **sodium hypochlorite** [692], **potassium ferricyanide** [924], and potassium ruthenate [196] (equation 513).

[443]

$C_6H_{13}CH_2NH_2$ $\xrightarrow{Pb(OAc)_4,\ C_6H_6,\ C_5H_5N,\ 60\ °C,\ 30\ min}$ 65% $C_6H_{13}CN$

         $\xrightarrow{K_3Fe(CN)_6}$ 55%

[924]

513

Hexylamine is converted into capronitrile in 73% yield on refluxing with nickel peroxide in benzene for 1.5 h [936]. Octylamine is dehydrogenated to octanenitrile in 60% yield by treatment with sodium hypochlorite and tetrabutylammonium hydrogen sulfate in ethyl acetate at room temperature for 35 min [692]. The preparation of benzonitrile is shown in equation 514 [196, 381, 1136].

The oxidation of *o*-phenylenediamine with oxygen in the presence of cuprous chloride in pyridine at 25 °C furnishes mucononitrile (1,4-dicyano-1,3-butadiene) in 73–77% yield [64].

Biochemical oxidation with ***Hygrophorus conicus*** converts tryptamine into 2-indolone-3-acetic acid in 43% yield (equation 515) [1069].

[381]   AgO, H_2O, 25-70 °C   514

                                                        77%*

$C_6H_5CH_2NH_2$ O_2, Co_2O_3, C_6H_6, reflux 30 min   85% C_6H_5CN

[1136]

[196] K_2RuO_4/$K_2S_2O_8$, RT, 24 h   66%

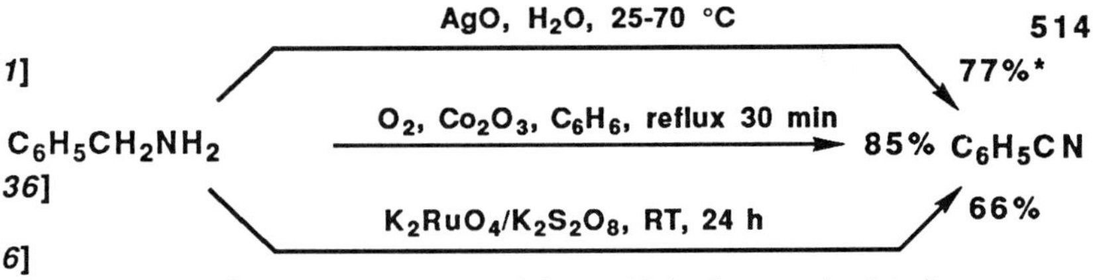

*In addition, 23% of benzaldehyde was isolated.

[1069]   515

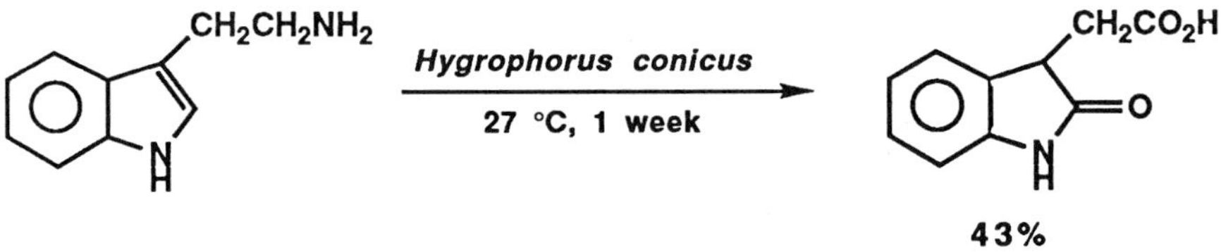

[indole-CH$_2$CH$_2$NH$_2$]   *Hygrophorus conicus*   [oxindole-CH$_2$CO$_2$H]

                             27 °C, 1 week

                                                                 43%

Oxidation of the Carbon Chain of Amines

N-benzoylated cyclic amines and bicyclic amines undergo transannular *hydroxylation* when incubated with *Sporotrichum sulfurescens*. 1-Benzoylpiperidine is converted into 1-benzoyl-4-hydroxypiperidine in approximately 20% yield [1081]. N-Benzoylhexamethyleneimine gives 4-hydroxy-3-keto and 4-keto derivatives as a result of transannular hydroxylation and subsequent oxidation (equation 516).

[1081]   516

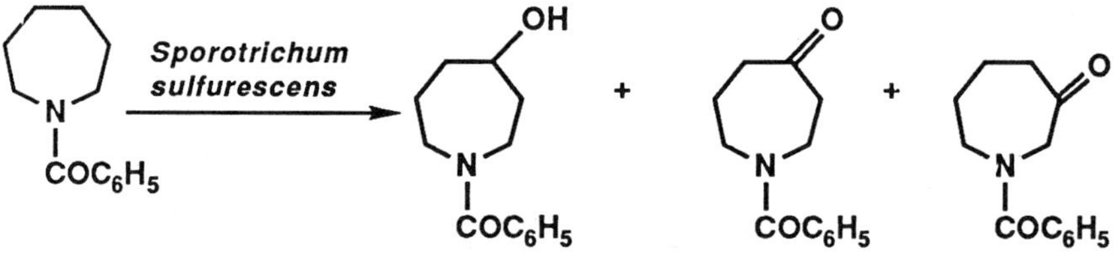

[N-COC$_6$H$_5$ azepane] → *Sporotrichum sulfurescens* → [4-OH-N-COC$_6$H$_5$] + [4-keto-N-COC$_6$H$_5$] + [3-keto-N-COC$_6$H$_5$]

Similar oxidations are achieved in bicyclic amides [1082], spirocyclic amides [1084], and acylated 1-adamantamines [1083]. N-Acetyladamantamine is hydroxylated on the tertiary as well as the secondary carbons to yield N-acetyl-1-adamantamine-4α-ol and N-acetyl-1-adamantamine-3-ol [1083]. The yields of all hydroxylated products are not very high (equations 517 and 518).

Tertiary amines containing a methyl group bound to carbon are oxidized to *formamides*. Dimethylaniline gives N-methylformanilide in 78% yield on treatment with **manganese dioxide** in chloroform for 18 h at room temperature [826]. Other reagents for this purpose are **potassium permanganate** in acetone and acetic acid at room temperature (yields 67–82%)

[1084] 517

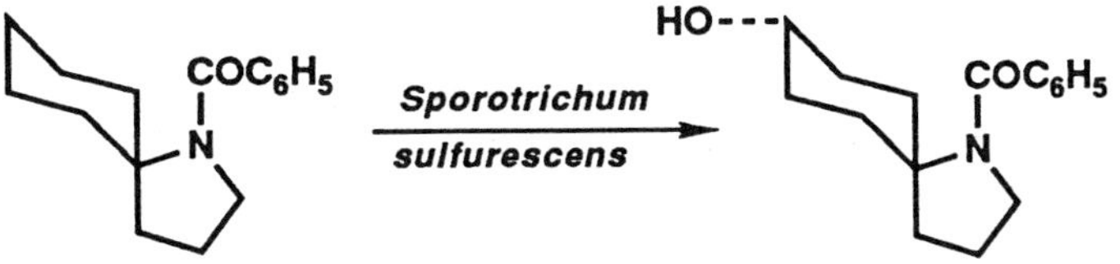
HO---
COC_6H_5
N
Sporotrichum
sulfurescens
COC_6H_5
N

54%

[1083] 518

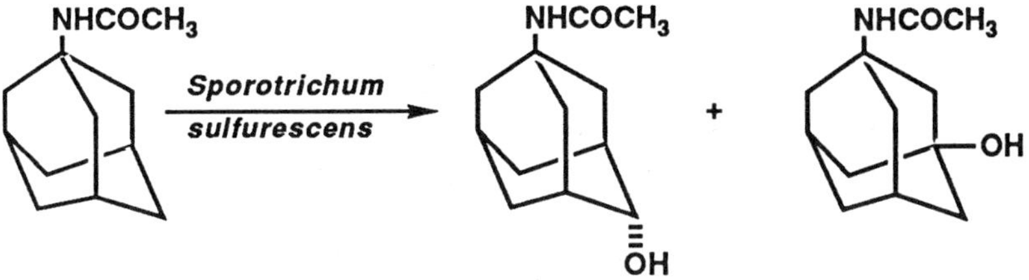
$NHCOCH_3$
Sporotrichum
sulfurescens
$NHCOCH_3$
OH
+
$NHCOCH_3$
OH

35% 7%

[854] and **diethyl azodicarboxylate** [980]. The formamide, on hydrolysis, is converted into a demethylated amine. Thus *N*-methylisopelletierine gives isopelletierine (equation 519) [980].

[980] 519

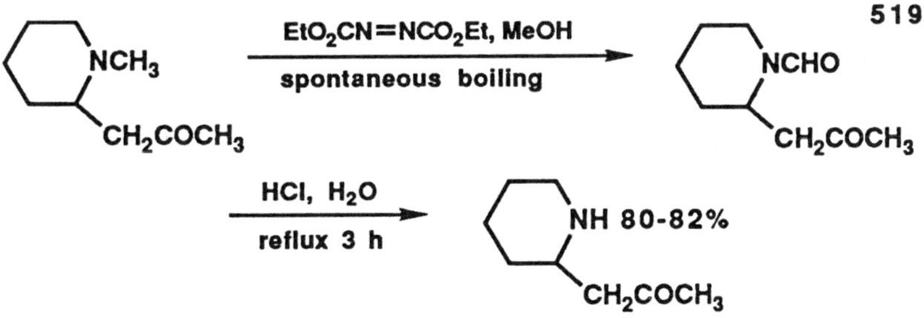
NCH_3
CH_2COCH_3
$EtO_2CN{=}NCO_2Et$, MeOH
spontaneous boiling
NCHO
CH_2COCH_3
HCl, H_2O
reflux 3 h
NH 80-82%
CH_2COCH_3

Such ***demethylation*** takes place often without interception of the formamide. Dicyclohexylmethylamine, on treatment with **potassium ferricyanide** in potassium hydroxide at room temperature overnight, is converted into dicyclohexylamine in 87% yield [926]. Other oxidants suitable for the demethylation of tertiary amines are mercuric acetate in 5% aqueous acetic acid at 100 °C [403], manganese dioxide in cyclohexane at 20 °C [812], and even oxygen, which, under irradiation at 20 °C in the presence of rose bengal in aqueous *tert*-butyl alcohol, converts codeine into norcodeine in 75% yield [46].

Tertiary amines with groups other than methyl are oxidized to ***amides***. *N,N*-Diethylaniline gives a 63% yield of *N*-ethylacetanilide on heating with benzyltriethylammonium permanganate in methylene chloride at 0–40 °C [904]. *p*-Bromobenzyldiphenylamine, on refluxing with potassium permanganate in acetone and pyridine, gives a 91% yield of diphenyl-*p*-bromobenzamide (equation 520) [855].

[855]   520

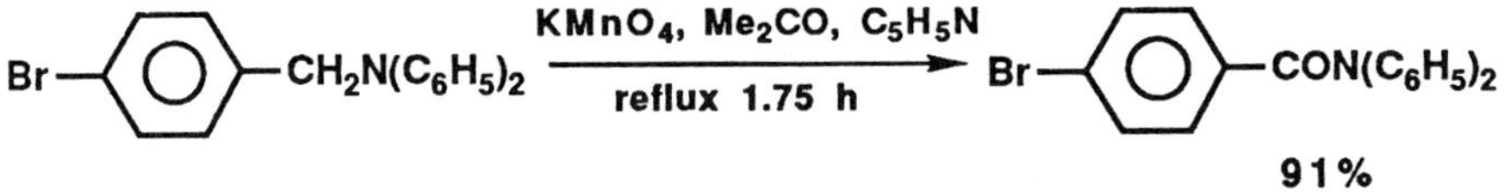

Br—C$_6$H$_4$—CH$_2$N(C$_6$H$_5$)$_2$ $\xrightarrow[\text{reflux 1.75 h}]{\text{KMnO}_4,\text{ Me}_2\text{CO, C}_5\text{H}_5\text{N}}$ Br—C$_6$H$_4$—CON(C$_6$H$_5$)$_2$

91%

In some cases, tertiary amines are cleaved to carbonyl compounds (equation 521) [818].

[818]   521

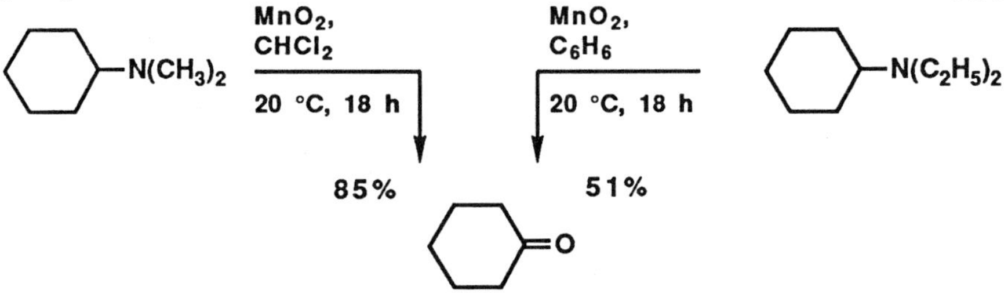

C$_6$H$_{11}$—N(CH$_3$)$_2$ $\xrightarrow[\text{20 °C, 18 h}]{\text{MnO}_2,\text{ CHCl}_2}$ cyclohexanone (85%) $\xleftarrow[\text{20 °C, 18 h}]{\text{MnO}_2,\text{ C}_6\text{H}_6}$ (51%) C$_6$H$_{11}$—N(C$_2$H$_5$)$_2$

Peroxybenzoic acid converts N-methylindoline into o-dimethylaminobenzaldehyde (equation 522) [311], and manganese dioxide opens the ring in dibenzylpiperazine to give N,N'-dibenzyl-N,N'-diformylethylenediamine (equation 523) [818].

[311]   522

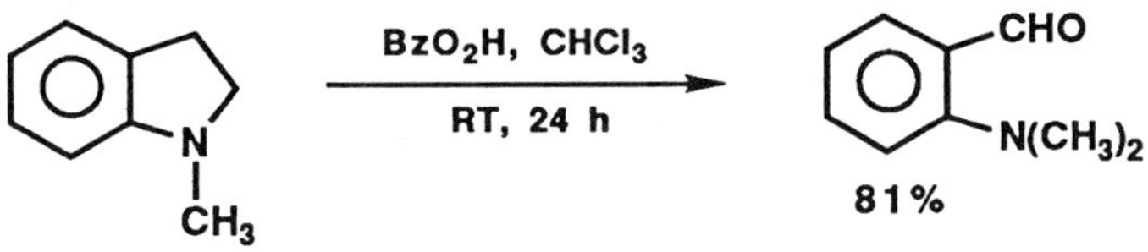

N-methylindoline $\xrightarrow[\text{RT, 24 h}]{\text{BzO}_2\text{H, CHCl}_3}$ 2-(dimethylamino)benzaldehyde

81%

[818]   523

C$_6$H$_5$CH$_2$N(CH$_2$CH$_2$)$_2$NCH$_2$C$_6$H$_5$ $\xrightarrow[\substack{\text{20 °C,}\\\text{18 h}}]{\text{MnO}_2,\text{ CH}_2\text{Cl}_2}$ C$_6$H$_5$CH$_2$N(CHO)CH$_2$CH$_2$N(CHO)CH$_2$C$_6$H$_5$

80%

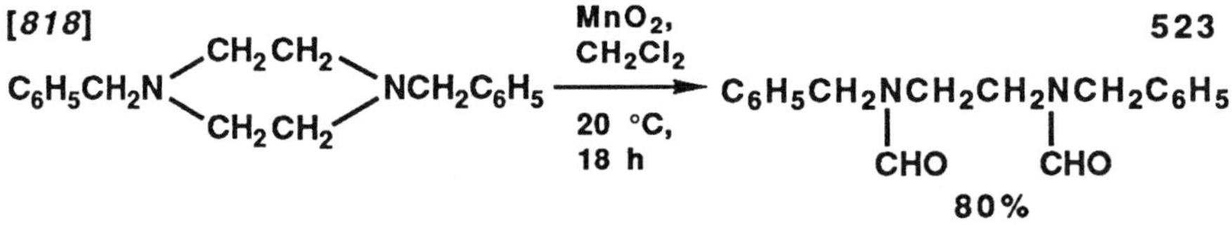

Tertiary *enamines* are oxidized to different compounds. N-Methylpyrrol gives 92% of N-methylsuccinimide by oxidation with peroxybenzoic acid in chloroform at room temperature [311]. N-Cyclohexenylmorpholine, with singlet oxygen, is oxidized to cyclohexanedione (equation 524) [42].

The oxidation of N-phenylacetylenyl-N,N-diethylamine yields the N,N-diethylamide of phenylglyoxylic acid (equation 525) [786].

Amides containing methylene groups linked to the nitrogen atoms are oxidized to **imides** by **ruthenium tetroxide,** which is usually generated in situ from ruthenium dioxide or ruthenium chloride and **sodium periodate**

[*42*] **524**

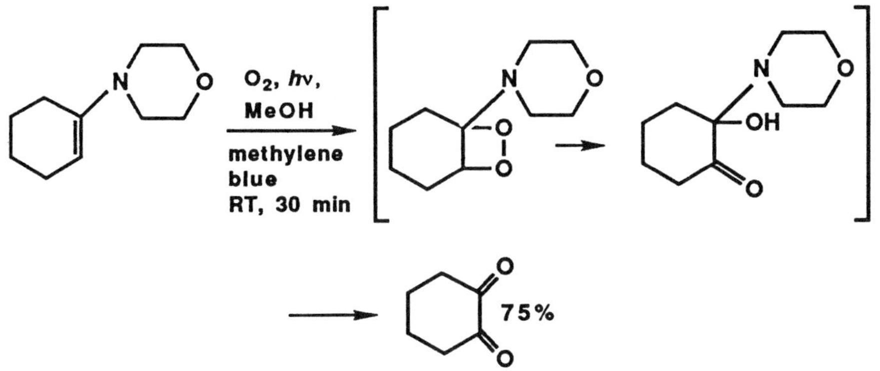

[*786*] **525**

$$C_6H_5C{\equiv}CN(C_2H_5)_2 \xrightarrow[\text{exothermic reaction, RT, 1.5 h}]{PhIO/(Ph_3P)_3RuCl_2,\ CH_2Cl_2} C_6H_5COCON(C_2H_5)_2 \quad 76\%$$

[*940, 944*]. *N*-Hexylenanthamide gives a 58% yield of *N*-caproylenanthamide [*940*] on treatment with ruthenium tetroxide in carbon tetrachloride at 10–15 °C. Ethyl *N*-acetylpipecolinate, when treated with ruthenium dioxide and sodium periodate in aqueous ethyl acetate at room temperature, gives a 95% yield of ethyl 1-acetyl-2-piperidone-6-carboxylate (equation 526) [*944*].

[*944*] **526**

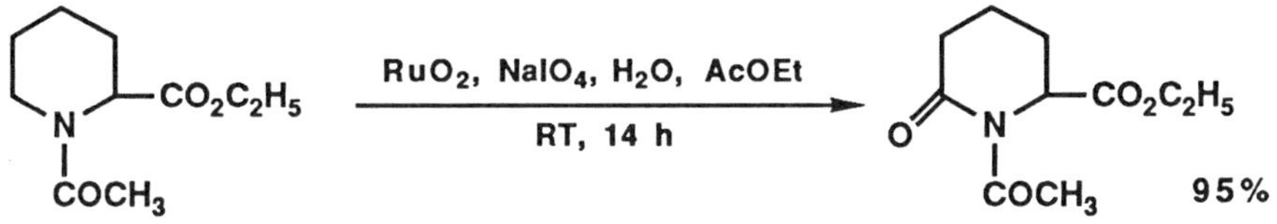

A similar oxidation converts ***lactams into imides***. The reaction can be carried out with **ruthenium tetroxide** [*940, 944*] or with *tert*-butyl hydroperoxide or peroxyacetic acid in the presence of manganese dichloride or diacetylacetonate [*225, 255*] (equation 527).

[*940*] **527**

[*225*]

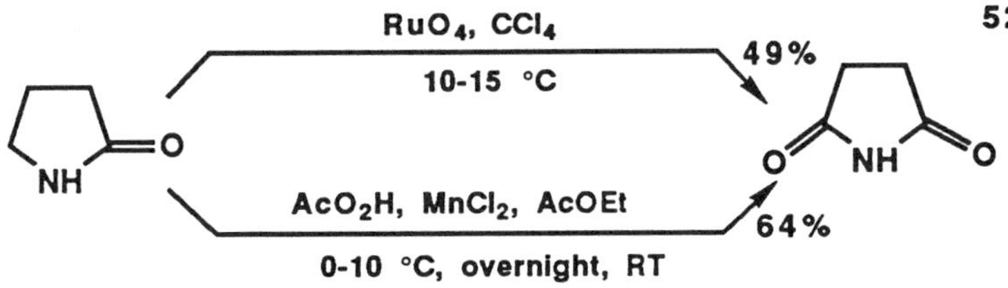

Oxidation of Aromatic Amines to Quinones

The presence of an amino group on an aromatic ring often results in oxidation of the ring to a ***quinone.*** The classical and industrial method is the treatment of anilines with **potassium dichromate** and sulfuric acid. Thus, aniline at room temperature is converted into *p*-benzoquinone in 86% yield [*647*], and 2,5-dimethylaniline at 80 °C gives a 55% yield of *p*-xyloquinone [*648*]. A specific reagent for such oxidations is the *Fremy salt, potassium nitrosodisulfonate* (equation 528) [*490*]. The oxidation of the amino group takes place even if it is acylated (equation 529) [*1190*].

[*490*] 528

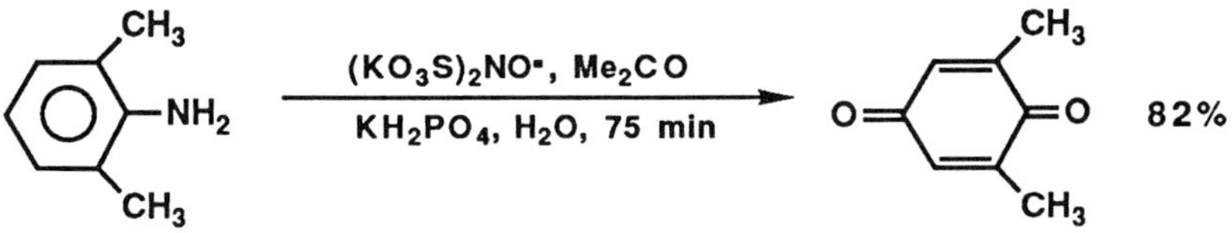
CH3
NH2
CH3
(KO3S)2NO•, Me2CO
KH2PO4, H2O, 75 min
CH3
O
O
CH3
82%

[*1190*] 529

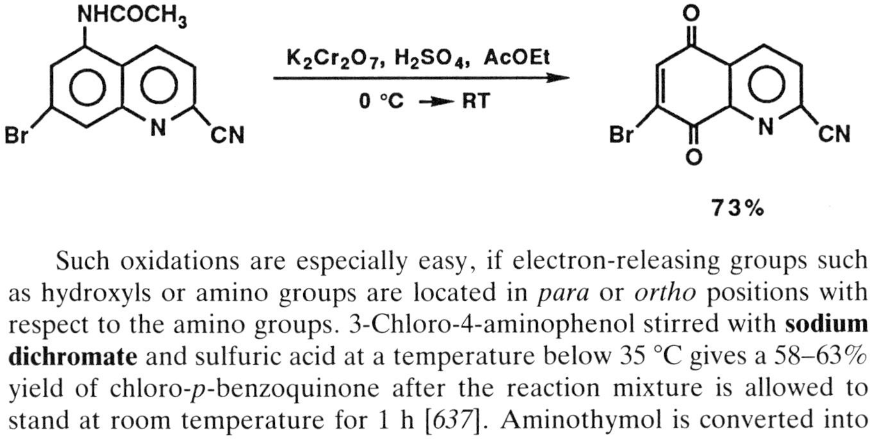
NHCOCH3
Br
N
CN
K2Cr2O7, H2SO4, AcOEt
0 °C → RT
O
Br
N
CN
O
73%

Such oxidations are especially easy, if electron-releasing groups such as hydroxyls or amino groups are located in *para* or *ortho* positions with respect to the amino groups. 3-Chloro-4-aminophenol stirred with **sodium dichromate** and sulfuric acid at a temperature below 35 °C gives a 58–63% yield of chloro-*p*-benzoquinone after the reaction mixture is allowed to stand at room temperature for 1 h [*637*]. Aminothymol is converted into thymoquinone on warming for 30 min with dilute sulfuric acid and sodium nitrite (equation 530) [*451*].

[*451*] 530

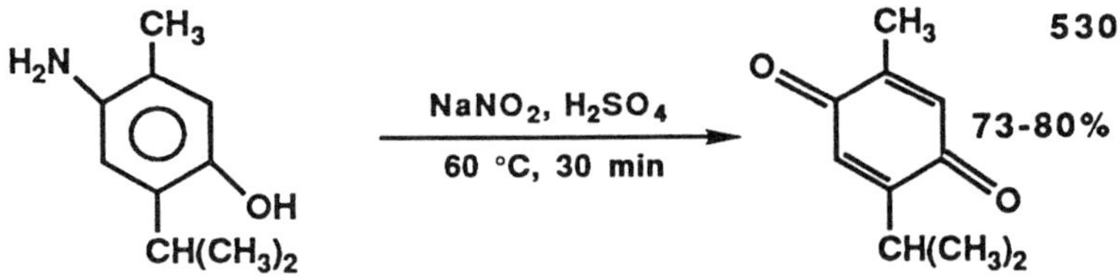
CH3
H2N
OH
CH(CH3)2
NaNO2, H2SO4
60 °C, 30 min
CH3
O
O
CH(CH3)2
73-80%

Diaminodurene in the form of its hexachlorostannate is oxidized to duroquinone with **ferric chloride** (equation 531) [*908*].

[908]   531

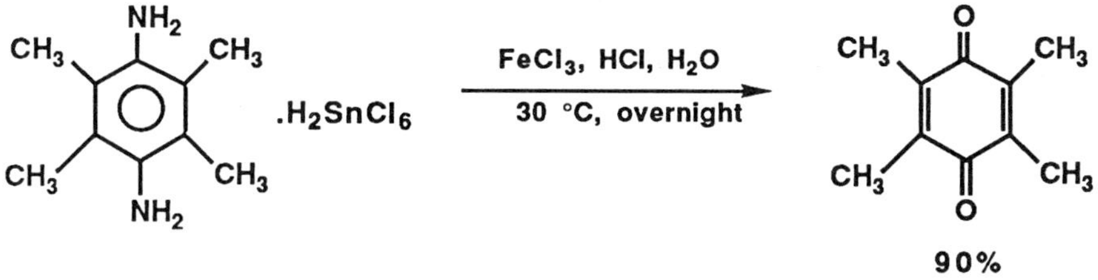

[Structure: 2,3,5,6-tetramethyl-1,4-diaminobenzene·H$_2$SnCl$_6$] → [Structure: 2,3,5,6-tetramethyl-1,4-benzoquinone]

Reagents: FeCl$_3$, HCl, H$_2$O, 30 °C, overnight

90%

Aminonaphthols are converted into naphthoquinones [*1191, 1192, 1193*] by nitric acid [*1193*], by potassium dichromate [*1191*], or by ferric chloride [*1192*] (equations 532 and 533).

[1191]   532

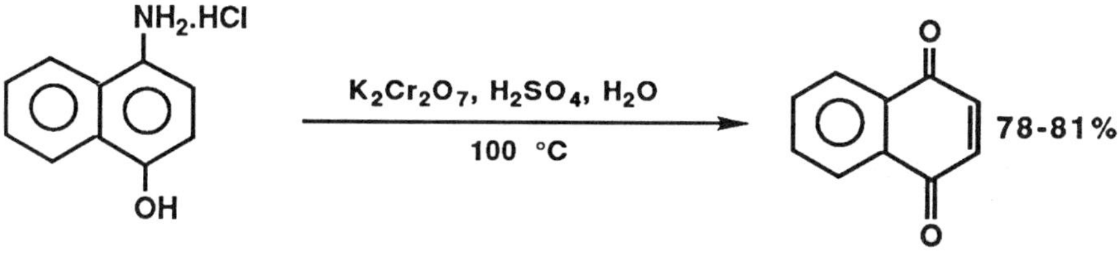

[Structure: 4-amino-1-naphthol hydrochloride] → [Structure: 1,4-naphthoquinone]

Reagents: K$_2$Cr$_2$O$_7$, H$_2$SO$_4$, H$_2$O, 100 °C

78–81%

[1192] X = H   533

[Structure: 1-amino-2-naphthol hydrochloride with substituent X] → [Structure: 1,2-naphthoquinone with X]

Reagents: FeCl$_3$, HCl, H$_2$O, 35 °C   93–94%
         HNO$_3$, H$_2$O, 30 °C   94–98%*

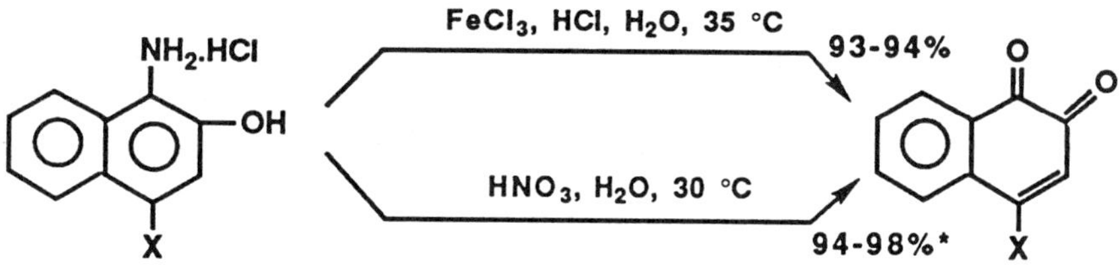

[1193] X = SO$_3$H

*The product was isolated as the NH$_4^+$ salt.

Several other examples of the oxidation of hydroxyamino and diamino aromatic compounds with silver oxide [*368*], lead dioxide [*368*], lead tetraacetate [*441*], and sodium hypochlorite [*694*] are documented in equations 534–536.

[694]   534

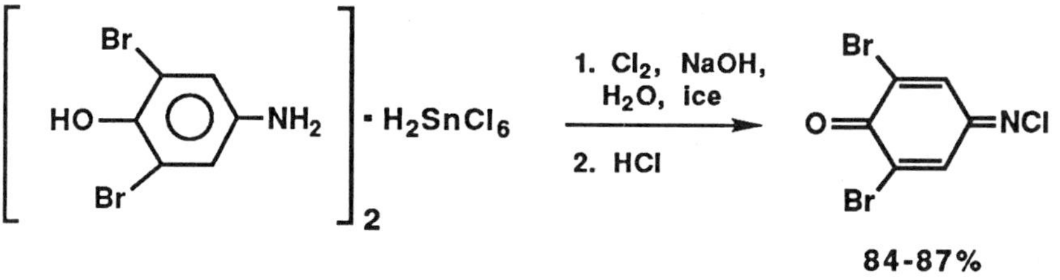

[Structure: (2,6-dibromo-4-aminophenol)$_2$·H$_2$SnCl$_6$] → [Structure: 2,6-dibromo-4-(N-chloroimino)cyclohexa-2,5-dien-1-one]

Reagents: 1. Cl$_2$, NaOH, H$_2$O, ice; 2. HCl

84–87%

[441]   535

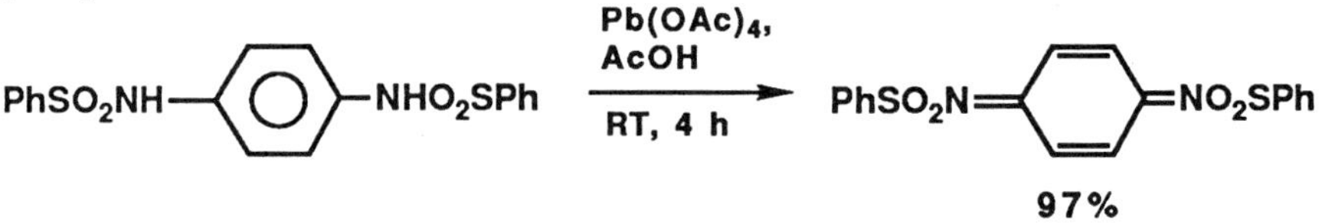

PhSO$_2$NH—⟨C$_6$H$_4$⟩—NHO$_2$SPh $\xrightarrow[\text{RT, 4 h}]{\text{Pb(OAc)}_4,\ \text{AcOH}}$ PhSO$_2$N=⟨⟩=NO$_2$SPh

                                                97%

[368]   Ag$_2$O, Na$_2$SO$_4$, Et$_2$O, RT, 1 h → 90%   536

HO—⟨⟩—N=N—⟨⟩—OH   O=⟨⟩=N—N=⟨⟩=O

    PbO$_2$, Na$_2$SO$_4$, Et$_2$O, RT, 3 h → 70%

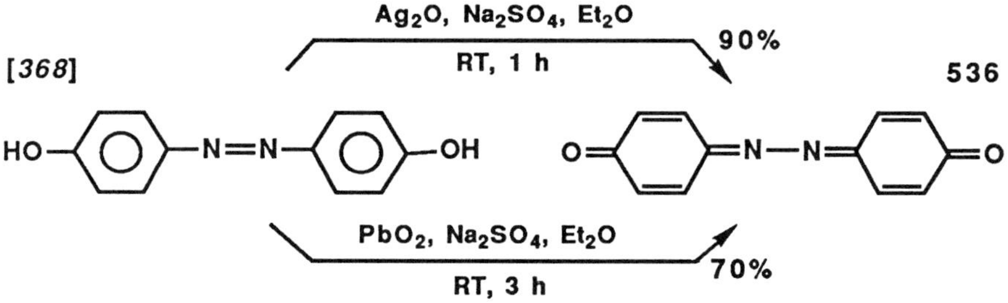

PHOSPHORUS AND ARSENIC COMPOUNDS

Like tertiary amines, *tertiary phosphines* are readily transformed into the corresponding oxides, *phosphine oxides*. Tributylphosphine in dichloromethane is oxidized with **ozone** adsorbed on silica gel at −70 °C to tributylphosphine oxide in 92% yield [*113*]. Other oxidants used to transform phosphines into phosphine oxides are **potassium peroxysulfate** [*205*], argentic oxide [*381*], **manganese dioxide** [*813*], and **barium manganate** [*833*] (equation 537).

                                                537

        $(C_6H_5)_3P$ ⟶ $(C_6H_5)_3P=O$

[113]	O$_3$, SiO$_2$, CH$_2$Cl$_2$, −65 °C	99%
[205]	2KHSO$_5$·KHSO$_4$·K$_2$SO$_4$, H$_2$O, EtOH, 70 °C, 4 h	100%
[381]	AgO, H$_2$O, RT, exothermic reaction	40%
[813]	MnO$_2$, petroleum ether, reflux 5 h	75%
[833]	BaMnO$_4$, C$_6$H$_6$, reflux 2 h	100%

A similar oxidation of trivalent phosphorus to pentavalent phosphorus occurs in trialkyl or triaryl phosphites. Triphenyl phosphite is converted by ozone in dichloromethane at 5–10 °C into triphenyl phosphate in 95% yield [*113*] and by argentic oxide in 30% yield [*381*] (equation 538).

The oxidation of *phosphites to phosphates* is very important in the chemistry of nucleoside phosphites, which must be handled under anhydrous conditions. For this purpose, **anhydrous hydrogen peroxide, bis(trimethylsilyl) peroxide, cumene hydroperoxide,** and *N*-methylmor-

[113] $(C_6H_5O)_3P$ →[O_3, CH_2Cl_2, 5-10 °C] $(C_6H_5O)_3P=O$ 95% **538**

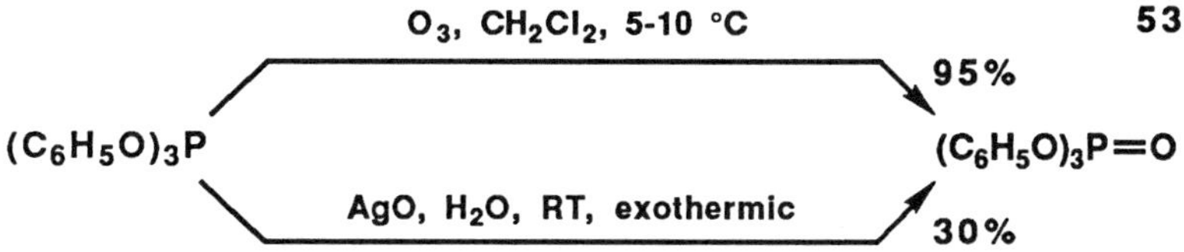

[381] $(C_6H_5O)_3P$ →[AgO, H_2O, RT, exothermic] $(C_6H_5O)_3P=O$ 30%

pholine N-oxide are used in solutions in dichloromethane and give 80–92% yields of the phosphates (equation 539) [228].

[228]

539

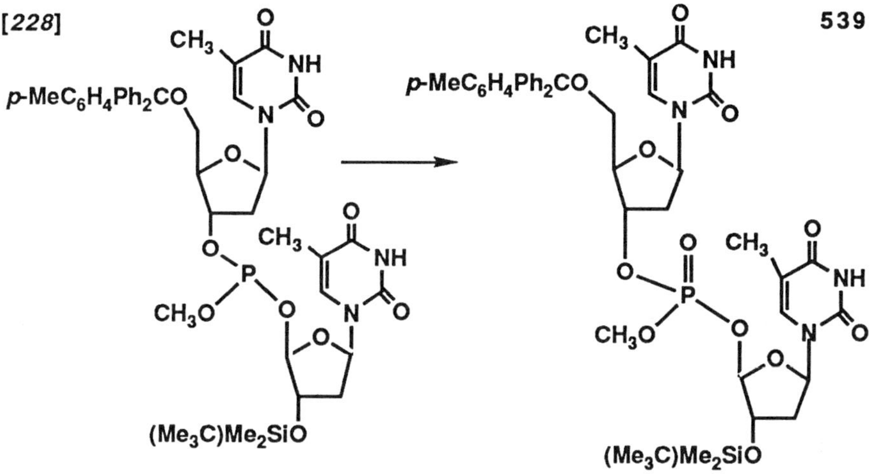

Reagent	Yield
$(Me_3SiO)_2$, CH_2Cl_2, 20 °C, 3 h	85%
$(Me_3SiO)_2$, $Me_3SiOSO_2CF_3$, Et_3N, CH_2Cl_2, -20 °C, 5 min	87%
$PhCMe_2OOH$, CH_2Cl_2, 0 °C, 40 min	92%
morpholine N-oxide (N-Me), CH_2Cl_2, 25 °C, 7 h	80%

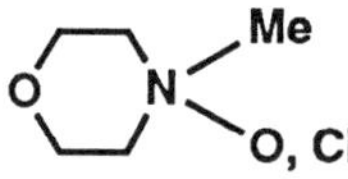

Trialkyl and triarylphosphine oxides are also prepared by *oxidation of trialkyl and triarylphosphine sulfides and selenides* (equation 540) [1007, 1194].

[1007]   **540**

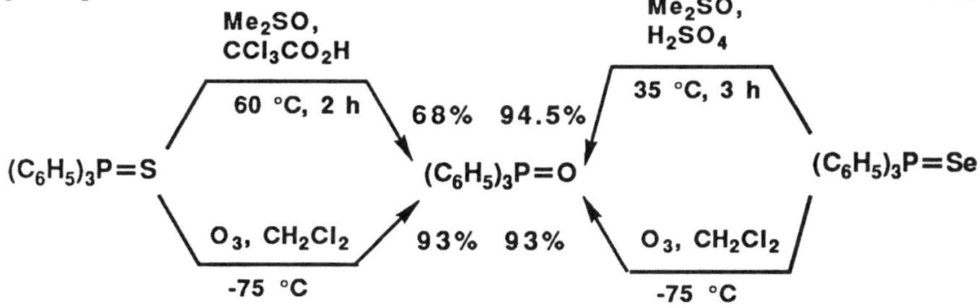

$(C_6H_5)_3P=S$ →[Me_2SO, CCl_3CO_2H, 60 °C, 2 h] 68% → $(C_6H_5)_3P=O$ 94.5% ←[Me_2SO, H_2SO_4, 35 °C, 3 h] $(C_6H_5)_3P=Se$

$(C_6H_5)_3P=S$ →[O_3, CH_2Cl_2, -75 °C] 93% → $(C_6H_5)_3P=O$ 93% ←[O_3, CH_2Cl_2, -75 °C] $(C_6H_5)_3P=Se$

[1194]

The same oxidation procedures apply to the conversion of **trialkyl or triaryl thiophosphates** and **selenophosphates** into **phosphates** [1007, 1194] (equation 541).

[1194]   541

$(C_2H_5O)_3P=S \xrightarrow[-75\ °C]{O_3,\ CH_2Cl_2} \quad (C_2H_5O)_3P=O \quad \xleftarrow[-75\ °C]{O_3,\ CH_2Cl_2} (C_2H_5O)_3P=Se$

                    95%   94%

Triphenylphosphine ylides are cleaved by sodium periodate to triphenylphosphine oxides and carbonyl compounds (equation 542) [1195].

[1195]   542

$C_6H_5COCH=P(C_6H_5)_3 \xrightarrow[\text{reflux 1 h}]{NaIO_4,\ H_2O} C_6H_5COCHO + (C_6H_5)_3P=O$

                                                            100%

Tertiary arsines are oxidized to **arsine oxides** by **hydrogen peroxide** (equation 543) [1196].

[1196]   543

$(C_6H_5)_3As \xrightarrow[\text{25-30 °C, 20-30 min + 30 min}]{30\%\ H_2O_2,\ Me_2CO} (C_6H_5)_3As=O \quad 84\text{-}87\%$

SULFUR COMPOUNDS

The oxidation of some sulfur-containing compounds, such as thioketones and thioamides (derivatives of ketones and acids, respectively), has been mentioned in earlier sections.

Oxidation of Thiols (Mercaptans)

Gentle oxidation of mercaptans and thiols leads to coupled products—**disulfides**. The oxidants used for this purpose are **air** [9], **hydrogen peroxide** [186], **dichromates** [663], **chlorochromates** [619], **manganese dioxide** [816], **copper permanganate** [896], **ferric chloride** [907], and **diethyl azodicarboxylate** [977] (equations 544–546).

[9]   544

$C_4H_9SH \xrightarrow[\text{RT, 4 h}]{\text{air, Al}_2O_3,\ C_6H_6} C_4H_9SSC_4H_9 \quad 89\%$

Oxidations 251

[816] MnO_2, $CHCl_3$ 545
                    ⎯⎯reflux 5-6 h⎯⎯ 66%
$CH_2=CHCH_2SH$ $CH_2=CHCH_2SSCH_2CH=CH_2$
                    ⎯⎯$EtO_2CN=NCO_2Et$⎯⎯ 93%
[977]   exothermic
                    reaction, RT,
                    a few minutes

                                                                              546
            C_6H_5SH ⎯⎯⎯⎯⎯⎯⎯⎯⎯⎯⎯⎯⎯⎯⎯⎯⎯ $C_6H_5SSC_6H_5$

[9]   air, Al_2O_3, C_6H_6, RT, 6 h   96%

[663]   (pyridine-3-CO_2H)$_2Cr_2O_7$, CH_2Cl_2, C_5H_5N, 0 °C, 1 h 90%
[619] Bu_4NCrO_3Cl, $CHCl_3$, RT, 1 h   87%

[896]   (pyridyl-pyridyl)$_2Cu(MnO_4)_4$, CH_2Cl_2, RT, 10 min   100%

In compounds containing oxidizable functional groups, these groups may be left intact [163, 186] or oxidized concurrently [907] (equations 547 and 548).

[186]   547

H_2N—⟨C$_6H_4$⟩—SNa ⎯⎯30% H_2O_2⎯⎯→ H_2N—⟨C$_6H_4$⟩—SS—⟨C$_6H_4$⟩—NH_2
                       65-70 °C,
                       2 h   58-64%

[907]   548

2-mercaptoresorcinol ⎯⎯2 N $FeCl_3$,⎯⎯→ disulfide-bis-quinone
                        EtOH, H_2O
                        RT   100%

More energetic oxidation *of thiols* leads *to sulfonic acids.* Thiophenol dissolved in a solution of potassium hydroxide in dimethylformamide is oxidized by **oxygen** at 23.5 °C over a 22-h period to benzenesulfonic acid in 91% yield [53]. Lauryl mercaptan gives dodecanesulfonic acid in 100% isolated yield upon treatment with potassium peroxymonosulfate, $2KHSO_5·KHSO_4·K_2SO_4$, at room temperature for 30 min [205].

Whereas 2-(2-pyridyl)ethanethiol is oxidized with hydrogen peroxide to 2-(2-pyridyl)ethyl disulfide, its oxidation by nitric acid yields 2-(2-pyridyl)ethanesulfonic acid (equation 549) [474].

[474]   549

$$\text{2-Py-CH}_2\text{CH}_2\text{SH}^* \xrightarrow[\text{RT, 30 min}]{30\% \text{ H}_2\text{O}_2, \text{ HCl}} \text{2-Py-CH}_2\text{CH}_2\text{SSCH}_2\text{CH}_2\text{-2-Py}$$

$$\text{2-Py-CH}_2\text{CH}_2\text{SH}^* \xrightarrow[\text{exothermic, 100 °C}]{\text{HNO}_3 \text{ (}d\text{ 1.4)}} \text{2-Py-CH}_2\text{CH}_2\text{SO}_3\text{H} \quad 90\%$$

*The starting material was the hydrochloride of the compound.

In 2-mercaptobenzothiazole, energetic oxidation with hydrogen peroxide leads to the replacement of the mercapto group by hydrogen, probably via oxidation to sulfinic acid (equation 550) [1197].

[1197]   550

$$\text{benzothiazole-2-SH} \xrightarrow[\substack{\text{cool to}\\ \text{65-70 °C,}\\ \text{heat to 80 °C}}]{\substack{\text{concentrated HCl,}\\ 30\% \text{ H}_2\text{O}_2}} \left[\text{benzothiazole-2-SO}_2\text{H} \right] \rightarrow \text{benzothiazole} \quad 81.5\%$$

The oxidation of **mercaptans** with **chlorine** in water gives **sulfonyl chlorides**. Ethyl mercaptan and pentyl mercaptan afford ethanesulfonyl chloride and pentanesulfonyl chloride in yields of 73 and 78% at 5 and 1–5 °C, respectively [683].

Sulfonyl chlorides result also from oxidations, with chlorine in aqueous media, of **isothiocyanates and isothioureas** (equation 551) [684].

[684]   551

$$\text{C}_6\text{H}_5\text{CH}_2\text{SCN} \xrightarrow[20\text{-}30 \text{ °C}]{\text{Cl}_2, \text{ H}_2\text{O}} \underset{73\%}{\text{C}_6\text{H}_5\text{CH}_2\text{SO}_2\text{Cl}} \xleftarrow[20\text{-}30 \text{ °C}]{\text{Cl}_2, \text{ H}_2\text{O}}{76\%} \text{C}_6\text{H}_5\text{CH}_2\text{SC}(=\text{NH})\text{NH}_2 \cdot \text{HCl}$$

Oxidation of Sulfides

Dialkyl **sulfides**, alkyl aryl sulfides, diaryl sulfides, and cyclic sulfides are oxidized to the corresponding **sulfoxides** or **sulfones**. Often, both products can be obtained on oxidation with the same oxidant, depending on whether one or two equivalents of the reagent are used [*164, 279*]. Some of the oxidants, however, selectively oxidize sulfides to sulfoxides only [*541, 770*] or to sulfones only [*207, 209, 589*]. The rate constant of the oxidation of diphenyl sulfides to diphenyl sulfoxides by peroxyacetic acid is several hundred times greater than that of the oxidation of sulfoxides to sulfones [*265*]. Oxidations at sulfur take precedence over oxidation at other functional groups.

Dibutyl sulfide is converted into dibutyl sulfoxide with one equivalent of **peroxytrifluoroacetic acid** and into dibutyl sulfone with two equivalents of peroxytrifluoroacetic acid [*279*]. On the other hand, with **manganese dioxide,** dibutyl sulfide yields dibutyl sulfoxide exclusively [*541*], and with **chromic acid,** it yields dibutyl sulfoxide, even when an excess of the oxidant is used and even when the reaction is carried out at 100 °C [*541*] (equation 552).

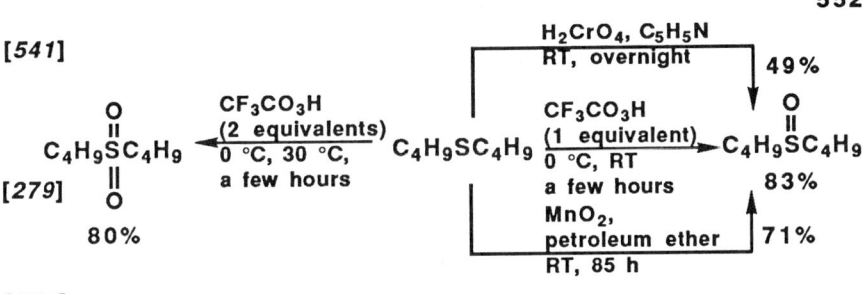

[*541*]

Organic peroxy acids, especially at low temperatures, oxidize sulfides to sulfoxides [*163, 324*], whereas tetrabutylammonium persulfate [*209*] at room temperature and hydrogen peroxide at higher temperatures yield sulfones [*163, 324*]. However, hydrogen peroxide in acetic anhydride at room temperature yields sulfoxides [*166*]. Under these conditions, double bonds resist epoxidation [*163, 166, 324*] (equation 553).

The oxidation reagents used most frequently for the conversion of sulfides into sulfoxides and sulfones are **hydrogen peroxide, peroxy acids,** and **periodates.** Periodates usually do not oxidize sulfoxides to sulfones [*770 771, 772, 776, 778*]. In addition, many other, even rather exotic, oxidants have been used especially for chemoselective oxidations of sulfides containing functional groups vulnerable to attack by peroxy compounds, such as double bonds and carbonyl groups.

Many examples with references are quoted in equations 554–557 to show various reaction conditions.

[*1198*]   554

$(C_6H_5CH_2)_2S \xrightarrow{\text{90\% AcO}_2\text{H, Et}_2\text{O}}_{2 \rightarrow 35\ °C,\ 30\ \text{min}} (C_6H_5CH_2)_2S=O$ 99%

[*695*]   555

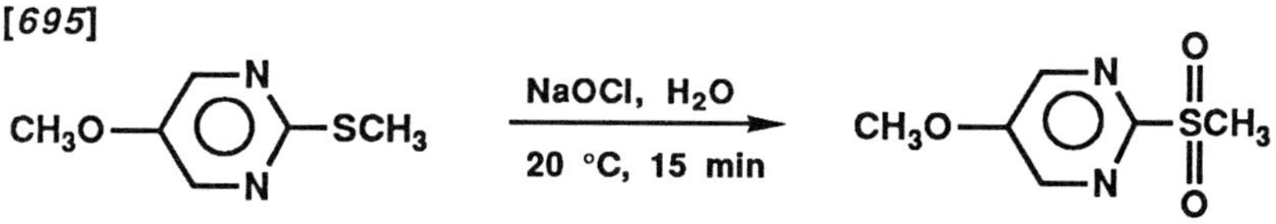

$CH_3O-\underset{N}{\overset{N}{\bigcirc}}-SCH_3 \xrightarrow{\text{NaOCl, H}_2\text{O}}_{20\ °C,\ 15\ \text{min}} CH_3O-\underset{N}{\overset{N}{\bigcirc}}-\overset{O}{\underset{O}{\overset{\|}{S}}}CH_3$ 88%

556

$C_6H_5SCH_3 \longrightarrow C_6H_5\overset{O}{\overset{\|}{S}}CH_3 \text{ or } C_6H_5\overset{O}{\underset{O}{\overset{\|}{\underset{\|}{S}}}}CH_3$

[*132*]	30% H_2O_2/TiCl$_3$, H_2O, MeOH, RT, 5 min	100%
[*207*]	2KHSO$_5$·KHSO$_4$·K$_2$SO$_4$ (Oxone) (3 equivalents), MeOH, H$_2$O, 0 °C, RT, 4 h	95%
[*209*]	2KHSO$_5$·KHSO$_4$·K$_2$SO$_4$ (Oxone), Bu$_4$NBr, CH$_2$Cl$_2$, RT	78%
[*224*]	*t*-BuOOH, Ti(O-i-Pr)$_4$, (*R,R*)-diethyl tartrate (1:1:2) (Sharpless reagent), CH$_2$Cl$_2$, −20 °C, 5 h	80% (ee 89%) (*R*)
[*710*]	*t*-BuOCl, MeOH, −78 °C, 15 min; NaHCO$_3$, RT, 2 h	90%
[*770*]	NaIO$_4$, H$_2$O, 0 °C, 3–12 h	99%
[*772*]	NaIO$_4$, H$_2$O, 0 °C, 15 h	91%
[*771*]	NaIO$_4$/Al$_2$O$_3$, EtOH, RT, 5 h	88%
[*210*] 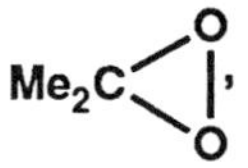	Me$_2$C⟨$\overset{O}{\underset{O}{}}$⟩, Me$_2$CO, RT	65%

[*1058*] *Corynebacterium equii* IFO 3730, 100% (ee 70%) (+) C$_{16}$H$_{34}$, 30 °C, 3 days*

*Exceptionally, the same microorganism converts decylphenyl sulfide to decyl phenyl sulfone in 88% yield with only 7% yield of the sulfoxide.

Oxidations 255

$$C_6H_5SC_6H_5 \longrightarrow C_6H_5\overset{O}{\underset{\|}{S}}C_6H_5 \text{ and/or } C_6H_5\overset{O}{\underset{\underset{O}{\|}}{\overset{\|}{S}}}C_6H_5 \qquad 557$$

Ref	Conditions	Sulfoxide	Sulfone
[132]	30% H$_2$O$_2$/TiCl$_3$, MeOH, H$_2$O, RT, 5 min	100%	
[194]	NaBO$_3$ (small excess), AcOH, 50-55 °C	71%	
[194]	NaBO$_3$ (large excess), AcOH, 50-55 °C		98%
[208]	2KHSO$_5$.KHSO$_4$.K$_2$SO$_4$ (2:1:1) (Oxone), Bu$_4$NBr, CH$_2$Cl$_2$, -10 °C →RT, 18 h	30%	+ 70%
[205]	KHSO$_5$.KHSO$_4$.K$_2$SO$_4$, H$_2$O, EtOH, H$_2$SO$_4$, AcOH RT, 5 h		97%
[426]	(NH$_4$)$_2$Ce(NO$_3$)$_6$ (4 mol), MeCN, H$_2$O, RT, 3 min	82%	
[506]	SO$_2$Cl$_2$, CH$_2$Cl$_2$, -30 to -40 °C, 9 min; EtOH	95%	
[739]	NaBrO$_2$, H$_2$O, RT, 30 min	78%	
[770]	NaIO$_4$, H$_2$O, 0 °C, 3-12 h	98%	
[778]	NaIO$_4$, C$_{16}$H$_{33}$PBu$_3$Br, CHCl$_3$, reflux 36 h	70%	
[942]	RuO$_4$, CCl$_4$, 0 °C, 14 h	32%	+ 42%

Sulfides containing sulfur in the ring are oxidized to *cyclic sulfoxides*. Ethylene sulfide, upon treatment with sodium periodate in aqueous methanol at 20–25 °C, is converted into ethylene sulfoxide in 65% yield [768]. With **peroxybenzoic acid** in dichloromethane, a 77% yield is obtained [309] (equation 558).

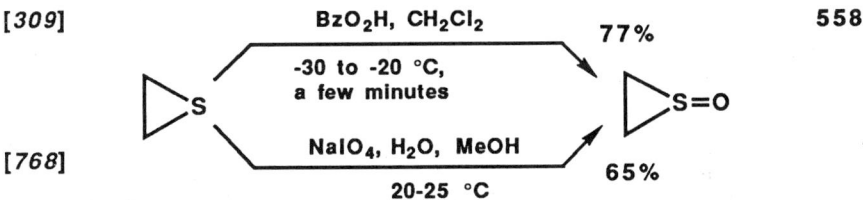

Tetrahydrothiophene (tetramethylene sulfide) gives tetramethylene sulfoxide in 90% yield on refluxing for 2 h with **tetrabutylammonium periodate** [776]. The oxidation of tetramethylene sulfide with one equivalent of **hydrogen peroxide** at room temperature affords tetramethylene sulfoxide; with two equivalents at refluxing temperature, tetramethylene sulfone is obtained (equation 559) [164].

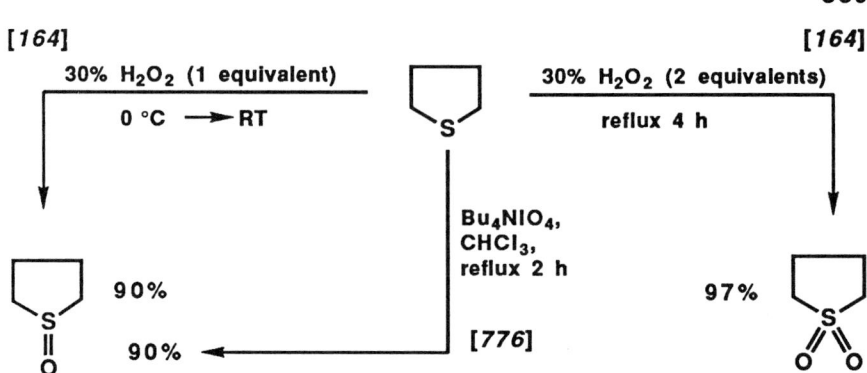

2-Methyl-4,5-dihydrothiophene-3-carboxylic acid yields the corresponding sulfoxide on treatment with **peroxyphthalic acid** and the sulfone on refluxing with hydrogen peroxide in acetic acid (equation 560) [*163*]. Neither oxidant affects the double bond under the conditions used.

Thiaxanthone is oxidized to thiaxanthone oxide in 78% yield by refluxing for 4 h with iodobenzene diacetate in acetic acid (equation 561)

[*795*]. 2,3-Dihydro-1-benzothiepin is converted into 2,3-dihydro-1-benzothiepin 1-oxide with sulfuryl chloride in 79% yield or with sodium periodate in 77% yield (equation 562) [*506*].

[506] 562

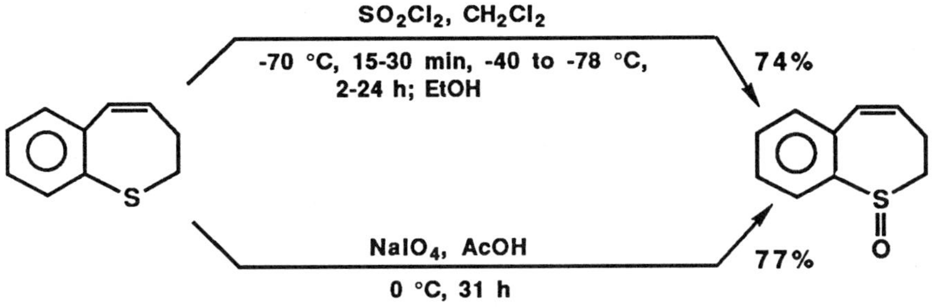

SO$_2$Cl$_2$, CH$_2$Cl$_2$
-70 °C, 15-30 min, -40 to -78 °C, 2-24 h; EtOH   74%

NaIO$_4$, AcOH
0 °C, 31 h   77%

Thiacyclohexane (thiane) is oxidized to thiane oxide by the enzyme *cyclohexanone oxygenase* [1034]. Thiacyclohexanes substituted at position 4 give two possible stereoisomers of the sulfoxides, *cis* and *trans*, in different ratios, depending on the oxidants used (equation 563) [1199].

[1199] 563

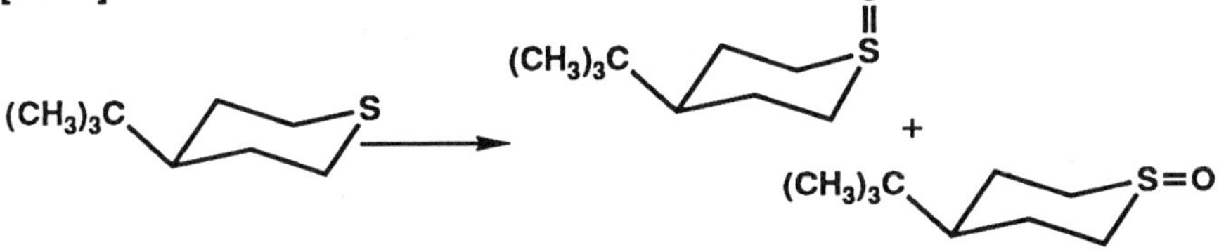

	cis	trans
30% H$_2$O$_2$, Me$_2$CO, 25 °C	37%	63%
30% H$_2$O$_2$, AcOH, 25 °C	35%	65%
t-BuOOH, C$_6$H$_6$, 50 °C	36%	64%
t-BuOOH, MeOH, 50 °C	27%	73%
m-ClC$_6$H$_4$CO$_3$H, CH$_2$Cl$_2$, 0 °C	36%	64%
N$_2$O$_4$, 0 °C	81%	19%
HNO$_3$, Ac$_2$O, 0 °C	67%	33%
H$_2$CrO$_4$, C$_5$H$_5$N, 25 °C	27%	73%
t-BuOCl, MeOH, -70 °C	100%	0%
NaIO$_4$, H$_2$O, 0 °C	75%	25%
PhIO, C$_6$H$_6$, 80 °C	46%	54%

1-Thiacyclooctan-5-one is converted into 1-thiacyclooctan-5-one 1-oxide by aqueous sodium periodate at 0 °C in 91% yield (equation 564) [767].

Optically active sulfoxides are obtained either by chemical enantio-selective oxidation with the **Sharpless reagent** (page 44) or by ***biochemical processes***.

[767]   NaIO$_4$, H$_2$O, 0 °C   564
91%

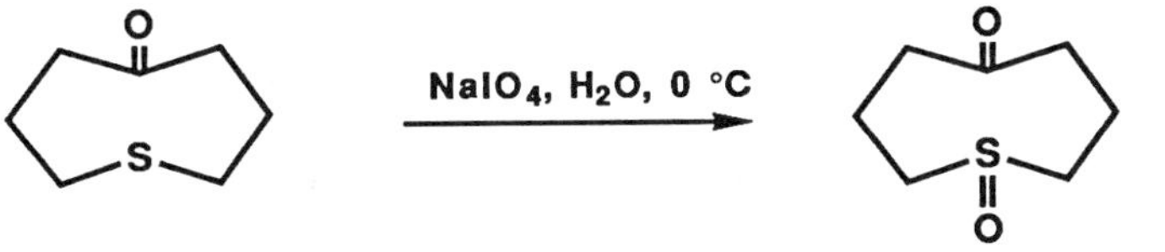

258   Oxidations in Organic Chemistry

The original Sharpless reagent, a mixture of tetraisopropyl orthotitanate, (R,R)-diethyl tartrate, and *tert*-butyl hydroperoxide in the ratio 1:1:2 in dry dichloromethane or 1,2-dichloroethane [*1025*], is modified by adding 1 mol of water [*224, 1029*]. Such a reagent gives higher enantiomeric excesses. Of many sulfide oxidations that have been carried out, the conversion of methyl *p*-tolyl sulfide into the sulfoxide is shown in equation 565.

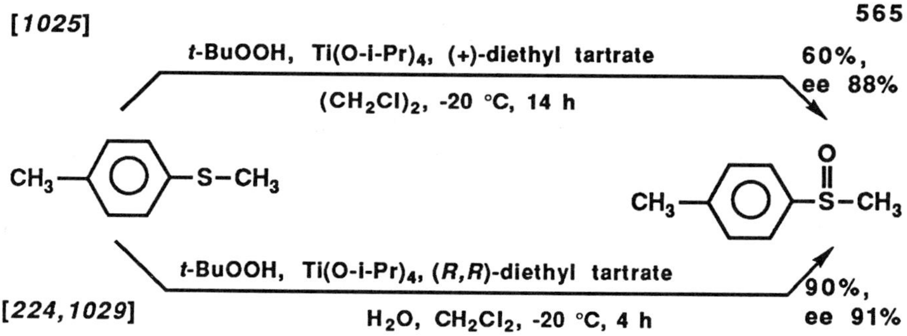

565

[*1025*]
$CH_3-C_6H_4-S-CH_3$ → $CH_3-C_6H_4-S(=O)-CH_3$
t-BuOOH, Ti(O-i-Pr)$_4$, (+)-diethyl tartrate
(CH$_2$Cl)$_2$, −20 °C, 14 h
60%, ee 88%

[*224, 1029*]
t-BuOOH, Ti(O-i-Pr)$_4$, (R,R)-diethyl tartrate
H$_2$O, CH$_2$Cl$_2$, −20 °C, 4 h
90%, ee 91%

Different microorganisms oxidize sulfides to sulfoxides, usually in low yields but with high enantiomeric excesses [*1046, 1047, 1068*] (equation 566).

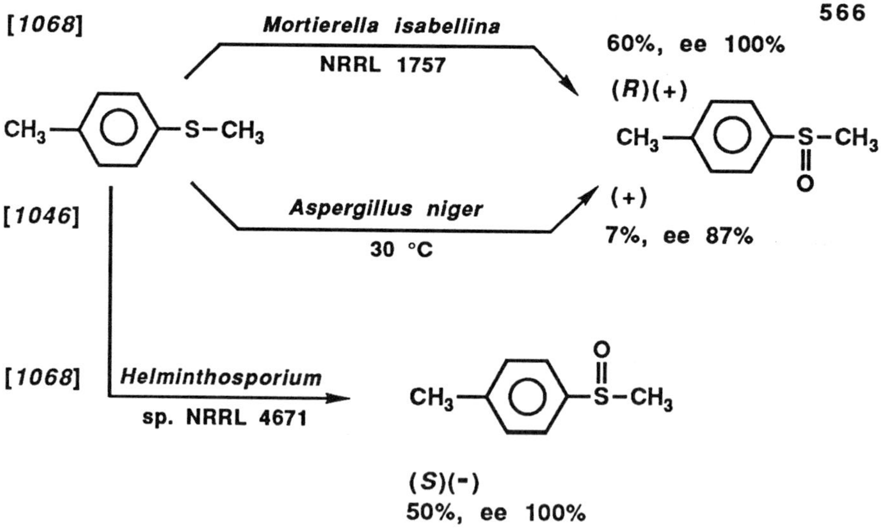

566

[*1068*] *Mortierella isabellina* NRRL 1757 → (R)(+) $CH_3-C_6H_4-S(=O)-CH_3$
60%, ee 100%

[*1046*] *Aspergillus niger* 30 °C → (+) 7%, ee 87%

[*1068*] *Helminthosporium* sp. NRRL 4671 → $CH_3-C_6H_4-S(=O)-CH_3$
(S)(−) 50%, ee 100%

Steroidal sulfides are converted into optically active sulfoxides by ***Calonectria decora*** [*1052*] and by ***Rhizopus stolonifer*** [*1048*]. ***Aspergillus niger*** transforms benzyl phenyl sulfide into optically active benzyl phenyl sulfoxide in 23% yield and into benzyl phenyl sulfone in 9% yield [*1048*].

The thermal decomposition of sulfoxides whose sulfur atom is attached to the α carbons of ketones or esters leads to α,β-unsaturated ketones or esters, respectively, via a *cis* elimination. The reaction is reminiscent of alkene formation by Cope elimination of dialkylhydroxylamines from tertiary amine oxides (equation 567) [*321*].

[*321*] 567

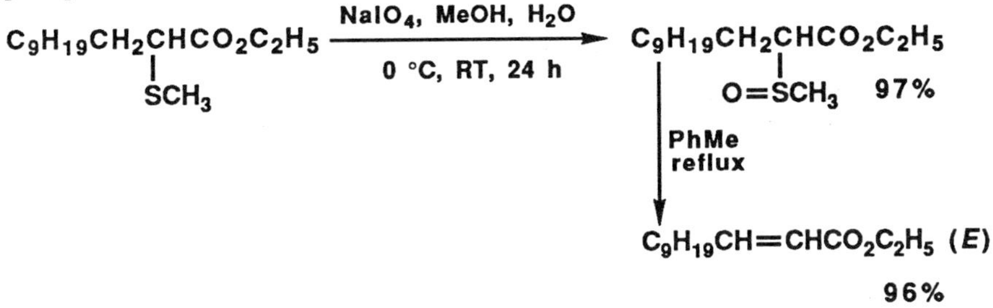

$$C_9H_{19}CH_2CHCO_2C_2H_5 \xrightarrow[0\ °C,\ RT,\ 24\ h]{NaIO_4,\ MeOH,\ H_2O} C_9H_{19}CH_2CHCO_2C_2H_5$$
$$|\qquad\qquad\qquad\qquad\qquad\qquad\qquad\qquad\qquad |$$
$$SCH_3 \qquad\qquad\qquad\qquad\qquad\qquad\qquad O=SCH_3\quad 97\%$$

PhMe reflux ↓

$$C_9H_{19}CH=CHCO_2C_2H_5\ (E)$$
96%

The oxidation of sulfides having two remote sulfur atoms may affect only one sulfur atom; monosulfoxide and monosulfone are obtained by the oxidation of 2,5-diphenyl-1,4-dithiadiene (equation 568) [*264*].

[*264*] 568

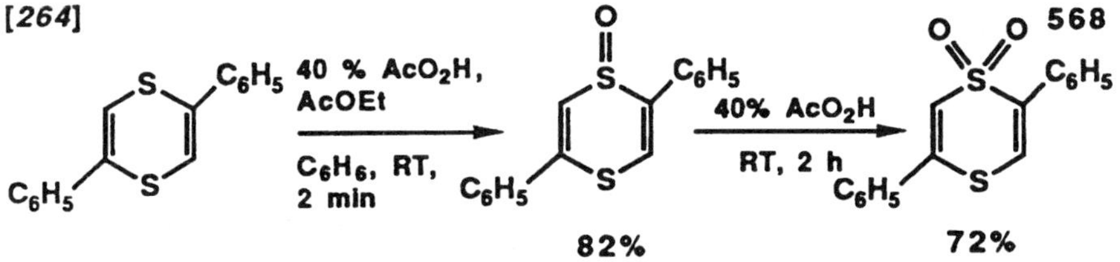

<chemical scheme: 2,5-diphenyl-1,4-dithiadiene → 40% AcO₂H, AcOEt, C₆H₆, RT, 2 min → monosulfoxide (82%) → 40% AcO₂H, RT, 2 h → sulfoxide-sulfone (72%)>

Oxidation of Thioacetals (Mercaptals and Mercaptoles)

The cyclic mercaptal of acetaldehyde (2-methyl-1,3-dithiane) is oxidized by **m-chloroperoxybenzoic acid** or **sodium periodate** to a mixture of diastereomeric *cis* and *trans* monosulfoxides (equation 569) [*323*].

[*323*] 569

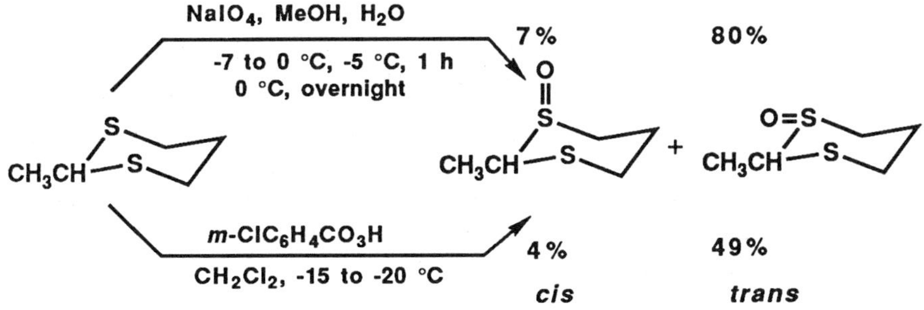

<chemical scheme: 2-methyl-1,3-dithiane
- NaIO₄, MeOH, H₂O, -7 to 0 °C, -5 °C, 1 h, 0 °C, overnight → cis (7%) + trans (80%)
- m-ClC₆H₄CO₃H, CH₂Cl₂, -15 to -20 °C → cis (4%) + trans (49%)>

260   Oxidations in Organic Chemistry

Acetone 2-chloroethylmercaptole is oxidized with **potassium permanganate** to the corresponding disulfone (sulfonal) (equation 570) [*846*].

[*846*]   570

$$(CH_3)_2C(SCH_2CH_2Cl)_2 \xrightarrow[C_6H_6,\ 25\text{-}30\ °C,\ 2\ h]{KMnO_4,\ H_2SO_4,\ H_2O} (CH_3)_2C(SO_2CH_2CH_2Cl)_2$$
$$81\%$$

Mercaptals of pentoses and hexoses are converted into sulfonals (disulfones), which easily undergo dehydration. The dimethylmercaptal of D-ribose or D-arabinose gives, on oxidation with peroxypropionic acid, 1,1-bis(methylsulfonyl)-D-*erythro*-pentos-1-ene (equation 571) [*340*].

[*340*]   571

$$\begin{array}{c} CH(SCH_3)_2 \\ \vdash \\ \vdash \\ \vdash \\ CH_2OH \end{array} \xrightarrow[\text{RT, 26 h}]{50\%\ EtCO_3H,\ Me_2CO} \begin{array}{c} CH(SO_2CH_3)_2 \\ \| \\ \vdash \\ \vdash \\ CH_2OH \end{array} \xleftarrow[\text{RT, 26 h}]{50\%\ EtCO_3H,\ Me_2CO} \begin{array}{c} CH(SCH_3)_2 \\ \vdash \\ \vdash \\ \vdash \\ CH_2OH \end{array}$$
$$86\%$$

Under controlled conditions, *thioacetals* can be transformed into *disulfoxides* or *disulfones* (equation 572) [*136*].

[*136*]   572

$$\begin{array}{c} CH(SOCH_3)_2 \\ \vdash \\ \vdash \\ \vdash \\ CH_2OH\ 32\% \end{array} \xleftarrow[\text{0 °C, 15 min}]{30\%\ H_2O_2,\ AcOH} \begin{array}{c} CH(SCH_3)_2 \\ \vdash \\ \vdash \\ \vdash \\ CH_2OH \end{array} \xrightarrow[\text{0 °C, 30 min}]{30\%\ H_2O_2,\ (NH_4)_2MoO_4} \begin{array}{c} CH(SO_2CH_3)_2 \\ \vdash \\ \vdash \\ \vdash \\ 74\%\ CH_2OH \end{array}$$

Hexylketene diethylmercaptole is oxidized with peroxyphthalic acid to 1,1-bis(ethanesulfonyl)-1-octene (equation 573) [*337*].

[*337*]   573

$$C_6H_{13}CH{=}C(SC_2H_5)_2 \xrightarrow[0\ °C,\ RT,\ 1\ day]{o\text{-}C_6H_4(CO_3H)(CO_2H),\ Et_2O} C_6H_{13}CH{=}C(SO_2C_2H_5)_2$$
$$70\text{-}95\%$$

In most instances, double and triple bonds remain unaffected by the oxidation of sulfides to sulfoxides or sulfones, even when hydrogen peroxide and peroxy acids are used [*163, 264, 322, 324, 506, 521*]. Evidently, the affinity of sulfur to oxygen is higher than that of the carbon–carbon multiple bonds (equations 574 and 575).

[521]

$$C_6H_5SCH_2CH=CH_2 \xrightarrow[<RT,\ 1\ h]{30\%\ H_2O_2/SeO_2,\ MeOH} \underset{\underset{O}{\|}}{\overset{\overset{O}{\|}}{C_6H_5S}}CH_2CH=CH_2 \quad \begin{array}{c}574\\ \\100\%\end{array}$$

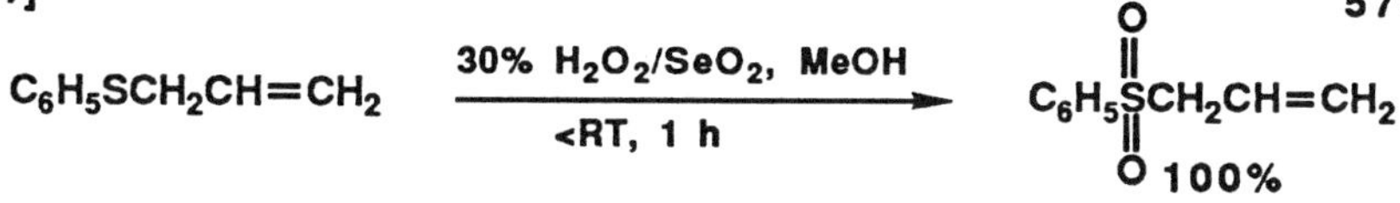

[322]

$$C_6H_5C{\equiv}CSCH_3 \xrightarrow[\substack{BzO_2H,\ CHCl_3 \\ -10\ °C;\ -23\ °C,\ 12\ h \\ \\ NaIO_4,\ MeCN,\ H_2O \\ -10\ °C;\ -5\ to\ -10\ °C,\ 12\ h}]{} \underset{O}{\overset{O}{\underset{\|}{C_6H_5C{\equiv}CSCH_3}}} \quad \begin{array}{c}575\\92\%\\ \\ \\46\%\end{array}$$

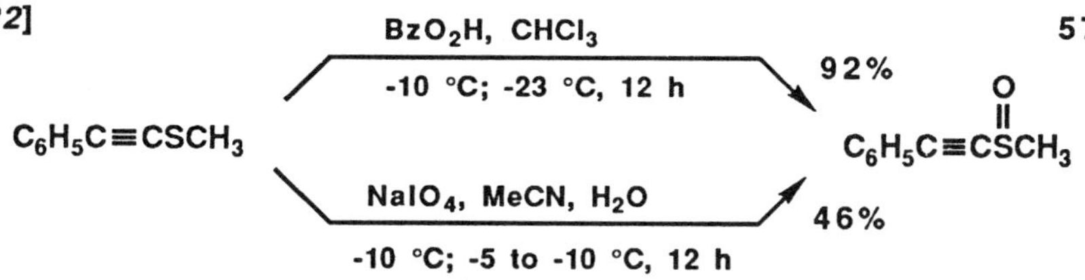

Other functional groups are also unaffected. 2-Phenylmercaptoethanol is oxidized at sulfur only. With hydrogen peroxide alone, it is converted into a sulfoxide. When catalysts such as tungstic acid or sodium orthovanadate are used, the product is the corresponding sulfone (equation 576) [135].

[135]   576

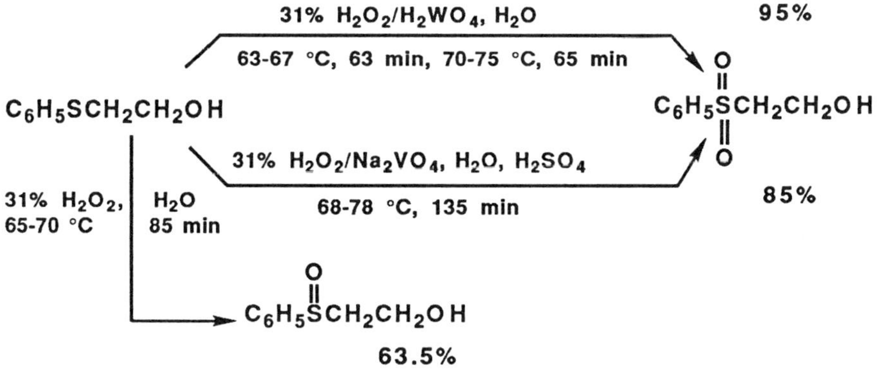

$$C_6H_5SCH_2CH_2OH \xrightarrow[\substack{31\%\ H_2O_2/H_2WO_4,\ H_2O \\ 63-67\ °C,\ 63\ min,\ 70-75\ °C,\ 65\ min \\ \\ 31\%\ H_2O_2/Na_2VO_4,\ H_2O,\ H_2SO_4 \\ 68-78\ °C,\ 135\ min}]{} \underset{\underset{O}{\|}}{\overset{\overset{O}{\|}}{C_6H_5S}}CH_2CH_2OH \quad \begin{array}{c}95\%\\ \\ \\85\%\end{array}$$

$$\xrightarrow[\substack{31\%\ H_2O_2, \\ 65-70\ °C}]{H_2O,\ 85\ min} \underset{O}{\overset{O}{\underset{\|}{C_6H_5S}}}CH_2CH_2OH \quad 63.5\%$$

On the other hand, secondary β-hydroxy sulfides yield β-keto sulfides on treatment with chloral on alumina (equation 577) [960]. This result is understandable because dehydrogenation, and not true oxidation, is involved.

[960]   577

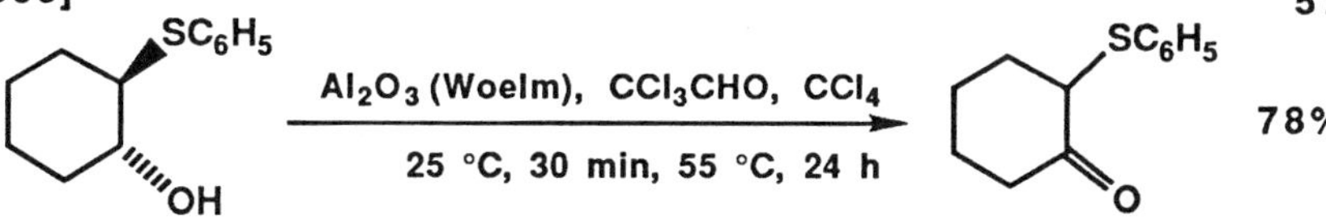

trans-2-(phenylthio)cyclohexanol $\xrightarrow[25\ °C,\ 30\ min,\ 55\ °C,\ 24\ h]{Al_2O_3\ (Woelm),\ CCl_3CHO,\ CCl_4}$ 2-(phenylthio)cyclohexanone   78%

Primary amino groups remain untouched when methionine is oxidized to methionine S-oxide in 97.5% yield by diethyl azodicarboxylate in aqueous ethanol at room temperature [979].

Oxidation of Sulfoxides

The oxidation of *sulfoxides to sulfones* is accomplished with **air** in the presence of noble metal salts [*10*] or with sodium permanganate [*837*] (equations 578 and 579).

[*10*] 578
$$(CH_3)_2S=O \xrightarrow[\text{heat, 16 h}]{\text{air, RhCl}_3\text{, i-PrOH/H}_2\text{O (9:1)}} (CH_3)_2S{\overset{O}{\underset{O}{\lessgtr}}}$$
80%

[*837*] 579
$$(C_6H_5CH_2)_2S=O \xrightarrow[\text{20 °C, 5 min}]{\text{NaMnO}_4\text{, dioxane, H}_2\text{O}} (C_6H_5CH_2)_2S{\overset{O}{\underset{O}{\lessgtr}}}$$
95–100%

Sodium permanganate and, especially, **osmium tetroxide** oxidize sulfoxides to sulfones but do not oxidize sulfides [*837*]. When a mixture of equimolar amounts of diphenyl sulfide, diphenyl sulfoxide, and osmium tetroxide is refluxed in ether for 48 h, the sulfoxide is oxidized in 96% yield to the sulfone, whereas the sulfide is completely recovered [*837*].

Certain unsaturated sulfoxides treated with a catalytic amount of osmium tetroxide and **trimethylamine oxide** as a reoxidant are not only oxidized to sulfones but also hydroxylated at the double bond. The sulfoxide group exerts a steric influence in the hydroxylation: two stereoisomers of a sulfoxide yield, by *syn* hydroxylation, two different stereoisomeric diols (equation 580) [*1102*].

[*1102*] 580

*The products were isolated as the diacetates.

Nitrogen analogues of sulfoxides, *sulfilimines*, when protected by a tosyl group at the nitrogen, are converted into nitrogen analogues of sulfones, *sulfoximines*, by **tetrabutylammonium hypochlorite**, which is prepared in situ from tetrabutylammonium bromide and an excess of sodium hypochlorite (equation 581) [1200].

[1200]

$$C_6H_5\underset{NSO_2C_7H_7}{\overset{\|}{S}}C_2H_5 \xrightarrow[\text{RT}]{\text{NaOCl, Bu}_4\text{NBr, H}_2\text{O,} \atop \text{AcOEt, CH}_2\text{Cl}_2} C_6H_5\underset{NSO_2C_7H_7}{\overset{\overset{O}{\|}}{\underset{\|}{S}}}C_2H_5 \quad 581$$

90%

Oxidation of Sulfones

Sulfones undergo peculiar reactions on treatment with carbon tetrachloride and potassium hydroxide in aqueous *tert*-butyl alcohol at 25–80 °C. Methyl phenyl sulfone yields phenyl trichloromethyl sulfone, which is hydrolyzed to benzenesulfonic acid [954]. Dibenzyl sulfone is quantitatively converted into *trans*-stilbene, and dicyclohexyl sulfone is converted into a mixture of bicyclohexylidene and 1,1-dichloro-2,3-dicyclohexylcyclopropane [954] (equations 582–584).

[954]   582

$$C_6H_5SO_2CH_3 \xrightarrow[\text{25-80 °C, 10-60 min}]{\text{CCl}_4,\text{ KOH, }t\text{-BuOH, H}_2\text{O}} C_6H_5SO_2CCl_3 \longrightarrow C_6H_5SO_3K$$

100%

[954]   583

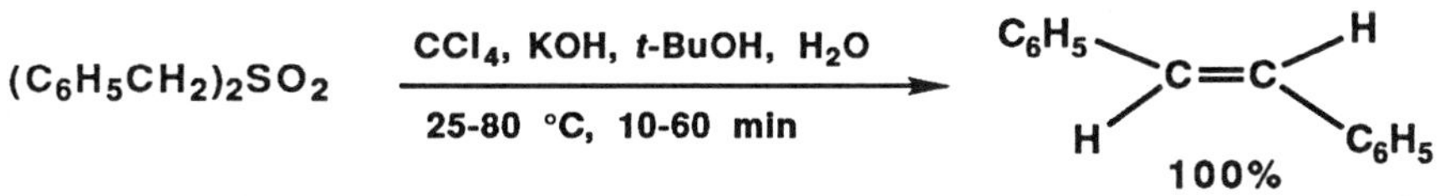

$$(C_6H_5CH_2)_2SO_2 \xrightarrow[\text{25-80 °C, 10-60 min}]{\text{CCl}_4,\text{ KOH, }t\text{-BuOH, H}_2\text{O}} \underset{H}{\overset{C_6H_5}{>}}C=C\underset{C_6H_5}{\overset{H}{<}}$$

100%

[954]   584

32%   60%

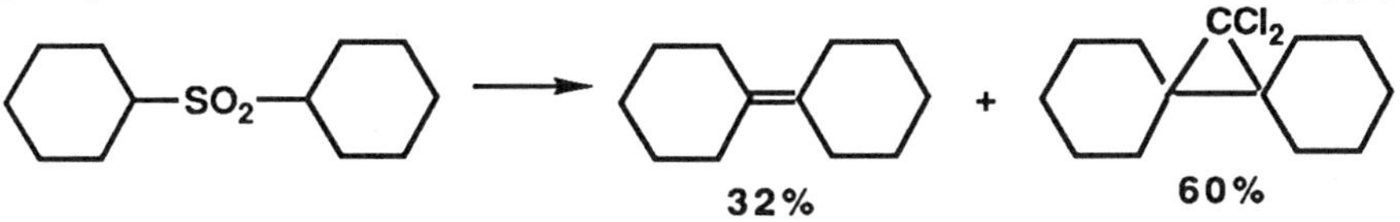

Oxidation of Disulfides

Disulfides are oxidized to **monosulfoxides (alkyl thiolsulfinates)** with peroxybenzoic acid (equation 585) [308].

Under anhydrous conditions, **chlorine** converts dimethyl disulfide into **methanesulfinyl chloride** by a sequence of reactions (equation 586) [687].

[308]   585

$$(CH_3)_3CSSC_2H_5 \xrightarrow[\text{0 °C, 30 min; RT, 1 h}]{\text{BzO}_2\text{H, CHCl}_3} (CH_3)_3C\overset{\overset{O}{\|}}{S}SC_2H_5$$

60%

[687]

$$CH_3SSCH_3 \xrightarrow[\substack{\text{0 to -10 °C} \\ \text{vacuum} \\ \text{distillation}}]{Cl_2,\ Ac_2O} \left[CH_3SCl \xrightarrow{Cl_2} CH_3SCl_3 \right] \xrightarrow{Ac_2O} \underset{+\ 2\ CH_3COCl}{CH_3SOCl\ 83\text{-}86\%}$$

586

The oxidation of disulfides with **oxygen** in solutions of potassium hydroxide in dimethylformamide at 23.5 °C gives *sulfonic acids* in 90% yields (equation 587) [53].

[53]

$$C_4H_9SSC_4H_9 \xrightarrow[23.5\ °C]{O_2,\ KOH,\ DMF} C_4H_9SO_3K\ \ 90\%$$

587

2,2'-Diaminodiphenyl disulfide, when treated with **peroxysulfuric acid,** is oxidized to *o*-anilinesulfonic acid in 80% yield (equation 588) [187].

[187]

30% H_2O_2, 97% H_2SO_4, 30-40 °C, 1.5 h, 25 °C, 1 h → 80%

588

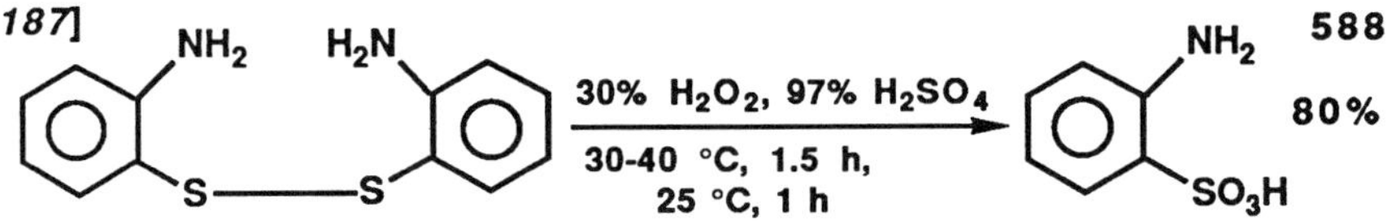

Chlorine in the presence of water converts disulfides into *sulfonyl chlorides* (equation 589) [685].

[685]

Cl_2, HCl, HNO_3, H_2O, 70 °C, 1.5 h → 84%

589

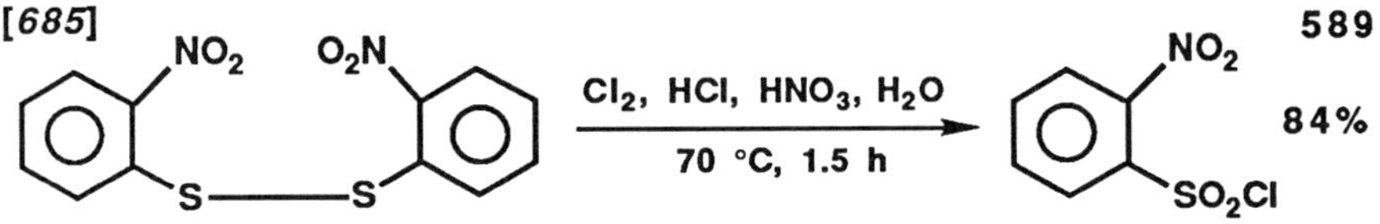

Sulfonic acids and sulfonyl chlorides can also be prepared from *sulfinic acids* by oxidation with **sodium permanganate** (equation 590) [836] and **chlorine** in water (equation 591) [686], respectively.

[836]

$$C_{12}H_{25}\text{-}C_6H_4\text{-}SO_2H \xrightarrow[\substack{35\text{-}40\ °C, \\ \text{RT overnight}}]{NaMnO_4,\ NaOH} C_{12}H_{25}\text{-}C_6H_4\text{-}SO_3Na\ \ 70\%$$

590

[686]

$$o\text{-}CH_3C_6H_4SO_2H \xrightarrow[30\ °C]{Cl_2,\ KOH,\ H_2O} o\text{-}CH_3C_6H_4SO_2Cl\ \ 57\%$$

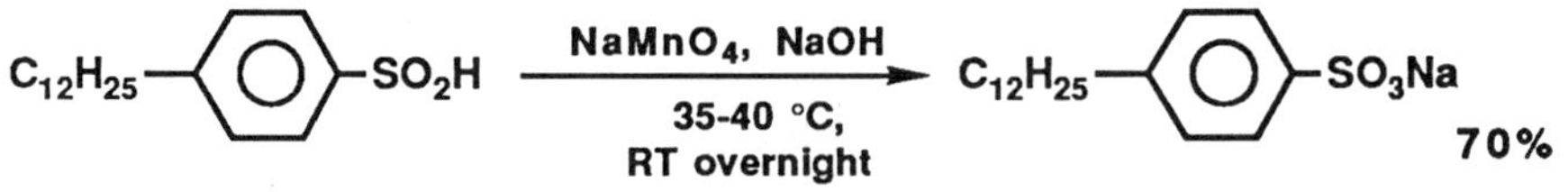

591

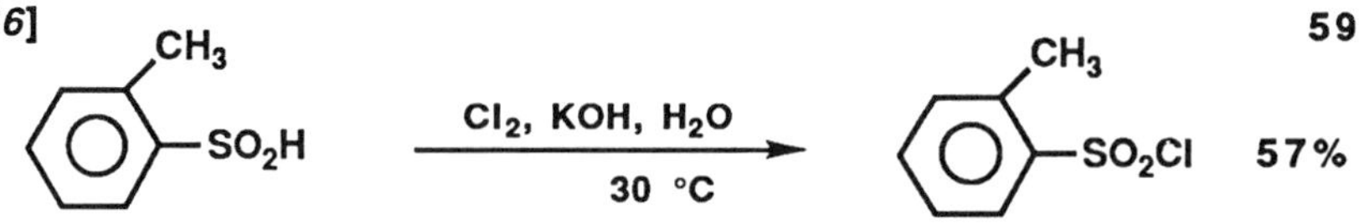

SELENIDES

Like sulfides, *selenides* can be oxidized to the corresponding oxides, *selenoxides,* or to *selenones.* Oxidation to selenoxides is much faster and, with some oxidants, final. Reaction conditions are specified in equation 592 for the oxidation of methyl phenyl selenide [*325, 711, 773, 1034*].

592

$$C_6H_5SeCH_3 \longrightarrow C_6H_5\overset{O}{\underset{\parallel}{S}e}CH_3 \text{ or } C_6H_5\overset{O}{\underset{\underset{O}{\parallel}}{\overset{\parallel}{S}e}}CH_3$$

[*325*]	CF_3CO_3H, CH_2Cl_2, 20 °C, 30 min		70%
[*325*]	m-$ClC_6H_4CO_3H$, CH_2Cl_2, 20 °C, 2 h		40%
[*711*]	t-BuOCl, MeCN, C_5H_5N, MeOH, -25 °C, 40 min	80%	
[*325*]	$KMnO_4$, CH_2Cl_2, 20 °C, 30 min		75%
[*773*]	$NaIO_4$, MeOH, H_2O, 0 °C	79%	
[*1034*]	cyclohexanone oxygenase (*Acinetobacter* sp.)	*	

*The yield is not reported.

Diphenyl selenide is oxidized with **peroxyacetic acid** at room temperature to diphenyl selenoxide hydrate, $C_6H_5Se(OH)_2$, in 43% yield after 2 h [*1198*]. Benzyl phenyl selenide is oxidized to benzyl phenyl selenoxide by **sodium periodate** in aqueous methanol at 0 °C in 95% yield and by **iodobenzene dichloride** in aqueous pyridine at −40 °C in 85% yield [*773*].

Selenoxides are useful intermediates in the preparation of α,β-unsaturated carbonyl compounds and esters. The treatment of aldehydes, ketones, or esters with benzeneselenyl chloride, C_6H_5SeCl, followed by the oxidation of the selenides to selenoxides by **hydrogen peroxide, peroxy acids,** or **sodium periodate,** gives α,β-unsaturated aldehydes, ketones, or esters. Thus, dehydrogenation with the formation of a carbon–carbon double bond is accomplished under very mild conditions [*167, 169*] (equation 593).

[*167*]   **593**

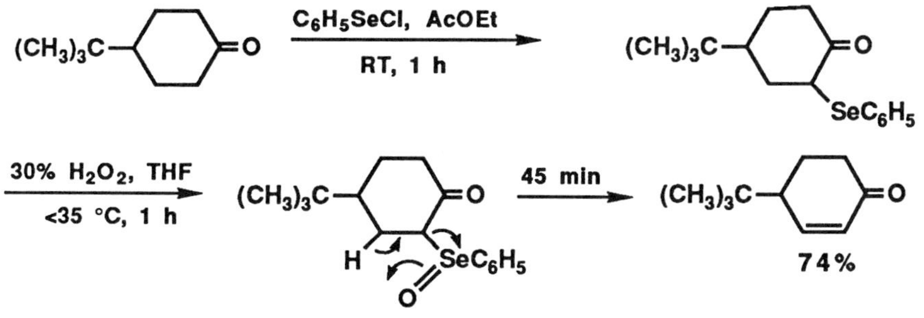

$(CH_3)_3C\text{-cyclohexanone} \xrightarrow{\underset{RT, 1\ h}{C_6H_5SeCl,\ AcOEt}} (CH_3)_3C\text{-cyclohexanone-}SeC_6H_5$

$\xrightarrow{\underset{<35\ °C,\ 1\ h}{30\%\ H_2O_2,\ THF}} (CH_3)_3C\text{-cyclohexanone intermediate} \xrightarrow{45\ min} (CH_3)_3C\text{-cyclohexenone}$ 74%

Another synthetic application of selenoxides is the conversion of epoxides into allylic alcohols (equation 594) [*168*]. Dehydrogenation occurs almost exclusively away from the hydroxyl group.

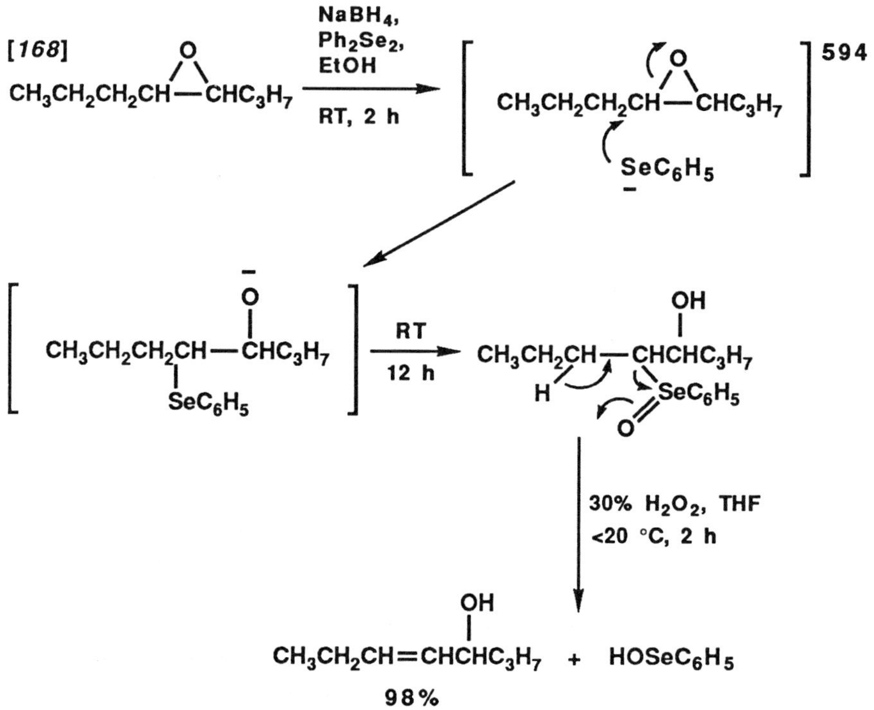

[*168*]

$$CH_3CH_2CH_2CH\overset{O}{-}CHC_3H_7 \xrightarrow[\text{RT, 2 h}]{\text{NaBH}_4,\text{Ph}_2\text{Se}_2,\text{EtOH}} \left[CH_3CH_2CH_2CH\overset{\overset{O^-}{|}}{-}CHC_3H_7\;\;\underset{SeC_6H_5}{} \right] \quad 594$$

$$\left[CH_3CH_2CH_2CH\overset{\overset{O^-}{|}}{\underset{SeC_6H_5}{-}}CHC_3H_7 \right] \xrightarrow[\text{12 h}]{\text{RT}} CH_3CH_2CH-CHCHC_3H_7 \text{ (OH, SeC}_6\text{H}_5\text{)}$$

↓ 30% H$_2$O$_2$, THF
<20 °C, 2 h

$$CH_3CH_2CH=CHCHC_3H_7 \text{ (OH)} + HOSeC_6H_5$$

98%

IODO COMPOUNDS

Iodo compounds can be oxidized to ***iodoso compounds*** or to ***iodoxy compounds.*** Iodobenzene treated with **peroxyacetic acid** at 30 °C gives an 83–91% yield of iodosobenzene diacetate [*266*]. 2-Iodo-4,5-dinitrobenzoic acid, when heated for 3 min at 100 °C with fuming **nitric acid,** is transformed into 2-iodoso-4,5-dinitrobenzoic acid in 92% yield [*475*]. Under more energetic conditions, the oxidation products are iodoxy compounds [*204, 205, 267*] (equation 595).

o-Iodosobenzoic acid is converted into *o*-iodoxybenzoic acid in 81% yield by heating at 100 °C with potassium permanganate in dilute sodium hydroxide [*893*].

Both iodoso and iodo compounds are oxidizing agents. The newest iodine-containing oxidant is the **Dess–Martin periodinane,** whose structure

Oxidations

[266] 40% AcO₂H, AcOH   [204] K₂S₂O₈, H₂SO₄   595
C₆H₅I(OCOCH₃)₂ ◀─────────── C₆H₅I ─── RT, 2.5 h ──▶ C₆H₅IO₂   88%
83-91%   30 °C, 1 h   KHSO₅·KHSO₄·K₂SO₄*
                                   H₂SO₄, ice, 4 h
                                   40% AcO₂H, AcOH   94%
                          [267]   35 °C, 30 min; H₂O
                                  35-100 °C, 2 min
                                  100 °C, 45 min

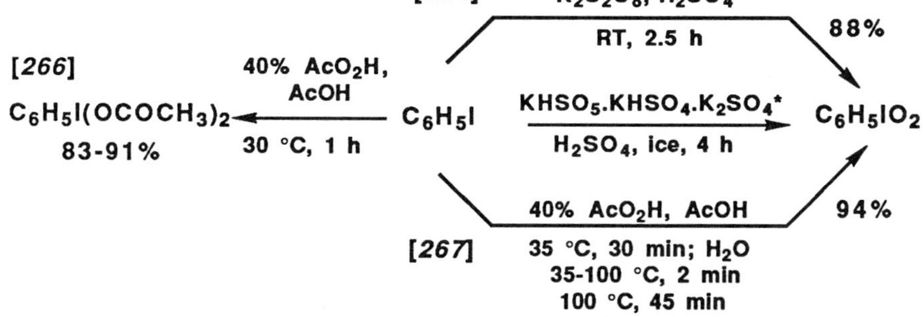

*Oxidation with Oxone is described in reference 205 and gives a 72% yield of iodoxybenzene.

and preparation are shown in equation 596 [801]. This reagent oxidizes alcohols to aldehydes or ketones under mild conditions [801].

[801]

<chemical scheme: 2-iodobenzoic acid → (KBrO₃, H₂SO₄) → cyclic iodinane with I(=O)(OH) 93% → (Ac₂O, AcOH, 100 °C, 40 min) → Dess-Martin type periodinane with OCOCH₃ groups 93%>   596

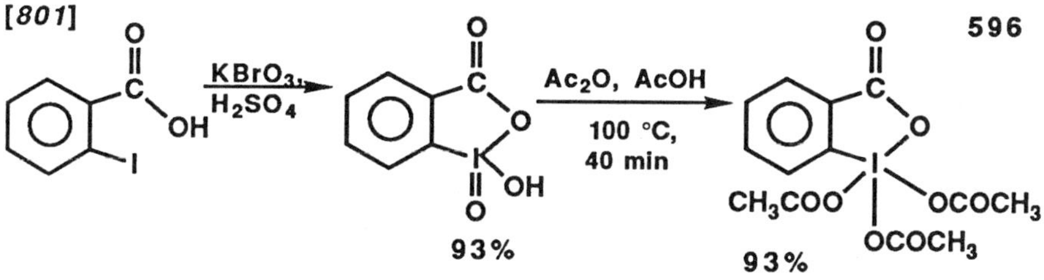

BORON COMPOUNDS

The oxidation of carbon–boron bond converts **boranes** into alkyl or aryl **borates,** which may be hydrolyzed subsequently to **alcohols** and boric acid [991]. The oxidation is carried out with **hydrogen peroxide** [183, 1201, 1202] or **trimethylamine oxide** [991, 992]. Phenylboronic acids are oxidized to phenols *biochemically* [1034].

A general scheme for the conversion of alkylboranes into alcohols is given in equation 597 [991].

[991]   597
$(C_4H_9)_3B \xrightarrow{Me_3NO} (C_4H_9)_2BOC_4H_9 \xrightarrow{Me_3NO}$

$C_4H_9B(OC_4H_9)_2 \xrightarrow{Me_3NO} B(OC_4H_9)_3$

The transformation of tributyl borane with trimethylamine oxide in chloroform into dibutyl borinate is exothermic and rapid at 0 °C, and the transformation of dibutyl borinate into butyl boronate takes place at 25 °C. However, the final conversion of the butyl boronate into tributyl borate requires refluxing in chloroform for 24 h [*991*]. The alcohols are isolated by transesterification of the alkyl borates, for example, by distillation with decanol [*991*].

To prepare alcohols by the oxidation of alkyl boranes, borane, diborane, alkyl and dialkyl boranes, and dialkoxy boranes are added to alkenes or alkynes. The addition takes place in the anti-Markovnikov mode because boron is more electrophilic than hydrogen.

Borane, which is used as a complex with tetrahydrofuran [*992*] or dimethyl sulfide [*611, 992*] or generated in situ from lithium borohydride with boron trifluoride etherate [*646*] or sodium borohydride with aluminum chloride [*184*], reacts with 3 mol of an alkene to form a tertiary borane. The oxidation with alkaline hydrogen peroxide [*183, 992, 1201*] or with trimethylamine oxide [*991, 992*] yields an alcohol (equations 598 and 599).

[*992*]   598

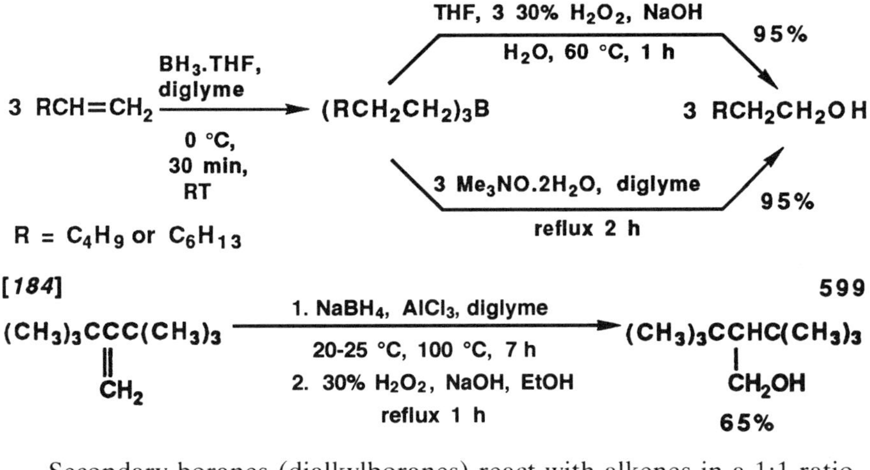

$$3\ RCH=CH_2 \xrightarrow[\substack{0\ °C,\\ 30\ min,\\ RT}]{BH_3 \cdot THF,\ diglyme} (RCH_2CH_2)_3B$$

$$(RCH_2CH_2)_3B \xrightarrow{\substack{THF,\ 3\ 30\%\ H_2O_2,\ NaOH \\ H_2O,\ 60\ °C,\ 1\ h}} 3\ RCH_2CH_2OH \quad 95\%$$

$$(RCH_2CH_2)_3B \xrightarrow{\substack{3\ Me_3NO \cdot 2H_2O,\ diglyme \\ reflux\ 2\ h}} 95\%$$

R = C_4H_9 or C_6H_{13}

[*184*]   599

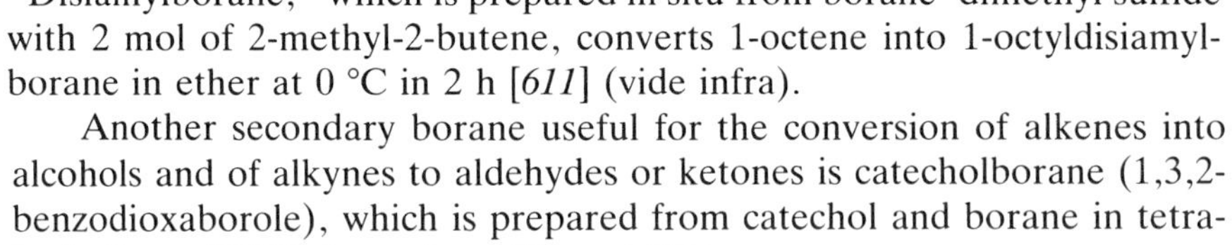

$$(CH_3)_3CCC(CH_3)_3 \underset{CH_2}{\parallel} \xrightarrow[\substack{2.\ 30\%\ H_2O_2,\ NaOH,\ EtOH \\ reflux\ 1\ h}]{1.\ NaBH_4,\ AlCl_3,\ diglyme,\ 20\text{-}25\ °C,\ 100\ °C,\ 7\ h} (CH_3)_3CCHC(CH_3)_3 \underset{CH_2OH}{|}$$

65%

Secondary boranes (dialkylboranes) react with alkenes in a 1:1 ratio. "Disiamylborane," which is prepared in situ from borane–dimethyl sulfide with 2 mol of 2-methyl-2-butene, converts 1-octene into 1-octyldisiamylborane in ether at 0 °C in 2 h [*611*] (vide infra).

Another secondary borane useful for the conversion of alkenes into alcohols and of alkynes to aldehydes or ketones is catecholborane (1,3,2-benzodioxaborole), which is prepared from catechol and borane in tetrahydrofuran [*1201*] (equations 600 and 601).

[*1201*]   600

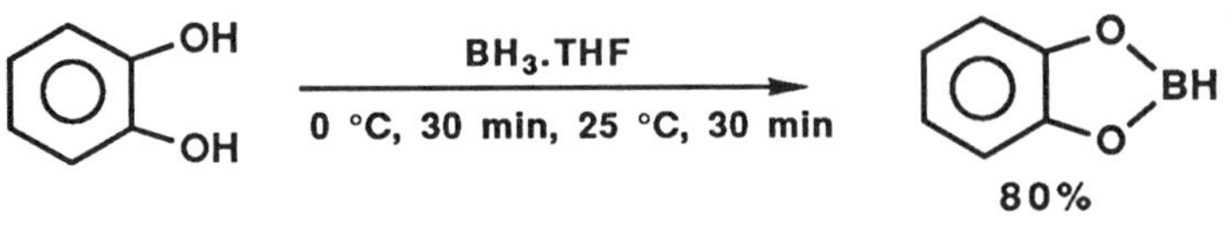

$$\text{catechol} \xrightarrow[0\ °C,\ 30\ min,\ 25\ °C,\ 30\ min]{BH_3 \cdot THF} \text{catecholborane}$$

80%

[1201]

601

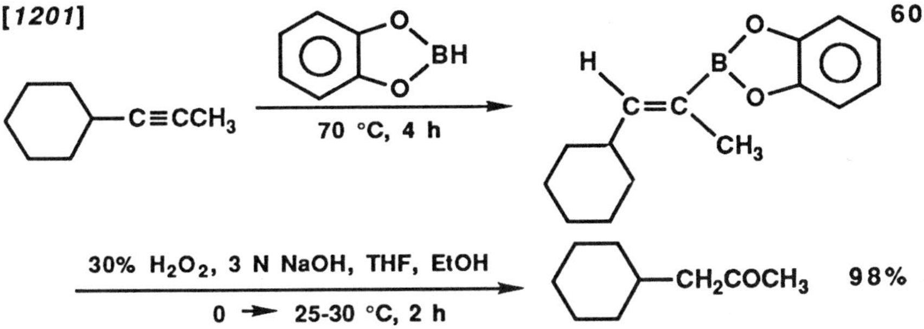

[Reaction: cyclohexyl-C≡CCH₃ + benzodioxaborole-BH, 70 °C, 4 h → vinyl boronate intermediate; then 30% H₂O₂, 3 N NaOH, THF, EtOH, 0 → 25–30 °C, 2 h → cyclohexyl-CH₂COCH₃, 98%]

If the oxidation of the boranes is carried out by hexavalent chromium compounds, the products are aldehydes (equation 602) [*611*] or ketones (equation 603) [*646*].

[*611*] **602**

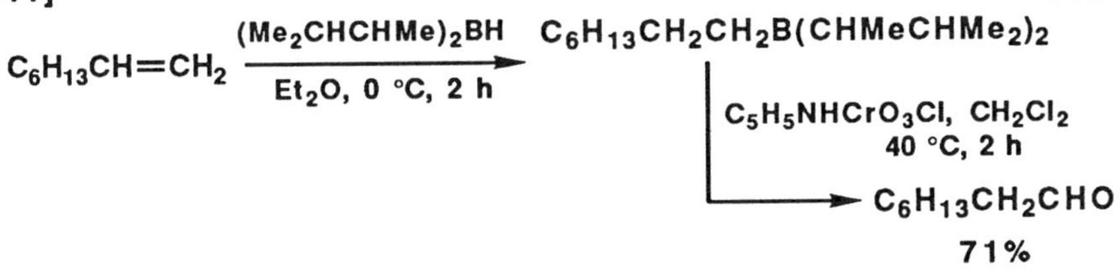

$$C_6H_{13}CH=CH_2 \xrightarrow[\text{Et}_2\text{O, 0 °C, 2 h}]{(Me_2CHCHMe)_2BH} C_6H_{13}CH_2CH_2B(CHMeCHMe_2)_2$$

$$\xrightarrow[\text{40 °C, 2 h}]{C_5H_5NHCrO_3Cl, CH_2Cl_2} C_6H_{13}CH_2CHO$$

71%

[*646*] **603**

[Reaction: 1-methylcyclohexene → LiBH₄, BF₃, Et₂O, 25–30 °C, 2.25 h → 2-methylcyclohexyl-BH₂ → Na₂Cr₂O₇, H₂SO₄, H₂O, 25–30 °C, 15 min; reflux 2 h → 2-methylcyclohexanone, 78%]

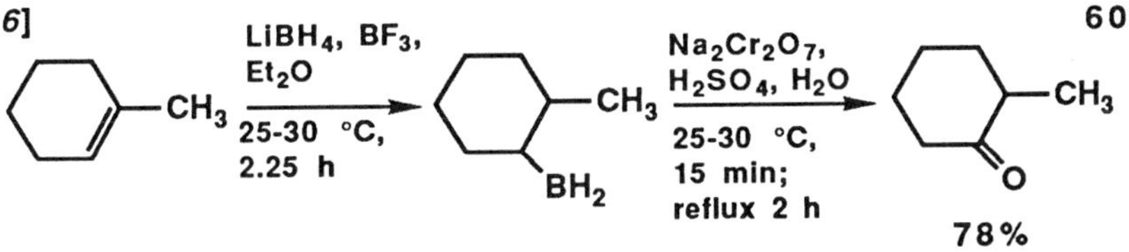

Aromatic boranes are synthesized from trimethyl borate and arylmagnesium halides [*185*] or from diborane and arylmercury halides [*183*], or diarylmercury compounds [*183*]. Their oxidation furnishes phenols in variable yields. A large excess (up to 16 mol) of BH₃ is necessary for high yields [*183*] (equations 604 and 605).

[*185*] **604**

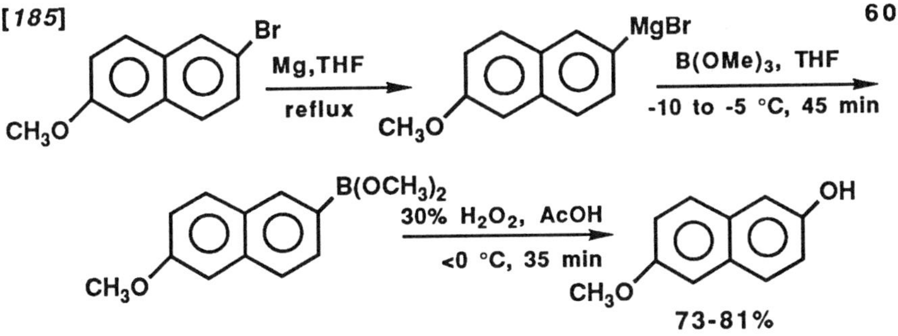

[Reaction: 6-methoxy-2-bromonaphthalene → Mg, THF, reflux → 6-methoxy-2-naphthyl-MgBr → B(OMe)₃, THF, −10 to −5 °C, 45 min → 6-methoxy-2-naphthyl-B(OCH₃)₂ → 30% H₂O₂, AcOH, <0 °C, 35 min → 6-methoxy-2-naphthol, 73–81%]

[183]

C_6H_5HgBr →
1. 8 B_2H_6, THF, RT, 1 h
2. H_2O_2, OH⁻

↓ 72% C_6H_5OH ↓ 97.5%

1. 8 B_2H_6, THF, RT, 1 h
2. H_2O_2, OH⁻
→ $(C_6H_5)_2Hg$ 605

Phenols are also obtained by oxidation of phenylboronic acids with **cyclohexanone oxygenase** produced by *Acinetobacter* strain NCIB 9871 (equation 606) [*1034*].

[*1034*] $C_6H_5B(OH)_2$ —cyclohexanone oxygenase→ C_6H_5OH 606

Alkyldichloroboranes are oxidized by oxygen to alkyl hydroperoxides (equation 607) [*51*].

[*51*] $C_6H_{13}BCl_2$ —O_2, Et_2O, -180 °C, 20 min; H_2O, RT→ $C_6H_{13}OOH$ 94%   607

SILICON AND TIN COMPOUNDS

The silicon–silicon bond is oxidatively cleaved in phenylpentamethyldisilane, which is converted by **peroxybenzoic acid** into phenyldimethylsilyl trimethylsilyl ether in 86% yield (equation 608) [*1203*].

[*1203*] $C_6H_5(CH_3)_2SiSi(CH_3)_3$ —BzO_2H, CH_2Cl_2, RT, overnight→ $C_6H_5(CH_3)_2SiOSi(CH_3)_3$ 86%   608

The oxidative cleavage of the *carbon–tin bond* takes place in the reaction of alkylstannanes with **lead tetraacetate**. Phenoxymethyltributylstannane, when refluxed in dichloromethane with an equimolar amount of lead tetraacetate for 32 h, is converted into phenoxymethyl acetate in 58% yield (equation 609) [*1204*].

[*1204*] $C_6H_5OCH_2Sn(C_4H_9)_3$ —$Pb(OAc)_4$, CH_2Cl_2, reflux 32 h→ $C_6H_5OCH_2OCOCH_3$ 58%   609

ORGANOMAGNESIUM AND ORGANOMERCURY COMPOUNDS

The reaction of **2-bromomagnesiumthiophene** with **tert-butyl peroxybenzoate** gives 2-*tert*-butoxythiophene, which, on acid-catalyzed thermal decomposition, yields **2-hydroxythiophene** (equation 610) [*235*].

[*235*]   610

thiophene-MgBr $\xrightarrow[\text{overnight}]{\text{BzO}_2\text{OCMe}_3, \ 0\ °\text{C, 45 min}}$ thiophene-OC(CH$_3$)$_3$ (70-76%) $\xrightarrow[\text{5-10 min}]{\text{TosOH}, \ 155\ °\text{C}}$ thiophene-OH (89-94%)

Secondary **organomercury chlorides** are oxidized with ozone to **ketones** in fair yields. Primary organomercurials usually give mixtures of acids resulting from deeper oxidation (equation 611) [*114*].

[*114*]   C$_6$H$_{11}$-HgCl $\xrightarrow[\text{10 °C}]{\text{O}_3,\ \text{CH}_2\text{Cl}_2}$ cyclohexanone   60%   611

Safety Advisory

THE SAFETY INFORMATION is included within this book by the American Chemical Society as a precaution to the readers.

The following chapter discusses the preparation and the use of various oxidizing agents. Strong oxidizing agents are among the most hazardous materials used in the laboratory. The extreme caution required when preparing and using these reagents cannot be overemphasized. The hazards generally associated with the preparation, handling, and use of oxidizing agents are explosion, fire, and uncontrolled and violent exothermic reaction resulting in possible equipment failure and physical injury to the operator. To minimize the potential for any of these occurrences, chemists must prudently exercise all safe practices at their disposal.

All new reactions should be conducted on the smallest possible scale with clean glassware that has been carefully inspected for defects. If at all possible, all reactions should be conducted in an adequate laboratory fume hood with the sash closed. This measure will provide the operator with a physical barrier against fires, explosions, and chemical splashes, as well as provide for the removal of potentially toxic, flammable, or offensive vapors. The laboratory should possess the basic safety facilities necessary to handle and use oxidizing agents, including approved and periodically inspected and tested eye wash stations, safety showers, and fire suppression equipment. To keep personal exposure to a minimum, chemists must wear appropriate and adequate personal protective equipment when conducting oxidation experiments. Essential safety apparel includes chemical splash goggles, full face shields, chemically impervious gloves, and laboratory coats or aprons.

Before following the procedures outlined in this chapter, readers should refer to the original reference for any specific handling practice that may be cited. Chemists should also refer to the manufacturer's material safety data sheet for any additional handling precautions.

Although many oxidation reactions and oxidizing agents have some degree of hazard associated with their use, the procedures marked with the hazard symbol 🔥 may require special caution.

Chapter 4
Preparative Procedures

1. PREPARATION OF ANHYDROUS HYDROGEN PEROXIDE IN ETHER [126]

A 2.5 M solution of anhydrous hydrogen peroxide in ether is prepared by repeated extraction of 30% aqueous hydrogen peroxide with ether. The ether extracts are dried over anhydrous magnesium sulfate, and the concentration of hydrogen peroxide is determined by iodometric titration.

2. PREPARATION OF 90% HYDROGEN PEROXIDE AND OF 90% PEROXYACETIC ACID [1198]

In an *absolutely clean* apparatus consisting of a sidearm flask fitted with a ground-glass capillary, a short column, and a condenser fitted with a ground-glass thermometer, 29% hydrogen peroxide is brought to boiling. First, water distills at 29 °C at 18 mm of Hg. Then, at 55 °C at 18 mm of Hg, hydrogen peroxide condenses as an oily liquid sticking to the walls. The residue in the flask is 90% hydrogen peroxide.

The mixing of 64 g (1.69 mol) of 90% hydrogen peroxide, 88 g (0.86 mol) of acetic anhydride, and 18 g of concentrated sulfuric acid and distillation in the apparatus just described gives 122 g (85%) of 90% peroxyacetic acid distilling at 25 °C at 12 mm of Hg and 10 g (7.8%) of 69% peroxyacetic acid distilling at 22 °C at 10 mm of Hg.

All the operations must be carried out with strict safety measures.

3. PREPARATION OF THE JONES REAGENT [585]

Chromium trioxide (267 g, 2.67 mol) is dissolved in 400 mL of water, and 230 mL of concentrated sulfuric acid is added with cooling. The cold solution is diluted with water up to 1 L to form an 8 N reagent. The reagent is added dropwise to a solution of the compound to be oxidized in acetone (distilled from potassium permanganate) at 20 °C.

Oxidants . . . Use Caution

4. PREPARATION OF CHROMIUM TRIOXIDE–PYRIDINE COMPLEX, CrO$_3$·2C$_5$H$_5$N (COLLINS REAGENT) [593]

A 3-L, three-necked flask equipped with a sealed mechanical stirrer and a drying tube filled with Drierite (calcium sulfate) is charged with 946 mL of dry pyridine and cooled in an acetone–ice bath to –15 to –18 °C. Chromium trioxide (90 g, 0.9 mol) dried previously over phosphorus pentoxide for 12 h under reduced pressure is added portionwise through the neck by using glazed paper cones (a new cone for each added portion) over a period of 5 min while the flask contents are stirred intensively. With these precautions, ignition of the reaction mixture, which would occur had pyridine been added to chromium trioxide, is avoided. After the addition of the chromium trioxide, stirring is continued for 6 h, during which time no further ice is added to the acetone bath. The initially formed bright-yellow viscous mixture turns into a bright-red crystalline slurry.

The complex CrO$_3$·2C$_5$H$_5$N is filtered with suction through a sintered-glass funnel, washed immediately with 300 mL of dry reagent-grade petroleum ether, and stored immediately in a desiccator over phosphorus pentoxide at 15–30 mm of Hg. The yield is 180 g (80%). The entire operation starting with the filtration should be performed within less than 3 min to minimize hydration of the extremely hygroscopic complex, which, when stored in this way, is still active after 10 days. The dry complex does not cause any fires.

5. PREPARATION OF PYRIDINIUM CHLOROCHROMATE ADSORBED ON ALUMINA [604]

To a solution of 6 g (0.06 mol) of chromium trioxide in 11 mL of 6 N hydrochloric acid is added 4.75 g (0.06 mol) of pyridine within 10 min at 40 °C. The mixture is kept at 10 °C until a yellow-orange solid forms and then reheated to 40 °C to dissolve the solid. Alumina (50 g) is added with stirring at 40 °C. After evaporation in a rotary evaporator, the orange solid is dried in vacuo for 2 h at room temperature. The reagent can be kept under vacuum for several weeks in the dark without losing its activity.

6. PREPARATION OF TETRABUTYLAMMONIUM CHROMATE [618]

To a stirred solution of 1.0 g (10 mmol) of chromium trioxide in 25 mL of water, an aqueous solution of 2.92 g (10.5 mmol) of tetrabutylammonium chloride in 50 mL of water is rapidly added at room temperature. A yellow-orange solid precipitates immediately. The mixture is cooled to 0 °C, and the solid is filtered with suction through a sintered-glass funnel. It is carefully washed with cold water, dried under vacuum over phosphorus pentoxide, and stored over calcium chloride. The yield of the tetrabutylammonium chromate is 2.76 g (77%).

7. PREPARATION OF ACTIVE MANGANESE DIOXIDE [805]

A solution of 1110 g (5 mol) of manganese sulfate tetrahydrate in 1500 mL of water and a solution of 1170 mL of 40% sodium hydroxide (16 mol) are added simultaneously over a period of 1 h to a hot stirred solution of 960 g (6.1 mol) of potassium permanganate in 6 L of water. Manganese dioxide starts precipitating as a fine brown solid. Stirring is continued for an additional hour, the mixture is centrifuged, and the solid is washed with water until the washings are colorless. The solid is dried in an oven at 100–120 °C and ground to a fine powder before use. The yield is 920 g (95%).

8. DEHYDROGENATION BY 2,3-DICHLORO-5,6-DICYANO-*p*-BENZOQUINONE (DDQ) [*970*]

A solution of 0.65 g (0.005 mol) of **tetralin** and 1.14 g (0.005 mol) of 2,3-dichloro-5,6-dicyano-*p*-benzoquinone in 5 mL of benzene is refluxed for 45 min, during which period the initially red solution becomes colorless. A solution of an additional 1.14 g (0.005 mol) of the quinone in 2 mL of benzene is added, and the refluxing is continued for an additional 75 min. After dilution with light petroleum, the solution is filtered, passed through a column of alumina, and evaporated to give 0.42 g **(70%)** of **naphthalene,** mp 79–80 °C. The petroleum-insoluble residue yields 0.7 g **(61%)** of colorless 2,3-dichloro-5,6-dicyanohydroquinone, mp 263 °C (dec), after crystallization from aqueous ethanol.

9. DEHYDROGENATION VIA SELENOXIDES [*167*]

To a solution of 5.50 g (14.2 mmol) of **3-cholestanone** in 125 mL of ethyl acetate is added 3.30 g (17.2 mmol) of benzeneselenyl chloride (C_6H_5SeCl), and the red-orange solution is stirred for 1 h until it turns pale yellow. Water (25 mL) is added to the stirred mixture, the aqueous layer is drained, and to the organic phase is added 55 mL of tetrahydrofuran and, dropwise, 3.5 mL (34.3 mmol) of 30% hydrogen peroxide while the mixture is stirred for an additional hour at a temperature below 35 °C. The reaction mixture is washed with water and a solution of sodium carbonate, dried, and evaporated to give 5.25 g (96%) of crude product, which on recrystallization from ethanol furnishes 2.43 g **(45%)** of **1-cholesten-3-one,** mp 95–97 °C.

10. OXYGENATION WITH SINGLET OXYGEN

Photochemical Generation [*33*]

To a solution of 6.4 g (0.097 mol) of **cyclopentadiene** in 1.8 L of distilled methanol cooled to 0 °C is added 5 g of thiourea and 0.2 g of rose bengal. Oxygen is bubbled through the solution for 5 min, and the mixture is cooled with ice and water to 15 °C and irradiated with a 450-W high-pressure mercury immersion lamp (Hanovia 679 A36, cooled by running water) with a Pyrex filter while the passage of oxygen is continued for 2.5 h. After 12 h at room temperature in the dark, the solvent is evaporated in vacuo, the residue is treated with water, and the mixture is extracted with benzene until the aqueous layer becomes clear. The aqueous layer is then evaporated, and the residue is fractionated in vacuo to give 5.7 g **(60%)** of *cis*-**2-cyclopentene-1,4-diol** distilling at 100–102 °C at 1.3 mm of Hg, mp 59–60 °C.

Oxidants . . . Use Caution

Generation from Hydrogen Peroxide and Sodium Hypochlorite [13]

A solution of 0.17 g (0.44 mmol) of **tetraphenylcyclopentadienone** in 125 mL of dioxane cooled in ice water is first treated with 5.0 mL (44 mmol) of 8.8 M hydrogen peroxide and subsequently with a solution of 32.8 mL (22 mmol) of 0.67 M sodium hypochlorite added dropwise with stirring. Extraction of the solution with benzene and evaporation of the solvent gives 0.085 g **(50%)** of crystalline *cis*-**dibenzoylstilbene**, mp 215.9–216.3 °C.

Generation from Triphenyl Phosphite Ozonide [19]

A solution of 9.3 g (0.03 mol) of **triphenyl phosphite** in 100 mL of dichloromethane is ozonized to form **triphenyl phosphite ozonide**. After the excess ozone is purged with nitrogen, a cold solution of 1.60 g (0.02 mol) of **1,3-cyclohexadiene** in 45 mL of dichloromethane is added from a dropping funnel while the mixture is stirred by a stream of nitrogen and cooled with a dry-ice–acetone bath. After the removal of the bath, the mixture is allowed to warm to room temperature. The solution is concentrated on a rotary evaporator, and the residue is distilled at less than 0.1 mm of Hg in a short-path still to give 1.51 g **(67.4%)** of pale-yellow semisolid **3,6-endooxocyclohexene,** which, after three recrystallizations from pentane, melts at 90–91 °C.

11. OZONIZATION

Ozonolysis of Double Bonds [81]

In a 2-L round-bottom reactor cooled to -20 °C, 120.4 g (0.40 mol) of **methyl oleate** is dissolved in 1800 mL of reagent-grade methanol, and oxygen containing 1.092 mmol of ozone per liter is passed through the solution at the rate of 1.94 L/min. After 195 min, ozone is no longer absorbed, as indicated by the liberation of iodine from a solution of potassium iodide at the reactor exit. Glacial acetic acid (150 mL) is added; the solution is warmed to 30 °C, and zinc dust is added, a small amount at a time, until a total of 60 g (0.92 mol) has been used, while the temperature is maintained at 30–35 °C by cooling. The mixture is filtered with suction, and the filtrate is distilled on a steam bath until about one-half of the methanol has been removed. Then 500 mL of dichloromethane and 500 mL of water are added, the aqueous layer is separated, and the dichloromethane layer is washed with water (each wash is back-washed with a small amount of dichloromethane) until all the acid has been removed. The methylene chloride is distilled off on a steam bath, and the residue is fractionated through a small Vigreux column with a nitrogen capillary ebullator. **Pelargonaldehyde** distills at 37–47 °C at 0.35 mm of Hg. The total yield (43.9 g in the main fraction and 2.27 g in the dry-ice trap) is 46.17 g **(81%)**. **Methyl 9-oxononanoate** distills at 94–96 °C at 0.75 mm of Hg (65.3 g) and at 96–120 °C at 0.30 mm of Hg (4.87 g). The crude yield is **94%** (an **85.4%** yield of the pure compound is obtained on redistillation).

12. ELECTROLYTIC DECARBOXYLATIVE COUPLING OF CARBOXYLIC ACIDS [118]

The electrolytic cell is a water-jacketed tube 6 in. (15 cm) long and 1.5 in. (3.8 cm) in diameter and is stoppered with a rubber stopper fitted with a dropping funnel, a reflux condenser, and two platinum foil electrodes (5 by 2.5 cm) placed about 1–2 mm apart. The cell is charged with a cooled solution of 0.1 g (0.0043 mol) of sodium in 80 mL of methanol, and 10 g (0.083 mol) of **5-fluorovaleric acid** is added. A direct current of 1.5 A at 110 V is passed through the cell for 3 h while the temperature is maintained below 50 °C. The methanolic solution is diluted with water, neutralized with acetic acid, and extracted several times with ether. The combined ether extracts are washed twice with a 5% solution of sodium carbonate, dried over anhydrous sodium sulfate or magnesium sulfate overnight, and evaporated. The residue is refluxed for 1 h with 50 mL of concentrated hydrochloric acid to remove esters formed during the electrolysis. The mixture is cooled and extracted with ether, and the extracts are washed with a 5% solution of sodium carbonate and with water. After the extracts have been dried and the solvent has been evaporated, 2.8 g **(45%)** of **1,8-difluorooctane,** bp 75–75.5 °C at 13 mm of Hg, is obtained.

13. OXIDATION WITH HYDROGEN PEROXIDE

Conversion of Alkenes into Alcohols by Hydroboration Followed by Oxidation [*184*]

To a solution of 9.5 g (0.25 mol) of sodium borohydride in 250 mL of purified diglyme is added 70 g (0.50 mol) of 90–92% pure **1,1-di-*tert*-butylethylene** followed by a solution of 11.2 g (0.084 mol) of anhydrous aluminum chloride in 50 mL of diglyme. After the mixture has been stirred for 1 h at 20–25 °C and 7 h on a steam bath, most of the diglyme is removed under reduced pressure, and the residue is treated with excess hydrochloric acid. The organic material (73.8 g), which is isolated by the usual procedure, is dissolved in 100 mL of ethanol containing 8 g (0.2 mol) of sodium hydroxide. To this solution, 68 g of 30% hydrogen peroxide (0.6 mol) is added at a rate sufficient to maintain reflux. After 1 h, 350 mL of water is added, and the organic product is isolated in the usual manner after excess hydrogen peroxide has been destroyed with sodium bisulfite. Fractional distillation gives 51.5 g **(65%)** of **2,2-di-*tert*-butylethanol,** bp 105–110 °C at 29 mm of Hg, mp 52–54 °C (54–55 °C after recrystallization from low-boiling petroleum ether).

anti Hydroxylation of a Double Bond [*131*]

A solution of 8.2 g (0.1 mol) of **cyclohexene** in 45 mL of acetic acid is stirred for 10 h at 50 °C with a solution of 0.2 g of WO$_3$ in 13 g (0.11 mol) of 30% hydrogen peroxide. Some solvent is removed under reduced pressure, the residue is refluxed for 1 h with an excess of 2 N sodium hydroxide, and the product is obtained by continuous extraction with ether. Distillation of the extract gives a **71%** yield of ***trans*-1,2-cyclohexanediol** boiling at 130–140 °C at 30 mm of Hg.

14. OXIDATION WITH *tert*-BUTYL HYDROPEROXIDE

Oxidation of Tertiary Amines to Amine Oxides [*226*]

To a stirred mixture of 106 g (0.5 mol) of **dimethyldodecylamine** and 3 g of vanadium pentoxide in 400 mL of benzene, 60 g (0.5 mol) of 75% *tert*-butyl hy-

Oxidants . . . Use Caution

droperoxide is added at 25 °C over a period of 30 min. After having been stirred for 4 h at 40–45 °C, the warm mixture is filtered, the filtrate is evaporated in a vacuum of a water aspirator at a temperature not exceeding 70 °C, and the residue is diluted with 2 volumes of ethyl acetate. The precipitated crystals are filtered with suction, washed with petroleum ether, and dried in vacuo to yield 92 g **(80%)** of **dimethyldodecylamine oxide,** mp 115–117 °C (120–121 °C after recrystallization from ethyl acetate).

Asymmetric Oxidation of Sulfides to Sulfoxides [224]

Isopropyl orthotitanate, Ti(O-i-Pr)$_4$, (1.49 mL, 5 mmol) and diethyl (R,R)-tartrate (1.71 mL, 10 mmol) are dissolved in 50 mL of dichloromethane at room temperature under nitrogen. Water (0.09 mL, 5 mmol) is injected by means of a microsyringe, the mixture is stirred for 15–20 min until it becomes homogeneous, and finally, 0.7 g (5 mmol) of **methyl p-tolyl sulfide** is added. The solution is stirred at −20 °C, and 5.5 mmol of a 2 M solution of *tert*-butyl hydroperoxide in dichloromethane is introduced. After 4 h, 10 mol equiv of water is added dropwise via a microsyringe at −20 °C, and vigorous stirring is maintained for 1 h at −20 °C and for an additional 1 h at room temperature. The white gel is filtered after the addition of a small amount of alumina and washed thoroughly with dichloromethane. The filtrate is stirred with 5% sodium hydroxide and brine for 1 h, and the layers formed are separated. The organic phase is dried over anhydrous sodium sulfate and evaporated to give the crude product free of sulfone. Chromatography on silica gel and elution with cyclohexane–ethyl acetate (1:1) gives 0.70 g **(90%)** of **methyl p-tolyl sulfoxide,** $[\alpha]_D^{20}$ +131° (c 2, acetone) (90% enantiomeric excess).

15. OXIDATION WITH POTASSIUM PEROXYSULFATE

Selective Oxidation of Sulfides to Sulfones [207]

A solution of 18.44 g of Oxone (49.5% KHSO$_5$) in 40 mL of water is added to a cooled (0 °C) solution of 1.24 g (0.01 mol) of **methyl phenyl sulfide** in 40 mL of methanol. The resulting cloudy slurry is stirred for 4 h at room temperature, diluted with water, and extracted 3 times with chloroform. The combined extracts are washed with water and brine, dried over anhydrous sodium sulfate, and concentrated under reduced pressure to give a quantitative yield of methyl phenyl sulfone as a white solid. Recrystallization yields **95%** of pure **methyl phenyl sulfone,** mp 85–88.5 °C.

16. OXIDATION WITH ORGANIC PEROXY ACIDS

Oxidation of Primary Aromatic Amines to Nitroso Compounds with Peroxyacetic Acid [153]

A solution of 16.2 g (0.10 mol) of **2,6-dichloroaniline** in a mixture of 400 mL of glacial acetic acid and 80 mL (0.70 mol) of 30% hydrogen peroxide is allowed to stand at room temperature for 48 h. The straw-colored crystals are filtered with suction and recrystallized from a minimum amount of glacial acetic acid to give

16.1 g **(91.3%)** of almost white **2,6-dichloronitrosobenzene,** melting at 173–175 °C to a pale-green liquid.

Epoxidation of Alkenes with Peroxytrifluoroacetic Acid [283]

To a dispersion of 8.2 mL (0.3 mol) of 90% hydrogen peroxide in 50 mL of dichloromethane cooled in an ice bath is added over a 10-min period 50.8 mL (0.36 mol) of trifluoroacetic anhydride. The solution of peroxytrifluoroacetic acid so obtained is stirred in the cold for 15 min and transferred to a pressure-equalizing dropping funnel, from which it is added over a 30-min period with good stirring to a mixture containing 95 g (0.9 mol) of sodium carbonate, 14.0 g (0.2 mol) of **1-pentene,** and 200 mL of dichloromethane in a flask fitted with an ice-cooled reflux condenser. During the addition, the solvent boils vigorously. After the addition has been completed, the mixture is refluxed for 30 min. The mixture is centrifuged for 15 min at 3000 rpm, the salt is triturated with 300 mL of dichloromethane, and the resulting mixture is centrifuged again. The combined dichloromethane solutions are fractionated in an efficient column (90 by 1.2 cm) packed with 4-mm glass helices and equipped with a variable-reflux-ratio head. After the removal of most of the solvent, the residue is fractionated through a microcolumn to give 14.0 g **(81%)** of **1,2-epoxypentane,** bp 89–90 °C.

Oxidation of Ketones to Esters (Baeyer–Villiger Reaction) with Peroxybenzoic Acid [305]

Cyclohexyl methyl ketone (10.2 g, 0.081 mol) is mixed with 0.11 mol of peroxybenzoic acid in a standardized chloroform solution, and the reaction mixture is allowed to stand at room temperature for 77 h. The benzoic acid and the unreacted peroxybenzoic acid are removed by extraction with 1 M sodium bicarbonate. The chloroform solution is washed with water, and the aqueous solutions are extracted with ether. The combined chloroform and ether solutions are dried, and the solvents are removed by evaporation. Distillation of the residual oil at 63–64 °C at 13 mm of Hg yields 8.3 g **(72%)** of **cyclohexyl acetate.**

17. OXIDATION WITH CUPRIC ACETATE

Coupling of Terminal Alkynes [58]

To a solution of 2 g (0.01 mol) of cupric acetate in 200 mL of pyridine–methanol–ether (1:1:4), 0.400 g (0.0048 mol) of **2-ethyl-3-butyn-2-ol** is added, and the mixture is refluxed for 20 min. After evaporation of most of the solvent in vacuo, the residue is acidified and extracted with ether, and the extract is evaporated to give 0.350 g **(88%)** of **2,7-dimethylocta-3,5-diyne-2,7-diol,** mp 128–130 °C.

18. OXIDATION WITH SILVER OXIDE (IN SITU) [365]

To a solution of 25.5 g (0.15 mol) of silver nitrate in 50 mL of water in a 300-mL bottle, a solution of 5.5 g (0.05 mol) of **3-cyclohexenecarboxaldehyde** in

10 mL of ethanol is added followed by a solution of 17 g (0.3 mol) of potassium hydroxide in 50 mL of water over a period of 1 h. The bottle is cooled with ice and water, and the mixture is stirred intensively. After the mixture has been shaken at room temperature for an additional 2–3 h, the deposited silver is filtered off and washed with a small amount of water, and the filtrate is cooled and acidified with dilute sulfuric acid. The mixture is extracted with ether, and the ether extract is washed with water, dried with sodium sulfate, and distilled to give 4 g **(62.5%)** of **3-cyclohexenecarboxylic acid,** bp 125–126 °C at 13 mm of Hg.

19. OXIDATION WITH MERCURIC OXIDE

Oxidation of Hydrazones to Diazo Compounds [*387*]

A mixture of 13 g (0.066 mol) of **benzophenone hydrazone,** 15 g of anhydrous sodium sulfate, 35 g (0.16 mol) of yellow (or red) mercuric oxide, 200 mL of dry ether, and 5 mL (10 mL if red mercuric oxide is used) of ethanol saturated with potassium hydroxide is shaken for 75 min in a pressure bottle wrapped in a wet towel. The reaction mixture is filtered, and the solvent is evaporated under reduced pressure at room temperature. The dark red oil thus obtained is dissolved in petroleum ether (bp 30–60 °C), and the solution is filtered again. Removal of the solvent from the filtrate under reduced pressure at room temperature gives an oil, which crystallizes when frozen in a stoppered flask in dry ice. After the flask has warmed to room temperature, the dark-red crystals are spread over a porous plate to give 11.4 g **(89%)** of **diphenyldiazomethane,** mp 29–30 °C.

Oxidation of Vicinal Dihydrazones to Alkynes [*394*]

In a three-necked flask equipped with a stirrer and a reflux condenser with a water-separating adapter, 23 g (0.106 mol) of mercuric oxide, 1 g of finely ground potassium hydroxide, and 20 g of anhydrous sodium sulfate in 100 mL of toluene are heated to a gentle reflux. While the suspension is stirred vigorously, 11.0 g (0.056 mol) of finely powdered **1,2-cyclodecanedione dihydrazone** is added in six portions over a period of 30 min. After refluxing for 2 h, the black mixture is filtered. The yellow toluene solution is passed through a column of 10 g of aluminum oxide (activity II–III), and the colorless filtrate is evaporated and distilled to give 5.12 g **(67%)** of **cyclodecyne,** bp 78.5 °C at 12 mm of Hg (after fractionation through a Vigreux column).

20. OXIDATION WITH THALLIUM TRINITRATE

Oxidation of Acetylenes to α-Diketones [*413*]

A solution of 1.78 g (0.01 mol) of **diphenylacetylene** in 20 mL of glyme is added to a solution of 8.9 g (0.02 mol) of thallium trinitrate in 10 mL of water containing 5 mL of 70% perchloric acid. The mixture is gently refluxed for 3 h. After the mixture has been cooled, thallium mononitrate is filtered off, and the filtrate is diluted with 100 mL of water. The mixture is extracted with two 25-mL portions of chloroform. The chloroform extract is dried with anhydrous sodium sulfate, the solvent is evaporated, and the residue is freed from traces of inorganic thallium salts by passage through a short column (2 by 10 cm) of acid-washed alumina. Elution with benzene–chloroform (1:1) and evaporation of the eluent gives 1.78 g **(85%)** of **benzil,** mp 93–94 °C.

21. OXIDATION WITH CERIC AMMONIUM NITRATE

Oxidation of Sulfides to Sulfoxides [*426*]

Ceric ammonium nitrate (1.10 g, 0.002 mol) is added in one portion to a stirred solution of 0.093 g (0.0005 mol) of **diphenyl sulfide** in 8 mL of 75% aqueous acetonitrile. After 3 min at room temperature, the mixture is quenched with 5 mL of water, and the product is obtained by extraction with two 20-mL portions of ether and evaporation of the solvents. **Diphenyl sulfoxide,** mp 68–70 °C, is obtained in **82%** yield.

22. OXIDATION WITH LEAD TETRAACETATE

Cleavage of Vicinal Diols in Anhydrous Medium [*1155*]

Under a nitrogen atmosphere, 20 g (0.17 mol) of ***cis*-1,2-cyclohexanediol** is dissolved in 200 mL of dry benzene. Anhydrous potassium carbonate (50 g) is added, and the mixture is stirred vigorously while 76 g (0.17 mol) of lead tetraacetate is added in 5-g portions over a period of 1 h. After an additional hour of stirring, the mixture is filtered with suction, and the salts are extracted with benzene. The combined benzene solutions are dried over anhydrous sodium sulfate and evaporated in vacuo. The residue is distilled under a nitrogen atmosphere to give 13.4 g **(68%)** of colorless **adipic aldehyde,** bp 68–70 °C at 3 mm of Hg.

23. OXIDATION WITH NITRIC ACID

Oxidation of Alcohols to Acids [*1206*]

In a 250-mL flask fitted with a magnetic stirring bar and a reflux condenser, nitric acid prepared by diluting 40 mL (0.89 mol) of concentrated nitric acid (d 1.4) with 40 mL of water is heated almost to boiling. **Cyclohexanol** (10 g, 0.1 mol) is added dropwise from a separatory funnel through the reflux condenser while the mixture is stirred without further heating. Each drop of cyclohexanol brings the mixture to boiling, and additional cyclohexanol is added only when the effervescence has subsided. After the addition has been completed, the mixture is gently refluxed for an additional 30 min and then transferred into a porcelain dish and evaporated to one-half of its volume on a steam bath in a hood. After cooling, crystals of **adipic acid** are filtered with suction and recrystallized from a minimum amount of water. The yield is 7–8 g **(48–55%),** mp 151 °C.

Oxidants . . . Use Caution

24. OXIDATION WITH AMMONIUM NITRATE

Oxidation of α-Hydroxy Ketones to α-Diketones [476]

In a 100-mL flask fitted with a reflux condenser are placed 0.1 g (0.0005 mol) of cupric acetate monohydrate (as a catalyst), 5 g (0.0625 mol) of ammonium nitrate, 10.6 g (0.05 mol) of **benzoin,** and 35 mL of 80% aqueous acetic acid. The mixture is heated with occasional shaking of the flask. Soon a vigorous evolution of nitrogen commences. After a 1.5-h reflux, the solution is cooled and seeded with a crystal of benzil (or by scratching the walls of the flask with a glass rod) until crystallization takes place. Water is added to the mixture to precipitate the residual benzil. The crystals are filtered with suction and washed with water containing about 20% of ethanol. The yield of **benzil** is **quantitative.** The melting point of the product after recrystallization from 70% ethanol is 95 °C.

25. OXIDATION WITH LEAD NITRATE

Conversion of Haloalkyl Group into Carbonyl [477]

To 110 g (1.0 mol) of *o*-**fluorotoluene,** 58 mL (182 g, 1.14 mol) of bromine is added over a period of 1 h at 105–120 °C while the mixture is irradiated with a 500-W bulb from a distance of 5–10 cm. After the addition of bromine has been completed, the *o*-**fluorobenzyl bromide** thus formed is heated in the course of 15 min to 150 °C. The cooled mixture is added in two portions to a solution of 175 g (0.53 mol) of lead nitrate in 1000 mL of boiling water, and the mixture is heated under reflux with occasional shaking. After the first 0.5 h of heating, nitrogen oxides start to evolve, and crystals of lead bromide begin to deposit. After 5–6 h of refluxing, the mixture is steam distilled. The distillate (94 g) is treated with a 5 N solution of sodium bisulfite. The aldehyde–bisulfite compound thus formed (139.5 g, 61.2%) is decomposed with a solution of sodium carbonate, and the aldehyde is steam distilled. The organic layer of the steam distillate is dried and distilled to give 55 g of *o*-**fluorobenzaldehyde**, bp 172–174 °C, n_D^{20} 1.5216. This amount represents a **78.5%** yield based on the aldehyde–bisulfite compound or a **48%** yield based on *o*-**fluorotoluene.**

26. OXIDATION WITH SULFUR

Willgerodt Oxidation of Ketones to Carboxylic Acids (Kindler Modification) [504]

A mixture of 373 g (2.2 mol) of **methyl 2-naphthyl ketone,** 105 g (3.3 mol) of sulfur, and 290 g (3.3 mol) of morpholine is refluxed gently for 14 h and briskly for 2 h in a hood. While still hot, the mixture is poured into 1.2 L of warm ethanol. After the mixture has been cooled, 534 g **(89.6%)** of pale-buff crystals of **2-naphthylacetic acid thiomorpholide,** mp 100–106 °C, is obtained. A mixture of 388 g (1.61 mol) of the thiomorpholide, 800 mL of acetic acid, 120 mL of concentrated sulfuric acid, and 180 mL of water is refluxed for 5 h and then poured into 6 L of water. A considerable amount of tar is left in the flask. After the mixture has been allowed to stand overnight, the solid is collected, washed with water, and digested with a solution of 150 g of sodium hydroxide in 3 L of water. The dark insoluble substance is removed by filtration, and 2-naphthylacetic acid is precipitated as a cream-colored product by acidification. After this material has been washed with water, dried, and digested with several portions of a mixture of benzene with petroleum ether, **2-naphthylacetic acid** is obtained as a purplish-tinged solid, mp 138–141 °C, in **89.5%** yield. Recrystallization from benzene raises the melting point to 142.2–143 °C.

27. OXIDATION WITH SELENIUM DIOXIDE

Oxidation of Ketones to α-Diketones [*516*]

A solution of 72.9 g (0.66 mol) of selenium dioxide, 10 mL of water, and 107.1 g (0.66 mol) of **2-acetylmesitylene** in 600 mL of dioxane is stirred and refluxed for 5 h. After separation from selenium by filtration, the dioxane is evaporated, and the residue is distilled at 130–134 °C at 20 mm of Hg to give an **82.5%** yield of **mesitylglyoxal** as a heavy yellow oil, bp 105–106 °C at 4 mm of Hg.

28. OXIDATION WITH CHROMIUM TRIOXIDE

Oxidation of Alcohols to Aldehydes and Ketones with Chromic Acid Adsorbed on Silica Gel [*537*]

To a solution of 2.0 g (0.02 mol) of chromium trioxide in 50 mL of water, 20 g of silica gel is added with stirring. The mixture is evaporated in vacuo, and the yellow solid is dried overnight at 100 °C. If kept under vacuum in the dark, the reagent retains its activity for at least a week.

The reagent (3 g) is suspended in 5 mL of ether, the mixture is stirred magnetically, and a solution of 0.158 g (1 mmol) of **1-decanol** is added. After 5 min, the solid is filtered off and washed with three 10-mL portions of ether. The combined filtrates are evaporated to give 0.135 g **(86%)** of **decanal,** mp 102–104 °C.

Oxidation of Acetylenic Alcohols to Ketones by the Jones Reagent [*578*]

A solution of 10.3 g (0.103 mol) of chromium trioxide in 30 mL of water and 8.7 mL of concentrated sulfuric acid is added over a period of 2 h to a stirred solution of 15 g (0.126 mol) of **3-octyn-2-ol** in 30 mL of acetone at 5–10 °C. After being stirred for an additional 30 min, the mixture is diluted with water to 250 mL, and the ketone is isolated by extraction with ether. Distillation of the ether solution gives 11.5 g **(73.6%)** of **3-octyn-2-one,** bp 70.5–71.5 at 11 mm of Hg.

29. OXIDATION WITH CHROMIUM TRIOXIDE–PYRIDINE COMPLEX

Oxidation of Primary Alcohols to Aldehydes [*598*]

A solution of 5 g (0.043 mol) of **2-ethyl-2-methyl-1-butanol** in 5 mL of dry pyridine is added to a stirred slurry of chromium trioxide–pyridine complex (*Pro-*

Oxidants . . . Use Caution

cedure 4) in 50 mL of pyridine, and the mixture is allowed to stand overnight at room temperature. After a 3-h reflux, the mixture is poured into cold dilute sulfuric acid, and the product is extracted with ether. Evaporation of the dried ether extract and fractionation of the residue under atmospheric pressure yields 4.4 g **(88%)** of **diethylmethylacetaldehyde,** bp 132–133 °C.

30. OXIDATION WITH CHROMIUM TRIOXIDE–3,5-DIMETHYLPYRAZOLE COMPLEX

Oxidation of Secondary Alcohols to Ketones [*602*]

3,5-Dimethylpyrazole (0.580 g, 0.006 mol) is added to a suspension of 0.600 g (0.006 mol) of chromium trioxide in 20 mL of dichloromethane, and the mixture is stirred under argon at room temperature for 15 min. To the dark-red solution, 0.263 g (0.0022 mol) of **α-phenylethanol** in 2 mL of dichloromethane is added in one portion, and the reaction mixture is stirred at room temperature for 30 min. The progress of the reaction is monitored by vapor-phase chromatography with a 6-ft (~1.8-m) 5% Carbowax 20 M (a polyethylene glycol compound) column. The solvent is removed under reduced pressure, the residue is extracted with 50 mL of ether, and the resulting mixture is filtered. The residue after concentration is dissolved in pentane, the solution is filtered through a short column packed with silica, and the filtrate is evaporated to give 0.260 g **(100%)** of **acetophenone.**

31. OXIDATION WITH PYRIDINIUM CHLOROCHROMATE (PCC)

Oxidation of Primary Alcohols to Aldehydes [*605*]

To a stirred suspension of 32.3 g (0.1 mol) of pyridinium chlorochromate in 200 mL of methylene chloride is added rapidly at room temperature a solution of 11.6 g (0.1 mol) of **1-heptanol** in 100–150 mL of dichloromethane. A black precipitate is formed shortly. After 2 h, the mixture is diluted with 5 volumes of anhydrous ether, the solvents are decanted, and the black solid is washed twice with ether. The organic solution is filtered through Florisil (magnesium silicates), and the solvent is evaporated at reduced pressure. **Heptanal** (10.1 g), bp 153 °C, is obtained in **78%** yield.

32. OXIDATION WITH PYRIDINIUM CHLOROCHROMATE ADSORBED ON ALUMINA

Oxidation of Unsaturated Primary Alcohols to Unsaturated Aldehydes [*604*]

A solution of 0.51 g (3.8 mmol) of **cinnamyl alcohol** in 10 mL of hexane is stirred with 9.3 g (7.6 mmol) of the pyridinium chlorochromate reagent (*Procedure 5*) at room temperature for 24 min. The solid is filtered off and washed with three 10 mL-portions of ether. The combined filtrates are evaporated, and the residue is distilled in vacuo to give 0.42 g **(84%)** of **cinnamaldehyde,** bp 130 °C at 20 mm of Hg.

Oxidants . . . Use Caution

33. OXIDATION WITH TETRABUTYLAMMONIUM CHROMATE

Oxidation of Secondary Alcohols to Ketones [618]

A solution of 1.58 g (5.68 mmol) of tetrabutylammonium chloride in 28 mL of water is added rapidly with stirring to a solution of 0.54 g (5.4 mmol) of chromium trioxide in 14 mL of water at room temperature. The tetrabutylammonium chromate is extracted with 200 mL of chloroform. The chloroform solution is concentrated to 6 mL, and a solution of 0.5 g (2.71 mmol) of **benzhydrol** in 4 mL of chloroform is added with stirring. After 3 h at 60 °C, the mixture is diluted with ether, and the solution is poured into 1 N sodium hydroxide. The ether layer is washed with a saturated solution of sodium chloride, dried with anhydrous sodium sulfate, and evaporated to give 0.450 g **(91%)** of **benzophenone,** mp 45–47 °C.

34. OXIDATION WITH SODIUM DICHROMATE

Oxidation of Secondary Alcohols to Ketones [642]

In a 2-L, three-necked flask fitted with a mechanical stirrer, a dropping funnel, and a reflux condenser, a solution of 120 g (0.400 mol, 20% excess) of sodium dichromate dihydrate and 135 g (1.33 mol) of 96% sulfuric acid in 500 mL of water is added over a 40-min period to a well-stirred mixture of 128.0 g (1.00 mol) of **4-ethylcyclohexanol** in 200 mL of water. Within 2 min, the mixture becomes greenish-black, and the temperature rises from 30 to 68 °C during the addition of the first half of the oxidizing solution. Immediately after the addition has been completed, the temperature starts dropping and, within 5 min, decreases to 55 °C. The mixture is cooled and extracted twice with 400 mL of ether–pentane (3:1). The extracts are washed several times with water, dried, and evaporated to yield 113.6 g **(90%)** of **4-ethylcyclohexanone,** bp 109–112 °C at 50 mm of Hg.

35. OXIDATION WITH SODIUM HYPOCHLORITE

Epoxidation of Unsaturated Ketones [691]

To a solution of 1.0 g (0.048 mol) of **benzalacetophenone** in 7.5 mL of pyridine is added 11 mL of a fresh 5.25% solution of sodium hypochlorite (Clorox). An exothermic reaction is accompanied by the almost immediate fading of the yellow color of the solution. When the mixture becomes almost colorless, 25 mL of water is added. The precipitated white crystals of the epoxide are filtered with suction, washed thoroughly with water, and recrystallized from ethanol to give 1.0 g **(94%)** of **1,3-diphenyl-2,3-epoxy-1-propanone,** mp 89–90 °C.

36. OXIDATION WITH SODIUM CHLORITE

Oxidation of Aldehydes to Carboxylic Acids [1162]

Solutions of 1.52 g (0.01 mol) of **vanillin** in 19 mL of *tert*-butyl alcohol, of 1.43 g (0.013 mol) of resorcinol (as chlorine scavenger) in an acetate buffer of

Oxidants . . . Use Caution

pH 3.93, and of 13 mL (0.012 mol) of sodium chlorite in water are mixed. The precipitate of vanillic acid (0.84 g) is filtered with suction. The filtrate is extracted with dichloromethane, and an additional crop (0.52 g) is obtained by extraction with a solution of sodium hydrogen carbonate followed by acidification. The total yield of **vanillic acid** is **81%**.

37. OXIDATION WITH SODIUM HYPOBROMITE

Conversion of Methyl Ketones into Carboxylic Acids with One Less Carbon [736]

A solution of sodium hypobromite, prepared by dissolving 42 g (1.05 mol) of sodium hydroxide in 200 mL of water and adding 15 mL (47 g, 0.29 mol) of bromine at 0 °C, is added over a 30-min period to a stirred solution of 15.0 g (0.066 mol) of **4-(*p*-methoxyphenyl)acetophenone** in 150 mL of dioxane. During the addition, the temperature is allowed to rise to 35–40 °C. After being stirred for an additional 15 min, the mixture is treated with enough sodium bisulfite to destroy the excess of sodium hypobromite. Water (1 L) is added, and 200 mL of water is distilled off to remove the bromoform and some dioxane. Acidification of the hot solution and subsequent cooling yields 13.8 g **(91%)** of **4-(*p*-methoxyphenyl)benzoic acid**, mp 247–248 °C.

38. OXIDATION WITH *N*-BROMOACETAMIDE

Selective Oxidation of Secondary Alcohols in Preference to Primary Alcohols [743]

To a solution of 13.05 g (0.095 mol) of *N*-bromoacetamide in 233 mL of methanol, 3.6 mL of pyridine, and 12.7 mL of water, 15.00 g (0.023 mol) of **20-cyano-3α,21-dihydroxy-17-pregnen-11-one** is added. The mixture, while protected from light, is stirred overnight at room temperature. After this time, a considerable amount of crystals accumulates. To destroy the excess of *N*-bromoacetamide, 4.2 mL of allyl alcohol is added, followed by 4.4 mL of 6 N hydrochloric acid to neutralize the pyridine. Crystallization is completed by the addition of 900 mL of water over a period of 20 min. After the mixture has been left in an ice bath for 30 min, the crystals are filtered with suction, washed with water, and dried to furnish 14.3 g **(96.5%)** of **20-cyano-21-hydroxy-17-pregnene-3,11-dione**, mp 254.1–255.1 °C.

39. OXIDATION WITH ACETYL HYPOIODITE

syn Hydroxylation of Double Bonds (Woodward Method) [782]

A mixture of 0.82 g (0.01 mol) of **cyclohexene**, 3.67 g (0.022 mol) of silver acetate, and 2.54 g (0.01 mol) of iodine in 65 mL of glacial acetic acid is shaken for 4.5 h at room temperature. Wet acetic acid (10 mL), containing 0.2 mL (0.011 mol) of water, is added, and the mixture is refluxed for 1 h. After the mixture has been cooled, the precipitated silver salts are filtered off and washed with a small

Oxidants . . . Use Caution

amount of acetic acid. The combined filtrate and washings are evaporated at 100 °C under reduced pressure. The residue is diluted with water, the solution is extracted with ether, and the ether extract is washed with concentrated ammonium hydroxide and with water. After the removal of ether, the residue is refluxed for 1 h with an excess of 3 N aqueous potassium hydroxide. The mixture is neutralized with concentrated hydrochloric acid and evaporated to dryness, and the residue is extracted with chloroform. Evaporation of the solvent gives 0.83 g **(66%)** of crude *cis*-**1,2-cyclohexanediol**. Recrystallization from ethyl acetate yields 0.52 g **(41%)** of pure product, mp 94–97 °C.

40. OXIDATION WITH PERIODIC ACID

Cleavage of Vicinal Diols to Carbonyl Compounds in Aqueous Medium [759]

A solution of 5.4 g (0.015 mol) of **glucose phenylosazone** in 2 L of warm 66% ethanol is cooled to room temperature and treated with a solution of 10 g (0.044 mol) of paraperiodic acid in 70 mL of water. The yellow-orange precipitate appears at once. After dilution of the mixture with 500 mL of water, the precipitate is filtered with suction, and the crystals are washed with 66% ethanol and dried in vacuo. Two recrystallizations of the residue (3.4 g, **85%**) from 66% ethanol give yellow-orange needles of **mesoxaldehyde osazone,** mp 198 °C (dec).

41. OXIDATION WITH SODIUM PERIODATE

Oxidation of Sulfides to Sulfoxides [770]

Thioanisole (methyl phenyl sulfide) (12.4 g, 0.1 mol) is added to 210 mL (0.105 mol) of a 0.5 M solution of sodium metaperiodate at 0 °C, and the mixture is stirred in an ice bath overnight. The precipitated sodium iodate is filtered, and the filtrate is extracted with chloroform. The extract is dried over anhydrous magnesium sulfate, and the solvents are evaporated under reduced pressure. The residue (13.9 g, **99%** yield) is distilled at 83–85 °C at 0.1 mm of Hg to give pure **methyl phenyl sulfoxide,** mp 29–30 °C.

42. OXIDATION WITH ALUMINA-SUPPORTED SODIUM PERIODATE

Preparation of the Supported Oxidant [771]

Acidic aluminum oxide (Merck 90) for column chromatography (100–325 mesh) (33.4 g) is added in one portion to a magnetically stirred solution of 21.4 g (0.1 mol) of **sodium metaperiodate** in 60 mL of water at 60 °C. The mixture is stirred for 20 min at 60 °C and then dried in a rotary evaporator. The white residue is heated at 120 °C for 16 h to a constant weight. The concentration of the oxidant is 3 mmol/g of alumina.

Oxidants . . . Use Caution

Oxidation of Sulfides to Sulfoxides [771]

To a solution of 6.2 g (0.05 mol) of **methyl phenyl sulfide** in 50 mL of 95% ethanol is added 54.8 (0.1 mol) of the supported oxidant in one portion at room temperature. The mixture is stirred vigorously until the sulfide has been completely consumed, as detected by gas-liquid chromatography with a 5% Carbowax 20 M (a polyethylene glycol compound) column (5 h). After filtration of the solid and removal of most of the ethanol by evaporation of the filtrate, 25 mL of dichloromethane is added. The solution is dried with anhydrous sodium sulfate and evaporated to dryness to give 5.54 g **(88%)** of **methyl phenyl sulfoxide,** bp 98–102 °C at 1 mm of Hg.

43. OXIDATION WITH SODIUM PERIODATE AND POTASSIUM PERMANGANATE

Cleavage of Alkenes by Lemieux–Rudloff Reagents [763]

A mixture of 0.2824 g (1 mmol) of **oleic acid,** 0.414 g (3 mmol) of potassium carbonate, 1.712 g (8 mmol) of sodium metaperiodate, and 0.021 g (0.134 mmol) of potassium permanganate in 400 mL of water is kept at 20 °C for 20 h. The solution is strongly acidified with 10% sulfuric acid and extracted with ether. Evaporation of the extract gives 0.354 g of a residue, which on trituration with petroleum ether gives 0.161 g of oily pelargonic acid and 0.193 g of crystalline **azelaic acid,** mp 102–104.5 °C. Pure **pelargonic acid** is obtained by vacuum distillation, and pure azelaic acid (mp 106.5–107.5 °C), by recrystallization, first from water and then from ethyl acetate. Yields of both crude acids are **quantitative,** but the yields of purified products are not stated.

44. OXIDATION WITH MANGANESE DIOXIDE

Oxidation of Allylic Alcohols to α,β-Unsaturated Aldehydes [805]

Cinnamyl alcohol (5 g, 0.037 mol) is stirred for 30 min with a suspension of 50 g (0.57 mol) of active manganese dioxide (***Procedure 7***) in 250 mL of carbon tetrachloride. The mixture is filtered, and the filtrate is evaporated. The residue is treated with a solution of 7.5 g (0.067 mol) of semicarbazide hydrochloride and 9 g (0.066 mol) of sodium acetate trihydrate in aqueous ethanol. The mixture is heated to 60 °C and kept at room temperature for 2 h. The **semicarbazone of cinnamaldehyde,** mp 212–215 °C, is obtained in **71.5%** yield.

Oxidation of Vicinal Diols to Ketones [817]

A solution of 1 g (0.0058 mol) of *cis*-**9,10-decalindiol** in 50 mL of dichloromethane is stirred at room temperature with 20 g of active manganese dioxide for 1 h. The mixture is filtered through Celite (diatomaceous earth), and the filtrate is evaporated in a rotary evaporator under vacuum generated by a water aspirator.

Oxidants . . . Use Caution

The residue is distilled to give 0.9 g **(90%)** of **1,6-cyclodecanedione,** mp 92–95 °C, on crystallization from acetone.

45. OXIDATION WITH POTASSIUM PERMANGANATE

Cleavage of Alkenes to Carboxylic Acids [869]

In a 5-L, three-necked round-bottom flask equipped with a gas-tight (Trubore) stirrer, an efficient reflux condenser connected to a dry-ice trap, a dropping funnel, and a thermometer reaching almost to the bottom of the flask are placed 460 g (2.6 mol) of potassium permanganate, 315 g (5.5 mol) of potassium hydroxide, and 3500 mL of water. The solids are dissolved by heating to 60 °C and constant stirring. **2,3-Dichlorohexafluoro-2-butene** (473 g, 2.03 mol) is added dropwise at such a rate as the capacity of the reflux condenser permits. After the addition has been completed, the solution is heated until the temperature of the liquid reaches 95 °C (8–10 h). The mixture is then cooled to 40 °C, and a stream of sulfur dioxide is passed through the liquid with stirring and cooling below 60 °C until the permanganate color just fades. The solution is acidified with just enough of 50% sulfuric acid to neutralize the potassium hydroxide used, and sulfur dioxide is passed again through the mixture until all the manganese dioxide is dissolved. Continuous extraction with ether and fractionation of the extract through an efficient column yield 480 g of the 80:20 azeotropic mixture of trifluoroacetic acid and water boiling at 103–105 °C at 745 mm of Hg. The yield of **trifluoroacetic acid** is **87%.**

Oxidation of Acetylenes to α-Diketones [864]

Into a solution of 5.6 g (0.02 mol) of **stearolic acid** in 3 L of water containing 1.5 g (0.027 mol) of potassium hydroxide, carbon dioxide is introduced until pH 7.5 is reached. A solution of 6.3 g (0.04 mol) of potassium permanganate in 300 mL of water is added all at once. The temperature of the reaction mixture is kept at about 25 °C, and the pH is maintained between 7.0 and 7.5 by the gradual bubbling of carbon dioxide into the mixture as the oxidation proceeds. After 1 h, the excess of potassium permanganate is destroyed by the addition of sodium bisulfite, and hydrochloric acid is added to precipitate the product. **9,10-Diketostearic acid** is filtered with suction and recrystallized from 200 mL of absolute ethanol at 0 °C. The total yield is 5.7–6.0 g **(92–96%)** of the product, mp 84.5–85.0 °C.

Oxidation of Secondary Alcohols to Carboxylic Acids [1147]

To a vigorously stirred mixture of 120 g (0.42 mol) of crystalline sodium carbonate (decahydrate) in 500 mL of water and 60 g (0.6 mol) of **cyclohexanol,** 270 g (1.71 mol) of potassium permanganate is added portionwise while the mixture is cooled to maintain the temperature between 15 and 30 °C. After removal of the manganese dioxide by filtration and acidification of the filtrate with 120 g of concentrated hydrochloric acid, 61 g **(70%)** of **adipic acid,** mp 151 °C, is obtained.

Oxidants . . . Use Caution

46. OXIDATION WITH POTASSIUM-PERMANGANATE-COATED MOLECULAR SIEVES

Oxidation of Secondary Alcohols to Ketones [*863*]

A 2-L, round-bottom flask is charged with 500 mL of a 0.06 M aqueous solution of potassium permanganate, and 20 g of Linde 13X molecular sieves (1/16-in. [0.16-cm] pellets) is added in one portion. The water is removed under reduced pressure in a rotary evaporator, and the coated pellets are separated from the nonadsorbed potassium permanganate by screening (20 mesh). The reagent contains 0.27 mmol of potassium permanganate per gram.

A suspension of 15.0 g (4.05 mol) of the reagent in 20 mL of freshly distilled benzene containing 0.250 g (1.36 mmol) of **cyclododecanol** is heated to 70 °C for 1.5 h. The solution is filtered through Celite (diatomaceous earth), the pellets are washed with 70 mL of benzene, and the combined filtrates are evaporated under reduced pressure to yield 0.226 g **(90%)** of **cyclododecanone** as colorless crystals, mp 56–59 °C. The pure compound melts at 59–61 °C.

47. OXIDATION WITH TETRABUTYLAMMONIUM PERMANGANATE (PURPLE BENZENE)

Cleavage of Alkenes to Carboxylic Acids [*845*]

Potassium permanganate (4.8 g, 0.03 mol) and 50 mL of water are vigorously stirred for 10 min, and the mixture is cooled in a water bath. Then 30 mL of benzene, 0.5 g (0.0016 mol) of tetrabutylammonium bromide, and 2 g (0.011 mol) of *trans*-**stilbene** are added. The mixture is stirred for 3 h at room temperature and then treated sequentially with sodium bisulfite and acid. The benzene layer is separated, dried, and evaporated to give 2.49 g **(92%)** of **benzoic acid.**

48. OXIDATION WITH FERRIC CHLORIDE

Dehydrogenation of Hydroquinones to Quinones and of Thiols to Disulfides [*907*]

To a solution of 0.285 g (0.002 mol) of **2-mercaptohydroquinone** in 10 parts of alcohol is added 3.5 mL (0.007 mol) of a 2 N solution of ferric chloride. The **disulfide of *p*-benzoquinone** deposits in the form of a yellow precipitate. It is collected and recrystallized from glacial acetic acid to give yellow crystals, mp 178 °C, in **quantitative** yield.

49. OXIDATION WITH OSMIUM TETROXIDE

syn Hydroxylation of Double Bond with an Equivalent Amount of Osmium Tetroxide [*949*]

1,2-Dimethylcyclopentene (6.4 g, 0.067 mol) is added portionwise to a solution of 17 g (0.089 mol) of osmium tetroxide in 300 mL of ether with 11 mL of pyridine as a catalyst. After 1 h, the dark deposit (20.5 g) is collected, dissolved in chloroform, and decomposed by shaking with a solution of 30 g of mannitol in 300 mL of a 10% solution of potassium hydroxide (added in four portions during a 24-h period). The combined aqueous solutions are continuously extracted with chloroform, the chloroform and pyridine are distilled off through a small column, and the viscous residue is distilled at 90 °C at 9 mm of Hg to give 6.1 g **(70%)** of *cis*-**1,2-dimethyl-1,2-cyclopentanediol**, bp 86 °C at 8 mm of Hg (mp 25–25.2 °C after recrystallization from ethyl acetate).

syn Hydroxylation of Double Bond with a Catalytic Amount of Osmium Tetroxide [*130*]

A solution of hydrogen peroxide in *tert*-butyl alcohol is prepared by adding 400 mL of pure *tert*-butyl alcohol (free of isobutylene) to 100 mL of 30% hydrogen peroxide. The solution is treated with small portions of anhydrous sodium sulfate until two layers separate. The alcohol layer is removed and dried with anhydrous sodium sulfate and finally with anhydrous calcium sulfate (Drierite). The solution contains 6.3% of hydrogen peroxide in *tert*-butyl alcohol.

To 6.1 g (0.105 mol) of **allyl alcohol** is added 54.6 mL (0.1 mol) of 6.3% hydrogen peroxide in *tert*-butyl alcohol followed by 1 mL of a 0.5% solution of osmium tetroxide in *tert*-butyl alcohol. The reaction is exothermic and requires cooling with tap water. After 3 h, the reaction mixture is fractionated to remove the solvent, the catalyst, and the unreacted allyl alcohol (1.7 g). The residue is 4.2 g **(60%)** of **glycerol**.

50. OXIDATION WITH RUTHENIUM TETROXIDE

Oxidation of Amides to Imides [*1205*]

A solution of 2.4 g (0.012 mol) of **ethyl 1-acetylpiperidine-2-carboxylate** in 40 mL of ethyl acetate is added to a mixture of 0.240 g of hydrated ruthenium dioxide and 120 mL of 10% aqueous solution of sodium periodate (0.056 mol). The mixture is vigorously stirred at room temperature for 14 h. The organic layer is separated, the aqueous layer is extracted with three 40-mL portions of ethyl acetate, and the combined organic solutions are stirred with 2 mL of isopropyl alcohol for 2–3 h to destroy the excess of ruthenium tetroxide. The black ruthenium dioxide that precipitates from the solution is filtered, and the filtrate is washed with 40 mL of water, dried with anhydrous sodium sulfate, and evaporated in vacuo to give 2.42 g **(95%)** of crude **ethyl 1-acetyl-6-oxopiperidine-2-carboxylate.**

All operations should be done *in the hood,* because ruthenium tetroxide possesses an unpleasant ozone-like smell. Commercially available ruthenium dioxide may differ in its properties, especially in its reaction with sodium periodate, depending on the way it is prepared and on the content of water in its hydrated form [*1207*].

51. OXIDATION WITH DIETHYL AZODICARBOXYLATE

Demethylation of Tertiary Amines [*980*]

N-**Methylisopelletierine** (6 g, 0.039 mol) and diethyl azodicarboxylate (6.8 g, 0.039 mol) are dissolved in 15 mL of methanol. The heat of the reaction brings the solution to boiling so that moderate cooling with water has to be applied. After the reaction is over, the methanol is evaporated. The residue, a thick amber-colored syrup, is then refluxed with 80 mL of 1 N hydrochloric acid for 3 h while formaldehyde is evolved and the solution darkens. After the solution has been allowed to stand for 2 days, diethyl hydrazodicarboxylate, mp 131 °C, settles at the bottom of flask. The isopelletierine is liberated by concentrated alkali and extracted with ether. The solution is dried with potassium carbonate. Distillation of the solution yields 4.4–4.5 g **(80–82%)** of **isopelletierine,** bp 101–102 °C at 12 mm of Hg.

52. OXIDATION WITH POTASSIUM NITROSODISULFONATE (FREMY SALT)

Conversion of Aromatic Amines into Quinones [490]

A solution of 0.605 g (0.005 mol) of **2,6-dimethylaniline** in 25 mL of acetone is mixed with a solution of 3 g (0.01 mol) (10% excess) of potassium nitrosodisulfonate in 50 mL of 0.6 M potassium dihydrogen phosphate and 100 mL of water. After 75 min, the oxidant has reacted. The solution is extracted several times with a total of 120–150 mL of chloroform. The combined extracts are dried with anhydrous sodium sulfate, and the solvent is evaporated in vacuo at 40–50 °C to yield as a residue 0.560 g **(82%)** of **2,6-dimethyl-*p*-benzoquinone,** mp 63–65 °C (dec).

53. OXIDATION WITH *p*-NITROSODIMETHYLANILINE

Oxidation of Activated Methylene Group to Carbonyl [986]

To a boiling solution of 24 g (0.24 mol) of **acetylacetone** and 36 g (0.24 mol) of *p*-nitrosodimethylaniline in 120 mL of ethanol is added 4.4 mL (0.15 mol) of a solution of sodium hydroxide (*d* 1.36) all at once. After about 1 min, the color of the solution changes, and the mixture keeps on boiling even after removal of the water bath. If the boiling becomes too vigorous, the mixture is cooled intermittently without suppressing the boiling. The reaction is finished when the reaction mixture becomes bright red. The mixture is cooled and mixed with 400 mL of ether, which causes the precipitation of sodium acetate, resulting from the cleavage of the acetylacetone. The mixture is filtered, and the filtrate is treated with a solution of 150 mL of dilute sulfuric acid (*d* 1.16) in 50 mL of water and, finally, extracted with six 300–500-mL portions of ether. The combined ether extracts are distilled and then evaporated in vacuo to remove the alcohol. The residue is fractionated at 12 mm of Hg at 54–55 °C to give 15 g **(55%)** of **triketopentane (pentanetrione).**

54. SOMMELET OXIDATION

Oxidation of Aralkyl Halides to Aldehydes [956]

Equivalent quantities of **2,4-bis(chloromethyl)anisole** and hexamethylenetetramine (small excess) in chloroform are mixed and refluxed for 2–3 h until precipitation of the salt is complete. The crystals are filtered with suction to give an essentially quantitative yield.

Oxidants . . . Use Caution

A solution of 59 g (0.12 mol) of the salt in 500 mL of water is refluxed for 4 h and filtered while hot. The filtrate is cooled in an ice bath. The white needles that separate are collected and washed with cold water to give 14 g **(70%)** of **4-methoxyisophthalaldehyde,** mp 117–118 °C. One recrystallization from 60% alcohol gives 12.6 g (63%) of the compound, mp 119–120 °C.

55. OXIDATION WITH DIMETHYL SULFIDE AND N-CHLOROSUCCINIMIDE

Oxidation of Secondary Alcohols to Ketones [1138]

To a stirred solution of 0.400 g (3.0 mmol) of N-chlorosuccinimide in 10 mL of analytical-grade toluene is added, at 0 °C, 0.3 mL (4.1 mmol) of dimethyl sulfide under argon. A white precipitate is formed immediately after the addition. The mixture is cooled to −25 °C with carbon tetrachloride–dry ice. A solution of 0.312 g (2.0 mmol) of **4-*tert*-butylcyclohexanol** (a mixture of *cis* and *trans* isomers) in 2 mL of toluene is added dropwise while the mixture is stirred for 2 h at −25 °C. Then a solution of 0.303 g (3.0 mmol) of triethylamine in 0.5 mL of toluene is dropped in, the cooling bath is removed, and after 5 min, 20 mL of ether is added. The organic layer is washed with 5 mL of water, dried with anhydrous magnesium sulfate, and evaporated to give 0.310 g **(~100%)** of **4-*tert*-butylcyclohexanone** as white plates, mp 44–47 °C.

56. OXIDATION WITH DIMETHYL SULFOXIDE

Oxidation of Acetylenic Alcohols to Acetylenic Aldehydes [1015]

To a refluxing solution of 4 g (0.033 mol) of **4,6-octadiyn-1-ol** and 21 g (0.10 mol) of dicyclohexylcarbodiimide in 200 mL of absolute ether, a solution of 1.7 g of crystalline phosphoric acid in 100 mL (110 g, 1.41 mol, 43× molar excess) of dimethyl sulfoxide is added dropwise. After an additional heating for 4 h, 100 mL of 4 N sulfuric acid is added. The precipitated crystals of N,N'-dicyclohexylurea are filtered with suction and washed with ether. The filtrate is washed until neutral with a solution of sodium hydrogen carbonate. The dried ether solution is evaporated, and the residue is distilled at 60 °C at 0.001 mm of Hg to give a **73%** yield of **4,6-octadiynal** as a colorless oil.

Oxidation of α-Bromo Ketones to Dicarbonyl Compounds [1003]

A solution of 15.98 g (0.065 mol) of ***p*-bromophenacyl bromide** in 100 mL (110 g, 1.41 mol, 38.5 equiv) of dimethyl sulfoxide is kept at room temperature for 9 h. It is then poured into an ice–water mixture and extracted with ether. The extracts are washed with water, dried with anhydrous magnesium sulfate, and evaporated in vacuo. The residual pasty pale-yellow solid is recrystallized from butyl ethyl ether to give 11.2 g **(84%)** of ***p*-bromophenylglyoxal hydrate,** mp 123–124 °C.

Oxidants . . . Use Caution

57. BIOCHEMICAL OXIDATION

Asymmetric Oxidation of Sulfide to Sulfoxide [*1046*]

Mycelia of *Aspergillus niger* from two malt slopes are transferred to four 500-mL Erlenmeyer flasks, each containing 100 mL of Czapek Dox liquid medium and 2 mL of corn steep liquor (Dista Products Ltd). The four flasks are used to inoculate 57 1-L flasks, each containing 100 mL of Czapek Dox medium at pH 4–5. The flasks are shaken on a platform shaker (180 rpm) at 30 °C for 2 days. **Benzyl *tert*-butyl sulfide** (30 mg dissolved in 1 mL of ethanol) is injected into each flask, and the flasks are shaken for 4 days at 30 °C. The mycelia are filtered off and extracted with boiling dichloromethane. The filtrate is continuously extracted with dichloromethane for 6 days. Evaporation of the extract and chromatography of the residues on deactivated alumina give sulfide, sulfoxide, and sulfone in this sequence. The sulfoxide fraction is recrystallized to give a **61%** yield of **(–)-benzyl *tert*-butyl sulfoxide** of 77% optical purity, mp 75–76 °C, [α] –265 ° (after six recrystallizations from pentane). *tert*-Butyl *p*-tolyl sulfide furnishes *tert*-butyl *p*-tolyl sulfoxide in 24% yield and 94% optical purity.

Oxidants . . . Use Caution

CORRELATION TABLES

THE PURPOSE of the correlation tables is to show what oxidants are suitable for the oxidation of compounds to their various oxidation products and where in the book such oxidations can be found. Abbreviations used in the tables are explained in footnotes or in the list of abbreviations on pages 319 and 320.

The absence of a reagent in the column "Oxidants" means that the reagent was not used in examples shown in this book. It does not mean that oxidants not listed here cannot be used successfully for a particular oxidation.

Numbers in italics in the "Page" column refer to pages in Chapter 4, Preparative Procedures.

CORRELATION TABLES

Table 1

DEHYDROGENATION

Starting Compound	Product	Oxidants	Page
−C=C−CH−CH−	−C=C−C=C−	$Hg(OAc)_2$, H_2O_2, DDQ* $PhIO_2/Ph_2Se_2$, m-$HO_2C_6H_4IO_2/Ph_2Se_2$ $H_2O_2/PhSeCl$, biochemical oxidation	47,48 48,49 49,*275*
Ar−CH−CH−	Ar−C=C−	ZnO-Cr_2O_3, o-$C_6Cl_4O_2$, DDQ	49
−CH−NH−	−C=N−	$Hg(OAc)_2$, Ag_2O, t-BuOCl	49,50
cyclohexane, decalin	benzene, naphthalene	Pt, Pd, S, Se, o-$C_6Cl_4O_2$ DDQ*, $AlCl_3$	50-52 *275*
piperidine, dihydropyridine	pyridine	HNO_3, As_2O_3, MnO_2 $K_3Fe(CN)_6$, $PhNO_2$	52
Ar−CH_2−	Ar−CH−CH−Ar	O_2/KOH, MnO_2, $KMnO_4$	53
Ar−H	Ar−Ar	$AlCl_3$, H_2CrO_4, H_2SO_4/Fe_2O_3, H_5IO_6, $FeCl_3$ $K_3Fe(CN)_6$, biochemical oxidation	53-55

*DDQ is 2,3-dichloro-5,6-dicyano-p-benzoquinone.

Table 2

OXIDATION OF ALKANES AND CYCLOALKANES

Starting Compound	Product	Oxidants	Page
$-CH_3$	$-CO_2H$	biochemical oxidation	58
$-CH_2-$	$-CH(OH)-$	biochemical oxidation	58
$-CH(-)-$	$-C(OH)(-)-$	O_3, CrO_3, $PhCH_2NEt_3MnO_4$, $p\text{-}O_2NC_6H_4CO_3H$	58, 59
$-CH(-)-$	$-C(OOH)(-)-$	O_2	60

Table 3

OXIDATION OF ALKENES AND CYCLOALKENES

Starting Compound	Product	Oxidants	Page
$-\underset{|}{C}=\underset{|}{C}-$	epoxide $-\underset{|}{C}-\underset{|}{C}-$ with O bridge	O_2, O_3, H_2O_2, t-BuOOH RCO_3H, Me_2CO_2* $Tl(OAc)_3$, CrO_3, $CrO_2(NO_3)_2$ $NaOCl$, $Ca(OCl)_2$, NBS** I_2/Ag_2O, $I(OCOCF_3)_3$ electrolysis, biochemical oxidation	60-62 61-64,279 61,62 61,63 62,63 61,62
	$-\underset{|}{C}-\underset{|}{C}-$ with O-O bridge	O_2, $(PhO)_3PO_3$	64-65
	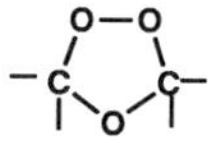 1,2,4-trioxolane (ozonide)	O_3	65-67
	OH OH $-\underset{|}{C}-\underset{|}{C}-$	$KMnO_4$, OsO_4 $Ph_3MePMnO_4$, $PhCH_2NEt_3MnO_4$ $OsO_4 + H_2O_2$, $NaClO_3$, or $KClO_3$ $OsO_4 + AgClO_3$, $Ba(ClO_3)_2$ or R_3NO $I_2 + AgOAc$, $AgOBz$ or $TlOAc + H_2O$ biochemical oxidation	68,71-73,290 72,73 68,69,291 68,69,72,73 70,286 71
	OH $-\underset{|}{C}-\underset{|}{C}-$ OH	H_2O_2/V_2O_5, SeO_2, MoO_3, WO_3 RCO_3H, $I_2 + AgOAc$, $AgOBz$ or $TlOAc$	69,72,277 69-72
	Br $-\underset{|}{C}-\underset{|}{C}-$ OH	$AcNHBr$, NBS**	73,74
	OH $-\underset{|}{C}-\underset{|}{C}-$ NHTos	$TosNClNa/OsO_4$	74
	RCOO $-\underset{|}{C}-\underset{|}{C}-OCOR$	$Tl(OAc)_3$, $Pb(OAc)_4$, $Bz_2O_2 + I_2$ $AgOAc$, $AgOBz$ or $TlOAc + I_2$	74,75
	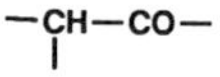 $-CH-CO-$	air, electrolysis, H_2O_2, HCO_3H CrO_3, $CrO_2Cl_2 + Zn$, DDQ*	75,76
	Hal $-\underset{|}{C}-CO-$	CrO_2Cl_2, $Ag_2CrO_4 + I_2$	76

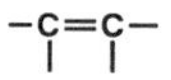

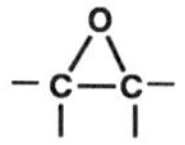

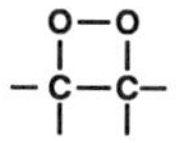

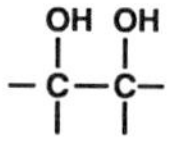

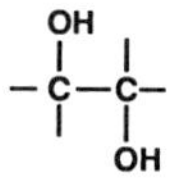

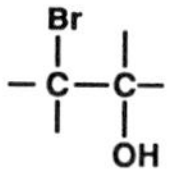

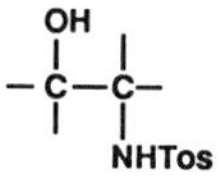

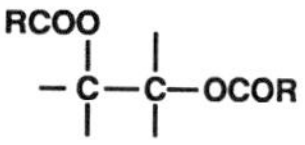

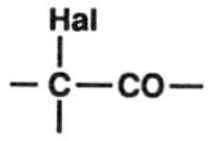

* Me_2CO_2 is dimethyldioxirane $Me_2C{<}^O_O$.

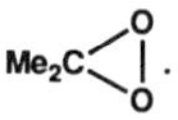

**NBS is N-bromosuccinimide.

Continued on next page

Table 3 (continued)

Starting Compound	Product	Oxidants	Page
—C=C— | |	OH | —C—CO— |	$KMnO_4$	76,77
	—CO—CO—	SeO_2, $KMnO_4$	76
	OH OH | | —CH + CH— | |	$O_3/LiAlH_4$	77,78
	OH | —CH + OC— | |	$O_3/NaBH_4$	77,78
	** —CO + OC— | |	O_3/Pd or Raney Ni O_3/Zn, O_3/(MeO)$_3$P, O_3/Me$_2$S O_2/hv, H_2O_2, CrO_3 $CrO_2(OAc)_2$, $NaIO_4$ $OsO_4/NaIO_4$, $KMnO_4$	77,78 276 78-80
	*** —CO$_2$H + HO$_2$C—	O_3/H_2O_2 or RCO_3H; Ag_2O, HNO_3 CrO_3, $K_2Cr_2O_7$, $NaMnO_4$, $KMnO_4$ Bu_4MnO_4, $NaIO_4/KMnO_4$, RuO_4	81-84 289 288,290
—C=C—CH— | | |	OOH | —C—C=C— | | |	O_2/hv, (PhO)$_3$PO$_3$, H_2O_2/NaOCl	84,85
	OH | —C=C—C— | | |	SeO_2	85,86
	OCOR | —C=C—C— | | |	t-BuO$_2$Ac, t-BuO$_2$Bz, Hg(OAc)$_2$ Pb(OAc)$_4$	86
	—C=C—CO | |	O_2/Cu(II), HgBr$_2$/hv, CrO_3 PhIO$_2$/(RSe)$_2$	86,87

* DDQ is 2,3-dichloro-5,6-dicyano-*p*-benzoquinone.
** If the vinylic carbon is linked to hydrogen, the product of the reductive ozonolysis is an aldehyde.
***Carboxylic acids are formed only when hydrogen or halogen are present at the carbon of the double bond.

Table 4

OXIDATION OF DIENES

Starting Compound	Product	Oxidants	Page
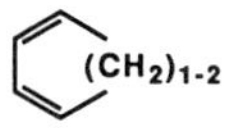 (CH$_2$)$_{1-2}$ (cyclic diene)	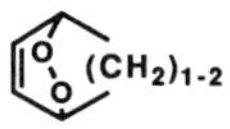 epoxide (CH$_2$)$_{1-2}$	O$_2$/hv, H$_2$O$_2$/NaOCl, H$_2$O$_2$/Br$_2$/KOH RCO$_3$K/OH$^-$, (PhO)$_3$PO$_3$	87-89
	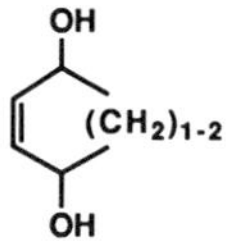 diol (CH$_2$)$_{1-2}$ with two OH	O$_2$hv H$_2$O$_2$/OsO$_4$, KMnO$_4$	275 89
	(CH$_2$)$_2$ with CH$_2$OH, CH$_2$OH	O$_3$/LiAlH$_4$	90

Table 5

OXIDATION OF ALKYNES

Starting Compound	Product	Oxidants	Page
—C≡CH	—C≡C—C≡C—	O$_2$/Cu(I), Cu(II)	90,91 279
	—CO$_2$H	Tl(NO$_3$)$_3$, KMnO$_4$, PhIO, C$_6$F$_5$I(O$_2$CR)$_2$	91,92
—C≡C—	OH —C—CO— |	Tl(NO$_3$)$_3$	91
	—CO—CO—	O$_3$, Tl(NO$_3$)$_3$, SeO$_2$, ZnCr$_2$O$_7$ Mo(O$_2$)$_2$, KMnO$_4$, Zn(MnO$_4$)$_2$, OsO$_4$ RuO$_4$, DMSO, PhIO	91,92 289

Table 6

OXIDATION OF AROMATIC HYDROCARBONS AND NONFUNCTIONALIZED HETEROCYCLES

Starting Compound	Product	Oxidants	Page
benzene, pyridine	*phenol, hydroxypyridine	KOH CF_3CO_3H, biochemical oxidation	92 93
	aryl-OCOR	RCO_3H, $Pb(OCOR)_4$, $(i\text{-}PrO_2)_2CO$ $Bz_2O_2 + I_2$	93,94
	**quinone	O_2, H_2O_2, RCO_3H, $Ce(SO_4)_2$ SeO_2, CrO_3, $Na_2Cr_2O_7$, $K_2Cr_2O_7$ $ZnCr_2O_7$, HIO_4, $Mn_2(SO_4)_3$	94-96
	—CHO	O_3	96,97
	—CO_2H	O_2, O_3/H_2O_2, electrolysis, RCO_3H HNO_3, CrO_3, $Na_2Cr_2O_7$, $KMnO_4$ $RuO_4/NaOCl$ or $NaIO_4$	96-98
$ArCH_2$—	ArCH(OOH)—	$O_2/h\nu$	99
	ArCH(OH)—, ArCH(OCOR)—	H_2O_2, $Hg(OAc)_2$, $(NH_4)_2Ce(NO_3)_6$ $Pb(OAc)_4$	100,101
	ArCHO, ArCO—	O_2, AgO, $HgBr_2$, $(NH_4)_2Ce(NO_3)_6$, $K_2S_2O_8$ SeO_2, CrO_3, CrO_2Cl_2, $Na_2Cr_2O_7$, $K_2Cr_2O_7$ HIO_4, $NaIO_4$, MnO_2, $KMnO_4$, RONO	101-105
	$ArCO_2H$	O_2, O_3, electrolysis, $KHSO_5$ HNO_3, CrO_3, $Na_2Cr_2O_7$ NaOCl, $KMnO_4$, NiO_2, biochemical oxidation	105-109

* This product is obtained only when the ring contains nitro groups.
**This product is obtained only with condensed aromatic hydrocarbons.

Table 7
OXIDATION OF HALOGEN DERIVATIVES AND TOSYLATES

Starting Compound	Product	Oxidants	Page
$-CH_2Hal^*$	$-CHO$	R_3NO, DMSO (Me_2SO)	109,110
$-CH_2OSO_2C_7H_7$	$-CHO$	DMSO	109,110
$-C{=}C{-}CH_2Hal$	$-CHO$	H_2CrO_4, $Na_2Cr_2O_7$, $C_6H_{12}N_4$	109-111
$ArCH_2Hal$	$ArCHO$	$Pb(NO_3)_2$, H_2CrO_4, $K_2Cr_2O_7$ Bu_4NIO_4, $C_6H_{12}N_4$, RNO $Me_2C{=}N{\diagup}^O_{\diagdown ONa}$	110-112,*282,292*
$-CH_2Hal$	$-CO_2H$	$N_2O_4^{**}$, biochemical oxidation	112,113
$-CHHal{-}$	$-CO{-}$	CrO_3 $Na_2Cr_2O_7$, Bu_4NOCl	112
$-C{=}C{-}CHHal$	$-C{=}C{-}CO$	$C_6H_{12}N_4$	110
$ArCHHal{-}$	$ArCO{-}$	CrO_3, Bu_4NOCl	112
(pentafluorophenyl, H)	(tetrafluoro-methyl-benzoquinone)	HNO_3	113,114

* Hal = Cl, Br, or I.
**N_2O_4 was also used to convert the groups $-CHF_2$ and $-CF_2I$ to $-CO_2H$.

Table 8

OXIDATION OF PRIMARY ALCOHOLS

Starting Compound	Product	Oxidants	Page
$-CH_2OH$	$-CHO$ *	Cu, CuO, $CuCr_2O_4$ O_2/Cu, Ag, Co_2O_3, Pt, PtO_2	114,115,123,126
$-C=C-CH_2OH$	$-C=C-CHO$ *	AgO, $(NH_4)_2Ce(NO_3)_6$ $Pb(OAc)_4$, N_2O_4	115,124,126, 283,284
$ArCH_2OH$	ArCHO *	CrO_3**, H_2CrO_4 $Na_2Cr_2O_7$, $K_2Cr_2O_7$ $(R_4N)_2CrO_4$	115-118,123-126
$ArC=C-CH_2OH$	$ArC=CCHO$ *	$(C_5H_5NH)_2Cr_2O_7$, $Ag_2Cr_2O_7$ CrO_2Cl_2, $(t$-Bu$)_2CrO_4$, KOCl t-BuOCl, NBS, NCS	119,123,125,126
		MnO_2, $KMnO_4$, $BaMnO_4$ K_2FeO_4, NiO_2, RuO_4 $C_6Cl_4O_2$, DDQ, $EtO_2CN=NCO_2Et$ $Me_2S + Cl_2$ or NBS, DMSO $(PhSeO)_2O$, periodinane***	119,120,124-126,288 120,125-127 120,125,126 120-123,125,126,293 123,124,126
$-CH_2OH$	$-CO_2H$	O_2/Pt, PtO_2, electrolysis KOH, NaOH/Ag_2O, HNO_3	127,128,130,281
$-C=C-CH_2OH$	$-C=C-CO_2H$	CrO_3**, $ZnCr_2O_7$ $(t$-Bu$)_2CrO_4$, $(C_5H_5NH)_2Cr_2O_7$	128,130
$ArCH_2OH$	$ArCO_2H$ *	MnO_2, $NaMnO_4$, $KMnO_4$, $Cu(MnO_4)_2$ Bu_4NMnO_4, NiO_2, Na_2RuO_4, K_2RuO_4	129,130
$ArC=C-CH_2OH$	$ArC=C-CO_2H$	biochemical oxidation	130
$-CH_2OH$	$-CO_2R$	MnO_2, NaCN	131
	$-CONH_2$	NiO_2, NH_3	132
$2-CH_2OH$	$-COOCH_2-$	$KHSO_5$, $Na_2Cr_2O_7$, $(t$-Bu$)_2CrO_4$ $Br_2 + NaBrO_3$, MnO_2	131

* In the preparations of the compounds marked with asterisks, more specific oxidants are shown in the text.

** CrO_3 includes its complexes such as $CrO_3 \cdot 2C_5H_5N$ and $C_5H_5NHCrO_3Cl$.

***Periodinane (Dess-Martin reagent) is

Table 9

OXIDATION OF SECONDARY ALCOHOLS

Starting Compound	Product		Oxidants	Page
$-\underset{\underset{OH}{\mid}}{CH}-$	$-CO-$	*	Cu, $CuCr_2O_4$, Raney Ni $O_2/h\nu$, O_2/Cu, Ag, Pd, Pt, PtO_2	132,133,147,149
$-\underset{\underset{OH}{\mid}}{\underset{\mid}{C}}=\underset{\mid}{C}-\underset{\mid}{CH}-$	$-\underset{\mid}{C}=\underset{\mid}{C}-CO-$		electrolysis, H_2O_2, $KHSO_5$, RCO_3H CuO, Ag_2O, Ag_2CO_3, AgO $(NH_4)_2Ce(NO_3)_6$, $Ce(SO_4)_2$ $Pb(OAc)_4$, N_2O_4	133,147,148 133,149
$ArCH-$ $\mid$ OH	$ArCO-$	*	CrO_3**, H_2CrO_4, $Na_2Cr_2O_7$, $K_2Cr_2O_7$ $ZnCr_2O_7$, $(Bu_4N)_2CrO_4$ Bu_4NHCrO_3Cl, CrO_2Cl_2	133,147-149 *283,284,285*
$Ar\underset{\mid}{C}=\underset{\mid}{C}-\underset{\underset{OH}{\mid}}{CH}-$	$Ar\underset{\mid}{C}=\underset{\mid}{C}-CO-$		Cl_2, Br_2, NaOCl, $Ca(OCl)_2$ Bu_4NOCl, t-BuOCl $NaBrO_2$, $NaBrO_3$	138-140,147-149
			MnO_2, $NaMnO_4$, $KMnO_4$, $BaMnO_4$ $(Bu_4N)MnO_4$	140-142,147-149,*290*
			K_2FeO_4, RuO_4, Na_2RuO_4, K_2RuO_4	142,148,149
			Oppenauer oxidation, DDQ	142-144,149,261
			$EtO_2CN=NCO_2Et$	144,149
			NCS, Me_2S + NCS, DMSO	144-146,147,148,*293*
			Ph_3CBF_4, $(PhSeO)_2O$	146-149
			biochemical oxidation	146,216
$-\underset{\underset{OH}{\mid}}{CH}-$	$-\underset{\underset{OH}{\mid}}{\overset{\overset{OOH}{\mid}}{C}}-$		$O_2/h\nu$	150
	$-CO_2H$		HNO_3, CrO_3, $KMnO_4$	150,*289*

* In the preparation of the compounds marked with asterisks, more specific oxidants are shown in the text.
** CrO_3 includes all its complexes such as $CrO_3 \cdot 2C_5H_5N$ and $C_5H_5NHCrO_3Cl$.

Table 10

OXIDATION OF TERTIARY ALCOHOLS

Starting Compound	Product	Oxidants	Page
$-\underset{\mid}{\overset{\mid}{C}}-OH$	$-\underset{\mid}{\overset{\mid}{C}}-OOH$	H_2O_2/H_2SO_4	150,151
	$-CO-$	$Pb(OAc)_4$, AcOI	151
$-\underset{\mid}{\overset{\mid}{C}}-\underset{\mid}{\overset{\mid}{C}}-OH$	$-CO_2H$ + $-CO-$	CrO_3	151

Table 11
OXIDATION OF UNSATURATED ALCOHOLS AT DOUBLE BONDS

Starting Compound	Product	Oxidants	Page
—C=C—OH	—C(OAc)—C(OAc)—OH	Pb(OAc)$_4$	151, 152
—C=C~~~C—OH	epoxide —C—C~~~C—OH	t-BuOOH, RCO$_3$H Sharpless reagent CrO$_3$, Na$_2$CrO$_4$	152–154

Table 12
SELECTIVE OXIDATION OF ALCOHOLS AND DIOLS

Starting Compound	Product	Oxidants	Page

OH OH
—C~~~C—H
     H
	OH
	—CO~~~C—H
	     H
		Ce(SO$_4$)$_2$, (NH$_4$)$_2$Ce(NO$_3$)$_6$
		CrO$_3$, Cl$_2$, NaOCl, AcNHBr, BaMnO$_4$
		Me$_2$S + Cl$_2$ or NCS, DDQ
		biochemical oxidation
			155
			286

	O~~~ —O~~~CO |	CuCr$_2$O$_4$, Ag$_2$CO$_3$ ring +N=O Cl$^-$, biochemical oxidation	157–159
OH OH —C—C—H H H	—CO + CHO	(NH$_4$)$_2$Ce(NO$_3$)$_6$ Pb(OAc)$_4$ NaBiO$_3$, CrO$_3$, K$_2$Cr$_2$O$_7$ HIO$_4$, NaIO$_4$, KIO$_4$ MnO$_2$, KMnO$_4$ PhIO, PhI(OAc)$_2$, PhIO$_2$	159–162, 281 286 288
	—CO$_2$H + CO$_2$H	O$_2$/Co, electrolysis	163

Table 13

OXIDATION OF PHENOLS

Starting Compound	Product	Oxidants	Page
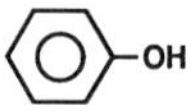 C₆H₅–OH	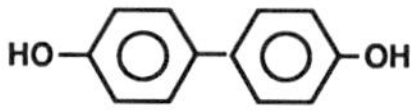 HO–C₆H₄–C₆H₄–OH	$K_2Cr_2O_7$, $FeCl_3$	163
	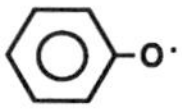 C₆H₅–O·	$K_3Fe(CN)_6$	163
	 HO–C₆H₄–OH	$K_2S_2O_8$	163, 164
	 O=C₆H₄=O (quinone)	HgO, $Hg(OCOCF_3)_2$, $(NH_4)_2Ce(NO_3)_6$ CrO_3, Br_2, DDQ $(KO_3S)_2NO\cdot$, biochemical oxidation + H_2O_2	164, 165
 HO–C₆H₄–OH	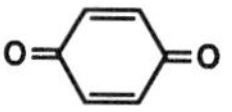 O=C₆H₄=O	Ag_2O, HgO, $(NH_4)_2Ce(NO_3)_6$ PbO_2, $Pb(OAc)_4$, N_2O_4, $Na_2Cr_2O_7$ $NaClO_3$, $NaIO_3$, $BaMnO_4$, $FeCl_3$ $Fe_2(SO_4)_3$	165-168 *290*
	 HO_2C–CH=CH–CO_2H	RCO_3H	168

Table 14

OXIDATION OF ETHERS AND EPOXIDES

Starting Compound	Product	Oxidants	Page
—CH$_2$O—	—CH(OOH)—	O$_2$	168,169
	—COO—	O$_3$, CrO$_3$, ZnCr$_2$O$_7$, Br$_2$, KMnO$_4$, Zn(MnO$_4$)$_2$, PhCH$_2$NEt$_3$MnO$_4$, RuO$_4$	169,170
—C=C—O—	epoxide —C(—O—)C—O—	RCO$_3$H	170
	—C(—)—CO—O—	HNO$_3$, C$_5$H$_5$NHCrO$_3$Cl	171
—C=C—C—O—	epoxide —C(—O—)C—C—O—	RCO$_3$H	171
	—C=C—CHO	SeO$_2$	171,172
—C≡C—O—	—CO—COO—	PhIO	172
—CH$_2$—OSi—	—CO—O—	CrO$_3$, NBS/hν	172
—CH—OSi—	—CO—	NBS	172
—C=C—OSi—	—C(OH)—CO—	O$_2$, RCO$_3$H, CrO$_2$Cl$_2$	172,173
epoxide —C(—O—)C—	—C(OOH)—C(OH)—	H$_2$O$_2$	173,174
	—CO—C(OH)—	DMSO	173,174
	—CO + OC—	HIO$_4$	173,174

Table 15

OXIDATION OF ALDEHYDES AND THEIR DERIVATIVES

Starting Compound	Product	Oxidants	Page
—CHO	—CO$_2$H	O$_2$, H$_2$O$_2$, ROOH, H$_2$SO$_5$, Ag$_2$O AgO, HNO$_3$, NaOH, KOH K$_2$Cr$_2$O$_7$, NaClO$_2$, MnO$_2$ KMnO$_4$, R$_4$NMnO$_4$, NiO$_2$, RuO$_4$ K$_2$RuO$_4$, DMSO, PhIO	174-180 279 285
	—CO$_3$H	O$_2$	180
	—OCHO ⟶ —OH	H$_2$O$_2$, RCO$_3$H	180, 181
—C=C—CHO	—C(O)C—CHO (epoxide)	H$_2$O$_2$, t-BuOOH	182
—C(OH)⁓CHO	—C(OH)⁓CO$_2$H	O$_2$, Br$_2$	182, 183
	—CO⁓CO$_2$H	KMnO$_4$, biochemical oxidation	182, 183
	HO$_2$C⁓CO$_2$H	HNO$_3$	183, 184
—CH(OR)$_2$	—CO$_2$R	O$_3$, RCO$_3$H	184, 185
—C=C—CH(OR)$_2$	—C(O)C—CH(OR)$_2$ (epoxide)	RCO$_3$H	184, 185
	—C=C—CO$_2$R	RCO$_3$H	184, 185
	—C(OH)—C(OH)—CH(OR)$_2$	KMnO$_4$	185

Table 16

OXIDATION OF KETONES AND THEIR DERIVATIVES

Starting Compound	Product	Oxidants	Page
$\|$ CO $\|$ —CH $\|$	$\left(\begin{array}{c}\|\\-C-OH\\\|\\-C-O-\end{array}\right)_2$	H_2O_2	186
—CO—	—COO— and/or —OCO—	H_2O_2/H^+, $(Me_3SiO)_2$, PhSe(O)OOH H_2SO_5, $KHSO_5$, RCO_3H $(NH_4)_2Ce(NO_3)_6$, biochemical oxidation	186-190 195, *279*
—C=C—CO—	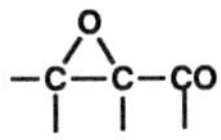 $-\overset{O}{\overset{/\backslash}{C-C}}-CO$	H_2O_2, NaOCl	191, 192, *285*
	—C=C—OCO—	$(Me_3SiO)_2$, RCO_3H	191, 192 194, 195
—CO—CO—	—CO—O—CO—	H_2O_2, H_2SO_5, RCO_3H	193, 194
$\|$ —C$\sim\sim$CO— $\|$ H	$\|$ —C$\sim\sim$CO— $\|$ OH	CrO_2Cl_2, MoO_5, ArIO, biochemical oxidation	196-199
$\|$ —C—CO— $\|$ H	$\|$ —C—CO— $\|$ OAc	$Pb(OAc)_4$	199
—CH$_2$—CO—	—CO—CO—	SeO_2, H_2SeO_3, CrO_3 RNO, DMSO	199, 200, *283* 201, *292*
—CH—CO— $\|$ Br	—CO—CO—	RNO, DMSO	201, 202, *293*
—CO—CH$_3$	—CH$_2$—CO$_2$H		
		$Tl(NO_3)_3$, S/$(NH_4)_2$S, S/HN$\diagup$O CCl$_4$/KOH	202-206 *282*
	—CO—CO$_2$H	KMnO$_4$	206
	—CO$_2$H	HNO_3, $KMnO_4$, $Na_2Cr_2O_7$ NaOCl, KOCl, Cl$_2$/NaOH Br$_2$/NaOH, I$_2$/KOH, CCl$_4$/KOH	205-210 *286*
—CO—C— $\|$	—CO$_2$H —+—CO— or —CO$_2$H + —CO$_2$H	O_2, KO_2, HNO_3, CrO_3, $KMnO_4$	211, 212

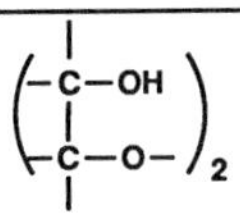

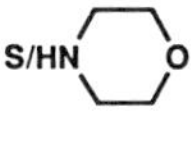

Continued on next page

Table 16 (continued)

Starting Compound	Product	Oxidants	Page
—C=C—CO—	epoxide —C(−O−)C—CO—	H_2O_2, t-BuOOH, RCO_3H, NaOCl, $Ca(OCl)_2$, NBS	212-214
	—C=C—OCO—	RCO_3H	212-214
	or epoxide —C(−O−)C—OCO—		192
	—C(OH)—C(OH)—CO—	OsO_4	214
	—CO + HO_2C—CO—	O_3	215
—CH—C=C—CO—	—C(OH)—C=C—CO—	biochemical oxidation	214
—CH(OH)—CO—	—CO~~CO—	CrO_3, DMSO, biochemical oxidation	215,216
—CH(OH)—CO—	—CO—CO—	O_2, $CuSO_4$, $Cu(OAc)_2$, $(NH_4)_2Ce(NO_3)_6$, $NH_4NO_3/Cu(OAc)_2$, Bi_2O_3, $NaBrO_3$, biochemical oxidation	217,218, 282
—CO—CO—	—CO_2H + HO_2C—	RCO_3H	218,219
—C=N—	oxaziridine —C(−O−)N—	RCO_3H	219
—C=NOH	—CH—NO_2	MnO_2	219
—C=NOH, —C=NNH—	—CO	O_3, $(NH_4)_2Ce(NO_3)_6$, $ZnCr_2O_7$, $(PhSeO)_2O_2$, MoF_3, $MoOCl_3$, WF_6	219,220
—C=NNH_2	—C=N=N	Ag_2O, HgO, $Pb(OAc)_4$, NiO_2	219-221,*280*
—C(=NNH_2)—C(=NNH_2)—	—C≡C—	HgO	221,222,*280*

Table 17

OXIDATION OF CARBOXYLIC ACIDS AND THEIR DERIVATIVES

Starting Compound	Product	Oxidants	Page
$-CO_2H$	$-CO_3H$	H_2O_2	222
$-\underset{H}{\overset{\|}{C}}-CO_2R$	$-\overset{\|}{\underset{\|}{C}}-CO_2R$ $-\overset{\|}{\underset{\|}{C}}-CO_2R$	1. LDA*, 2. $FeCl_3$	223
	$-\overset{\|}{\underset{OH}{C}}-CO_2R$	1. LDA*, 2. MoO_5; camphorylsulfonyloxaziridine; $PhI(OAc)_2$	223
$-\overset{H}{\underset{H}{\overset{\|}{\underset{\|}{C}}}}-CO_2R$	$-CO-CO_2R$	air, N_2O_3	223
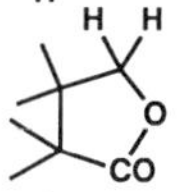 (bicyclic lactone structure with H H and CO)	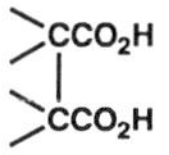 $>CCO_2H$ $>CCO_2H$	HNO_3, $KMnO_4$	224
$-\overset{\|}{CH}-CO_2H$	$-CHO$	Bu_4NIO_4	224
$2-CH_2-CO_2H$	$-CH_2-CH_2-$	electrolysis	224, 276
$-\underset{\|}{C}=\underset{\|}{C}\sim CO_2H$	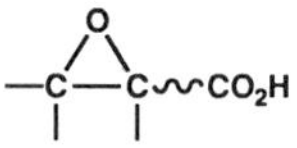 $-\overset{\|}{\underset{\|}{C}}\overset{O}{\overset{\triangle}{-}}\overset{\|}{\underset{\|}{C}}\sim CO_2H$	H_2O_2, RCO_3H, NBS, 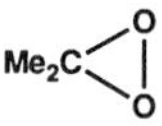 $Me_2C\overset{O}{\underset{O}{<}}$	224, 225
	$-\overset{OH}{\underset{\|}{C}}-\overset{OH}{\underset{\|}{C}}\sim CO_2H$	$KMnO_4$, OsO_4	225, 226
	$-\overset{\|}{\underset{OH}{C}}-\overset{OH}{\underset{\|}{C}}\sim CO_2H$	RCO_3H	225, 226
	$-C=\underset{OH}{\overset{\|}{C}}\sim CO_2H$	biochemical oxidation	225, 226
	$-COCO\sim CO_2H$	$KMnO_4$	226

*LDA is lithium diisopropylamide.

Continued on next page

Table 17 (continued)

Starting Compound	Product	Oxidants	Page
$-CH_2-C=C-CO_2R$ (with substituents on C=C)	$-CO-C=C-CO_2R$	CrO_3	226
$-C=C-CO_2H$ (with H, H)	$-CHO$	$KMnO_4$	226
	$-CHO + HO_2CCO_2H$	$KMnO_4/NaIO_4$	226, 227
$-C\equiv C\sim\sim CO_2H$	$-COCO\sim\sim CO_2H$	$KMnO_4$	226, 227
	$-CO_2H + HO_2C\sim\sim CO_2H$	O_3/H_2O_2, RCO_3H	226, 227
$-\underset{OH}{\overset{|}{C}}-CO_2H$	$-CO-CO_2H$	Br_2, $KMnO_4$, RuO_4	227, 228
	$-CHO, -CO-$	H_2O_2, R_4NIO_4, NBS, $Pb(OAc)_4$	228, 229
$-\underset{NH_2}{\overset{|}{C}}-CO_2H$	$-CHO$	NaOCl	229
	$-CO_2H$	H_2O_2, AgO	229
$-CO-CO_2H$	$-CO_2H$	H_2O_2	229
$-CO-CO_2R$	$-CO-OCO_2R$	RCO_3H	229
$-CONH_2$	$-NCO$	$Pb(OAc)_4$	230
$-CONHNH_2$	$-CHO$	$K_3Fe(CN)_6$	230
	$-CO_2H$	$(NH_4)_2Ce(NO_3)_6$	230
$-C=C-CN$	$-\overset{O}{\overset{/\backslash}{C-C}}-CN$	H_2O_2	230

Table 18

OXIDATION OF NITROGEN COMPOUNDS

Starting Compound	Product	Oxidants	Page
—CHNO$_2$	—CHO, —CO—	KMnO$_4$	230,231
—NO	—NO$_2$	H$_2$O$_2$, RCO$_3$H, K$_2$Cr$_2$O$_7$	231
—NHOH	—NO•	Ag$_2$O	232
	—NO (=O)	Ag$_2$O, Ag$_2$CO$_3$, Na$_2$Cr$_2$O$_7$ K$_2$Cr$_2$O$_7$, MnO$_2$ EtO$_2$CN=NCO$_2$Et	231,232
—N=N—	—N=N— (=O)	RCO$_3$H	232
—N$_3$	—NO$_2$	1. Ph$_3$P, 2. O$_3$	232
—NH—NH—	—N=N—	air, Ag$_2$O, CuO, HNO$_3$, Br$_2$, NBS MnO$_2$, EtO$_2$CN=NCO$_2$Et diphenylpicrylhydrazyl	233
—NH$_2$	—N=N—	O$_2$, NaBO$_3$, MnO$_2$ BaMnO$_4$, AgMnO$_4$ NiO$_2$	
	—N=N— (—O)	H$_2$O$_2$, RCO$_3$H	234
	—NO	H$_2$SO$_5$, KHSO$_5$, RCO$_3$H	235,278
	—NO$_2$	HBO$_3$, H$_2$SO$_5$, RCO$_3$H, Me$_2$C(—O—O—) KMnO$_4$, biochemical oxidation	235,236
—NH—	—N(OH)—	H$_2$O$_2$	236
—N(|)—	—N(=O)(|)—	H$_2$O$_2$, t-BuOOH, RCO$_3$H, Me$_2$C(—O—O—)	236-239 277
—CH$_2$—NH$_2$	—CH=NH (→ —CH=O)	Na$_2$S$_2$O$_8$, NaOCl, KMnO$_4$, BaMnO$_4$ Zn(MnO$_4$)$_2$, K$_2$FeO$_4$, CH$_2$O/C$_6$H$_{12}$N$_4$ NaNO$_2$/DMSO	239,240
—CH(|)—NH$_2$	—C(|)=NH (→ —C(|)=O)	Hg(OAc)$_2$, MnO$_2$	240,241

Continued on next page

Table 18 (continued)

Starting Compound	Product	Oxidants	Page
—CH$_2$—NH$_2$	—C≡N	AgO, Co$_2$O$_3$, NiO$_2$, Pb(OAc)$_4$, NaOCl K$_3$Fe(CN)$_6$, K$_2$RuO$_4$	241,242
	—CO$_2$H	biochemical oxidation	242
—CH—NH—	$\underset{\vert}{-\mathrm{C}}$=N—	Hg(OAc)$_2$, HNO$_3$, MnO$_2$	240,241
—CH$\sim\!\!\sim$N—	$\underset{\vert}{-\overset{\mathrm{OH}}{\mathrm{C}}}\!\sim\!\!\sim\!\underset{\vert}{\mathrm{N}}$—	biochemical oxidation	242,243
	—CO$\sim\!\!\sim\underset{\vert}{\mathrm{N}}$—	biochemical oxidation	242
—CH$_2\underset{\vert}{\mathrm{N}}$—	—CO$\underset{\vert}{\mathrm{N}}$—	RCO$_3$H, MnO$_2$, KMnO$_4$, K$_3$Fe(CN)$_6$ EtO$_2$CN=NCO$_2$Et	242-244 *291*
	—NH—	O$_2$/hν, Hg(OAc)$_2$, MnO$_2$, K$_3$Fe(CN)$_6$	243
	—CO—	RO$_3$H, MnO$_2$	244
—C≡C—$\underset{\vert}{\mathrm{N}}$—	—CO—CO—	O$_2$/hν	244,245
—C≡C—$\underset{\vert}{\mathrm{N}}$—	—COCO—$\underset{\vert}{\mathrm{N}}$—	PhIO	245
—CH—$\underset{\vert}{\mathrm{N}}$—CO—	—CO—$\underset{\vert}{\mathrm{N}}$—CO	t-BuOOH, RCO$_3$H, RuO$_4$	244,245,*291*
C$_6$H$_5$—NH$_2$	O=C$_6$H$_4$=O (benzoquinone)	Ag$_2$O, PbO$_2$, Pb(OAc)$_4$ NaNO$_2$/H$_2$SO$_4$, HNO$_3$ Na$_2$Cr$_2$O$_7$, K$_2$Cr$_2$O$_7$, NaOCl FeCl$_3$, (KO$_2$S)$_2$NO·	246-248,*292*

Table 19
OXIDATION OF COMPOUNDS OF PHOSPHORUS AND ARSENIC

Starting Compound	Product	Oxidants	Page
—P⟨	—P(=O)⟨	O_3, $KHSO_5$, AgO, MnO_2, $BaMnO_4$	248, 249
	—P(O₂)⟨ (cyclic peroxide)	O_3	2
—O—P(O)(O)—	—O—P(=O)(O)(O)—	O_3, H_2O_2, $(Me_3SiO)_2$, $ROOH$, AgO, R_3NO	248, 249
—P=S, —P=Se	—P=	O_3, DMSO	249, 250
—CH=P—	—CHO + O=P—	$NaIO_4$	250
—As⟨	—As=O	H_2O_2	250

Table 20

OXIDATION OF COMPOUNDS OF SULFUR AND SELENIUM

Starting Compound	Product	Oxidants	Page
—SH	—S—S—	air, H_2O_2, $(R_4N)_2Cr_2O_7$ R_4NCrO_3Cl, MnO_2, $Cu(MnO_4)_2$ $FeCl_3$, $EtO_2CN=NCO_2Et$	250-252 252,*290*
	—SO_2H	H_2O_2	251,252
	—SO_3H	O_2, $KHSO_5$, HNO_3	252
	—SO_2Cl	Cl_2/H_2O	
—SCN, —S—C(=NH)(NH$_2$)	—SO_2Cl	Cl_2, H_2O	252
—S—	—S(=O)—	H_2O_2, $NaBO_3$, $KHSO_5$, $(R_4N)_2S_2O_8$	253-259
		t-BuOOH, Me_2C(dioxirane), RCO_3H	261
		$(NH_4)_2Ce(NO_3)_6$, N_2O_4, HNO_3, SO_2Cl_2 H_2CrO_4, t-BuOCl	*281*
		$NaBrO_2$, $NaIO_4$, R_4NIO_4	*287,288*
		MnO_2, ArIO, ArI(OAc)$_2$, $EtO_2CN=NCO_2Et$	257,258
		Sharpless reagent, biochemical oxidation	*278,294*
	—S(=O)$_2$—	H_2O_2, $NaBO_3$, $KHSO_5$, $(R_4N)_2S_2O_8$, RCO_3H NaOCl, RuO_4	253-257 261,*278*
—C(SR)$_2$—	—C(SR)$_2$(=O)	H_2O_2, RCO_3H, $NaIO_4$	259,260
	—C(SR)$_2$(=O)$_2$	H_2O_2, RCO_3H, $KMnO_4$	260
—S(=O)—	—S(=O)$_2$—	air, $NaMnO_4$, OsO_4/R_3NO	262

Table 20 (continued)

Starting Compound	Product	Oxidants	Page
−C=C⌇S−	−S(OH)−S(OH)⌇S (with OH, OH on S's)	OsO_4/R_3NO	262
−CH(OH)⌇SR	−CO⌇SR	CCl_3CHO/Al_2O_3	261
−S(=N−)−	−S(=O)(=N−)−	R_4NOCl	262, 263
−S−S−	−S(=O)−S−	RCO_3H	263
	−SOCl	Cl_2/Ac_2O	264
	−SO$_3$H	O_2, H_2SO_5	264
	−SO$_2$Cl	Cl_2/H_2O	264
−SO$_2$H	−SO$_3$H	KOCl, NaMnO$_4$	264
−Se−	−Se(=O)−	H_2O_2, RCO_3H, t-BuOCl, NaIO$_4$, ArICl$_2$, biochemical oxidation	265, 266
	−Se(=O)(=O)−	RCO_3H, $KMnO_4$	265

Table 21

OXIDATION OF IODO COMPOUNDS

Starting Compound	Product	Oxidants	Page
—I	—IO	RCO_3H, HNO_3	266, 267
	—IO_2	$KHSO_5$, $K_2S_2O_8$, RCO_3H, $KBrO_3$, $KMnO_4$	266, 267

Table 22

OXIDATION OF BORON, SILICON, TIN, MAGNESIUM, AND MERCURY COMPOUNDS

Starting Compound	Product	Oxidants	Page
R_3B	—O—B—O— with O branch		267-270
	3 R-OH + H_3BO_3	H_2O_2, R_3NO biochemical oxidation	277
—Si—Si—	—Si—O—Si—	RCO_3H	270
—CH_2—Sn—	—CH_2OH (—CH_2OAc)	$Pb(OAc)_4$	270
—R—MgBr	ROH	BzO_2OCMe_3	271
—CH—HgX	—CO—	O_3	271

ABBREVIATIONS AND GENERIC NAMES

$[\alpha]_D^t$	specific rotation for the D line of sodium light at temperature t (°C)
Amberlyst	ion-exchange macroreticular resin suitable for aqueous or nonaqueous catalysis
bp	boiling point
c	concentration in grams per 100 ml
Carbitol	2-(2-ethoxyethoxy)ethanol or diethylene glycol monoethyl ether
Celite	silicon dioxide (95% SiO_2), diatomaceous earth, infusorial earth, siliceous earth, or kieselguhr
Cellosolve	2-ethoxyethanol
d	density
d^{18}	density (specific gravity) at 18 °C
d_{20}^{20}	density (specific gravity) at 20 °C related to the density of water at 20 °C
DCC	dicyclohexylcarbodiimide
DDQ	2,3-dichloro-5,6-dicyano-p-benzoquinone
dec	with decomposition
Diglyme	bis(2-methoxyethyl)ether or diethylene glycol dimethyl ether
DMF	dimethylformamide
DMSO	dimethyl sulfoxide
Enz-FAD	enzyme–flavin adenine dinucleotide
Florisil	activated magnesium silicate
HMPA	hexamethylphosphoramide or hexamethylphosphoric triamide
LDA	lithium diisopropylamide
Monoglyme	ethylene glycol dimethyl ether or 1,2-dimethoxyethane
mp	melting point
n_D^{20}	refractive index for the D line of sodium light at 20 °C
NADPH	reduced nicotinamide adenine dinucleotide phosphate
NBA	N-bromoacetamide

NBS	*N*-bromosuccinimide
NCS	*N*-chlorosuccinimide
PCC	pyridinium chlorochromate
rpm	revolutions per minute
RT	room temperature
THF	tetrahydrofuran
Tos	*p*-methylbenzenesulfonyl, *p*-toluenesulfonyl, or tosyl

REFERENCES

The numbers in square brackets at the end of the references refer to the page in the book where the reference is quoted.

[1.] House, H. O.   *Org. Synth., Collective Volume* **1963**, *4*, 367 [1, 53].
[2.] Bowden, K.; Heilbron, I.; Jones, E. R. H.; Sargent, K. H.   *J. Chem. Soc.* **1947**, 1579 [1, 4, 155].
[3.] Senseman, C. E.; Nelson, O. A.   *Ind. Eng. Chem.* **1923**, *15*, 521 [1, 4, 19, 95].
[4.] Church, J. M.; Lynn, L.   *Ind. Eng. Chem.* **1950**, *42*, 768 [1, 4, 114, 115, 174, 175].
[5.] Heyns, K.; Stöckel, O.   *Justus Liebigs Ann. Chem.* **1947**, *558*, 192 [1, 4, 174, 182, 183].
[6.] Heyns, K.; Heinemann, R.   *Justus Liebigs Ann. Chem.* **1947**, *558*, 187 [1, 4, 174, 182].
[7.] Moureu, C.; Mignonac, G.   *C. R. Hebd. Seances Acad. Sci.* **1920**, *170*, 258 [1, 114, 124, 126, 132].
[8.] Sergeev, P. G.; Sladkov, A. M.   *Zh. Obshch. Khim.* **1957**, *27*, 819 [1, 223].
[9.] Liu, K.-T.; Tong, Y.-C.   *Synthesis* **1978**, 669 [1, 4, 250, 251].
[10.] Trocha-Grimshaw, J.; Henbest, H. B.   *Chem. Commun.* **1968**, 1035 [1, 4, 262].
[11.] Friedman, N.; Gorodetsky, M.; Mazur, Y.   *Chem. Commun.* **1971**, 874 [1, 16, 86, 103, 104, 105].
[12.] Foote, C. S.   *Acc. Chem. Res.* **1968**, *1*, 104 [1].
[13.] Foote, C. S.; Wexler, S.; Ando, W.; Higgins, R.   *J. Am. Chem. Soc.* **1968**, *90*, 975 [1, 2, 3, 87, 88, 89, 276].
[14.] Foote, C. S.; Wexler, S.; Ando, W.   *Tetrahedron Lett.* **1965**, 4111 [1, 2, 87].
[15.] Foote, C. S.; Wexler, S.   *J. Am. Chem. Soc.* **1964**, *86*, 3879 [1, 2, 84].
[16.] McKeown, E.; Waters, W. A.   *J. Chem. Soc. B* **1966**, 1040 [1, 2, 87, 88].
[17.] Wasserman, H. H.; Scheffer, J. R.   *J. Am. Chem. Soc.* **1967**, *89*, 3073 [1, 2, 88, 89].
[18.] Moureu, C.; Dufraisse, C.; Dean, P. M.   *C. R. Hebd. Seances Acad. Sci.* **1926**, *182*, 1584 [1, 88, 89].
[19.] Murray, R. W.; Kaplan, M. L.   *J. Am. Chem. Soc.* **1969**, *91*, 5358 [1, 2, 84, 87, 88, 89, 276].
[20.] Turro, N. J.; Ito, Y.; Chow, M.-F.; Adam, W.; Rodriquez, O.; Yany, F.   *J. Am. Chem. Soc.* **1977**, *99*, 5836 [1, 3, 65].
[21.] Fenical, W.; Kearns, D. R.; Radlick, P.   *J. Am. Chem. Soc.* **1969**, *91*, 7771 [2, 3, 84, 85].
[22.] Farmer, E. H.; Sundralingam, A.   *J. Chem. Soc.* **1942**, 121 [2, 4, 84, 85].
[23.] Kharasch, M. S.; Burt, J. G.   *J. Org. Chem.* **1951**, *16*, 150 [2, 84, 85, 99].
[24.] Scheffer, J. R.; Ouchi, M. D.   *Tetrahedron Lett.* **1970**, 223 [2, 88].
[25.] Schultz, A. G.; Schlessinger, R. H.   *Tetrahedron Lett.* **1970**, 2731 [2, 64, 65, 80, 85].
[26.] Schenck, G. O.; Gollnick, K.; Neumüller, O. A.   *Justus Liebigs Ann. Chem.* **1957**, *603*, 46 [2, 84].
[27.] Hock, H.; Lang, S.   *Chem. Ber.* **1942**, *75*, 1051 [2, 85, 99].

[28.] Hock, H.; Lang, S. *Chem. Ber.* **1943,** *76,* 169 [2, 99].
[29.] Hock, H.; Lang, S. *Chem. Ber.* **1944,** *77,* 257 [2, 99].
[30.] Verbeek, J.; Berends, W.; van Beek, H. C. A. *Rec. Trav. Chim. Pays-Bas* **1976,** *95,* 285 [2].
[31.] Koch, E.; Schenck, G. O. *Chem. Ber.* **1966,** *99,* 1984 [2].
[32.] Kondo, K.; Matsumoto, M. *Chem. Commun.* **1972,** 1332 [2, 87].
[33.] Kaneko, C.; Sugimoto, A.; Tanaka, S. *Synthesis* **1974,** 876 [2, 87, 88, 275].
[34.] Skold, C. N.; Schlessinger, R. H. *Tetrahedron Lett.* **1970,** 791 [2].
[35.] Windaus, A.; Brunken, J. *Justus Liebigs Ann. Chem.* **1928,** *460,* 225 [2].
[36.] Bartlett, P. D.; Schaap, A. P. *J. Am. Chem. Soc.* **1970,** *92,* 3223 [3, 64].
[37.] Mazur, S.; Foote, C. S. *J. Am. Chem. Soc.* **1970,** *92,* 3225 [3, 65].
[38.] Rio, G.; Berthelot, J. *Bull. Soc. Chim. Fr.* **1971,** 3555 [3, 64, 65].
[39.] Fenical, W.; Kearns, D. R.; Radlick, P. *J. Am. Chem. Soc.* **1969,** *91,* 3396 [3, 64].
[40.] Bartlett, P. D.; Schaap, A. P. *J. Am. Chem. Soc.* **1970,** *92,* 3223 [3, 64].
[41.] Mazur, S.; Foote, C. S. *J. Am. Chem. Soc.* **1970,** *92,* 3225 [3].
[42.] Wasserman, H. H.; Terao, S. *Tetrahedron Lett.* **1975,** 1735 [3, 244, 245].
[43.] Shimizu, N.; Bartlett, P. D. *J. Am. Chem. Soc.* **1976,** *98,* 4193 [3, 4].
[44.] Schenck, G. O.; Becker, H.-D.; Schulte-Elte, K.-H.; Krauch, C. H. *Chem. Ber.* **1963,** *96,* 509 [3, 150, 168, 169].
[45.] Wasserman, H. H.; Van Verth, J. E. *J. Am. Chem. Soc.* **1974,** *96,* 585 [3, 133, 147, 149].
[46.] Lindner, J. H. E.; Kuhn, H. J.; Gollnick, K. *Tetrahedron Lett.* **1972,** 1705 [3, 243].
[47.] Hock, H.; Lang, S. *Chem. Ber.* **1943,** *76,* 1130 [4, 99].
[48.] Criegee, R. *Chem. Ber.* **1944,** *77,* 22 [4, 60].
[49.] Hock, H.; Depke, F.; Knauel, G. *Chem. Ber.* **1950,** *83,* 238 [4].
[50.] Knight, H. B.; Swern, D. *Org. Synth., Collective Volume* **1963,** *4,* 895 [4, 99].
[51.] Midland, M. M.; Brown, H. C. *J. Am. Chem. Soc.* **1973,** *95,* 4069 [4, 270].
[52.] Wallace, T. J.; Pobiner, H.; Schriesheim, A. *J. Org. Chem.* **1965,** *30,* 3768 [4, 211].
[53.] Wallace, T. J.; Schriesheim, A. *Tetrahedron Lett.* **1963,** 1131 [4, 251, 264].
[54.] Rust, F. F.; Vaughan, W. E. *Ind. Eng. Chem.* **1949,** *41,* 2595 [4, 58, 60].
[55.] Sneeden, R. P. A.; Turner, R. B. *J. Am. Chem. Soc.* **1955,** *77,* 190 [4, 115, 123, 124, 147].
[56.] Heyns, K.; Blazejewicz, L. *Tetrahedron* **1960,** *9,* 67 [4, 115, 124, 127, 130, 133, 147].
[57.] Cameron, R. E.; Bocarsly, A. B. *J. Am. Chem. Soc.* **1985,** *107,* 6117 [4, 133].
[58.] Eglinton, G.; Galbraith, A. R. *J. Chem. Soc.* **1959,** 889 [4, 15, 90, 91, 155, 279].
[59.] Hay, A. S. *J. Org. Chem.* **1960,** *25,* 1275 [4, 90].
[60.] Clement, W. H.; Selwitz, C. M. *J. Org. Chem.* **1964,** *29,* 241 [4, 75].
[61.] Volger, H. C.; Brackman, W. *Rec. Trav. Chim. Pays-Bas* **1965,** *84,* 579 [4, 86].
[62.] Clarke, H. T.; Dreger, E. E. *Org. Synth., Collective Volume* **1932,** *1,* 87 [4, 217].
[63.] Harrisson, R. J.; Moyle, M. *Org. Synth., Collective Volume* **1963,** *4,* 493 [4, 16, 174].

[64.] Tsuji, J.; Takayanagi, H. Org. Synth., Collective Volume **1988**, *6*, 662 [4, 241].
[65.] Tsuji, J.; Nagashima, H.; Nemoto, H. Org. Synth. **1984**, *62*, 9 [4, 75].
[66.] Jones, G. E.; Kendrick, D. A.; Holmes, A. B. Org. Synth. **1987**, *65*, 52 [4, 90].
[67.] Ito, S.; Matsumoto, M. J. Org. Chem. **1983**, *48*, 1133 [4, 211, 212].
[68.] Hay, A. S.; Eustance, J. W.; Blanchard, H. S. J. Org. Chem. **1960**, *25*, 616 [4, 105, 106].
[69.] de Vries, G.; Schors, A. Tetrahedron Lett. **1968**, 5689 [4, 28, 163].
[70.] Blackburn, T. F.; Schwartz, J. Chem. Commun. **1977**, 157 [4, 133, 147].
[71.] Beroza, M.; Bierl, B. A. Microchim. Acta **1969**, 720 [4, 5].
[72.] Whitmore, F. C.; Church, J. M. J. Am. Chem. Soc. **1932**, *54*, 3710 [4, 5].
[73.] Smith, L. I.; Greenwood, F. L.; Hudrlik, O. Org. Synth., Collective Volume **1955**, *3*, 673 [4].
[74.] Smith, L. I. J. Am. Chem. Soc. **1925**, *47*, 1844, 1850 [4].
[75.] Criegee, R.; Blust, G.; Zinke, H. Chem. Ber. **1954**, *87*, 766 [5, 67].
[76.] Murray, R. W.; Youssefyeh, R. D.; Story, P. R. J. Am. Chem. Soc. **1967**, *89*, 2429 [5, 65, 66].
[77.] Asinger, F. Chem. Ber. **1942**, *75*, 656 [5, 6, 19, 67, 81, 82, 175, 179].
[78.] Criegee, R.; Schröder, G. Chem. Ber. **1960**, *93*, 689 [5, 67].
[79.] Schröder, G. Chem. Ber. **1962**, *95*, 733 [5, 66, 67].
[80.] Fischer, F. G.; Düll, H.; Ertel, L. Chem. Ber. **1932**, *65*, 1467 [5, 77, 79].
[81.] Pryde, E. H.; Anders, D. E.; Teeter, H. M.; Cowan, J. C. J. Org. Chem. **1960**, *25*, 618 [5, 6, 77, 78, 79, 81, 276].
[82.] White, R. W.; King, S. W.; O'Brian, J. L. Tetrahedron Lett. **1971**, 3587 [5, 77].
[83.] Odinokov, V. N.; Zhemaiduk, L. P.; Tolstikov, G. A. Zh. Org. Khim. **1978**, *14*, 54 (Engl. Transl. 48) [5, 81].
[84.] Criegee, R.; Lederer, M. Justus Liebigs Ann. Chem. **1953**, *583*, 1, 29 [5, 91, 92, 227].
[85.] Cook, N. C.; Whitmore, F. C. J. Am. Chem. Soc. **1941**, *63*, 3540 [5, 77].
[86.] Harries, C.; Haarmann, R. Chem. Ber. **1915**, *48*, 32 [5, 78, 81, 163].
[87.] Overend, W. G.; Stacey, M.; Wiggins, L. F. J. Chem. Soc. **1949**, 1358 [5, 8, 78, 81, 228, 229].
[88.] Lorenz, O. Anal. Chem. **1965**, *37*, 101 [5].
[89.] Knowles, W. S.; Thompson, Q. E. J. Org. Chem. **1960**, *25*, 1031 [5, 78, 79].
[90.] Parker, K. A.; Farmar, J. G. J. Org. Chem. **1986**, *51*, 4023 [5, 78].
[91.] Pappas, J. J.; Keaveney, W. P.; Gancher, E.; Berger, M. Tetrahedron Lett. **1966**, 4273 [5, 78].
[92.] Hudlický, T.; Luna, H.; Barbieri, G.; Kwart, L. D. J. Am. Chem. Soc. **1988**, *110*, 4735 [5, 46, 78, 90, 93].
[93.] Bailey, P. S.; Erickson, R. E. Org. Synth., Collective Volume **1973**, *5*, 489, 493 [5, 6].
[94.] Greenwood, F. L. J. Org. Chem. **1955**, *20*, 803 [5, 77].
[95.] Witkop, B.; Patrick, J. B. J. Am. Chem. Soc. **1952**, *74*, 3855 [5, 77, 78].
[96.] Diaper, D. G. M.; Mitchell, D. L. Can. J. Chem. **1960**, *38*, 1976 [5].
[97.] Fremery, M. I.; Fields, E. K. J. Org. Chem. **1963**, *28*, 2537 [5, 81].
[98.] Bailey, P. S. Ind. Eng. Chem. **1958**, *50*, 993 [5, 81].
[99.] Eichelberger, J. L.; Stille, J. K. J. Org. Chem. **1971**, *36*, 1840 [6, 81, 82].
[100.] Diaper, D. G. M. Can. J. Chem. **1955**, *33*, 1720 [6].

[101.] Gensler, W. J.; Schlein, H. N. *J. Am. Chem. Soc.* **1955,** *77,* 4846 [6, 11, 35, 92, 225, 226].
[102.] Warnell, J. L.; Shriner, R. L. *J. Am. Chem. Soc.* **1957,** *79,* 3165 [6, 81].
[103.] Hudlický, M.; Mareš, F. *Collect. Czech. Chem. Commun.* **1959,** *24,* 46 [6, 29, 81, 82, 208, 209].
[104.] Levine, A. A.; Cole, A. G. *J. Am. Chem. Soc.* **1932,** *54,* 338 [6, 96].
[105.] Cohen, Z.; Keinan, E.; Mazur, Y.; Varkony, T. H. *J. Org. Chem.* **1975,** *40,* 2141 [6].
[106.] Cohen, Z.; Varkony, H.; Keinan, E.; Mazur, Y. *Org. Synth., Collective Volume* **1988,** *6,* 43 [6, 59].
[107.] Bailey, P. S.; Lane, A. G. *J. Am. Chem. Soc.* **1967,** *89,* 4473 [6, 61, 62].
[108.] Durland, J. R.; Adkins, H. *J. Am. Chem. Soc.* **1939,** *61,* 429 [6].
[109.] Criegee, R.; Höver, H. *Chem. Ber.* **1960,** *93,* 2521 [6, 96, 97].
[110.] Fischer, F. G. *Justus Liebigs Ann. Chem.* **1929,** *476,* 233 [6, 169].
[111.] Deslongchamps, P.; Moreau, C. *Can. J. Chem.* **1971,** *49,* 2465 [6, 184, 185].
[112.] Dick, C. R.; Hanna, R. F. *J. Org. Chem.* **1964,** *29,* 1218 [6, 13, 14, 61, 180].
[113.] Thompson, Q. E. *J. Am. Chem. Soc.* **1961,** *83,* 845 [6, 248, 249].
[114.] Pike, P. E.; Marsh, P. G.; Erickson, R. E.; Waters, W. L. *Tetrahedron Lett.* **1970,** 2679 [6, 271].
[115.] Fichter, F.; Ackermann, F. *Helv. Chim. Acta* **1919,** *2,* 583 [6].
[116.] Kaulen, J.; Schäfer, H. J. *Synthesis* **1979,** 513 [6, 127, 130].
[117.] Kulka, M. *J. Am. Chem. Soc.* **1946,** *68,* 2472 [6, 98, 105, 108, 109].
[118.] Pattison, F. L. M.; Stothers, J. B.; Woolford, R. G. *J. Am. Chem. Soc.* **1956,** *78,* 2255 [6, 276].
[119.] Yoshida, J.; Hashimoto, J.; Kawabata, N. *J. Org. Chem.* **1982,** *47,* 3575 [6, 61, 62].
[120.] Tsuji, J.; Minato, M. *Tetrahedron Lett.* **1987,** *28,* 3683 [6, 75].
[121.] Shono, T.; Matsumura, Y.; Hayashi, J.; Mizoguchi, M. *Tetrahedron Lett.* **1979,** 165 [6, 131, 133, 147].
[122.] Ruholl, H.; Schäfer, H. J. *Synthesis* **1988,** 54 [6, 163].
[123.] Levin, A. I.; Chechina, O. N.; Sobolov, S. V. *Zh. Obshch. Khim.* **1965,** *35,* 1778; *Chem. Abstr.* **1966,** *64,* 1943a [6].
[124.] Shanley, E. S.; Greenspan, F. P. *Ind. Eng. Chem.* **1947,** *39,* 1536 [7].
[125.] Smith, C. W.; Holm, R. T. *J. Org. Chem.* **1957,** *22,* 746 [7, 8, 175, 182].
[126.] Saito, I.; Nagata, R.; Yuba, K.; Matsuura, T. *Tetrahedron Lett.* **1983,** *24,* 1737 [7, 273].
[127.] Cookson, P. G.; Davies, A. G.; Fazal, N. *J. Organomet. Chem.* **1975,** *99,* C31 [7, 10].
[128.] Matsumoto, M.; Kobayashi, H. *Heterocycles* **1986,** *24,* 2443 [7, 8, 187].
[129.] Payne, G. B.; Deming, P. H.; Williams, P. H. *J. Org. Chem.* **1961,** *26,* 659 [7, 8, 60, 62, 63, 234, 238].
[130.] Milas, N. A.; Sussman, S. *J. Am. Chem. Soc.* **1936,** *58,* 1302 [7, 8, 39, 69, 225, 291].
[131.] Mugdan, M.; Young, D. P. *J. Chem. Soc.* **1949,** 2988 [7, 8, 69, 71, 72, 277].
[132.] Watanabe, Y.; Numata, T.; Oae, S. *Synthesis* **1981,** 204 [7, 8, 254, 255].
[133.] Treibs, W. *Chem. Ber.* **1947,** *80,* 56 [7, 8].
[134.] Milas, N. A. *J. Am. Chem. Soc.* **1937,** *59,* 2342 [7, 8, 19, 78, 79].
[135.] Schultz, H. S.; Freyermuth, H. B.; Buc, S. R. *J. Org. Chem.* **1963,** *28,* 1140 [7, 8, 261].

[136.] Zinner, H.; Falk, K.-H.   *Chem. Ber.* **1955**, *88*, 566 [7, 260].
[137.] Church, J. M.; Blumberg, R.   *Ind. Eng. Chem.* **1951**, *43*, 1780 [7, 8].
[138.] Payne, G. B.; Williams, P. H.   *J. Org. Chem.* **1959**, *24*, 54 [8, 60, 225].
[139.] Yamaguchi, S.; Inoue, M.; Enomoto, S.   *Bull. Chem. Soc. Jpn.* **1986**, *59*, 2881 [8, 94, 95].
[140.] Roussel, M.; Mimoun, H.   *J. Org. Chem.* **1980**, *45*, 5387 [8, 75].
[141.] Criegee, R.   *Justus Liebigs Ann. Chem.* **1936**, *522*, 75 [8, 39, 79].
[142.] Weitz, E.; Scheffer, A.   *Chem. Ber.* **1921**, *54*, 2327 [8, 60, 212].
[143.] Rohrmann, E.; Jones, R. G.; Shonle, H. A.   *J. Am. Chem. Soc.* **1944**, *66*, 1856 [8, 17, 60, 212, 229].
[144.] Payne, G. B.; Williams, P. H.   *J. Org. Chem.* **1959**, *24*, 54 [8, 69].
[145.] Payne, G. B.   *J. Org. Chem.* **1959**, *24*, 2048 [8, 225].
[146.] Payne, G. B.   *J. Am. Chem. Soc.* **1958**, *80*, 6461; **1959**, *81*, 4901 [8, 60, 182].
[147.] Payne, G. B.; Williams, P. H.   *J. Org. Chem.* **1961**, *26*, 651 [8, 9, 60, 230].
[148.] Payne, G. B.   *Tetrahedron* **1962**, *18*, 763 [8, 12, 60, 187, 191].
[149.] Wasson, P. C.; House, H. O.   *Org. Synth., Collective Volume* **1963**, *4*, 552 [8, 212].
[150.] Linn, W. J.   *Org. Synth., Collective Volume* **1973**, *5*, 1007 [8, 230].
[151.] Felix, D.; Wintner, C.; Eschenmoser, A.   *Org. Synth., Collective Volume* **1988**, *6*, 679 [8, 212, 213].
[152.] Milas, N. A.; Maloney, L. S.   *J. Am. Chem. Soc.* **1940**, *62*, 1841 [8, 69, 89].
[153.] Holmes, R. R.; Bayer, R. P.   *J. Am. Chem. Soc.* **1960**, *82*, 3454 [8, 12, 235, 236, 278].
[154.] Dunstan, W. R.; Goulding, E.   *J. Chem. Soc.* **1899**, *75*, 1004 [8, 236].
[155.] Linsker, F.; Evans, R. L.   *J. Am. Chem. Soc.* **1946**, *68*, 403 [8, 239].
[156.] Cope, A. C.; Lee, H.-H.   *J. Am. Chem. Soc.* **1957**, *79*, 964 [8, 237].
[157.] Atkinson, C. M.; Simpson, J. C. E.   *J. Chem. Soc.* **1947**, 1649 [8].
[158.] Chivers, G. E.; Suschitzky, H.   *J. Chem. Soc. C* **1971**, 2867 [8, 12, 13, 237, 238].
[159.] Coeur, A.; Alary, J.   *Bull. Soc. Chim. Fr.* **1964**, 2412 [8].
[160.] Hardegger, E.; Nikles, E.   *Helv. Chim. Acta* **1957**, *40*, 2428 [8, 29, 208, 209, 238].
[161.] Cope, A. C.; Ciganek, E.   *Org. Synth., Collective Volume* **1963**, *4*, 612 [8, 237].
[162.] Taylor, E. C.; Crovetti, A. J.   *Org. Synth., Collective Volume* **1963**, *4*, 704 [8, 238].
[163.] Korte, F.; Löhmer, K. H.   *Chem. Ber.* **1957**, *90*, 1290 [8, 14, 251, 253, 256, 260].
[164.] Tarbell, D. S.; Weaver, C.   *J. Am. Chem. Soc.* **1941**, *63*, 2939 [8, 253, 255, 256].
[165.] Gilman, H.; Nobis, J. F.   *J. Am. Chem. Soc.* **1945**, *67*, 1479 [8].
[166.] Sviridova, A. V.; Laba, V. I.; Prilezhaeva, E. N.   *Zh. Org. Khim.* **1971**, *7*, 2480 (Engl. Transl. 2577) [8, 253].
[167.] Sharpless, K. B.; Lauer, R. F.; Teranishi, A. Y.   *J. Am. Chem. Soc.* **1973**, *95*, 6137 [8, 43, 44, 48, 49, 265, 275].
[168.] Sharpless, K. B.; Lauer, R. F.   *J. Am. Chem. Soc.* **1973**, *95*, 2697 [8, 266].
[169.] Ley, S. V.; Murray, P. J.; Palmer, B. D.   *Tetrahedron* **1985**, *41*, 4765 [8, 265].
[170.] Dobrowsky, A.   *Monatsh. Chem.* **1955**, *86*, 325 [8].

[*171.*] Barger, G. *J. Chem. Soc.* **1918,** *113,* 218 [8, 16, 166, 167, 180, 181].
[*172.*] Dakin, H. D. *Org. Synth., Collective Volume* **1932,** *1,* 149 [8, 180, 181].
[*173.*] Surrey, A. R. *Org. Synth., Collective Volume* **1955,** *3,* 759 [8, 180, 181].
[*174.*] Greenspan, F. P.; Gall, R. J.; MacKellar, D. G. *J. Org. Chem.* **1955,** *20,* 215 [8, 14, 61, 222, 225].
[*175.*] Payne, G. B. *Org. Synth., Collective Volume* **1973,** *5,* 805 [8, 222].
[*176.*] Silbert, L. S.; Siegel, E.; Swern, D. *Org. Synth., Collective Volume* **1973,** *5,* 904 [8, 13, 222].
[*177.*] Ross, H.; Hüttel, R. *Chem. Ber.* **1956,** *89,* 2641 [8, 150].
[*178.*] Bhatt, M. V.; Rao, G. V.; Rao, K. S. R. *Chem. Commun.* **1971,** 822 [8, 42, 173, 174].
[*179.*] Milas, N. A.; Mageli, O. L. *J. Am. Chem. Soc.* **1952,** *74,* 1471 [8, 150].
[*180.*] Hawkins, E. G. E.; Large, R. *J. Chem. Soc., Perkin Trans. 1* **1974,** 2561 [8, 140, 186].
[*181.*] Kharasch, M. S.; Sosnovsky, G. *J. Org. Chem.* **1958,** *23,* 1322 [8, 186].
[*182.*] Badovskaya, L. A.; Krapivin, G. D.; Kalyugina, T. Y.; Kul'nevich, V. G.; Muzychenko, G. F. *Zh. Org. Khim.* **1975,** *11,* 2446 (Engl. Transl. 2508) [8].
[*183.*] Breuer, S. W.; Leatham, M. J.; Thorpe, F. G. *Chem. Commun.* **1971,** 1475 [8, 267, 268, 269, 270].
[*184.*] Newman, M. S.; Arkell, A.; Fukunaga, T. *J. Am. Chem. Soc.* **1960,** *82,* 2498 [8, 22, 117, 128, 268, 277].
[*185.*] Kidwell, R. L.; Murphy, M.; Darling, S. D. *Org. Synth., Collective Volume* **1973,** *5,* 918 [8, 269].
[*186.*] Price, C. C.; Stacy, G. W. *Org. Synth., Collective Volume* **1955,** *3,* 86 [8, 250, 251].
[*187.*] Chao, T. H.; Hardy, W. B. *Chem. Ind. (London)* **1965,** 81 [8, 264].
[*188.*] Dakin, H. D. *J. Biol. Chem.* **1909,** *5,* 409 [8, 229].
[*189.*] Snyder, H. R.; Buck, J. S.; Ide, W. S. *Org. Synth., Collective Volume* **1943,** *2,* 333 [8, 229].
[*190.*] Ho, T.-L.; Olah, G. A. *Synthesis* **1976,** 807 [8].
[*191.*] Hudlický, M. unpublished results [8].
[*192.*] Lissel, M.; Dehmlow, E. V. *Tetrahedron Lett.* **1978,** 3689 [8, 211].
[*193.*] Santurri, P.; Robbins, F.; Stubbings, R. *Org. Synth., Collective Volume* **1973,** *5,* 341 [8, 234].
[*194.*] McKillop, A.; Tarbin, J. A. *Tetrahedron Lett.* **1983,** *24,* 1505 [8, 235, 236, 255].
[*195.*] Meisenheimer, J.; Hesse, E. *Chem. Ber.* **1919,** *52,* 1161 [8, 235].
[*196.*] Schröder, M.; Griffith, W. P. *Chem. Commun.* **1979,** 58 [8, 38, 130, 142, 148, 241, 242].
[*197.*] Baker, W.; Brown, N. C. *J. Chem. Soc.* **1948,** 2303 [8, 163, 164].
[*198.*] Bacon, R. G. R.; Stewart, D. *J. Chem. Soc. C* **1966,** 1384 [8, 239, 240].
[*199.*] Langley, W. D. *Org. Synth., Collective Volume* **1955,** *3,* 334 [8, 235].
[*200.*] Nishihara, A.; Kubota, I. *J. Org. Chem.* **1968,** *33,* 2525 [8, 175, 182].
[*201.*] Büchi, G.; Jeger, O. *Helv. Chim. Acta* **1949,** *32,* 538 [8, 188].
[*202.*] Atkinson, C. M.; Brown, C. W.; McIntyre, J.; Simpson, J. C. E. *J. Chem. Soc.* **1954,** 2023 [8].
[*203.*] Wiley, R. H.; Hartman, J. L. *J. Am. Chem. Soc.* **1951,** *73,* 494 [8, 235, 236].
[*204.*] Bamberger, E.; Hill, A. *Chem. Ber.* **1900,** *33,* 533 [8, 266].
[*205.*] Kennedy, R. J.; Stock, A. M. *J. Org. Chem.* **1960,** *25,* 1901 [9, 105, 106, 131, 133, 175, 179, 188, 195, 235, 248, 251, 255, 266].
[*206.*] Pearl, I. A. *Org. Synth., Collective Volume* **1963,** *4,* 972 [9, 16, 175, 180].

[207.] Trost, B. M.; Curran, D. P. *Tetrahedron Lett.* **1981**, *22*, 1287 [9, 253, 254, 278].
[208.] Evans, T. L.; Grade, M. M. *Synth. Commun.* **1986**, *16*, 1207 [9, 255].
[209.] Trost, B. M.; Braslau, R. *J. Org. Chem.* **1988**, *53*, 532 [9, 253, 254].
[210.] Murray, R. W.; Jeyaraman, R. *J. Org. Chem.* **1985**, *50*, 2847 [9, 61, 62, 63, 64, 225, 237, 238, 254].
[211.] Murray, R. W.; Jeyaraman, R.; Mohan, L. *Tetrahedron Lett.* **1986**, *27*, 2335 [9, 235].
[212.] Milas, N. A.; Surgenor, D. M. *J. Am. Chem. Soc.* **1946**, *68*, 205 [9].
[213.] Hill, J. G.; Rossiter, B. E.; Sharpless, K. B. *J. Org. Chem.* **1983**, *48*, 3607 [9].
[214.] Sharpless, K. B.; Verhoeven, T. R. *Aldrichimica Acta* **1979**, *12*, 63 [9].
[215.] Sharpless, K. B.; Michaelson, R. C. *J. Am. Chem. Soc.* **1973**, *95*, 6136 [9, 61, 152].
[216.] Itoh, T.; Jitsukawa, K.; Kaneda, K.; Teranishi, S. *J. Am. Chem. Soc.* **1979**, *101*, 159 [9, 19, 44, 61, 152, 153].
[217.] Indictor, N.; Brill, W. F. *J. Org. Chem.* **1965**, *30*, 2074 [9, 61, 62].
[218.] Rossiter, B. E.; Verhoeven, T. R.; Sharpless, K. B. *Tetrahedron Lett.* **1979**, 4733 [9, 153].
[219.] Payne, G. B. *J. Org. Chem.* **1960**, *25*, 275 [9, 182].
[220.] Yang, N. C.; Finnegan, R. A. *J. Am. Chem. Soc.* **1958**, *80*, 5845 [9, 61, 212, 213].
[221.] Katsuki, T.; Sharpless, K. B. *J. Am. Chem. Soc.* **1980**, *102*, 5974 [9, 44, 45, 61, 154].
[222.] Rossiter, B. E.; Katsuki, T.; Sharpless, K. B. *J. Am. Chem. Soc.* **1981**, *103*, 464 [9, 44, 45, 61, 154].
[223.] Katsuki, T.; Sharpless, K. B. *J. Am. Chem. Soc.* **1980**, *102*, 5974 [9, 44, 45, 61].
[224.] Pitchen, P.; Duñach, E.; Deshmukh, M. N.; Kagan, H. B. *J. Am. Chem. Soc.* **1984**, *106*, 8188 [9, 10, 44, 45, 254, 258, 278].
[225.] Doumaux, A. R., Jr.; Trecker, D. J. *J. Org. Chem.* **1970**, *35*, 2121 [9, 12, 245].
[226.] Kuhnen, L. *Chem. Ber.* **1966**, *99*, 3384 [9, 237, 277].
[227.] Sheng, M. N.; Zajacek, J. G. *Org. Synth., Collective Volume* **1988**, *6*, 501 [9, 237].
[228.] Hayakawa, Y.; Uchiyama, M.; Noyori, R. *Tetrahedron Lett.* **1986**, *27*, 4191 [9, 10, 42, 249].
[229.] Price, C. C.; Krebs, E. *Org. Synth., Collective Volume* **1955**, *3*, 649 [10, 14].
[230.] Perret, A.; Perrot, R. *Helv. Chim. Acta* **1945**, *28*, 558 [10, 74].
[231.] Doyle, M. P.; Patrie, W. J.; Williams, S. B. *J. Org. Chem.* **1979**, *44*, 2955 [10, 96].
[232.] Kurz, M. E.; Kovacic, P. *J. Am. Chem. Soc.* **1967**, *89*, 4960 [10, 94].
[233.] Kharasch, M. S.; Sosnovsky, G.; Yang, N. C. *J. Am. Chem. Soc.* **1959**, *81*, 5819 [10, 86].
[234.] Pedersen, K.; Jakobson, P.; Lawesson, S.-O. *Org. Synth., Collective Volume* **1973**, *5*, 70 [10, 86].
[235.] Frisell, C.; Lawesson, S.-O. *Org. Synth., Collective Volume* **1973**, *5*, 642 [10, 271].
[236.] Kanemoto, S.; Oshima, K.; Matsubara, S.; Takai, K.; Nozaki, H. *Tetrahedron Lett.* **1983**, *24*, 2185 [10, 25, 125].
[237.] Suzuki, M.; Takada, H.; Noyori, R. *J. Org. Chem.* **1982**, *47*, 902 [10, 187, 191, 194, 195].

[238.] Sydnes, L. K.; Burkow, I. C.; Hansen, S. H.   *Tetrahedron* **1985**, *41*, 5703 [10, 17, 101].
[239.] Greenspan, F. P.; MacKellar, D. G.   *Anal. Chem.* **1948**, *20*, 1061 [10, 12].
[240.] Swern, D.; Billen, G. N.; Scanlan, J. T.   *J. Am. Chem. Soc.* **1946**, *68*, 1504 [11, 69].
[241.] Swern, D.; Billen, G. N.; Findley, T. W.; Scanlan, J. T.   *J. Am. Chem. Soc.* **1945**, *67*, 1786 [11, 12].
[242.] Adkins, H.; Roebuck, A. K.   *J. Am. Chem. Soc.* **1948**, *70*, 4041 [11, 71, 72].
[243.] Mancera, O.; Rosenkranz, G.; Djerassi, C.   *J. Org. Chem.* **1951**, *16*, 192 [11].
[244.] Roebuck, A.; Adkins, H.   *Org. Synth., Collective Volume* **1955**, *3*, 217 [11, 71, 72].
[245.] Goldschmidt, S.; Veer, W. L. C.   *Rec. Trav. Chim. Pays-Bas* **1948**, *67*, 489 [11].
[246.] Horan, J. E.; Schiessler, R. W.   *Org. Synth., Collective Volume* **1973**, *5*, 647 [11, 75, 76].
[247.] Roberts, S. M.; Suschitzky, H.   *J. Chem. Soc. C* **1969**, 1485 [11, 237, 238, 239].
[248.] Roberts, S. M.; Suschitzky, H.   *J. Chem. Soc. C* **1968**, 1537 [11, 13, 237, 238, 239].
[249.] Greenspan, F. P.   *Ind. Eng. Chem.* **1947**, *39*, 847 [12, 95, 163, 168, 234].
[250.] Dittmann, W.; Stürzenhofecker, F.   *Justus Liebigs Ann. Chem.* **1965**, *688*, 57 [12, 61].
[251.] Reif, D. J.; House, H. O.   *Org. Synth., Collective Volume* **1963**, *4*, 860 [12, 61, 62, 63].
[252.] Korach, M.; Nielsen, D. R.; Rideout, W. H.   *Org. Synth., Collective Volume* **1973**, *5*, 414 [12].
[253.] Emmons, W. D.   *J. Am. Chem. Soc.* **1957**, *79*, 5528 [12, 13, 235, 236].
[254.] Payne, G. B.; Williams, P. H.   *J. Org. Chem.* **1959**, *24*, 284 [12, 191, 192, 212].
[255.] Doumaux, A. R., Jr.; McKeon, J. E.; Trecker, D. J.   *J. Am. Chem. Soc.* **1969**, *91*, 3992 [12, 245].
[256.] O'Connor, W. F.; Moriconi, E. J.   *Ind. Eng. Chem.* **1953**, *45*, 277 [12, 98].
[257.] Emmons, W. D.   *J. Am. Chem. Soc.* **1956**, *78*, 6208 [12].
[258.] Wenkert, E.; Rubin, M.   *Nature (London)* **1952**, *170*, 708 [12, 191, 192].
[259.] Böeseken, J.; Cohen, W. D.; Kip, C. J.   *Rec. Trav. Chim. Pays-Bas* **1936**, *55*, 815 [12, 180, 181].
[260.] Reissenweber, G.; Mangold, D.   *Angew. Chem., Int. Ed. Engl.* **1980**, *19*, 222 [12, 193, 194].
[261.] Doering, v. W. E.; Speers, L.   *J. Am. Chem. Soc.* **1950**, *72*, 5515 [12, 188, 195].
[262.] Starcher, P. S.; Phillips, B.   *J. Am. Chem. Soc.* **1958**, *80*, 4079 [12, 186, 188, 189, 195].
[263.] Mosher, H. S.; Turner, L.; Carlsmith, A.   *Org. Synth., Collective Volume* **1963**, *4*, 828 [12, 238].
[264.] Szmant, H. H.; Alfonso, L. M.   *J. Am. Chem. Soc.* **1957**, *79*, 205 [12, 259, 260].
[265.] Böeseken, M. J.; Arrias, E.   *Rec. Trav. Chim. Pays-Bas* **1935**, *54*, 711 [12, 253].
[266.] Sharefkin, J. G.; Saltzman, H.   *Org. Synth., Collective Volume* **1973**, *5*, 658, 660 [12, 31, 266, 267].

[267.] Sharefkin, J. G.; Saltzman, H.   Org. Synth., Collective Volume **1973**, *5*, 665 [12, 32, 266].
[268.] Böeseken, J.; Slooff, G.   Rec. Trav. Chim. Pays-Bas **1930**, *49*, 95 [12, 92, 226, 227].
[269.] Leffler, J. E.   J. Org. Chem. **1951**, *16*, 1785 [12, 218].
[270.] Böeseken, J.; Jacobs, J.   Rec. Trav. Chim. Pays-Bas **1936**, *55*, 804 [12, 218, 219].
[271.] Böeseken, J.; Slooff, G.   Rec. Trav. Chim. Pays-Bas **1930**, *49*, 91 [12, 193, 218].
[272.] Böeseken, J.; Slooff, G.   Rec. Trav. Chim. Pays-Bas **1930**, *49*, 100 [12, 97, 98].
[273.] Kobrina, L. S.; Akulenko, N. V.; Yakobson, G. G.   Zh. Org. Khim. **1972**, *8*, 2165 (Engl. Transl. 2209) [12, 163, 168, 218].
[274.] Page, G. A.; Tarbell, D. S.   Org. Synth., Collective Volume **1963**, *4*, 136 [12, 163, 168].
[275.] Lassak, E. V.; Pinhey, J. T.; Simes, J. J. H.   Aust. J. Chem. **1973**, *26*, 1051 [12, 22, 47].
[276.] Corey, E. J.; Barrette, E.-P.; Magriotis, P. A.   Tetrahedron Lett. **1985**, *26*, 5855 [12, 133, 147].
[277.] Heywood, D. L.; Phillips, B.   J. Org. Chem. **1960**, *25*, 1699 [12, 184, 185].
[278.] Taylor, E. C.; McKillop, A.   J. Org. Chem. **1965**, *30*, 3153 [12, 13, 231].
[279.] Venier, C. G.; Squires, T. G.; Chen, Y.-Y.; Hussmann, G. P.; Shei, J. C.; Smith, B. F.   J. Org. Chem. **1982**, *47*, 3773 [12, 13, 253].
[280.] Hawthorne, M. F.   Anal. Chem. **1956**, *28*, 540 [12, 13, 189].
[281.] Emmons, W. D.   J. Am. Chem. Soc. **1954**, *76*, 3470 [12, 13, 236].
[282.] Hawthorne, M. F.; Emmons, W. D.   J. Am. Chem. Soc. **1958**, *80*, 6398 [12, 13, 187, 188].
[283.] Emmons, W. D.; Pagano, A. S.   J. Am. Chem. Soc. **1955**, *77*, 89 [12, 61, 188, 189, 279].
[284.] Emmons, W. D.; Lucas, G. B.   J. Am. Chem. Soc. **1955**, *77*, 2287 [12, 13, 187, 188, 189, 195].
[285.] Emmons, W. D.; Ferris, A. F.   J. Am. Chem. Soc. **1953**, *75*, 4623; **1954**, *76*, 3468, 3470 [12, 13, 236].
[286.] Sauers, R. R.; Ubersax, R. W.   J. Org. Chem. **1965**, *30*, 3939 [12, 13, 14, 187, 190].
[287.] McKittrick, B. A.; Ganem, B.   Tetrahedron Lett. **1985**, *26*, 4895 [12, 13, 14, 61, 171].
[288.] Emmons, W. D.; Pagano, A. S.; Freeman, J. P.   J. Am. Chem. Soc. **1954**, *76*, 3472 [12, 69].
[289.] Emmons, W. D.; Lucas, G. B.   J. Am. Chem. Soc. **1955**, *77*, 2287 [13].
[290.] Pagano, A. S.; Emmons, W. D.   Org. Synth., Collective Volume **1973**, *5*, 367 [13, 236].
[291.] Emmons, W. D.   J. Am. Chem. Soc. **1954**, *76*, 3468 [13, 231].
[292.] DiCosimo, R.; Szabo, H.-C.   J. Org. Chem. **1986**, *51*, 1365 [13, 93, 94].
[293.] Braun, G.   Org. Synth., Collective Volume **1932**, *1*, 431 [13].
[294.] Moyer, J. R.; Manley, N. C.   J. Org. Chem. **1964**, *29*, 2099 [13].
[295.] Swern, D.; Findley, T. W.; Scanlan, J. T.   J. Am. Chem. Soc. **1944**, *66*, 1925 [13, 61, 225, 235].
[296.] Bartlett, P. D.; Bavley, A.   J. Am. Chem. Soc. **1938**, *60*, 2416 [13, 34, 35, 61, 68].
[297.] Stevens, C. L.; Tazuma, J.   J. Am. Chem. Soc. **1954**, *76*, 715 [13, 61, 170].
[298.] Henbest, H. B.; Wilson, R. A. L.   J. Chem. Soc. **1957**, 1958 [13, 152].

[299.] Naves, Y.-R.; Schwarzkopf, O.; Lewis, A. D.   *Helv. Chim. Acta* **1947**, *30*, 880 [13, 212, 213].
[300.] Hibbert, H.; Burt, P.   *Org. Synth., Collective Volume* **1932**, *1*, 494 [13, 61, 62, 63].
[301.] Emmons, W. D.; Pagano, A. S.   *Org. Synth., Collective Volume* **1973**, *5*, 191 [13].
[302.] Ogata, Y.; Sawaki, Y.   *J. Org. Chem.* **1969**, *34*, 3985 [13, 180, 181].
[303.] Friess, S. L.   *J. Am. Chem. Soc.* **1949**, *71*, 2571 [13, 186, 189, 195, 208].
[304.] Friess, S. L.; Farnham, N.   *J. Am. Chem. Soc.* **1950**, *72*, 5518 [13, 187, 189].
[305.] Friess, S. L.; Pinson, R., Jr.   *J. Am. Chem. Soc.* **1952**, *74*, 1302 [13, 189, 279].
[306.] Mislow, K.; Brenner, J.   *J. Am. Chem. Soc.* **1953**, *75*, 2318 [13, 187].
[307.] Yokoyama, T.; Nohara, F.   *Bull. Chem. Soc. Jpn.* **1965**, *38*, 1498 [13, 191, 192].
[308.] Small, L. D.; Bailey, J. H.; Cavallito, C. J.   *J. Am. Chem. Soc.* **1947**, *69*, 1710 [13, 263].
[309.] Kondo, K.; Negishi, A.; Fukuyama, M.   *Tetrahedron Lett.* **1969**, 2461 [13, 255].
[310.] Braun, G.   *J. Am. Chem. Soc.* **1929**, *51*, 228 [13, 28, 39, 69, 225].
[311.] Nabih, I.; Helmy, E.   *J. Pharm. Sci.* **1967**, *56*, 649 [13, 244].
[312.] Brough, J. N.; Lythgoe, B.; Waterhouse, P.   *J. Chem. Soc.* **1954**, 4069 [13, 232].
[313.] McDonald, R. N.; Steppel, R. N.; Dorsey, J. E.   *Org. Synth., Collective Volume* **1988**, *6*, 276 [13].
[314.] Frimer, A. A.   *Synthesis* **1977**, 578 [13, 170].
[315.] Camps, F.; Coll, J.; Messeguer, A.; Pericàs, M. A.   *Tetrahedron Lett.* **1981**, *22*, 3895 [13, 14, 61, 62, 63, 180, 181].
[316.] Paquette, L. A.; Barrett, J. H.   *Org. Synth., Collective Volume* **1973**, *5*, 467 [13, 61].
[317.] Brown, H. C.; Kabalka, G. W.; Rathke, M. W.   *J. Am. Chem. Soc.* **1967**, *89*, 4530 [13].
[318.] Godfrey, I. M.; Sargent, M. V.; Elix, J. A.   *J. Chem. Soc., Perkin Trans. 1* **1974**, 1353 [13, 180, 181].
[319.] Robinson, C. H.; Milewich, L.; Hofer, P.   *J. Org. Chem.* **1966**, *31*, 524 [13, 235].
[320.] Craig, J. C.; Purushothaman, K. K.   *J. Org. Chem.* **1970**, *35*, 1721 [13, 237].
[321.] Trost, B. M.; Salzmann, T. N.; Hiroi, K.   *J. Am. Chem. Soc.* **1976**, *98*, 4887 [13, 259].
[322.] Russel, G. A.; Ochrymowycz, L. A.   *J. Org. Chem.* **1970**, *35*, 2106 [13, 30, 260, 261].
[323.] Carey, F. A.; Dailey, O. D., Jr.; Hernandez, O.; Tucker, J. R.   *J. Org. Chem.* **1976**, *41*, 3975 [13, 30, 259].
[324.] Paquette, L. A.; Carr, R. V. C.   *Org. Synth.* **1985**, *64*, 157 [13, 253, 260].
[325.] Krief, A.; Dumont, W.; Denis, J.-N.; Évrard, G.; Norberg, B.   *Chem. Commun.* **1985**, 569 [13, 35, 265].
[326.] Cella, J. A.; McGrath, J. P.; Regen, S. L.   *Tetrahedron Lett.* **1975**, 4115 [13, 147].
[327.] Grieco, P. A.; Oguri, T.; Yokoyama, Y.   *Tetrahedron Lett.* **1978**, 419 [13, 184, 185].
[328.] Tanaka, S.; Yamamoto, H.; Nozaki, H.; Sharpless, K. B.; Michaelson, R. C.; Cutting, J. D.   *J. Am. Chem. Soc.* **1974**, *96*, 5254 [14, 19, 21, 61, 153].

[329.] Müller, W.; Schneider, H.-J.   *Angew. Chem., Int. Ed. Engl.* **1979**, *18*, 407 [14, 58].
[330.] Royals, E. E.; Harrell, L. L., Jr.   *J. Am. Chem. Soc.* **1955**, *77*, 3405 [14, 61, 62, 63].
[331.] Böhme, H.   *Org. Synth., Collective Volume* **1955**, *3*, 619 [14, 222].
[332.] Karrer, P.; Stürzinger, H.   *Helv. Chim. Acta* **1946**, *29*, 1829 [14, 212, 213, 214].
[333.] House, H. O.; Gannon, W. F.   *J. Org. Chem.* **1958**, *23*, 879 [14, 218].
[334.] Karrer, P.; Haab, F.   *Helv. Chim. Acta* **1949**, *32*, 950 [14, 190, 193, 229].
[335.] Strating, J.; Thijs, L.; Zwanenburg, B.   *Tetrahedron Lett.* **1966**, 65 [14].
[336.] MacDonald, D. L.; Fischer, H. O. L.   *J. Am. Chem. Soc.* **1952**, *74*, 2087 [14].
[337.] Rinzema, L. C.; Stoffelsma, J.; Arens, J. F.   *Rec. Trav. Chim. Pays-Bas* **1959**, *78*, 354 [14, 260].
[338.] White, R. W.; Emmons, W. D.   *Tetrahedron* **1961**, *17*, 31 [14, 61, 62, 190, 195, 236].
[339.] Pollak, A.; Zupan, M.; Sket, B.   *Synthesis* **1973**, 495 [14, 237, 239].
[340.] Zinner, H.; Falk, K.-H.   *Chem. Ber.* **1956**, *89*, 2451 [14, 260].
[341.] Parker, W. E.; Ricciuti, C.; Ogg, C. L.; Swern, D.   *J. Am. Chem. Soc.* **1955**, *77*, 4037 [14].
[342.] Dzhemilev, U. M.; Vostrikov, N. S.; Moiseenkov, A. M.; Tolstikov, G. A.   *Izv. Akad. Nauk. SSSR* **1981**, 1320; *Chem. Abstr.* **1981**, *95*, 167877 [14].
[343.] Grieco, P. A.; Yokoyama, Y.; Gilman, S.; Ohfune, Y.   *Chem. Commun.* **1977**, 870 [14, 187, 188].
[344.] Neave, G. B.   *J. Chem. Soc.* **1912**, *101*, 513 [14, 132].
[345.] Davies, R. R.; Hodgson, H. H.   *J. Chem. Soc.* **1943**, 282 [15, 114, 124].
[346.] Weizmann, C.; Bergmann, E.; Sulzbacher, M.   *J. Org. Chem.* **1950**, *15*, 54 [15, 178].
[347.] Jallabert, C.; Riviere, H.   *Tetrahedron Lett.* **1977**, 1215 [15].
[348.] Tsuji, J.; Shimizu, I.; Yamamoto, K.   *Tetrahedron Lett.* **1976**, 2975 [15].
[349.] Sheikh, M. Y.; Eadon, G.   *Tetrahedron Lett.* **1972**, 257 [15, 115, 123, 124, 133, 148].
[350.] Lochte, H. L.; Noyes, W. A.; Bailey, J. R.   *J. Am. Chem. Soc.* **1922**, *44*, 2556 [15, 233].
[351.] Hartman, W. W.; Dickey, J. B.   *J. Am. Chem. Soc.* **1933**, *55*, 1228 [15, 217].
[352.] Gardner, T. S.; Wenis, E.; Lee, J.   *J. Org. Chem.* **1958**, *23*, 823 [15, 217].
[353.] Hauser, F. M.; Ellenberger, S. R.   *Synthesis* **1987**, 723 [15, 101, 102].
[354.] Dunbar, R. E.; Arnold, M. R.   *J. Org. Chem.* **1945**, *10*, 501 [15, 114, 132].
[355.] Longley, R. I., Jr.; Emerson, W. S.   *Org. Synth., Collective Volume* **1963**, *4*, 677 [15, 131, 158].
[356.] Nes, W. R.   *J. Org. Chem.* **1958**, *23*, 899 [15, 132].
[357.] Campbell, I. D.; Eglinton, G.   *Org. Synth., Collective Volume* **1973**, *5*, 517 [15, 90].
[358.] Ruggli, P.; Zeller, P.   *Helv. Chim. Acta* **1945**, *28*, 741 [15, 217].
[359.] Blomquist, A. T.; Goldstein, A.   *Org. Synth., Collective Volume* **1963**, *4*, 838 [15, 217].
[360.] Belli, A.; Giordano, C.   *Synthesis* **1980**, 477 [15].
[361.] Kaeding, W. W.; Collins, G. R.   *J. Org. Chem.* **1965**, *30*, 3750 [15, 93].
[362.] Pearl, I. A.   *J. Org. Chem.* **1947**, *12*, 85 [15, 16, 175, 177].
[363.] Pearl, I. A.   *J. Org. Chem.* **1947**, *12*, 79 [15, 177, 180].

[364.] Pearl, I. A. J. Am. Chem. Soc. **1946**, *68*, 1100 [15, 177, 180].
[365.] Fiesselmann, H. Chem. Ber. **1942**, *75*, 881 [16, 175, 279].
[366.] Pearl, I. A. J. Am. Chem. Soc. **1946**, *68*, 429 [16, 127, 175, 180].
[367.] Campaigne, E.; LeSuer, W. M. Org. Synth., Collective Volume **1963**, *4*, 919 [16, 175, 176].
[368.] Willstätter, R.; Benz, M. Chem. Ber. **1906**, *39*, 3482 [16, 17, 166, 167, 247, 248].
[369.] König, K.-H.; Schulze, W.; Möller, G. Chem. Ber. **1960**, *93*, 554 [16, 18, 167].
[370.] Reimlinger, H. Chem. Ber. **1964**, *97*, 3493 [16, 220, 221].
[371.] Boyer, J. H.; Borgers, R.; Wolford, L. T. J. Am. Chem. Soc. **1957**, *79*, 678 [16].
[372.] Schmitz, E.; Ohme, R. Org. Synth., Collective Volume **1973**, *5*, 897 [16, 233].
[373.] Willstätter, R.; Kubli, H. Chem. Ber. **1908**, *41*, 1936 [16, 231].
[374.] Wieland, H.; Roth, K. Chem. Ber. **1920**, *53*, 210 [16, 232].
[375.] Fryer, R. I.; Archer, G. A.; Brust, B.; Zally, W.; Sternbach, L. H. J. Org. Chem. **1965**, *30*, 1308 [16, 20, 49].
[376.] Rapoport, H.; Reist, H. N. J. Am. Chem. Soc. **1955**, *77*, 490 [16, 133].
[377.] Fetizon, M.; Golfier, M.; Louis, J.-M. Tetrahedron **1975**, *31*, 171 [16, 157, 158].
[378.] Maassen, J. A.; De Boer, T. J. Rec. Trav. Chim. Pays-Bas **1971**, *90*, 373 [16, 231, 232].
[379.] Noyes, A. A.; DeVault, D.; Coryell, C. D.; Deahl, T. J. J. Am. Chem. Soc. **1937**, *59*, 1326 [16].
[380.] Syper, L. Tetrahedron Lett. **1967**, 4193 [16, 101, 103, 104, 105, 115, 124, 126, 147].
[381.] Clarke, T. G.; Hampson, N. A.; Lee, J. B.; Morley, J. R.; Scanlon, B. Tetrahedron Lett. **1968**, 5685 [16, 229, 241, 242, 248, 249].
[382.] Corey, E. J.; Gilman, N. W.; Ganem, B. E. J. Am. Chem. Soc. **1968**, *90*, 5616 [16, 33, 131, 175, 176, 179].
[383.] McKillop, A.; Young, D. W. Synth. Commun. **1977**, *7*, 467 [16, 164, 165].
[384.] Jacini, G.; Bacchetti, T. Gazz. Chim. Ital. **1950**, *80*, 757 [16, 166].
[385.] Ried, W.; Junker, P. Angew. Chem. **1967**, *79*, 622 [16, 221].
[386.] Day, A. C.; Raymond, P.; Southam, R. M.; Whiting, M. C. J. Chem. Soc. C **1966**, 467 [16, 220].
[387.] Miller, J. B. J. Org. Chem. **1959**, *24*, 560 [16, 220, 221, 280].
[388.] Baltzly, R.; Mehta, N. B.; Russell, P. B.; Brooks, R. E.; Grivsky, E. M.; Steinberg, A. M. J. Org. Chem. **1961**, *26*, 3669 [16, 220, 221].
[389.] Nenitzescu, C. D.; Solomonica, E. Org. Synth., Collective Volume **1943**, *2*, 496 [16, 221].
[390.] Smith, L. I.; Howard, K. L. Org. Synth., Collective Volume **1955**, *3*, 351 [16, 220, 221].
[391.] Smith, L. I.; Hoehn, H. H. Org. Synth., Collective Volume **1955**, *3*, 356 [16, 220, 221].
[392.] Andrews, S. D.; Day, A. C.; Raymond, P.; Whiting, M. C. Org. Synth., Collective Volume **1970**, *50*, 27 [16, 220].
[393.] Blomquist, A. T.; Burge, R. E., Jr.; Sucsy, A. C. J. Am. Chem. Soc. **1952**, *74*, 3636 [16, 221, 222].
[394.] Prelog, V.; Schenker, K.; Günthard, H. H. Helv. Chim. Acta **1952**, *35*, 1598 [16, 221, 222, 280].
[395.] Prelog, V.; Schenker, K.; Küng, W. Helv. Chim. Acta **1953**, *36*, 471 [16, 20, 221].

[396.] Cope, A. C.; Smith, D. S.; Cotter, R. J.   *Org. Synth., Collective Volume* **1963**, *4*, 377 [16, 221, 222].
[397.] Overberger, C. G.; Palmer, L. C.; Marks, B. S.; Byrd, N. R.   *J. Am. Chem. Soc.* **1955**, *77*, 4100 [16].
[398.] Overberger, C. G.; Lombardino, J. G.; Hiskey, R. G.   *J. Am. Chem. Soc.* **1957**, *79*, 6430 [16].
[399.] Windaus, A.; Linsert, O.   *Justus Liebigs Ann. Chem.* **1928**, *465*, 148 [16, 47, 48].
[400.] Zürcher, A.; Heusser, H.; Jeger, O.; Geistlich, P.   *Helv. Chim. Acta* **1954**, *37*, 1562 [16, 47].
[401.] Leonard, N. J.; Hay, A. S.; Fulmer, R. W.; Gash, V. W.   *J. Am. Chem. Soc.* **1955**, *77*, 439 [16, 49, 50].
[402.] Leonard, N. J.; Hauck, F. P., Jr.   *J. Am. Chem. Soc.* **1957**, *79*, 5279 [16, 49].
[403.] Leonard, N. J.; Morrow, D. F.   *J. Am. Chem. Soc.* **1958**, *80*, 371 [16, 243].
[404.] Rappoport, Z.; Winstein, S.; Young, W. G.   *J. Am. Chem. Soc.* **1972**, *94*, 2320 [16, 86, 101].
[405.] Theis, R. J.; Dessy, R. E.   *J. Org. Chem.* **1966**, *31*, 624 [16, 220, 221].
[406.] Newman, M. S.; Arkell, A.   *J. Org. Chem.* **1959**, *24*, 385 [16, 22, 218, 220, 221].
[407.] Scholl, R.; Schwarzer, G.   *Chem. Ber.* **1922**, *55*, 324 [17, 54].
[408.] Baker, W.; Glockling, F.; McOmie, J. F. W.   *J. Chem. Soc.* **1951**, 1118 [17, 50, 52, 54].
[409.] Crouse, D. J.; Wheeler, M. M.; Goemann, M.; Tobin, P. S.; Basu, S. K.; Wheeler, D. M. S.   *J. Org. Chem.* **1981**, *46*, 1814 [17].
[410.] Cambie, R. C.; Rutledge, P. S.   *Org. Synth., Collective Volume* **1988**, *6*, 348 [17, 71, 72].
[411.] Anderson, C. B.; Winstein, S.   *J. Org. Chem.* **1963**, *28*, 605 [17, 74, 75].
[412.] Kruse, W.; Bednarski, T. M.   *J. Org. Chem.* **1971**, *36*, 1154 [17, 61].
[413.] McKillop, A.; Oldenziel, O. H.; Swann, B. P.; Taylor, E. C.; Robey, R. L.   *J. Am. Chem. Soc.* **1973**, *95*, 1296 [17, 91, 92, 280].
[414.] McKillop, A.; Swann, B. P.; Taylor, E. C.   *J. Am. Chem. Soc.* **1971**, *93*, 4919 [17, 206].
[415.] Trahanovsky, W. S.; Young, L. B.   *J. Org. Chem.* **1966**, *31*, 2033 [17, 100].
[416.] Baciocchi, E.; Mandolini, L.; Rol, C.   *J. Org. Chem.* **1980**, *45*, 3906 [17, 32].
[417.] Syper, L.   *Tetrahedron Lett.* **1966**, 4493 [17, 101, 103, 104].
[418.] Trahanovsky, W. S.; Young, L. B.   *J. Chem. Soc.* **1965**, 5777 [17].
[419.] Young, L. B.; Trahanovsky, W. S.   *J. Org. Chem.* **1967**, *32*, 2349 [17].
[420.] Trahanovsky, W. S.; Young, L. B.; Brown, G. L.   *J. Org. Chem.* **1967**, *32*, 3865 [17, 115, 124].
[421.] Ho, T.-L.   *Synthesis* **1978**, 936 [17, 29, 115].
[422.] Ho, T.-L.   *Synthesis* **1973**, 347 [17, 101, 103, 104, 166, 190, 219].
[423.] Laatsch, H.   *Justus Liebigs Ann. Chem.* **1986**, *1655*, 1669 [17, 165].
[424.] Trahanovsky, W. S.; Young, L. H.; Bierman, M. H.   *J. Org. Chem.* **1969**, *34*, 869 [17, 159, 161].
[425.] Ho, T.-L.   *Synthesis* **1972**, 560 [17, 217].
[426.] Ho, T.-L.; Wong, C. M.   *Synthesis* **1972**, 561 [17, 255, 281].
[427.] Bird, J. W.; Diaper, D. G. M.   *Can. J. Chem.* **1969**, *47*, 145 [17, 219].
[428.] Ho, T.-L.; Ho, H. C.; Wong, C. M.   *Synthesis* **1972**, 562 [17, 230].
[429.] Periasamy, M.; Bhatt, M. V.   *Synthesis* **1977**, 330 [17, 94, 95, 96].

[*430.*] deJonge, C. R. H. I.; van Dort, H. M.; Vollbracht, L.   *Tetrahedron Lett.* **1970,** 1881 [17, 164, 165].
[*431.*] Mitchell, P. W. D.   *Can. J. Chem.* **1963,** *41,* 550 [17, 166].
[*432.*] Todd, D.; Martell, A. E.   *Org. Synth., Collective Volume* **1973,** *5,* 617 [17].
[*433.*] Vargha, L.; Remenyi, M.   *J. Chem. Soc.* **1951,** 1068 [17, 159, 160].
[*434.*] Cason, J.   *Org. Synth., Collective Volume* **1955,** *3,* 3 [17, 18, 100, 101].
[*435.*] Partch, R. E.   *J. Am. Chem. Soc.* **1967,** *89,* 3662 [17, 18, 93, 94, 100, 101].
[*436.*] Criegee, R.   *Justus Liebigs Ann. Chem.* **1930,** *481,* 263 [18, 74].
[*437.*] Cavill, G. W. K.; Solomon, D. H.   *J. Chem. Soc.* **1954,** 3943 [18, 100].
[*438.*] Cavill, G. W. K.; Solomon, D. H.   *J. Chem. Soc.* **1955,** 4426 [18, 199].
[*439.*] Cocker, J. D.; Henbest, H. B.; Phillips, G. H.; Slater, G. P.; Thomas, D. A.   *J. Chem. Soc.* **1965,** 6 [18, 199].
[*440.*] Brown, B. R.; Todd, A. R.   *J. Chem. Soc.* **1954,** 1280 [18, 167].
[*441.*] Adams, R.; Nagarkatti, A. S.   *J. Am. Chem. Soc.* **1950,** *72,* 4601 [18, 247, 248].
[*442.*] Partch, R. E.   *Tetrahedron Lett.* **1964,** 3071 [18, 115, 124, 126, 133, 149].
[*443.*] Stojiljković, A.; Andrejević, V.; Mihailović, M. L.   *Tetrahedron* **1967,** *23,* 721 [18, 241].
[*444.*] Mihailović, M. L.; Stojiljković, A.; Andrejević, V.   *Tetrahedron Lett.* **1965,** 461 [18, 241].
[*445.*] Middleton, W. J.; Gale, D. M.   *Org. Synth., Collective Volume* **1988,** *6,* 161 [18, 220].
[*446.*] Criegee, R.   *Chem. Ber.* **1931,** *64,* 260 [18, 161].
[*447.*] Grob, C. A.; Schiess, P. W.   *Helv. Chim. Acta* **1960,** *43,* 1546 [18, 151].
[*448.*] Wolf, F. J.; Weijlard, J.   *Org. Synth., Collective Volume* **1963,** *4,* 124 [18, 159, 160].
[*449.*] Campbell, J. R.; Kalman, J. R.; Pinhey, J. T.; Sternhell, S.   *Tetrahedron Lett.* **1972,** 1763 [18, 93, 94].
[*450.*] Dox, A. W.   *Org. Synth., Collective Volume* **1932,** *1,* 266 [18, 223].
[*451.*] Kremers, E.; Wakeman, N.; Hixon, R. M.   *Org. Synth., Collective Volume* **1932,** *1,* 511 [18, 246].
[*452.*] Dickinson, J. D.; Eaborn, C.   *J. Chem. Soc.* **1959,** 2337 [18, 103, 104].
[*453.*] Park, J. R.; Partington, J. R.   *J. Chem. Soc.* **1924,** *125,* 72 [19].
[*454.*] Field, B. O.; Grundy, J.   *J. Chem. Soc.* **1955,** 1110 [19, 115].
[*455.*] Severson, W. A.; Brice, T. J.   *J. Am. Chem. Soc.* **1958,** *80,* 2313 [19, 113].
[*456.*] Grundy, J.   *J. Chem. Soc.* **1957,** 5087 [19, 133, 148, 149].
[*457.*] Brook, A. G.   *J. Chem. Soc.* **1952,** 5040 [19, 166].
[*458.*] Singer, A.; McElvain, S. M.   *Org. Synth., Collective Volume* **1943,** *2,* 214 [19, 52, 241].
[*459.*] Sucharda, E.   *Chem. Ber.* **1925,** *58,* 1727 [19, 98].
[*460.*] Valkanas, G.; Hopff, H.   *J. Chem. Soc.* **1963,** 3475 [19, 105, 106].
[*461.*] Hopff, H.; Valkanas, G.   *J. Org. Chem.* **1962,** *27,* 2923 [19, 105, 106].
[*462.*] Bengtsson, E. B.   *Acta Chem. Scand.* **1953,** *7,* 774; **1954,** *8,* 842; **1955,** *9,* 832 [19, 105, 106, 108, 109].
[*463.*] Zaugg, H. E.; Rapala, R. T.   *Org. Synth., Collective Volume* **1955,** *3,* 820 [19, 105, 106].
[*464.*] Tuley, W. F.; Marvel, C. S.   *Org. Synth., Collective Volume* **1955,** *3,* 822 [19, 105, 106].
[*465.*] Degering, E. F.; Boatright, L. G.   *J. Am. Chem. Soc.* **1950,** *72,* 5137 [19, 127].

[466.] Coates, R. R.; Cook, J. W.   *J. Chem. Soc.* **1942,** 559 [19, 50].
[467.] Kent, W. H.; Tollens, B.   *Justus Liebigs Ann. Chem.* **1885,** *227,* 221 [19, 128, 184].
[468.] Ellis, B. A.   *Org. Synth., Collective Volume* **1932,** *1,* 18 [19, 150].
[469.] Powell, S. G.; Huntress, E. H.; Hershberg, E. B.   *Org. Synth., Collective Volume* **1932,** *1,* 168 [19, 127].
[470.] Moureu, C.; Chaux, R.   *Org. Synth., Collective Volume* **1932,** *1,* 166 [19, 176].
[471.] Boedtker, E.   *J. Pharm Chim.* **1932,** *15,* 225 [19, 211].
[472.] Dean, F. H.; Amin, J. H.; Pattison, F. L. M.   *Org. Synth., Collective Volume* **1973,** *5,* 580 [19, 224].
[473.] Hudlický, M.; Bell, H. M.   *J. Fluorine Chem.* **1974,** *4,* 149 [19].
[474.] Bauer, L.; Gardella, L. A., Jr.   *J. Org. Chem.* **1961,** *26,* 82 [19, 252].
[475.] Goldstein, H.; Jaunin, R.   *Helv. Chim. Acta* **1951,** *34,* 2222 [19, 266].
[476.] Weiss, M.; Appel, M.   *J. Am. Chem. Soc.* **1948,** *70,* 3666 [19, 217, 282].
[477.] Hudlický, M.   *Collect. Czech. Chem. Commun.* **1960,** *25,* 1199 [19, 101, 111, 282].
[478.] Balandin, A. A.; Zelinskii, N. D.; Marukyan, G. M.; Bogdanova, O. K.   *J. Appl. Chem. USSR* **1941,** *14,* 161 (French 172); *Chem. Abstr.* **1942,** *36,* 417 [19, 49].
[479.] Downs, C. R.   *J. Soc. Chem. Ind.* **1926,** *45,* 188T [19, 97].
[480.] Morgan, M. S.; Tipson, R. S.   *J. Am. Chem. Soc.* **1946,** *68,* 1569 [19, 52].
[481.] Rigby, W.   *J. Chem. Soc.* **1951,** 793 [19, 217].
[482.] Uskoković, M.; Gut, M.; Trachtenberg, E. N.; Klyne, W.; Dorfman, R. I.   *J. Am. Chem. Soc.* **1960,** *82,* 4965 [19, 22, 39, 159, 162].
[483.] Rigby, W.   *J. Chem. Soc.* **1950,** 1907 [19, 159, 161].
[484.] Heiduschka, A.; Ripper, J.   *Chem. Ber.* **1923,** *56,* 1736 [20, 127].
[485.] Wohl, A.   *Chem. Ber.* **1899,** *32,* 3486 [20, 92].
[486.] Tschitschibabin, A. E.   *Chem. Ber.* **1923,** *56,* 1879 [20, 92].
[487.] Saá, J. M.; Morey, J.; Costa, A.   *Tetrahedron Lett.* **1986,** *27,* 5125 [20, 164].
[488.] Teuber, H.-J.   *Org. Synth., Collective Volume* **1988,** *6,* 480 [20, 164].
[489.] Wehrli, P. A.; Pigott, F.   *Org. Synth., Collective Volume* **1988,** *6,* 1010 [20, 164].
[490.] Teuber, H.-J.; Hasselbach, M.   *Chem. Ber.* **1959,** *92,* 674 [20, 246, 292].
[491.] Hershberg, E. B.; Fieser, L. F.   *Org. Synth., Collective Volume* **1943,** *2,* 423 [20, 50].
[492.] Weiss, R.   *Org. Synth., Collective Volume* **1955,** *3,* 729 [20, 50, 51].
[493.] Prelog, V.; Komzak, A.; Moor, K.   *Helv. Chim. Acta* **1942,** *25,* 1654 [20, 50, 51, 52].
[494.] Newman, M. S.; Bye, T. S.   *J. Am. Chem. Soc.* **1952,** *74,* 905 [20, 50, 51].
[495.] Ruzicka, L.; Waldmann, H.   *Helv. Chim. Acta* **1933,** *16,* 842 [20, 38, 50, 51].
[496.] Linstead, R. P.; Michaelis, K. O. A.   *J. Chem. Soc.* **1940,** 1134 [20, 38, 50, 51].
[497.] Marvel, C. S.; Brooks, L. A.   *J. Am. Chem. Soc.* **1941,** *63,* 2630 [20, 38, 50, 51].
[498.] Carmack, M.; Spielman, M. A.   *Org. React. (N.Y.)* **1946,** *3,* 83 [20].
[499.] Fieser, L. F.; Kilmer, G. W.   *J. Am. Chem. Soc.* **1940,** *62,* 1354 [20, 203, 204, 205].
[500.] King, J. A.; McMillan, F. H.   *J. Am. Chem. Soc.* **1946,** *68,* 525, 1369 [20].

[501.] Hartmann, M.; Bosshard, W.   *Helv. Chim. Acta* **1941,** *24E,* 28 [20, 203, 204, 205].
[502.] Carmack, M.; DeTar, D. F.   *J. Am. Chem. Soc.* **1946,** *68,* 2029 [20, 203].
[503.] Malan, R. L.; Dean, P. M.   *J. Am. Chem. Soc.* **1947,** *69,* 1797 [20, 35, 105, 108, 109, 203, 205].
[504.] Newman, M. S.   *J. Org. Chem.* **1944,** *9,* 518 [20, 203, 204, 282].
[505.] Campaigne, E.; Budde, W. M.; Schaeffer, G. F.   *Org. Synth., Collective Volume* **1963,** *4,* 31 [20, 101, 102].
[506.] Traynelis, V. J.; Yoshikawa, Y.; Tarka, S. M.; Livingston, J. R., Jr.   *J. Org. Chem.* **1973,** *38,* 3986 [20, 255, 256, 257, 260].
[507.] Ronzio, A. R.; Waugh, T. D.   *Org. Synth., Collective Volume* **1955,** *3,* 438 [20, 21].
[508.] Hach, C. C.; Banks, C. V.; Diehl, H.   *Org. Synth., Collective Volume* **1963,** *4,* 229 [20, 21, 200].
[509.] Postowsky, J. J.; Lugowkin, B. P.   *Chem. Ber.* **1935,** *68,* 852 [20, 76, 91, 92, 95, 103, 104].
[510.] Riley, H. L.; Friend, N. A. C.   *J. Chem. Soc.* **1932,** 2342 [20].
[511.] Kariyone, K.; Yazawa, H.   *Tetrahedron Lett.* **1970,** 2885 [20, 172].
[512.] Coxon, J. M.; Dansted, E.; Hartshorn, M. P.   *Org. Synth., Collective Volume* **1988,** *6,* 946 [20, 86].
[513.] Kaplan, H.   *J. Am. Chem. Soc.* **1941,** *63,* 2654 [20, 101, 102].
[514.] Glenn, R. A.; Bailey, J. R.   *J. Am. Chem. Soc.* **1941,** *63,* 639 [20, 24, 101, 102].
[515.] Burgher, A.; Modlin, L. R., Jr.   *J. Am. Chem. Soc.* **1940,** *62,* 1079 [20, 101, 102].
[516.] Gray, A. R.; Fuson, R. C.   *J. Am. Chem. Soc.* **1934,** *56,* 739 [21, 200, 283].
[517.] Bockstahler, E. R.; Wright, D. L.   *J. Am. Chem. Soc.* **1949,** *71,* 3760 [21, 199, 200].
[518.] Corey, E. J.; Schaefer, J. P.   *J. Am. Chem. Soc.* **1960,** *82,* 918 [21, 199, 200].
[519.] Riley, H. L.; Morley, J. F.; Friend, N. A. C.   *J. Chem. Soc.* **1932,** 1875 [21].
[520.] Riley, H. A.; Gray, A. R.   *Org. Synth., Collective Volume* **1943,** *2,* 509 [21, 200].
[521.] Drabowicz, J.; Lyzwa, P.; Mikolajczyk, M.   *Phosphorus Sulfur* **1983,** *17,* 169 [21, 260, 261].
[522.] Watanabe, H.   *Chem. Pharm. Bull.* **1957,** *5,* 426 [21].
[523.] Jacques, J.; Ourisson, G.; Sandris, C.   *Bull. Soc. Chim. Fr.* **1955,** 1293 [21, 199].
[524.] Heller, M.; Bernstein, S.   *J. Org. Chem.* **1961,** *26,* 3876 [21, 199].
[525.] Barton, D. H. R.; Brewster, A. G.; Hui, R. A. H. F.; Lester, D. J.; Ley, S. V.; Back, T. G.   *Chem. Commun.* **1978,** 952 [21, 122, 123, 149].
[526.] Barton, D. H. R.; Morzycki, J. W.; Motherwell, W. B.   *Chem. Commun.* **1981,** 1044 [21, 32, 48].
[527.] Barton, D. H. R.; Lester, D. J.; Ley, S. V.   *Chem. Commun.* **1977,** 445 [21, 219].
[528.] Olah, G. A.; Welch, J.; Prakash, G. K. S.; Ho, T.-L.   *Synthesis* **1976,** 808 [21, 219, 220].
[529.] Olah, G. A.; Welch, J.   *Synthesis* **1976,** 809 [21, 219, 220].
[530.] Guziec, F. S., Jr.; Luzzio, F. A.   *J. Org. Chem.* **1982,** *47,* 1787 [21, 23, 118, 124, 125].

[531.] Vedejs, E.; Engler, D. A.; Telschow, J. E.   *J. Org. Chem.* **1978,** *43*, 188 [21, 196, 223].
[532.] Vedejs, E.; Larsen, S.   *Org. Synth.* **1985,** *64*, 127 [21].
[533.] Vedejs, E.   *J. Am. Chem. Soc.* **1974,** *96*, 5944 [21].
[534.] Ballistreri, F. P.; Failla, S.; Tomaselli, G. A.; Curci, R.   *Tetrahedron Lett.* **1986,** *27*, 5139 [21, 92].
[535.] Schlecht, M. F.; Kim, H.-j.   *Tetrahedron Lett.* **1986,** *27*, 4889 [21, 154].
[536.] Brown, H. C.; Garg, C. P.; Liu, K.-T.   *J. Org. Chem.* **1971,** *36*, 387 [21, 22, 136, 148].
[537.] Santaniello, E.; Ponti, F.; Manzocchi, A.   *Synthesis* **1978,** 534 [21, 22, 117, 283].
[538.] Flatt, S. J.; Fleet, G. W. J.; Taylor, B. J.   *Synthesis* **1979,** 815 [21, 22, 117, 125, 135, 136, 147, 149].
[539.] Harrison, I. T.; Harrison, S.   *Chem. Commun.* **1966,** 752 [21, 22, 169, 170].
[540.] Cardillo, G.; Orena, M.; Sandri, S.   *Tetrahedron Lett.* **1976,** 3985 [22, 110, 111].
[541.] Edwards, D.; Stenlake, J. B.   *J. Chem. Soc.* **1954,** 3272 [22, 33, 253].
[542.] Snatzke, G.   *Chem. Ber.* **1961,** *94*, 729 [22, 135, 136].
[543.] Cardillo, G.; Orena, M.; Sandri, S.   *Synthesis* **1976,** 394 [22, 117, 124, 125, 126, 135, 147, 148, 149].
[544.] Roček, J.; Riehl, Sr. A.   *J. Org. Chem.* **1967,** *32*, 3569 [22].
[545.] Bingham, R. C.; Schleyer, v. P. R.   *J. Org. Chem.* **1971,** *36*, 1198 [22, 59].
[546.] Landa, S.; Vais, J.; Burkhard, J.   *Z. Chem.* **1967,** *7*, 233 [22, 59].
[547.] Lieberman, S. V.; Connor, R.   *Org. Synth., Collective Volume* **1943,** *2*, 441 [22, 103].
[548.] Tsang, S. M.; Wood, E. H.; Johnson, J. R.   *Org. Synth., Collective Volume* **1955,** *3*, 641 [22, 103].
[549.] Nishimura, T.   *Org. Synth., Collective Volume* **1963,** *4*, 713 [22, 103].
[550.] Wertheim, E.   *J. Am. Chem. Soc.* **1933,** *55*, 2540 [22, 105, 107].
[551.] Goldstein, H.; Voegeli, R.   *Helv. Chim. Acta* **1942,** *25*, 475 [22, 105, 106].
[552.] Rosenthal, D.; Grabowich, P.; Sabo, E. F.; Fried, J.   *J. Am. Chem. Soc.* **1963,** *85*, 3971 [22, 86, 87].
[553.] Nakayama, M.; Shinke, S.; Matsushita, Y.; Ohira, S.; Hayashi, S.   *Bull. Chem. Soc. Jpn.* **1979,** *52*, 184 [22, 86, 226].
[554.] Speer, J. H.; Dabovich, T. C.   *Org. Synth., Collective Volume* **1955,** *3*, 39 [22, 200].
[555.] Holmgren, A. V.; Wenner, W.   *Org. Synth., Collective Volume* **1963,** *4*, 23 [22, 169, 200].
[556.] Bamberger, E.; Burgdorf, C.   *Chem. Ber.* **1890,** *23*, 2433 [22, 96].
[557.] Smith, L. I.; Webster, I. M.   *J. Am. Chem. Soc.* **1937,** *59*, 662 [22, 94, 95].
[558.] Fieser, L. F.; Campbell, W. P.; Fry, E. M.; Gates, M. D., Jr.   *J. Am. Chem. Soc.* **1939,** *61*, 3216 [22].
[559.] Smith, L. I.; Byers, D. J.   *J. Am. Chem. Soc.* **1941,** *63*, 612 [22, 164].
[560.] Pacaud, R. A.; Allen, C. F. H.   *Org. Synth., Collective Volume* **1943,** *2*, 336 [22, 112].
[561.] Holloway, F.; Cohen, M.; Westheimer, F. H.   *J. Am. Chem. Soc.* **1951,** *51*, 65 [22, 134].
[562.] Awasthy, A. K.; Roček, J.   *J. Am. Chem. Soc.* **1969,** *91*, 991 [22].
[563.] Roček, J.; Drozd, J. C.   *J. Am. Chem. Soc.* **1970,** *92*, 6668 [22].

[564.] Hickinbottom, W. J.; Moussa, G. E. M. *J. Chem. Soc.* **1957,** 4195 [22, 61, 62, 80].
[565.] Rogers, H. R.; McDermott, J. X.; Whitesides, G. M. *J. Org. Chem.* **1975,** *40,* 3577 [22, 75].
[566.] Moussa, G. E. M.; Basyouni, M. N.; Shaban, M. E.; Youssef, A. M. *J. Appl. Chem. Biotechnol.* **1978,** *28,* 875 [22].
[567.] Sarel, S.; Yanuka, Y. *J. Org. Chem.* **1959,** *24,* 2018 [22, 30, 82, 83].
[568.] Schlecht, M. F.; Kim, H.-j. *Tetrahedron Lett.* **1985,** *26,* 127 [22, 154].
[569.] Riegel, B.; Moffett, R. B.; McIntosh, A. V. *Org. Synth., Collective Volume* **1955,** *3,* 234 [22, 82, 83].
[570.] Hasan, F.; Roček, J. *J. Am. Chem. Soc.* **1975,** *97,* 1444 [22].
[571.] Cainelli, G.; Cardillo, G.; Orena, M.; Sandri, S. *J. Am. Chem. Soc.* **1976,** *98,* 6737 [22, 117, 123, 135, 136, 147, 148, 149].
[572.] Harding, K. E.; May, L. M.; Dick, K. F. *J. Org. Chem.* **1975,** *40,* 1664 [22, 117, 124, 125, 126].
[573.] Sauer, J. C. *Org. Synth., Collective Volume* **1963,** *4,* 813 [22, 117].
[574.] Pattison, F. L. M.; Stothers, J. B.; Woolford, R. G. *J. Am. Chem. Soc.* **1956,** *78,* 2255 [22, 128, 224].
[575.] Schubert, A.; Langbein, G.; Siebert, R. *Chem. Ber.* **1957,** *90,* 2576 [22, 45, 46, 197].
[576.] Price, C. C.; Karabinos, J. V. *J. Am. Chem. Soc.* **1940,** *62,* 1159 [22, 135].
[577.] Roček, J.; Radkowsky, A. E. *J. Am. Chem. Soc.* **1968,** *90,* 2986 [22, 134].
[578.] Bowden, K.; Heilbron, I. M.; Jones, E. R. H.; Weedon, B. C. L. *J. Chem. Soc.* **1946,** 39 [22, 135, 136, 283].
[579.] Djerassi, C.; Engle, R. R.; Bowers, A. *J. Org. Chem.* **1956,** *21,* 1547 [22, 136, 215].
[580.] Schaeffer, J. R.; Snoddy, A. D. *Org. Synth., Collective Volume* **1963,** *4,* 19 [22, 150].
[581.] Eisenbraun, E. J. *Org. Synth., Collective Volume* **1973,** *5,* 310 [22, 135].
[582.] Rasmusson, G. H.; House, H. O.; Zaweski, E. F.; DePuy, C. H. *Org. Synth., Collective Volume* **1973,** *5,* 324 [22, 155].
[583.] Kleinfelter, D. C.; Schleyer, v. P. R. *Org. Synth., Collective Volume* **1973,** *5,* 852 [22].
[584.] Meinwald, J.; Crandall, J.; Hymans, W. E. *Org. Synth., Collective Volume* **1973,** *5,* 866 [22].
[585.] Bowers, A.; Halsall, T. G.; Jones, E. R. H.; Lemin, A. J. *J. Chem. Soc.* **1953,** 2548 [22, 273].
[586.] Bal, B. S.; Kochhar, K. S.; Pinnick, H. W. *J. Org. Chem.* **1981,** *46,* 1492 [22, 169, 170].
[587.] Ip, D.; Roček, J. *J. Org. Chem.* **1979,** *44,* 312 [22].
[588.] Tavares, D. F.; Borger, J. P. *Can. J. Chem.* **1966,** *44,* 1323 [22, 159].
[589.] Buu-Hoi, N. H.; Lecocq, J. J. *Bull Soc. Chim. Fr.* **1946,** 139 [22, 253].
[590.] Liu, H.-J.; Han, I.-S. *Synth. Commun.* **1985,** *15,* 759 [22, 172].
[591.] Sisler, H. H.; Bush, J. D.; Accountius, O. E. *J. Am. Chem. Soc.* **1948,** *70,* 3827 [22].
[592.] Poos, G. I.; Arth, G. E.; Beyler, R. E.; Sarett, L. H. *J. Am. Chem. Soc.* **1953,** *75,* 422 [22, 137].
[593.] Dauben, W. G.; Lorber, M.; Fullerton, D. S. *J. Org. Chem.* **1969,** *34,* 3587 [22, 23, 86, 87, 274].

References

[594.] Collins, J. C.; Hess, W. W. *Org. Synth., Collective Volume* **1988**, *6*, 644 [22, 117].
[595.] Collins, J. C.; Hess, W. W.; Frank, F. J. *Tetrahedron Lett.* **1968**, 3363 [22, 117, 118, 125, 137, 147, 148, 149].
[596.] Ratcliffe, R.; Rodehorst, R. *J. Org. Chem.* **1970**, *35*, 4000 [22, 117, 123, 125, 126].
[597.] Singh, R. P.; Subbarao, H. N.; Dev, S. *Tetrahedron* **1979**, *35*, 1789 [22, 23, 117, 124, 125, 126, 137, 148].
[598.] Sarel, S.; Newman, M. S. *J. Am. Chem. Soc.* **1956**, *78*, 5416 [22, 283].
[599.] Holum, J. R. *J. Org. Chem.* **1961**, *26*, 4814 [22, 23, 117, 124, 126, 137, 147, 148, 149].
[600.] Ratcliffe, R. W. *Org. Synth., Collective Volume* **1988**, *6*, 373 [22, 117].
[601.] Shaw, J. E.; Sherry, J. J. *Tetrahedron Lett.* **1971**, 4379 [23, 91].
[602.] Corey, E. J.; Fleet, G. W. J. *Tetrahedron Lett.* **1973**, 4499 [23, 118, 125, 126, 137, 138, 147, 148, 149, 175, 284].
[603.] Corey, E. J.; Schmidt, G. *Tetrahedron Lett.* **1979**, 399 [23, 25, 116, 118, 125, 128, 137].
[604.] Cheng, Y.-S.; Liu, W.-L.; Chen, S.-H. *Synthesis* **1980**, 223 [23, 118, 126, 138, 274, 284].
[605.] Corey, E. J.; Suggs, J. W. *Tetrahedron Lett.* **1975**, 2647 [23, 118, 138, 149, 284].
[606.] Parish, E. J.; Chitrakorn, S.; Wei, T.-Y. *Synth. Commun.* **1986**, *16*, 1371 [23, 86, 87].
[607.] Rathore, R.; Saxena, N.; Chandrasekaran, S. *Synth. Commun.* **1986**, *16*, 1493 [23, 103, 104].
[608.] Piancatelli, G.; Scettri, A.; D'Auria, M. *Tetrahedron Lett.* **1977**, 3483 [23, 171].
[609.] Rollin, P.; Sinaÿ, P. *Carbohydr. Res.* **1981**, *98*, 139 [23, 171].
[610.] Willis, J. P.; Gogins, K. A. Z.; Miller, L. L. *J. Org. Chem.* **1981**, *46*, 3215 [23, 29, 173].
[611.] Brown, H. C.; Kulkarni, S. U.; Rao, C. G. *Synthesis* **1980**, 151 [23, 268, 269].
[612.] Fréchet, J. M. J.; Warnock, J.; Farrall, M. J. *J. Org. Chem.* **1978**, *43*, 2618 [23, 118, 138, 147, 148].
[613.] Langheld, K. *Chem. Ber.* **1909**, *42*, 392 [23, 268].
[614.] Cardillo, G.; Orena, M.; Sandri, S. *Chem. Commun.* **1976**, 190 [23, 110, 111].
[615.] den Hollander, A. J. *Rec. Trav. Chim. Pays-Bas* **1920**, *39*, 481 [23, 166].
[616.] Cardillo, G.; Shimizu, M. *J. Org. Chem.* **1977**, *42*, 4268 [23, 76].
[617.] Brunelet, T.; Jouitteau, C.; Gelbard, G. *J. Org. Chem.* **1986**, *51*, 4016 [24, 117, 123, 135, 136, 147, 148].
[618.] Cacchi, S. *Synthesis* **1979**, 356 [24, 116, 125, 126, 274, 285].
[619.] Santaniello, E.; Milani, F.; Casati, R. *Synthesis* **1983**, 749 [24, 118, 126, 138, 149, 250, 251].
[620.] Wiberg, K. B.; Foster, G. *J. Am. Chem. Soc.* **1961**, *83*, 423 [24, 58].
[621.] Hasan, F.; Roček, J. *J. Am. Chem. Soc.* **1974**, *96*, 534 [24, 134, 136].
[622.] Huntress, E. H.; Hershberg, E. B.; Cliff, I. S. *J. Am. Chem. Soc.* **1931**, *53*, 2720 [24, 103].
[623.] Gilman, H.; Eisch, J. *J. Am. Chem. Soc.* **1957**, *79*, 4423 [24].
[624.] Nightingale, D. V.; Wagner, W. S. *J. Org. Chem.* **1960**, *25*, 32 [24, 105, 107, 108].

[625.] Allen, C. F. H.; Van Allan, J. A. *Org. Synth., Collective Volume* **1955**, *3*, 1 [24, 103, 105].
[626.] Rieveschl, G., Jr.; Ray, F. E. *Org. Synth., Collective Volume* **1955**, *3*, 420 [24, 210].
[627.] Fieser, L. F. *Org. Synth., Collective Volume* **1963**, *4*, 189 [24, 136].
[628.] Fieser, L. F. *Org. Synth., Collective Volume* **1963**, *4*, 195 [24, 136].
[629.] Rao, Y. S.; Filler, R. *J. Org. Chem.* **1974**, *39*, 3305 [24, 136, 148].
[630.] Pletcher, D.; Tait, S. J. D. *J. Chem. Soc., Perkin Trans. 2* **1979**, 788 [24, 95, 123, 125, 137, 148].
[631.] Vingiello, F. A.; Van Oot, J. G. *J. Am. Chem. Soc.* **1951**, *73*, 5070 [24].
[632.] Reitsema, R. H.; Allphin, N. L. *J. Org. Chem.* **1962**, *27*, 27 [24, 107].
[633.] Friedman, L.; Fishel, D. L.; Shechter, H. *J. Org. Chem.* **1965**, *30*, 1453 [24, 105, 107].
[634.] Kamm, O.; Matthews, A. O. *Org. Synth., Collective Volume* **1932**, *1*, 392 [24, 105, 106].
[635.] Clarke, H. T.; Hartman, W. W. *Org. Synth., Collective Volume* **1932**, *1*, 543 [24, 29].
[636.] Viet, E. B. *Org. Synth., Collective Volume* **1932**, *1*, 482 [24, 166].
[637.] Harman, R. E. *Org. Synth., Collective Volume* **1963**, *4*, 148 [24, 246].
[638.] Wächter, R. *Angew. Chem.* **1955**, *67*, 305 [24, 96].
[639.] Lee, D. G.; Spitzer, U. A. *J. Org. Chem.* **1970**, *35*, 3589 [24, 116, 125, 126].
[640.] Robertson, G. R. *Org. Synth., Collective Volume* **1932**, *1*, 138 [24, 131].
[641.] Brown, H. C.; Garg, C. P. *J. Am. Chem. Soc.* **1961**, *83*, 2952 [24, 136, 148].
[642.] Hussey, A. S.; Baker, R. H. *J. Org. Chem.* **1960**, *25*, 1434 [24, 136, 285].
[643.] Conant, J. B.; Quayle, O. R. *Org. Synth., Collective Volume* **1932**, *1*, 211 [24, 136].
[644.] Sandborn, L. T. *Org. Synth., Collective Volume* **1932**, *1*, 340 [24, 136].
[645.] Coleman, G. H.; McCloskey, C. M.; Stuart, F. A. *Org. Synth., Collective Volume* **1955**, *3*, 668 [24, 231, 232].
[646.] Brown, H. C.; Garg, C. P. *J. Am. Chem. Soc.* **1961**, *83*, 2951 [24, 268, 269].
[647.] Schniter, K. *Chem. Ber.* **1887**, *20*, 2282 [24, 94, 246].
[648.] Kehrmann, F.; Stiller, T. E. *Chem. Ber.* **1912**, *45*, 3346 [24, 246].
[649.] Linstead, R. P.; Levine, P. *J. Am. Chem. Soc.* **1942**, *64*, 2022 [24, 96].
[650.] Wendband, R.; LaLonde, J. *Org. Synth., Collective Volume* **1963**, *4*, 713 [24, 96].
[651.] Hutchins, R. O.; Natale, N. R.; Cook, W. J. *Tetrahedron Lett.* **1977**, 4167 [24, 116, 137, 149].
[652.] Pletcher, D.; Tait, S. J. D. *Tetrahedron Lett.* **1978**, 1601 [24].
[653.] Hurd, C. D.; Meinert, R. N. *Org. Synth., Collective Volume* **1943**, *2*, 541 [24, 116].
[654.] Rahman, M.; Roček, J. *J. Am. Chem. Soc.* **1971**, *93*, 5455 [24, 134].
[655.] Roček, J.; Radkowsky, A. E. *J. Am. Chem. Soc.* **1973**, *95*, 7123 [24, 134, 136].
[656.] Hurd, C. D.; Garrett, J. W.; Osborne, E. N. *J. Am. Chem. Soc.* **1933**, *55*, 1082 [24, 174, 176].
[657.] Bamberger, E. *Chem. Ber.* **1894**, *27*, 1548 [24, 231, 232].
[658.] Langley, W. D. *Org. Synth., Collective Volume* **1955**, *3*, 334 [24, 231].
[659.] Firouzabadi, H.; Sardarian, A.; Charibi, H. *Synth. Commun.* **1984**, *14*, 89 [25, 116].

[660.] Firouzabadi, H.; Sardarian, A. R.; Moosavipour, H.; Afshari, G. M. *Synthesis* **1986,** 285 [25, 91, 92, 95, 128, 130, 149, 169, 219].
[661.] Corey, E. J.; Schmidt, G. *Tetrahedron Lett.* **1980,** *21,* 731 [25].
[662.] Salmon, J. *Chem. Br.* **1982,** 703 [25].
[663.] López, C.; González, A.; Cossió, F. P.; Palomo, C. *Synth. Commun.* **1985,** *15,* 1197 [25, 95, 116, 126, 250, 251].
[664.] Singh, J.; Kalsi, P. S.; Jawanda, G. S.; Chhabra, B. R. *Chem. Ind. (London)* **1986,** 751 [26].
[665.] Sisler, H. H. *Inorg. Synth.* **1946,** *2,* 205 [26].
[666.] Wiberg, K. B.; Marshall, B.; Foster, G. *Tetrahedron Lett.* **1962,** 345 [26].
[667.] Richter, v. V. *Chem. Ber.* **1886,** *19,* 1060 [26, 101, 102].
[668.] Freeman, F.; DuBois, R. H.; Yamachika, N. J. *Tetrahedron* **1969,** *25,* 3441 [26].
[669.] Sharpless, K. B.; Teranishi, A. Y.; Bäckvall, J.-E. *J. Am. Chem. Soc.* **1977,** *99,* 3120 [26].
[670.] Sharpless, K. B.; Teranishi, A. Y. *J. Org. Chem.* **1973,** *38,* 185 [26, 76].
[671.] Freeman, F.; Cameron, P. J.; DuBois, R. H. *J. Org. Chem.* **1968,** *33,* 3970 [26].
[672.] Freeman, F.; DuBois, R. H.; McLaughlin, T. G. *Org. Synth., Collective Volume* **1988,** *6,* 1028 [26, 76].
[673.] Miller, v. W.; Rohde, G. *Chem. Ber.* **1890,** *23,* 1070 [26, 105].
[674.] Sharpless, K. B.; Akashi, K. *J. Am. Chem. Soc.* **1975,** *97,* 5927 [26, 119, 124, 125, 126].
[675.] San Filippo, J., Jr.; Chern, C.-I. *J. Org. Chem.* **1977,** *42,* 2182 [26, 119, 123, 125, 147, 149].
[676.] Lee, T. V.; Toczek, J. *Tetrahedron Lett.* **1982,** *23,* 2917 [26, 172, 173, 196].
[677.] Suga, T.; Kihara, K.; Matsuura, T. *Bull. Chem. Soc. Jpn.* **1965,** *38,* 893 [26, 119, 128, 131].
[678.] Suga, T.; Kihara, K.; Matsuura, T. *Bull. Chem. Soc. Jpn.* **1965,** *38,* 1141 [26, 119, 125, 200].
[679.] Miyaura, N.; Kochi, J. K. *J. Am. Chem. Soc.* **1983,** *105,* 2368 [26, 61, 62, 63].
[680.] Schildknecht, H.; Föttinger, W. *Justus Liebigs Ann. Chem.* **1962,** *659,* 20 [26, 79].
[681.] Wicha, J.; Zarecki, A. *Tetrahedron Lett.* **1974,** 3059 [27, 138, 156].
[682.] Neirabeyeh, M. A.; Ziegler, J.-C.; Gross, B. *Synthesis* **1976,** 811 [27, 28, 138, 139, 147, 148, 149].
[683.] Douglass, I. B.; Johnson, T. B. *J. Am. Chem. Soc.* **1938,** *60,* 1486 [27, 252].
[684.] Johnson, T. B.; Douglass, I. B. *J. Am. Chem. Soc.* **1939,** *61,* 2548 [27, 252].
[685.] Wertheim, E. *Org. Synth., Collective Volume* **1943,** *2,* 471 [27, 264].
[686.] Ullmann, F.; Lehner, A. *Chem. Ber.* **1905,** *38,* 729 [27, 264].
[687.] Douglass, I. B.; Norton, R. V. *Org. Synth., Collective Volume* **1973,** *5,* 709 [27, 264].
[688.] Mowry, D. T.; Ringwald, E. L. *J. Am. Chem. Soc.* **1950,** *72,* 2037 [27, 206, 207].
[689.] Krishman, S.; Kuhn, D. G.; Hamilton, G. A. *J. Am. Chem. Soc.* **1977,** *99,* 8121 [27, 61, 62, 64].
[690.] Cacchi, S.; La Torre, F. *Chem. Ind. (London)* **1986,** 286 [27, 112].
[691.] Marmor, S. *J. Org. Chem.* **1963,** *28,* 250 [27, 212, 213, 285].

[692.] Lee, G. A.; Freedman, H. H.   *Tetrahedron Lett.* **1976,** 1641 [27, 125, 139, 149, 239, 240, 241].
[693.] Stevens, R. V.; Chapman, K. T.; Weller, H. N.   *J. Org. Chem.* **1980,** *45,* 2030 [27, 139, 147, 148].
[694.] Hartman, W. W.; Dickey, J. B.; Stampfli, J. G.   *Org. Synth., Collective Volume* **1943,** *2,* 175 [27, 247].
[695.] Buděšínský, Z.; Přikryl, J.; Svátek, E.   *Collect. Czech. Chem. Commun.* **1964,** *29,* 2980 [27, 254].
[696.] Neiswender, D. D., Jr.; Moniz, W. B.; Dixon, J. A.   *J. Am. Chem. Soc.* **1960,** *82,* 2876 [27, 105, 107, 108, 207, 208, 210].
[697.] White, T.   *J. Chem. Soc.* **1943,** 238 [27, 207].
[698.] Newman, M. S.; Holmes, H. L.   *Org. Synth., Collective Volume* **1943,** *2,* 428 [27, 206, 207].
[699.] Smith, W. T.; McLeod, G. L.   *Org. Synth., Collective Volume* **1963,** *4,* 345 [27, 209].
[700.] Langheld, K.   *Chem. Ber.* **1909,** *42,* 2360 [27, 229].
[701.] Wolfe, S.; Hasan, S. K.; Campbell, J. R.   *Chem. Commun.* **1970,** 1420 [27, 38, 96, 97, 142, 148].
[702.] Meyers, C. Y.   *J. Org. Chem.* **1961,** *26,* 1046 [27, 119, 126].
[703.] Smith, L. I.; Prichard, W. W.; Spillane, L. J.   *Org. Synth., Collective Volume* **1955,** *3,* 302 [27, 207].
[704.] Zincke, T.   *Chem. Ber.* **1892,** *25,* 3599 [27, 61, 212, 213].
[705.] Schneider, M.; Weber, J.-V.; Faller, P.   *J. Org. Chem.* **1982,** *47,* 364 [27, 139, 148].
[706.] Teeter, H. M.; Bell, E. W.   *Org. Synth., Collective Volume* **1963,** *4,* 125 [27].
[707.] Poisel, H.; Schmidt, U.   *Chem. Ber.* **1975,** *108,* 2547 [27, 49].
[708.] Fonken, G. S.; Thompson, J. L.; Levin, R. H.   *J. Am. Chem. Soc.* **1955,** *77,* 172 [27].
[709.] Grob, C. A.; Schmid, H. J.   *Helv. Chim. Acta* **1953,** *36,* 1763 [27, 28, 119, 125, 139, 144, 148].
[710.] Skattebøl, L.; Boulette, B.; Solomon, S.   *J. Org. Chem.* **1967,** *32,* 3111 [27, 254].
[711.] Kobayashi, M.; Ohkubo, H.; Shimizu, T.   *Bull. Chem. Soc. Jpn.* **1986,** *59,* 503 [27, 265].
[712.] Dalcanale, E.; Montanari, F.   *J. Org. Chem.* **1986,** *51,* 567 [28, 176].
[713.] Wuts, P. G. M.; Bergh, C. L.   *Tetrahedron Lett.* **1986,** *27,* 3995 [28, 43, 178].
[714.] Hofmann, K. A.; Ehrhart, O.; Schneider, O.   *Chem. Ber.* **1913,** *46,* 1657 [28, 69].
[715.] Milas, N. A.; Terry, E. M.   *J. Am. Chem. Soc.* **1925,** *47,* 1412 [28, 39, 69, 225].
[716.] Milas, N. A.   *J. Am. Chem. Soc.* **1927,** *49,* 2005 [28, 39].
[717.] Bassignani, L.; Brandt, A.; Caciagli, V.; Re, L.   *J. Org. Chem.* **1978,** *43,* 4245 [28, 91].
[718.] Milas, N. A.   *Org. Synth., Collective Volume* **1943,** *2,* 302 [28].
[719.] Underwood, H. M., Jr.; Walsh, W. L.   *Org. Synth., Collective Volume* **1943,** *2,* 553 [28, 166].
[720.] Corey, E. J.; Kim, C. U.   *J. Am. Chem. Soc.* **1972,** *94,* 7586 [28, 42].
[721.] Corey, E. J.; Kim, C. U.; Misco, P. F.   *Org. Synth., Collective Volume* **1988,** *6,* 220 [28, 144].
[722.] Juenge, E. C.; Beal, D. A.   *Tetrahedron Lett.* **1968,** 5819 [28].
[723.] Herranz, E.; Sharpless, K. B.   *Org. Synth.* **1983,** *61,* 85 [28, 74].

[724.] Billimoria, J. D.; Maclagan, N. F. *J. Chem. Soc.* **1954,** 3257 [28].
[725.] Williams, D. M. *J. Chem. Soc.* **1931,** 2783 [28].
[726.] Blair, L. K.; Baldwin, J.; Smith, W. C., Jr. *J. Org. Chem.* **1977,** *42,* 1816 [28, 139, 148].
[727.] Oae, S.; Ohnishi, Y.; Kozuka, S.; Tagaki, W. *Bull. Chem. Soc. Jpn.* **1966,** *39,* 364 [28, 29].
[728.] Smith, B.; Hernestam, S. *Acta Chem. Scand.* **1954,** *8,* 1111 [28, 29].
[729.] Wineburg, J. P.; Abrams, C.; Swern, D. *J. Heterocycl. Chem.* **1975,** *12,* 749 [28, 228].
[730.] Clarke, H. T. *Org. Synth., Collective Volume* **1955,** *3,* 226 [28, 29].
[731.] Markees, D. G. *J. Org. Chem.* **1958,** *23,* 1490 [28, 29, 170].
[732.] Zincke, T.; Walter, W. *Justus Liebigs Ann. Chem.* **1904,** *334,* 367 [28, 164].
[733.] Overberger, C. G.; Huang, P.-t.; Berenbaum, M. B. *Org. Synth., Collective Volume* **1963,** *4,* 66 [28, 233].
[734.] Tsuda, Y.; Matsuhira, N.; Kanemitsu, K. *Chem. Pharm. Bull.* **1985,** *33,* 4095 [28, 139].
[735.] Komppa, G. *Chem. Ber.* **1899,** *32,* 1421 [29, 209].
[736.] Johnson, W. S.; Gutsche, C. D.; Offenhauer, R. D. *J. Am. Chem. Soc.* **1946,** *68,* 1648 [29, 207, 208, 286].
[737.] Sandborn, L. T.; Bousquet, E. W. *Org. Synth., Collective Volume* **1932,** *1,* 526 [29, 206, 207].
[738.] Calder, A.; Forrester, A. R.; Hepburn, S. P. *Org. Synth., Collective Volume* **1988,** *6,* 803 [29, 35, 235].
[739.] Kageyama, T.; Okawara, M. *Synthesis* **1983,** 815 [29, 139, 148, 255].
[740.] Ballard, D. A.; Dehn, W. M. *Org. Synth., Collective Volume* **1932,** *1,* 89 [29, 217].
[741.] Kanemoto, S.; Tomioka, H.; Oshima, K.; Nozaki, H. *Bull. Chem. Soc. Jpn.* **1986,** *59,* 105 [29, 133, 139, 156].
[742.] Farkas, L.; Schächter, O. *J. Am. Chem. Soc.* **1949,** *71,* 2827 [29, 131].
[743.] Jones, R. E.; Kocher, F. W. *J. Am. Chem. Soc.* **1954,** *76,* 3682 [29, 156, 157, 286].
[744.] Pinnick, H. W.; Lajis, N. H. *J. Org. Chem.* **1978,** *43,* 371 [29, 172].
[745.] Barakat, M. Z.; El-Wahab, M. F. A. *J. Am. Chem. Soc.* **1953,** *75,* 5731 [29, 228].
[746.] Guss, C. O.; Rosenthal, R. *J. Am. Chem. Soc.* **1955,** *77,* 2549 [29, 61, 62, 63, 73, 74, 212, 225].
[747.] Theilacker, W.; Fauser, K. *Justus Liebigs Ann. Chem.* **1939,** *539,* 103 [29].
[748.] McCarty, C. G.; Leeper, C. G. *J. Org. Chem.* **1970,** *35,* 4245 [29].
[749.] Smith, C. W.; Norton, D. G.; Ballard, S. A. *J. Am. Chem. Soc.* **1951,** *73,* 5273 [29, 35, 182, 183].
[750.] Morgan, K. J.; Bardwell, J.; Cullis, C. F. *J. Chem. Soc.* **1950,** 3190 [29, 30].
[751.] Parrilli, M.; Barone, G.; Adinolfi, M.; Mangoni, L. *Tetrahedron Lett.* **1976,** 207 [29, 61, 62, 63].
[752.] Heer, J.; Miescher, K. *Helv. Chim. Acta* **1945,** *28,* 157 [30, 210].
[753.] Evans, T. W.; Dehn, W. M. *J. Am. Chem. Soc.* **1930,** *52,* 3647 [30].
[754.] Bruce, J. M. *J. Chem. Soc.* **1959,** 2366 [30, 167, 168].
[755.] Bal, B.; Buse, C. T.; Smith, K.; Heathcock, C. H. *Org. Synth.* **1984,** *63,* 89 [30, 172, 173].
[756.] Dimant, E.; Banay, M. *J. Org. Chem.* **1960,** *25,* 475 [30, 160, 161].
[757.] Fatiadi, A. J. *J. Org. Chem.* **1967,** *32,* 2903 [30, 53].

[758.] Fatiadi, A. J. *Chem. Commun.* **1967,** 1087 [30, 94, 95].
[759.] Chargaff, E.; Magasanik, B. *J. Am. Chem. Soc.* **1947,** *69,* 1459 [30, 160, 287].
[760.] Dolby, L. J.; Booth, D. L. *J. Am. Chem. Soc.* **1966,** *88,* 1049 [30, 79, 80, 101, 102, 103].
[761.] Nagarkatti, J. P.; Ashley, K. R. *Tetrahedron Lett.* **1973,** 4599 [30, 173, 174].
[762.] Wolfrom, M. L.; Yosizawa, Z. *J. Am. Chem. Soc.* **1959,** *81,* 3477 [30, 160].
[763.] Lemieux, R. U.; Rudloff, v. E. *Can. J. Chem.* **1955,** *33,* 1701 [30, 34, 35, 226, 227, 288].
[764.] Lemieux, R. U.; Rudloff, v. E. *Can. J. Chem.* **1955,** *33,* 1710 [30, 74].
[765.] Pappo, R.; Allen, D. S., Jr.; Lemieux, R. U.; Johnson, W. S. *J. Org. Chem.* **1956,** *21,* 478 [30, 39, 79].
[766.] Yamagishi, A. *Chem. Commun.* **1986,** 290 [30].
[767.] Leonard, N. J.; Johnson, C. R. *J. Am. Chem. Soc.* **1962,** *84,* 3701 [30, 257].
[768.] Hartzell, G. E.; Paige, J. N. *J. Am. Chem. Soc.* **1966,** *88,* 2616 [30, 255].
[769.] Cook, M. J.; Tonge, A. P. *J. Chem. Soc., Perkin Trans. 2* **1974,** 767 [30].
[770.] Leonard, N. J.; Johnson, C. R. *J. Org. Chem.* **1962,** *27,* 282 [30, 253, 254, 255, 287].
[771.] Liu, K.-T.; Tong, Y.-C. *J. Org. Chem.* **1978,** *43,* 2717 [30, 254, 287, 288].
[772.] Johnson, C. R.; Keiser, J. E. *Org. Synth., Collective Volume* **1973,** *5,* 791 [30, 254].
[773.] Cinquini, M.; Colonna, S.; Giovini, R. *Chem. Ind. (London)* **1969,** 1737 [30, 31, 265].
[774.] Carlsen, P. H. J.; Katsuki, T.; Martin, V. S.; Sharpless, K. B. *J. Org. Chem.* **1981,** *46,* 3936 [30, 38, 83, 96, 97, 169, 170].
[775.] Fuson, R. C.; Tan, T.-L. *J. Am. Chem. Soc.* **1948,** *70,* 602 [30, 158].
[776.] Santaniello, E.; Manzocchi, A.; Farachi, C. *Synthesis* **1980,** 563 [30, 31, 228, 254, 255].
[777.] Santaniello, E.; Ponti, F.; Manzocchi, A. *Tetrahedron Lett.* **1980,** *21,* 2655 [30, 224].
[778.] Ferraboschi, P.; Azadani, M. N.; Santaniello, E.; Trave, S. *Synth. Commun.* **1986,** *16,* 43 [30, 31, 111, 228, 254, 255].
[779.] Cambie, R. C.; Chambers, D.; Rutledge, P. S.; Woodgate, P. D. *J. Chem. Soc., Perkin Trans. 1* **1978,** 1483 [31, 159].
[780.] Beebe, T. R.; Barnes, B. A.; Bender, K. A.; Halbert, A. D.; Miller, R. D.; Ramsay, M. L.; Ridenour, M. W. *J. Org. Chem.* **1975,** *40,* 1992 [31, 151].
[781.] Woodward, R. B.; Brutcher, F. V., Jr. *J. Am. Chem. Soc.* **1958,** *80,* 209 [31, 70, 71].
[782.] Gunstone, F. D.; Morris, L. J. *J. Chem. Soc.* **1957,** 487 [31, 70, 71, 72, 286].
[783.] Ellington, P. S.; Hey, D. G.; Meakins, G. D. *J. Chem. Soc. C* **1966,** 1327 [31, 71].
[784.] Späth, E.; Kuffner, F.; Kesztler, F. *Chem. Ber.* **1936,** *69,* 378 [31, 70, 71].
[785.] Linskeseder, M.; Zbiral, E. *Justus Liebigs Ann. Chem.* **1978,** 1076 [31, 61, 62].
[786.] Müller, P.; Godoy, J. *Tetrahedron Lett.* **1982,** *23,* 3661 [31, 172, 244, 245].
[787.] Müller, P.; Godoy, J. *Helv. Chim. Acta* **1981,** *64,* 2531 [31, 91, 92].
[788.] Müller, P.; Godoy, J. *Tetrahedron Lett.* **1981,** *22,* 2361 [31, 177, 179].

[789.] Criegee, R.; Beucker, H. *Justus Liebigs Ann. Chem.* **1939,** *541,* 218 [31, 32, 159, 160].
[790.] Moriarty, R. M.; Hou, K.-C.; Prakash, I.; Arora, S. K. *Org. Synth.* **1985,** *64,* 138 [31, 196].
[791.] Imamoto, T.; Koto, H. *Chem. Lett.* **1986,** 967 [31].
[792.] Askenasy, P.; Meyer, V. *Chem. Ber.* **1893,** *26,* 1354 [31].
[793.] Pausacker, K. H. *J. Chem. Soc.* **1953,** 107 [32].
[794.] Moriarty, R. M.; Hu, H. *Tetrahedron Lett.* **1981,** *22,* 2747 [32, 223].
[795.] Castrillón, J. P. A.; Szmant, H. H. *J. Org. Chem.* **1967,** *32,* 976 [32, 256].
[796.] Schmeisser, M.; Dahmen, K.; Sartori, P. *Chem. Ber.* **1967,** *100,* 1633 [32].
[797.] Moriarty, R. M.; Penmasta, R.; Awasthi, A. K.; Prakash, I. *J. Org. Chem.* **1988,** *53,* 6124 [32, 91, 92].
[798.] Barton, D. H. R.; Godfrey, C. R. A.; Morzycki, J. W.; Motherwell, W. B.; Stobie, A. *Tetrahedron Lett.* **1982,** *23,* 957 [32, 159, 160].
[799.] Barton, D. H. R.; Crich, D. *Tetrahedron* **1985,** *41,* 4359 [32, 86].
[800.] Linderman, R. J.; Graves, D. M. *Tetrahedron Lett.* **1987,** *28,* 4259 [32].
[801.] Dess, D. B.; Martin, J. C. *J. Org. Chem.* **1983,** *48,* 4155 [32, 123, 148, 267].
[802.] Periasamy, M.; Bhatt, M. V. *Tetrahedron Lett.* **1978,** 4561 [32, 94, 95, 96].
[803.] Bush, J. B., Jr.; Finkbeiner, H. *J. Am. Chem. Soc.* **1968,** *90,* 5903 [32, 74].
[804.] Heiba, E. I.; Dessau, R. M.; Williams, A. L.; Rodewald, P. G. *Org. Synth.* **1983,** *61,* 22 [32].
[805.] Attenburrow, J.; Cameron, A. F. B.; Chapman, J. H.; Evans, R. M.; Hems, B. A.; Jansen, A. B. A.; Walker, T. *J. Chem. Soc.* **1952,** 1094 [33, 119, 126, 274, 288].
[806.] Harfenist, M.; Bavley, A.; Lazier, W. A. *J. Org. Chem.* **1954,** *19,* 1608 [33, 119, 124, 126].
[807.] Mancera, O.; Rosenkranz, G.; Sondheimer, F. *J. Chem. Soc.* **1953,** 2189 [33, 119, 215].
[808.] Gritter, R. J.; Wallace, T. J. *J. Org. Chem.* **1959,** *24,* 1051 [33, 119, 124, 140].
[809.] Goldman, I. M. *J. Org. Chem.* **1969,** *34,* 1979 [33, 119, 140, 149].
[810.] Ball, S.; Goodwin, R. W.; Morton, R. A. *Biochem. J.* **1948,** *42,* 516 [33, 119].
[811.] Highet, R. J.; Wildman, W. C. *J. Am. Chem. Soc.* **1955,** *77,* 4399 [33, 119, 120].
[812.] Henbest, H. B.; Stratford, M. J. W. *J. Chem. Soc. C* **1966,** 995 [33, 243].
[813.] Barakat, M. Z.; Abdel-Wahab, M. F.; El-Sadr, M. M. *J. Chem. Soc.* **1956,** 4685 [33, 119, 140, 176, 179, 234, 248].
[814.] Pratt, E. F.; Suskind, S. P. *J. Org. Chem.* **1963,** *28,* 638 [33, 53, 103, 104].
[815.] Pratt, E. F.; Van de Castle, J. F. *J. Org. Chem.* **1961,** *26,* 2973 [33, 140, 147, 148, 149].
[816.] Papadopoulos, E. P.; Jarrar, A.; Issidorides, C. H. *J. Org. Chem.* **1966,** *31,* 615 [33, 231, 232, 250, 251].
[817.] Ohloff, G.; Giersch, W. *Angew. Chem.* **1973,** *85,* 401 [33, 159, 161, 162, 288].
[818.] Curragh, E. F.; Henbest, H. B.; Thomas, A. *J. Chem. Soc.* **1960,** 3559 [33, 244].
[819.] Thomas, D. A.; Warburton, W. K. *J. Chem. Soc.* **1965,** 2988 [33].

[*820.*] Murray, R. W.; Trozzolo, A. M. *J. Org. Chem.* **1961,** *26,* 3109 [33].
[*821.*] Curtin, D. Y.; Johnson, H. W., Jr.; Steiner, E. G. *J. Am. Chem. Soc.* **1955,** *77,* 4566 [33].
[*822.*] Adler, E.; Becker, H.-D. *Acta Chem. Scand.* **1961,** *15,* 849 [33, 140].
[*823.*] Canonica, L. *Gazz. Chim. Ital.* **1947,** *77,* 92 [33, 219].
[*824.*] Stork, G.; Tomasz, M. *J. Am. Chem. Soc.* **1964,** *86,* 475 [33, 140].
[*825.*] Pratt, E. F.; McGovern, T. P. *J. Org. Chem.* **1964,** *29,* 1540 [33, 52, 233, 240, 241].
[*826.*] Henbest, H. B.; Thomas, A. *J. Chem. Soc.* **1957,** 3032 [33, 242].
[*827.*] Henbest, H. B.; Stratford, M. J. W. *Chem. Ind. (London)* **1961,** 1170 [33].
[*828.*] Henbest, H. B.; Thomas, A. *Chem. Ind. (London)* **1956,** 1097 [33].
[*829.*] Rigby, W. *J. Chem. Soc.* **1956,** 2452 [33].
[*830.*] Kenyon, J.; Symons, M. C. R. *J. Chem. Soc.* **1953,** 2129 [33, 35].
[*831.*] Kenyon, J.; Symons, M. C. R. *J. Chem. Soc.* **1953,** 3580 [33, 35].
[*832.*] Firouzabadi, H.; Ghaderi, E. *Tetrahedron Lett.* **1978,** 839 [33, 120, 140, 155].
[*833.*] Firouzabadi, H.; Mostafavipoor, Z. *Bull. Chem. Soc. Jpn.* **1983,** *56,* 914 [33, 120, 125, 126, 140, 149, 166, 234, 239, 240, 248].
[*834.*] Birch, S. F.; Oldham, W. J.; Johnson, E. A. *J. Chem. Soc.* **1947,** 818 [34, 79, 80, 82].
[*835.*] Menger, F. M.; Lee, C. *Tetrahedron Lett.* **1981,** *22,* 1655 [34, 129, 130, 140, 147, 148].
[*836.*] Truce, W. E.; Lyons, J. F. *J. Am. Chem. Soc.* **1951,** *73,* 126 [34, 264].
[*837.*] Henbest, H. B.; Khan, S. A. *Chem. Commun.* **1968,** 1036 [34, 39, 262].
[*838.*] Burdon, J.; Tatlow, J. C. *J. Appl. Chem.* **1958,** *8,* 293 [34, 35, 82, 83].
[*839.*] Cornsworth, J. W. *Org. Synth., Collective Volume* **1963,** *4,* 467 [34, 35, 228].
[*840.*] Einhorn, A. *Chem. Ber.* **1884,** *17,* 119 [34, 35, 79, 81].
[*841.*] Sam, D. J.; Simmons, H. F. *J. Am. Chem. Soc.* **1972,** *94,* 4024 [34, 35, 82, 105, 106, 130, 142, 149].
[*842.*] Davey, W.; Gwilt, J. R. *J. Chem. Soc.* **1950,** 204 [34, 35, 79, 81, 226].
[*843.*] Krapcho, A. P.; Larson, J. R.; Eldridge, J. M. *J. Org. Chem.* **1977,** *42,* 3749 [34, 35, 82, 91, 92].
[*844.*] Menger, F. M.; Lee, C. *J. Org. Chem.* **1979,** *44,* 3446 [34, 35, 141, 147, 149].
[*845.*] Herriott, A. W.; Picker, D. *Tetrahedron Lett.* **1974,** 1511 [34, 35, 82, 129, 130, 290].
[*846.*] Büchi, J.; Füeg, H. R.; Aebi, A. *Helv. Chim. Acta* **1959,** *42,* 1368 [34, 35, 260].
[*847.*] Regen, S. L.; Koteel, C. *J. Am. Chem. Soc.* **1977,** *99,* 3837 [34].
[*848.*] Lee, D. G.; Chang, V. S. *Synthesis* **1978,** 462 [34, 35, 91].
[*849.*] Noureldin, N. A.; Lee, D. G. *Tetrahedron Lett.* **1981,** *22,* 4889 [34].
[*850.*] Rawalay, S. S.; Shechter, H. *J. Org. Chem.* **1967,** *32,* 3129 [34, 239, 240].
[*851.*] Shechter, H.; Rawalay, S. S. *J. Am. Chem. Soc.* **1964,** *86,* 1706 [34, 35, 239, 240].
[*852.*] Milewich, L.; Axelrod, L. R. *Org. Synth., Collective Volume* **1988,** *6,* 690 [34, 35, 82].
[*853.*] Wiberg, K. B.; Saegebarth, K. A. *J. Am. Chem. Soc.* **1957,** *79,* 2822 [34, 68, 79, 80].
[*854.*] Wiesner, K.; Jay, E. W. K.; Tsai, T. Y. R.; Demerson, C.; Jay, L.; Kanno, T.; Křepinský, J.; Vilím, A.; Wu, C. S. *Can. J. Chem.* **1972,** *50,* 1925 [34, 35, 243].

[855.] Forrest, J.; Tucker, S. H.; Whalley, M. *J. Chem. Soc.* **1951,** 303 [34, 35, 243, 244].
[856.] Srinivasan, N. S.; Lee, D. G. *J. Org. Chem.* **1979,** *44,* 1574 [34, 35, 91, 92].
[857.] Srinivasan, N. S.; Lee, D. G. *Synthesis* **1979,** 520 [34, 76, 77].
[858.] Klein, E.; Rojahn, W. *Tetrahedron* **1965,** *21,* 2353 [34, 35].
[859.] Kornblum, N.; Jones, W. J. *Org. Synth., Collective Volume* **1973,** *5,* 845 [34, 35, 235].
[860.] Sharpless, K. B.; Lauer, R. F.; Repič, O.; Teranishi, A. Y.; Williams, D. R. *J. Am. Chem. Soc.* **1971,** *93,* 3303 [34, 35, 82].
[861.] Jensen, H. P.; Sharpless, K. B. *J. Org. Chem.* **1974,** *39,* 2314 [34, 35, 76, 226].
[862.] Jacob, T. A.; Bachman, G. B.; Hass, H. B. *J. Org. Chem.* **1951,** *16,* 1572 [34, 35].
[863.] Regen, S. L.; Koteel, C. *J. Am. Chem. Soc.* **1977,** *99,* 3837 [34, 35, 120, 125, 126, 141, 147, 149, 290].
[864.] Khan, N. A.; Newman, M. S. *J. Org. Chem.* **1952,** *17,* 1063 [34, 35, 91, 92, 226, 227, 289].
[865.] Quici, S.; Regen, S. L. *J. Org. Chem.* **1979,** *44,* 3436 [34, 35, 141].
[866.] Winkler, W. *Chem. Ber.* **1948,** *81,* 256 [34, 35, 103].
[867.] Shechter, H.; Williams, F. T., Jr. *J. Org. Chem.* **1962,** *27,* 3699 [34, 35, 230, 231].
[868.] Ruhoff, R. *Org. Synth., Collective Volume* **1943,** *2,* 315 [34, 35, 176, 179].
[869.] Henne, A. L.; Trott, P. *J. Am. Chem. Soc.* **1947,** *69,* 1820 [35, 82, 83, 289].
[870.] Kaiser, E. M. *J. Am. Chem. Soc.* **1967,** *89,* 3659 [35, 53].
[871.] Witzemann, E. J.; Evans, W. L.; Hass, H.; Schroeder, E. F. *Org. Synth., Collective Volume* **1943,** *2,* 307 [35, 185].
[872.] Starks, C. M. *J. Am. Chem. Soc.* **1971,** *93,* 195 [35, 82].
[873.] Hill, J. W.; McEwen, W. L. *Org. Synth., Collective Volume* **1943,** *2,* 53 [35].
[874.] Raasch, M. S.; Castle, J. E. *Org. Synth., Collective Volume* **1973,** *5,* 393 [35, 82, 83, 84].
[875.] Gardner, J. H. *J. Am. Chem. Soc.* **1927,** *49,* 1831 [35, 98].
[876.] Jones, R. G.; McLaughlin, K. C. *Org. Synth., Collective Volume* **1963,** *4,* 824 [35, 98].
[877.] Crook, K. E.; McElvain, S. M. *J. Am. Chem. Soc.* **1930,** *52,* 4006 [35, 103, 104, 105].
[878.] Ullmann, F.; Uzbachian, J. B. *Chem. Ber.* **1903,** *36,* 1797 [35].
[879.] Brockman, H.; Kluge, F.; Muxfeldt, H. *Chem. Ber.* **1957,** *90,* 2302 [35, 224].
[880.] Bigelow, L. A. *J. Am. Chem. Soc.* **1922,** *44,* 2010 [35, 105, 106].
[881.] Whitmore, F. C.; Woodward, G. E. *Org. Synth., Collective Volume* **1932,** *1,* 159 [35, 105, 106].
[882.] Clarke, H. T.; Taylor, E. R. *Org. Synth., Collective Volume* **1943,** *2,* 135 [35, 105, 106].
[883.] Singer, A. W.; McElvain, S. M. *Org. Synth., Collective Volume* **1955,** *3,* 740 [35, 105, 108, 109].
[884.] Billman, J. H.; Parker, E. E. *J. Am. Chem. Soc.* **1944,** *66,* 538 [35, 129].
[885.] Shriner, R. L.; Kleiderer, E. C. *Org. Synth., Collective Volume* **1943,** *2,* 538 [35, 176, 180].
[886.] Heller, G. *Chem. Ber.* **1911,** *44,* 2418 [35, 227].

[887.] Corson, B. B.; Dodge, R. A.; Harris, S. A.; Hazen, R. K. *Org. Synth., Collective Volume* **1932,** *1,* 241 [35, 228].
[888.] Claus, A.; Wollner, R. *Chem. Ber.* **1885,** *18,* 1856 [35, 206].
[889.] Claus, A. *Chem. Ber.* **1886,** *19,* 230 [35, 206].
[890.] Claus, A.; Fickert, E. *Chem. Ber.* **1886,** *19,* 3182 [35, 206].
[891.] Koelsch, C. F. *Org. Synth., Collective Volume* **1955,** *3,* 791 [35, 105, 107, 108, 210].
[892.] Kornblum, N.; Clutter, R. J. *J. Am. Chem. Soc.* **1954,** *76,* 4494 [35, 235].
[893.] Hartmann, C.; Meyer, V. *Chem. Ber.* **1893,** *26,* 1727 [35, 266].
[894.] Noureldin, N. A.; Lee, D. G. *J. Org. Chem.* **1982,** *47,* 2790 [35, 129, 130, 141].
[895.] Firouzabadi, H.; Sardarian, A. R.; Naderi, M.; Vessal, B. *Tetrahedron* **1984,** *40,* 5001 [36, 79, 125, 126].
[896.] Firouzabadi, H.; Naderi, M.; Sardarian, A.; Vessal, B. *Synth. Commun.* **1983,** *13,* 611 [36, 250, 251].
[897.] Firouzabadi, H.; Vessal, B.; Naderi, M. *Tetrahedron Lett.* **1982,** *23,* 1847 [36, 234].
[898.] Wolfe, S.; Ingold, C. F. *J. Am. Chem. Soc.* **1983,** *105,* 7755 [36, 91, 169].
[899.] Sala, T.; Sargent, M. V. *Chem. Commun.* **1978,** 253 [36, 82, 129, 130, 142, 149, 176, 177].
[900.] Ogino, T.; Mochizuki, K. *Chem. Lett.* **1979,** 443 [36, 71, 72, 79, 80, 89].
[901.] Morris, J. A.; Mills, D. C. *Chem. Ind. (London)* **1978,** 446 [36, 142, 176].
[902.] Schmidt, H.-J.; Schäfer, H. J. *Angew. Chem., Int. Ed. Engl.* **1979,** *18,* 68 [36, 58, 59, 169].
[903.] Scholz, D. *Monatsh. Chem.* **1979,** *110,* 1471 [36, 176, 177, 179].
[904.] Schmidt, H.-J.; Schäfer, H. J. *Angew. Chem., Int. Ed. Engl.* **1981,** *20,* 109 [36, 243].
[905.] Reischl, W.; Zbiral, E. *Tetrahedron* **1979,** *35,* 1109 [37, 71, 72].
[906.] Stenberg, V. I.; Singh, S. P.; Narain, N. K.; Parmar, S. S. *J. Org. Chem.* **1977,** *42,* 171 [37].
[907.] Alcalay, W. *Helv. Chim. Acta* **1947,** *30,* 578 [37, 166, 250, 251, 290].
[908.] Smith, L. I. *Org. Synth., Collective Volume* **1943,** *2,* 254 [37, 246, 247].
[909.] Fieser, L. F. *Org. Synth., Collective Volume* **1943,** *2,* 430 [37].
[910.] Giovannini, E.; Portmann, P. *Helv. Chim. Acta* **1948,** *31,* 1381, 1392 [37].
[911.] Frazier, R. H., Jr.; Harlow, R. L. *J. Org. Chem.* **1980,** *45,* 5408 [37, 223].
[912.] Walder, H. *Chem. Ber.* **1882,** *15,* 2166 [37, 53].
[913.] Ullmann, F.; Mauthner, F. *Chem. Ber.* **1902,** *35,* 4302 [37].
[914.] Smith, L. I.; Wiley, P. F. *J. Am. Chem. Soc.* **1946,** *68,* 894 [37, 166].
[915.] Fierz-David, H. E.; Blangey, L.; Dübendorfer, H. *Helv. Chim. Acta* **1946,** *29,* 1661 [37, 53].
[916.] Szwarc, M.; Shaw, A. *J. Am. Chem. Soc.* **1951,** *73,* 1379 [37].
[917.] Audette, R. J.; Quail, J. W.; Smith, P. J. *Tetrahedron Lett.* **1971,** 279 [37, 120, 125, 126, 142, 148, 149, 239, 240].
[918.] Tsuda, Y.; Nakajima, S. *Chem. Lett.* **1978,** 1397 [37, 120, 126, 239, 240].
[919.] Merz, V.; Ris, C. *Chem. Ber.* **1887,** *20,* 1190 [37, 52].
[920.] Bower, J. D.; Ramage, G. R. *J. Chem. Soc.* **1957,** 4506 [37, 55].
[921.] Kalb, L.; Gross, O. *Chem. Ber.* **1926,** *59,* 727 [37].
[922.] Cook, C. D.; Woodworth, R. C. *J. Am. Chem. Soc.* **1953,** *75,* 6242 [37, 163].

[923.] Dimroth, K.; Berndt, A.; Perst, H.; Reichardt, C.   *Org. Synth., Collective Volume* **1973,** *5,* 1130 [37, 163].
[924.] Nikishin, G. I.; Troyanskii, E. I.; Joffe, V. A.   *Izv. Akad. Nauk SSSR* **1982,** 2758; *Chem. Abstr.* **1983,** *98,* 142911 [37, 241].
[925.] Thesing, J.; Müller, A.; Michel, G.   *Chem. Ber.* **1955,** *88,* 1027 [37].
[926.] Perrine, T. D.   *J. Org. Chem.* **1951,** *16,* 1303 [37, 243].
[927.] Krafft, M. E.; Crooks, W. J., III; Zorc, B.; Milczanowski, S. E.   *J. Org. Chem.* **1988,** *53,* 3158 [37, 132, 147].
[928.] Alder, K.; Wirtz, H.   *Justus Liebigs Ann. Chem.* **1956,** *601,* 138 [37, 132].
[929.] Belew, J. S.; Tek-Ling, C.   *Chem. Commun.* **1967,** 1100 [37, 120, 125].
[930.] Warrener, R. N.; Cain, E. N.   *Aust. J. Chem.* **1971,** *24,* 785 [37].
[931.] Hawkins, E. G. E.; Large, R.   *J. Chem. Soc., Perkin Trans. 1* **1974,** 280 [37].
[932.] Nakagawa, K.; Konaka, R.; Nakata, T.   *J. Org. Chem.* **1962,** *27,* 1597 [37, 120, 126, 130].
[933.] Ladner, D. W.   *Synth. Commun.* **1986,** *16,* 157 [37, 105, 108].
[934.] Nakagawa, K.; Mineo, S.; Kawamura, S.   *Chem. Pharm. Bull.* **1978,** *26,* 229 [37, 177, 179].
[935.] Nakagawa, K.; Onoue, H.; Minami, K.   *Chem. Commun.* **1966,** 730 [37, 220, 221].
[936.] Nakagawa, K.; Tsuji, T.   *Chem. Pharm. Bull.* **1963,** *11,* 296 [37, 234, 241].
[937.] Lee, D. G.; Hall, D. T.; Cleland, J. H.   *Can. J. Chem.* **1972,** *50,* 3741 [38, 130, 142, 148].
[938.] Gopal, H.; Gordon, A. J.   *Tetrahedron Lett.* **1971,** 2941 [38, 91, 92].
[939.] Guizard, C.; Cheradame, H.; Brunel, Y.; Beguin, C. G.   *J. Fluorine Chem.* **1979,** *13,* 175 [38, 83].
[940.] Berkowitz, L. M.; Rylander, P. N.   *J. Am. Chem. Soc.* **1958,** *80,* 6682 [38, 120, 125, 142, 148, 169, 177, 179, 245].
[941.] Caputo, J. A.; Fuchs, R.   *Tetrahedron Lett.* **1967,** 4729 [38, 96, 97, 228].
[942.] Djerassi, C.; Engle, R. R.   *J. Am. Chem. Soc.* **1953,** *75,* 3838 [38, 255].
[943.] Spitzer, U. A.; Lee, D. G.   *J. Org. Chem.* **1974,** *39,* 2468 [38, 97, 98].
[944.] Yoshifuji, S.; Tanaka, K.-I.; Kawai, T.; Nitta, Y.   *Chem. Pharm. Bull.* **1986,** *34,* 3873 [38, 245].
[945.] Linstead, R. P.; Millidge, A. F.; Thomas, S. L. S.; Walpole, A. L.   *J. Chem. Soc.* **1937,** 1146 [38, 50].
[946.] Linstead, R. P.; Thomas, S. L. S.   *J. Chem. Soc.* **1940,** 1127 [38, 50].
[947.] Linstead, R. P.; Michaelis, K. O. A.; Thomas, S. L. S.   *J. Chem. Soc.* **1940,** 1139 [38, 50].
[948.] Ainsworth, C.   *Org. Synth., Collective Volume* **1963,** *4,* 536 [38, 50].
[949.] Roček, J.; Westheimer, F. H.   *J. Am. Chem. Soc.* **1962,** *84,* 2241 [39, 68, 159, 161, 290].
[950.] Baran, J. S.   *J. Org. Chem.* **1960,** *25,* 257 [39, 214].
[951.] Hentges, S. G.; Sharpless, K. B.   *J. Am. Chem. Soc.* **1980,** *102,* 4263 [39, 44, 45, 71, 73].
[952.] Van Rheenen, V.; Cha, D. Y.; Hartley, W. M.   *Org. Synth., Collective Volume* **1988,** *6,* 342 [39, 42, 69, 71, 72].
[953.] Wiesner, K.; Chan, K. K.; Demerson, C.   *Tetrahedron Lett.* **1965,** 2893 [39, 159, 162].
[954.] Meyers, C. Y.; Malte, A. M.; Matthews, W. S.   *J. Am. Chem. Soc.* **1969,** *91,* 7510 [39, 127, 205, 206, 207, 263].
[955.] Dunn, F. W.; Waugh, T. D.; Dittmer, K.   *J. Am. Chem. Soc.* **1946,** *68,* 2118 [39, 111].

[956.] Wood, J. H.; Tung, C. C.; Perry, M. A.; Gibson, R. E. *J. Am. Chem. Soc.* **1950,** *72,* 2992 [39, 111, 112, 292].
[957.] Graymore, J.; Davies, D. D. *J. Chem. Soc.* **1945,** 293 [39, 239, 240].
[958.] Posner, G. H.; Chapdelaine, M. J. *Synthesis* **1977,** 555 [39, 143].
[959.] Posner, G. H.; Perfetti, R. B.; Runquist, A. W. *Tetrahedron Lett.* **1976,** 3499 [39, 143, 149].
[960.] Posner, G. H.; Chapdelaine, M. J. *Tetrahedron Lett.* **1977,** 3227 [39, 261].
[961.] Kleiderer, E. C.; Kornfeld, E. C. *J. Org. Chem.* **1948,** *13,* 455 [39, 143].
[962.] Davies, R. R.; Hodgson, H. H. *J. Soc. Chem. Ind., London* **1943,** 109 [39, 143].
[963.] Eastman, J. F.; Teranishi, R. *Org. Synth., Collective Volume* **1963,** *4,* 192 [39, 143].
[964.] Oppenauer, R. V. *Org. Synth., Collective Volume* **1955,** *3,* 207 [39, 143].
[965.] Corey, E. J.; Achiwa, K. *J. Am. Chem. Soc.* **1969,** *91,* 1429 [40].
[966.] Turner, A. B.; Ringold, H. J. *J. Chem. Soc. C* **1967,** 1720 [40, 47, 48].
[967.] Dost, N.; Van Nes, K. *Rec. Trav. Chim. Pays-Bas* **1951,** *70,* 403 [40].
[968.] Findlay, J. W.; Turner, A. B. *Org. Synth., Collective Volume* **1973,** *5,* 428 [40].
[969.] Barclay, B. M.; Campbell, N. *J. Chem. Soc.* **1945,** 530 [40, 52].
[970.] Braude, E. A.; Brook, A. G.; Linstead, R. P. *J. Chem. Soc.* **1954,** 3569 [40, 52, 275].
[971.] Becker, H.-D. *J. Org. Chem.* **1965,** *30,* 982 [40, 164, 165].
[972.] Becker, H.-D. *J. Org. Chem.* **1965,** *30,* 989 [40].
[973.] Braude, E. A.; Linstead, R. P.; Wooldridge, K. R. *J. Chem. Soc.* **1956,** 3070 [40, 120, 121, 125, 126].
[974.] Cacchi, S. *Synthesis* **1978,** 848 [40, 143, 144].
[975.] Burn, D.; Petrow, V.; Weston, G. O. *Tetrahedron Lett.* **1960,** 14 [40, 143, 144, 156].
[976.] Jasiczak, J.; Smoczkiewicz, M. A. *Synthesis* **1981,** 804 [40].
[977.] Yoneda, F.; Suzuki, K.; Nitta, Y. *J. Org. Chem.* **1967,** *32,* 727 [40, 120, 125, 144, 149, 233, 250, 251].
[978.] Taylor, E. C.; Yoneda, F. *Chem. Commun.* **1967,** 199 [40, 231, 232].
[979.] Axen, R.; Chaykovsky, M.; Witkop, B. *J. Org. Chem.* **1967,** *32,* 4117 [41, 261].
[980.] Hess, K. *Chem. Ber.* **1919,** *52,* 964 [41, 243, 291].
[981.] Jung, M. E.; Brown, R. W. *Tetrahedron Lett.* **1978,** 2771 [41, 146, 147, 148, 172].
[982.] Goldschmidt, S.; Renn, K. *Chem. Ber.* **1922,** *55,* 628 [41].
[983.] Braude, E. A.; Brook, A. G.; Linstead, R. P. *J. Chem. Soc.* **1954,** 3574 [41, 233].
[984.] Kröhnke, F. *Angew. Chem., Int. Ed. Engl.* **1963,** *2,* 380 [41, 112].
[985.] Kröhnke, F.; Börner, E. *Chem. Ber.* **1936,** *69,* 2006 [41, 201].
[986.] Sachs, F.; Röhmer, A. *Chem. Ber.* **1902,** *35,* 3307 [41, 201, 292].
[987.] Hass, H. B.; Bender, M. L. *J. Am. Chem. Soc.* **1949,** *71,* 1767 [42, 111].
[988.] Leopold, B. *Acta Chem. Scand.* **1950,** *4,* 1523 [42].
[989.] Pearl, I. A. *J. Am. Chem. Soc.* **1948,** *70,* 1746 [42, 81, 163].
[990.] Beard, H. G.; Hodgson, H. H. *J. Chem. Soc.* **1944,** 4 [42].
[991.] Soderquist, J. A.; Najafi, M. R. *J. Org. Chem.* **1986,** *51,* 1330 [42, 267, 268].
[992.] Kabalka, G. W.; Hedgecock, H. C., Jr. *J. Org. Chem.* **1975,** *40,* 1776 [42, 267, 268].

[993.] Franzen, V.   *Org. Synth., Collective Volume* **1973,** *5,* 872 [42, 109, 110].
[994.] Serra-Errante, G.; Sammes, P. G.   *Chem. Commun.* **1975,** 573 [42, 93, 238].
[995.] Miyazawa, T.; Endo, T.   *J. Org. Chem.* **1985,** *50,* 3930 [42, 157, 158].
[996.] Iwata, C.; Takemoto, Y.; Nakamura, A.; Imanishi, T.   *Tetrahedron Lett.* **1985,** *26,* 3227 [42, 173].
[997.] Wolfe, S.; Pilgrim, W. R.; Garrard, T. F.; Chamberlain, P.   *Can. J. Chem.* **1971,** *49,* 1099 [43, 91, 92].
[998.] Johnson, A. P.; Pelter, A.   *J. Chem. Soc.* **1964,** 520 [43, 109, 110].
[999.] Dave, P.; Byun, H.-S.; Engel, R.   *Synth. Commun.* **1986,** *16,* 1343 [43, 109, 110, 111].
[1000.] Kornblum, N.; Jones, W. J.; Anderson, G. J.   *J. Am. Chem. Soc.* **1959,** *81,* 4113 [43, 109, 110].
[1001.] Floyd, M. B.; Du, M. T.; Fabio, P. F.; Jacob, L. A.; Johnson, B. D.   *J. Org. Chem.* **1985,** *50,* 5022 [43, 200, 201].
[1002.] Schipper, E.; Cinnamon, M.; Rascher, L.; Chiang, Y. H.; Oroshnik, W.   *Tetrahedron Lett.* **1968,** 6201 [43, 201].
[1003.] Kornblum, N.; Powers, J. W.; Anderson, G. J.; Jones, W. J.; Larson, H. O.; Levand, O.; Weaver, W. M.   *J. Am. Chem. Soc.* **1957,** *79,* 6562 [43, 111, 201, 202, 293].
[1004.] Kornblum, N.; Frazier, H. W.   *J. Am. Chem. Soc.* **1966,** *88,* 865 [43, 202].
[1005.] Bauer, D. P.; Macomber, R. S.   *J. Org. Chem.* **1975,** *40,* 1990 [43, 202].
[1006.] Onodera, K.; Hirano, S.; Kashimura, N.   *J. Am. Chem. Soc.* **1965,** *87,* 4651 [43].
[1007.] Mikolajczyk, M.; Luczak, J.   *Chem. Ind. (London)* **1974,** 701 [43, 249, 250].
[1008.] Albright, J. D.; Goldman, L.   *J. Am. Chem. Soc.* **1965,** *87,* 4214 [43].
[1009.] Albright, J. D.; Goldman, L.   *J. Am. Chem. Soc.* **1967,** *89,* 2416 [43, 145, 146].
[1010.] Moffatt, J. G.   *Org. Synth., Collective Volume* **1973,** *5,* 242 [43, 122].
[1011.] Omura, K.; Sharma, A. K.; Swern, D.   *J. Org. Chem.* **1976,** *41,* 957 [43, 125, 126].
[1012.] Huang, S. L.; Omura, K.; Swern, D.   *Synthesis* **1978,** 297 [43, 122, 145].
[1013.] Amon, C. M.; Banwell, M. G.; Gravatt, G. L.   *J. Org. Chem.* **1987,** *52,* 4851 [43, 155].
[1014.] Trost, B. M.; Fray, M. J.   *Tetrahedron Lett.* **1988,** *29,* 2163 [43, 173, 174].
[1015.] Bohlmann, F.; Miethe, R.   *Chem. Ber.* **1967,** *100,* 3861 [43, 122, 293].
[1016.] Pfitzner, K. E.; Moffatt, J. G.   *J. Am. Chem. Soc.* **1965,** *87,* 5661 [43, 122, 144, 145, 215, 216].
[1017.] Onodera, K.; Hirano, S.; Kashimura, N.   *Carbohydr. Res.* **1968,** *6,* 276 [43].
[1018.] Parikh, J. R.; Doering, v. W. E.   *J. Am. Chem. Soc.* **1967,** *89,* 5505 [43, 145, 146, 215, 216].
[1019.] Corey, E. J.; Kim, C. U.   *Tetrahedron Lett.* **1974,** 287 [43, 155, 156].
[1020.] Corey, E. J.; Kim, C. U.   *Tetrahedron Lett.* **1973,** 919 [43, 122].
[1021.] Barton, D. H. R.; Garner, B. J.; Wightman, R. H.   *J. Chem. Soc.* **1964,** 1855 [43, 122, 123].
[1022.] Mancuso, A. J.; Huang, S.-L.; Swern, D.   *J. Org. Chem.* **1978,** *43,* 2480 [43, 122].
[1023.] Omura, K.; Swern, D.   *Tetrahedron* **1978,** *34,* 1651 [43, 121, 122, 125, 126, 145, 147, 148, 149].

[1024.] Smith, A. B., III; Leenay, T. L.; Liu, H.-J.; Nelson, L. A. K.; Ball, R. G. *Tetrahedron Lett.* **1988,** *29,* 49 [43, 122].
[1025.] Di Furia, F.; Modena, G.; Seraglia, R. *Synthesis* **1984,** 325 [44, 45, 258].
[1026.] Sharpless, K. B.; Behrens, C. H.; Katsuki, T.; Lee, A. W. M.; Martin, V. S.; Takatani, M.; Viti, S. M.; Walker, F. J.; Woodard, S. S. *Pure Appl. Chem.* **1983,** *55,* 589 [44, 45, 154].
[1027.] Sharpless, K. B.; Woodard, S. S.; Finn, M. G. *Pure Appl. Chem.* **1983,** *55,* 1823 [44, 45, 154].
[1028.] Hill, J. G.; Sharpless, K. B. *Org. Synth.* **1984,** *63,* 66 [44, 45].
[1029.] Pitchen, P.; Kagan, H. B. *Tetrahedron Lett.* **1984,** *25,* 1049 [44, 45, 258].
[1030.] Yamada, S.-i.; Mashiko, T.; Terashima, S. *J. Am. Chem. Soc.* **1977,** *99,* 1988 [44, 45].
[1031.] Michaelson, R. C.; Palermo, R. E.; Sharpless, K. B. *J. Am. Chem. Soc.* **1977,** *99,* 1990 [44, 45, 153].
[1032.] Davis, F. A.; Haque, M. S.; Ulatowski, T. G.; Towson, J. C. *J. Org. Chem.* **1986,** *51,* 2402 [44, 45, 223].
[1033.] Yamada, T.; Narasaka, K. *Chem. Lett.* **1986,** 131 [45, 71, 73].
[1034.] Branchaud, B. P.; Walsh, C. T. *J. Am. Chem. Soc.* **1985,** *107,* 2153 [45, 46, 177, 187, 190, 257, 265, 267, 270].
[1035.] Jones, J. B.; Jakovac, I. J. *Org. Synth.* **1984,** *63,* 10 [45, 46, 131, 159].
[1036.] Grunwald, J.; Wirz, B.; Scollar, M. P.; Klibanov, A. M. *J. Am. Chem. Soc.* **1986,** *108,* 6732 [45, 146].
[1037.] Brossi, A.; Ramel, A.; O'Brien, J.; Teitel, S. *Chem. Pharm. Bull.* **1973,** *21,* 1839 [45, 46, 53, 54, 55].
[1038.] Saunders, B. C.; Stark, B. P. *Tetrahedron* **1967,** *23,* 1867 [45, 46, 164, 165].
[1039.] Kazandjian, R. Z.; Klibanov, A. M. *J. Am. Chem. Soc.* **1985,** *107,* 5448 [45, 46, 163, 164].
[1040.] Stubbs, J. J.; Lockwood, L. B.; Roe, E. T.; Tabenkin, B.; Ward, G. E. *Ind. Eng. Chem.* **1940,** *32,* 1626 [45, 46, 177, 183].
[1041.] Reichstein, T. *Helv. Chim. Acta* **1934,** *17,* 996 [45, 46, 146, 156, 157].
[1042.] Taschner, M. J.; Black, D. J. *Abstracts of Papers,* 194th National Meeting of the American Chemical Society, New Orleans, LA; American Chemical Society: Washington, DC, 1987; ORGN 109 [45, 46].
[1043.] Schwab, J. M.; Li, W.-b.; Thomas, L. P. *J. Am. Chem. Soc.* **1983,** *105,* 4800 [45, 46, 190, 195].
[1044.] Ichimoto, I.; Fujii, K.; Sekido, F.; Nonomura, S.; Tatsumi, C. *Agric. Biol. Chem.* **1965,** *29,* 99 [45, 46, 130].
[1045.] Fried, J.; Thoma, R. W.; Genke, J. R.; Herz, J. E.; Donin, M. N.; Perlman, D. *J. Am. Chem. Soc.* **1952,** *74,* 3962 [45, 46, 197, 198].
[1046.] Auret, B. J.; Boyd, D. R.; Henbest, H. B.; Ross, S. *J. Chem. Soc. C* **1968,** 2371 [45, 46, 258, 294].
[1047.] Auret, B. J.; Boyd, D. R.; Henbest, H. B. *J. Chem. Soc. C* **1968,** 2374 [45, 46, 258].
[1048.] Dodson, R. M.; Newman, N.; Tsuchiya, H. M. *J. Org. Chem.* **1962,** *27,* 2707 [45, 46, 258].
[1049.] McCurdy, J. T.; Garrett, R. D. *J. Org. Chem.* **1968,** *33,* 660 [45, 46, 194].
[1050.] Tabenkin, B.; LeMahieu, R. A.; Berger, J.; Kierstead, R. W. *Appl. Microbiol.* **1969,** *17,* 714 [45, 46, 214].
[1051.] Herzog, H. L.; Gentles, M. J.; Charney, W.; Sutter, D.; Townley, E.; Yudis, M.; Kabasakalian, P.; Hershberg, E. B. *J. Org. Chem.* **1959,** *24,* 691 [45, 46, 197, 198].

[1052.] Holmlund, C. E.; Sax, K. J.; Nielsen, B. E.; Hartman, R. E.; Evans, R. H., Jr.; Blank, R. H. *J. Org. Chem.* **1962,** *27,* 1468 [45, 46, 258].
[1053.] Douros, J. D., Jr.; Frankenfeld, J. W. *Appl. Microbiol.* **1968,** *16,* 320 [45, 46, 58, 107].
[1054.] McAleer, W. J.; Kozlowski, M. A.; Stoudt, T. H.; Chemerda, J. M. *J. Org. Chem.* **1958,** *23,* 958 [45, 46, 130].
[1055.] Ohta, H.; Tetsukawa, H. *Chem. Commun.* **1978,** 849 [45, 46, 61, 62].
[1056.] Charney, W.; Nobile, A.; Federbush, C.; Sutter, D.; Perlman, P. L.; Herzog, H. L.; Payne, C. C.; Tully, M. E.; Gentles, M. J.; Hershberg, E. B. *Tetrahedron* **1962,** *18,* 591 [45, 46, 48, 49, 146, 215, 216].
[1057.] Herzog, H. L.; Payne, C. C.; Hughes, M. T.; Gentles, M. J.; Hershberg, E. B.; Nobile, A.; Charney, W.; Federbush, C.; Sutter, D.; Perlman, P. L. *Tetrahedron* **1962,** *18,* 581 [45, 46, 49].
[1058.] Ohta, H.; Okamoto, Y.; Tsuchihashi, G.-I. *Chem. Lett.* **1984,** 205 [45, 46, 254].
[1059.] Hanze, A. R.; Sebek, O. K.; Murray, H. C. *J. Org. Chem.* **1960,** *25,* 1968 [45, 46, 156, 157, 216].
[1060.] Mann, K. M.; Hanson, F. R.; O'Connell, P. W.; Anderson, H. V.; Brunner, M. P.; Karnemaat, J. N. *Appl. Microbiol.* **1955,** *3,* 14 [45, 46, 197, 198].
[1061.] O'Connell, P. W.; Mann, K. M.; Nielson, E. D.; Hanson, F. R. *Appl. Microbiol.* **1955,** *3,* 17 [45, 46, 197, 198].
[1062.] Lin, Y. Y.; Shibahara, M.; Smith, L. L. *J. Org. Chem.* **1969,** *34,* 3530 [45, 46, 197].
[1063.] Kurosawa, Y.; Hayano, M.; Bloom, B. M. *Agric. Biol. Chem.* **1961,** *25,* 838 [45, 46, 61].
[1064.] Meystre, C.; Vischer, E.; Wettstein, A. *Helv. Chim. Acta* **1955,** *38,* 381 [45, 46, 197, 198].
[1065.] Fried, J.; Thoma, R. W.; Klingsberg, A. *J. Am. Chem. Soc.* **1953,** *75,* 5764 [45, 46, 194, 195].
[1066.] Tamm, C.; Gubler, A.; Juhasz, G.; Weiss-Berg, E.; Zürcher, W. *Helv. Chim. Acta* **1963,** *46,* 889 [45, 46, 197, 199].
[1067.] Urech, J.; Vischer, E.; Wettstein, A. *Helv. Chim. Acta* **1960,** *43,* 1077 [45, 46, 49, 197, 199].
[1068.] Abushanab, E.; Reed, D.; Suzuki, F.; Sih, C. J. *Tetrahedron Lett.* **1978,** 3415 [46, 49, 258].
[1069.] Siehr, D. J. *J. Am. Chem. Soc.* **1961,** *83,* 2401 [46, 241, 242].
[1070.] Sih, C. J.; Lee, S. S.; Tsong, Y. Y.; Wang, K. C.; Chang, F. N. *J. Am. Chem. Soc.* **1965,** *87,* 2765 [46].
[1071.] Davis, J. B.; Raymond, R. L. *Appl. Microbiol.* **1961,** *9,* 383 [46, 58, 106, 107].
[1072.] Prairie, R. L.; Talalay, P. *Biochemistry* **1963,** *2,* 203 [46, 194, 195].
[1073.] Aberhart, D. J. *J. Org. Chem.* **1980,** *45,* 5218 [46, 71].
[1074.] Gibson, D. T.; Hensley, M.; Yoshioka, H.; Mabry, T. *J. Biochem.* **1970,** *9,* 1626 [46, 93].
[1075.] Peterson, D. H.; Murray, H. C. *J. Am. Chem. Soc.* **1952,** *74,* 1871 [46, 197].
[1076.] Peterson, D. H.; Murray, H. C.; Eppstein, S. H.; Reineke, L. M.; Weintraub, A.; Meister, P. D.; Leigh, H. M. *J. Am. Chem. Soc.* **1952,** *74,* 5933 [46, 197].
[1077.] Peterson, D. H.; Nathan, A. H.; Meister, P. D.; Eppstein, S. H.; Murray, H. C.; Weintraub, A.; Reineke, L. M.; Leigh, H. M. *J. Am. Chem. Soc.* **1953,** *75,* 419 [46, 197].

[*1078.*] Eppstein, S. H; Peterson, D. H.; Leigh, H. M.; Murray, H. C.; Weintraub, A.; Reineke, L. M.; Meister, P. D.   *J. Am. Chem. Soc.* **1953,** *75,* 421 [46, 197].
[*1079.*] Peterson, D. H.; Meister, P. D.; Weintraub, A.; Reineke, L. M.; Eppstein, S. H.; Murray, H. C.; Osborn, H. M. L.   *J. Am. Chem. Soc.* **1955,** *77,* 4428 [46, 197, 198].
[*1080.*] Fonken, G. S.; Herr, M. E.; Murray, H. C.; Reineke, L. M.   *J. Org. Chem.* **1968,** *33,* 3182 [46].
[*1081.*] Johnson, R. A.; Herr, M. E.; Murray, H. C.; Fonken, G. S.   *J. Org. Chem.* **1968,** *33,* 3187 [46, 242].
[*1082.*] Johnson, R. A.; Herr, M. E.; Murray, H. C.; Reineke, L. M.; Fonken, G. S.   *J. Org. Chem.* **1968,** *33,* 3195 [46, 242].
[*1083.*] Herr, M. E.; Johnson, R. A.; Murray, H. C.; Reineke, L. M.; Fonken, G. S.   *J. Org. Chem.* **1968,** *33,* 3201 [46, 242, 243].
[*1084.*] Johnson, R. A.; Herr, M. E.; Murray, H. C.; Fonken, G. S.   *J. Org. Chem.* **1970,** *35,* 622 [46, 242, 243].
[*1085.*] Kawai, S.; Oshima, T.; Egami, F.   *Biochem. Biophys. Acta* **1965,** *97,* 391 [46, 236].
[*1086.*] Jones, D. F.; Howe, R.   *J. Chem. Soc. C* **1968,** 2816 [46, 58, 112, 113].
[*1087.*] Jones, D. F.; Howe, R.   *J. Chem. Soc. C* **1968,** 2801 [46, 58, 112].
[*1088.*] Mamoli, L.; Vercellone, A.   *Chem. Ber.* **1938,** *71,* 1686 [46, 146, 215].
[*1089.*] Shieh, H. S.; Blackwood, A. C.   *Can. J. Biochem.* **1967,** *45,* 2045 [46, 225, 226].
[*1090.*] Sergienko, S. R.   *C. R. Acad. Sci. URSS* **1940,** *29,* 36; *Chem. Abstr.* **1941,** *35,* 3614 [49].
[*1091.*] Grundon, M. F.; Reynolds, B. E.   *J. Chem. Soc.* **1964,** 2445 [49, 240, 241].
[*1092.*] Newman, M. S.; Zahm, H. V.   *J. Am. Chem. Soc.* **1943,** *65,* 1097 [50].
[*1093.*] Linstead, R. P.; Michaelis, K. O. A.; Thomas, S. L. S.   *J. Chem. Soc.* **1940,** 1139 [50].
[*1094.*] Smith, C.; Lewcock, W.   *J. Chem. Soc.* **1912,** *101,* 1453 [52].
[*1095.*] Albert, H. E.   *J. Am. Chem. Soc.* **1954,** *76,* 4983 [53, 163].
[*1096.*] Grosse, A. V.; Morrell, J. C.; Mattox, W. J.   *Ind. Eng. Chem.* **1940,** *32,* 528 [55].
[*1097.*] Mareš, F.; Roček, J.   *Collect. Czech. Chem. Commun.* **1961,** *26,* 2370 [58].
[*1098.*] Heider, R. L. U.S. Patent 2 554 459, 1951; *Chem. Abstr.* **1951,** *45,* 9076e [60].
[*1099.*] Bach, R. D.; Knight, J. W.   *Org. Synth.* **1981,** *60,* 63 [60, 61].
[*1100.*] Pasetti, A.; Sianesi, D.   *Gazz. Chim. Ital.* **1968,** *98,* 265 [62].
[*1101.*] Bartlett, P. D.; Mendenhall, G. D.; Durham, D. L.   *J. Org. Chem.* **1980,** *45,* 4269 [64].
[*1102.*] Hauser, F. M.; Ellenberger, S. R.; Clardy, J. C.; Bass, L. S.   *J. Am. Chem. Soc.* **1984,** *106,* 2458 [71, 73, 262].
[*1103.*] Johnson, C. R.; Barbachyn, M. R.   *J. Am. Chem. Soc.* **1984,** *106,* 2459 [71, 73].
[*1104.*] Schmidt, E.; Knilling, v. W.; Ascherl, A.   *Chem. Ber.* **1926,** *59,* 1279 [73, 74].
[*1105.*] Herranz, E.; Sharpless, K. B.   *Org. Synth.* **1983,** *61,* 93 [74].
[*1106.*] Fischer, F. G.; Löwenberg, K.   *Chem. Ber.* **1933,** *66,* 665 [77, 79].
[*1107.*] Turner, R. B.; Mattox, V. R.; McGuckin, W. F.; Kendall, E. C.   *J. Am. Chem. Soc.* **1952,** *74,* 5814 [78, 80].

[1108.] Claus, R. E.; Schreiber, S. L.   *Org. Synth.* **1985,** *64* 150 [78, 79, 81].
[1109.] Wasserman, H. H.; Stiller, K.; Floyd, M. B.   *Tetrahedron Lett.* **1968,** 3277 [80].
[1110.] Bailey, P. S.; Bath, S. S.   *J. Am. Chem. Soc.* **1957,** *79,* 3120 [81].
[1111.] Grummitt, O.; Egan, R.; Buck, A.   *Org. Synth., Collective Volume* **1955,** *3,* 449 [81, 82].
[1112.] Schenck, G. O.; Schulte-Elte, K.-H.   *Justus Liebigs Ann. Chem.* **1958,** *618,* 185 [84].
[1113.] Kharasch, M. S.; Sosnovsky, G.   *J. Am. Chem. Soc.* **1958,** *80,* 756 [86].
[1114.] Whitham, G. H.   *J. Chem. Soc.* **1961,** 2232 [86].
[1115.] Schenck, G. O.; Kinkel, K. G.; Mertens, H.-J.   *Justus Liebigs Ann. Chem.* **1953,** *584,* 125 [87, 88].
[1116.] Ouannes, C.; Wilson, T.   *J. Am. Chem. Soc.* **1968,** *90,* 6527 [88].
[1117.] Dupont, G.; Dulou, R.; Lefort, D.   *Bull. Soc. Chim. Fr.* **1949,** 789 [91, 92, 227].
[1118.] McClure, J. D.; Williams, P. H.   *J. Org. Chem.* **1962,** *27,* 627 [93].
[1119.] Kovacic, P.; Reid, C. G.; Kurz, M. E.   *J. Org. Chem.* **1969,** *34,* 3302 [93, 94].
[1120.] Kovacic, P.; Reid, C. G.; Brittain, M. J.   *J. Org. Chem.* **1970,** *35,* 2152 [93, 94].
[1121.] Braude, E. A.; Fawcett, J. S.   *Org. Synth., Collective Volume* **1963,** *4,* 698 [94, 95].
[1122.] Hadler, H. I.; Raha, C. R.   *J. Org. Chem.* **1957,** *22,* 433 [95].
[1123.] Bailey, P. S.; Erickson, R. E.   *Org. Synth., Collective Volume* **1973,** *5,* 489 [96, 97].
[1124.] Bailey, P. S.; Erickson, R. E.   *Org. Synth., Collective Volume* **1973,** *5,* 493 [96, 97].
[1125.] Gardner, J. H.; Naylor, C. A., Jr.   *Org. Synth., Collective Volume* **1943,** *2,* 523 [97, 98].
[1126.] Heymann, H.; Trowbridge, L.   *J. Am. Chem. Soc.* **1950,** *72,* 84 [100].
[1127.] Gilliard, X.; Monnet, P.; Cartier, X. Ger. Patent 101 221; *Chem. Zentr.* **1899,** *I,* 960; Ger. Patent 107 722; *Chem. Zentr.* **1900,** *I,* 1113 [101].
[1128.] Emerson, W. S.; Deebel, G. F.   *Org. Synth., Collective Volume* **1963,** *4,* 579 [103].
[1129.] Clarke, H. T.; Hartman, W. W.   *Org. Synth., Collective Volume* **1932,** *1,* 543 [105, 106].
[1130.] Friedman, L.   *Org. Synth., Collective Volume* **1973,** *5,* 810 [105].
[1131.] Pamfilov, A. W.   *Chem. Zentr.* **1922,** *III,* 353 [105, 108].
[1132.] Angyal, S. J.; Rassack, R. C.   *J. Chem. Soc.* **1949,** 2700 [110, 111].
[1133.] Sartori, P.; Ahlers, K.; Frohn, H.-J.   *J. Fluorine Chem.* **1976,** *8,* 457 [113, 114].
[1134.] Yakobson, G. G.; Shteingarts, V. D.; Vorozhtsov, N. N.   *Zh. Vses. Khim. Ova. im. D. I. Mendeleeva* **1964,** *9,* 701; *Chem. Abstr.* **1965,** *62,* 9052c [113].
[1135.] Church, J. M.; Joshi, H. K.   *Ind. Eng. Chem.* **1951,** *43,* 1804 [114].
[1136.] Belew, J. S.; Garza, C.; Mathieson, J. W.   *Chem. Commun.* **1970,** 634 [115, 124, 126, 234, 241, 242].
[1137.] Bertrand, M.; Gil, G.; Viala, J.   *Tetrahedron Lett.* **1979,** 1595 [120, 132].
[1138.] Corey, E. J.; Kim, C. U.   *J. Am. Chem. Soc.* **1972,** *94,* 7586 [121, 123, 125, 293].
[1139.] Huang, S. L.; Omura, K.; Swern, D.   *J. Org. Chem.* **1976,** *41,* 3329 [122, 146].

[1140.] Kuwajima, I.; Shimizu, M.; Urabe, H.   *J. Org. Chem.* **1982,** *47,* 837 [122, 123, 124, 125, 126].
[1141.] Ahrens, H.; Korytnyk, W.   *J. Heterocycl. Chem.* **1967,** *4,* 625 [128].
[1142.] Wiberg, K. B.; Schäfer, H.   *J. Am. Chem. Soc.* **1967,** *89,* 455 [134].
[1143.] Rahman, M.; Roček, J.   *J. Am. Chem. Soc.* **1971,** *93,* 5462 [134].
[1144.] Kwart, H.; Francis, P. S.   *J. Am. Chem. Soc.* **1955,** *77,* 4907 [134, 135].
[1145.] Kwart, H.; Francis, P. S.   *J. Am. Chem. Soc.* **1959,** *81,* 2116 [135].
[1146.] Bruce, W. F.   *Org. Synth., Collective Volume* **1943,** *2,* 139 [136].
[1147.] Braun, v. J.; Lemke, G.   *Chem. Ber.* **1922,** *55,* 3526 [140, 150, 289].
[1148.] Djerassi, C.   *Org. React.* **1951,** *6,* 207 [143].
[1149.] Mancuso, A. J.; Brownfain, D. S.; Swern, D.   *J. Org. Chem.* **1979,** *44,* 4148 [145].
[1150.] Barbier, P.; Locquin, R.   *C. R. Hebd. Seances Acad. Sci.* **1913,** *156,* 1443 [151].
[1151.] Fuson, R. C.; Maynert, E. W.; Tan, T.-L.; Trumbull, E. R.; Wassmundt, F. W.   *J. Am. Chem. Soc.* **1957,** *79,* 1938 [151, 152].
[1152.] Stevens, R. V.; Chapman, K. T.; Stubbs, C. A.; Tam, W. W.; Albizati, K. F.   *Tetrahedron Lett.* **1982,** *23,* 4647 [156].
[1153.] English, J., Jr.; Griswold, P. H., Jr.   *J. Am. Chem. Soc.* **1948,** *70,* 1390 [160].
[1154.] Criegee, R.; Büchner, E.; Walther, W.   *Chem. Ber.* **1940,** *73,* 571 [160, 161].
[1155.] English, J., Jr.; Barber, G. W.   *J. Am. Chem. Soc.* **1949,** *71,* 3310 [161, 281].
[1156.] Hudlický, M.; Bell, H. M.   *J. Fluorine Chem.* **1975,** *6,* 201 [167].
[1157.] Toennies, G.   *J. Am. Chem. Soc.* **1937,** *59,* 552 [8].
[1158.] English, J., Jr.; Dayan, J. E.   *Org. Synth., Collective Volume* **1963,** *4,* 499 [171].
[1159.] Rubottom, G. M.; Gruber, J. M.; Juve, H. D., Jr.   *Org. Synth.* **1985,** *64,* 118 [173].
[1160.] Pearl, I. A.   *Org. Synth., Collective Volume* **1963,** *4,* 974 [176, 180].
[1161.] Cavalieri, L.; Pattison, D. B.; Carmack, M.   *J. Am. Chem. Soc.* **1945,** *67,* 1783 [176, 179, 204].
[1162.] Lindgren, B. O.; Nilsson, T.   *Acta Chem. Scand.* **1973,** *27,* 888 [176, 179, 180, 285].
[1163.] Hawkins, E. G. E.   *J. Chem. Soc.* **1950,** 2169 [177, 179].
[1164.] Jorissen, W. P.; Van Der Beek, P. A. A.   *Rec. Trav. Chim. Pays-Bas* **1926,** *45,* 245 [180].
[1165.] Hönig, M.; Ruzizcka, W.   *Chem. Ber.* **1930,** *63,* 1648 [182, 183].
[1166.] Doering, v. W. E.; Speers, L.   *J. Am. Chem. Soc.* **1950,** *72,* 5515 [187].
[1167.] Blomquist, A. T.; LaLancette, E. A.   *J. Am. Chem. Soc.* **1961,** *83,* 1387 [187].
[1168.] Baker, B. R.; Schaub, R. E.; Joseph, J. P.; McEvoy, F. J.; Williams, J. H.   *J. Org. Chem.* **1952,** *17,* 141 [194].
[1169.] King, J. A.; McMillan, F. H.   *J. Am. Chem. Soc.* **1946,** *68,* 525, 1369 [203, 204].
[1170.] Schwenk, E.; Papa, D.   *J. Org. Chem.* **1946,** *11,* 798 [203, 204].
[1171.] Staunton, J.; Eisenbraun, E. J.   *Org. Synth., Collective Volume* **1973,** *5,* 8 [207, 208].
[1172.] Levine, R.; Stephens, J. R.   *J. Am. Chem. Soc.* **1950,** *72,* 1642 [208].
[1173.] Morgan, K. J.; Bardwell, J.; Cullis, C. F.   *J. Chem. Soc.* **1950,** 3190 [210].
[1174.] Looft, E.   *Chem. Ber.* **1894,** *27,* 1542 [211].

[1175.] Roček, J.; Riehl, Sr. A. *J. Am. Chem. Soc.* **1967,** *89,* 6691 [211, 212].
[1176.] Turner, R. B. *J. Am. Chem. Soc.* **1950,** *72,* 579 [215].
[1177.] Buehler, C. A.; Harris, J. O. *J. Am. Chem. Soc.* **1950,** *72,* 5015 [218].
[1178.] Emmons, W. D. *J. Am. Chem. Soc.* **1957,** *79,* 5739 [219].
[1179.] Fleisher, G. A.; Kendall, E. C. *J. Org. Chem.* **1951,** *16,* 556 [219].
[1180.] McDonald, R. N.; Steppel, R. N.; Dorsey, J. E. *Org. Synth., Collective Volume* **1988,** *6,* 276 [222].
[1181.] Kuhn, R.; Roth, H. *Chem. Ber.* **1933,** *66,* 1274 [224].
[1182.] Baumgarten, H. E.; Staklis, A. *J. Am. Chem. Soc.* **1965,** *87,* 1141 [230].
[1183.] Corey, E. J.; Samuelsson, B.; Luzzio, F. A. *J. Am. Chem. Soc.* **1984,** *106,* 3682 [232].
[1184.] Kauer, J. C. *Org. Synth., Collective Volume* **1963,** *4,* 411 [233].
[1185.] Carpino, L. A.; Crowley, P. J. *Org. Synth., Collective Volume* **1973,** *5,* 160 [233].
[1186.] Di Nunno, L.; Florio, S.; Todesco, P. E. *J. Chem. Soc. C* **1970,** 1433 [234, 235].
[1187.] Cope, A. C.; Pike, R. A.; Spencer, C. F. *J. Am. Chem. Soc.* **1953,** *75,* 3212 [237].
[1188.] Boekelheide, V.; Linn, W. J. *J. Am. Chem. Soc.* **1954,** *76,* 1286 [238].
[1189.] Scheit, K. H.; Kampe, W. *Angew. Chem.* **1965,** *77,* 811 [239, 240].
[1190.] Rao, A. V. R.; Chavan, S. P.; Sivadasan, L. *Tetrahedron* **1986,** *42,* 5065 [246].
[1191.] Fieser, L. F. *Org. Synth., Collective Volume* **1932,** *1,* 383 [247].
[1192.] Fieser, L. F. *Org. Synth., Collective Volume* **1943,** *2,* 4300[247].
[1193.] Martin, E. L.; Fieser, L. F. *Org. Synth., Collective Volume* **1955,** *3,* 633 [247].
[1194.] Skowronska, A.; Krawczyk, E. *Synthesis* **1983,** 509 [249, 250].
[1195.] Bestman, H.-J.; Armsen, R.; Wagner, H. *Chem. Ber.* **1969,** *102,* 2259 [250].
[1196.] Shriner, R. L.; Wolf, C. N. *Org. Synth., Collective Volume* **1963,** *4,* 910 [250].
[1197.] Čech, J. *Collect. Czech. Chem. Commun.* **1949,** *14,* 558 [252].
[1198.] Arbusow, B. A. *J. Prakt. Chem.* **1931,** [2], *131,* 357 [254, 265, 273].
[1199.] Johnson, C. R.; McCants, D., Jr. *J. Am. Chem. Soc.* **1965,** *87,* 1109 [257].
[1200.] Akutagawa, K.; Furukawa, N. *J. Org. Chem.* **1984,** *49,* 2282 [263].
[1201.] Brown, H. C.; Gupta, S. K. *J. Am. Chem. Soc.* **1975,** *97,* 5249 [267, 268, 269].
[1202.] Zweifel, G.; Brown, H. C. *Org. React.* **1963,** *13,* 1 [267].
[1203.] Sakurai, H.; Imoto, T.; Hayashi, N.; Kumada, M. *J. Am. Chem. Soc.* **1965,** *87,* 4001 [270].
[1204.] Yamamoto, M.; Izukawa, H.; Saiki, M.; Yamada, K. *Chem. Commun.* **1988,** 560 [270].
[1205.] Yoshifuji, S.; Tanaka, K.-I.; Kawai, T.; Nitta, Y. *Chem. Pharm. Bull.* **1985,** *33,* 5515 [291].
[1206.] Lukeš, R. personal communication [281].
[1207.] Giddings, S.; Mills, A. *J. Org. Chem.* **1988,** *53,* 1103 [38].

BIBLIOGRAPHY

Review Articles in *Organic Reactions*, Wiley, New York

Angyal, S. J. "The Sommelet Reaction" **1954**, *8*, 197–217.
Behrman, E. J. "The Persulfate Oxidation of Phenols and Arylamines (The Elbs and the Boyland–Sims Oxidations)" **1988**, *35*, 421–511.
Carmack, M.; Spielman, M. A. "The Willgerodt Reaction" **1946**, *3*, 83–107.
Cason, J. "Synthesis of Benzoquinones by Oxidation" **1948**, *4*, 305–361.
Denny, R. W.; Nikon, A. "Sensitized Photooxygenation of Olefins" **1973**, *20*, 133–336.
Djerassi, C. "The Oppenauer Oxidation" **1951**, *6*, 207–272.
Hassall, C. H. "The Baeyer–Villiger Oxidation of Aldehydes and Ketones" **1957**, *9*, 73–106.
Jackson, E. L. "Periodic Acid Oxidation" **1944**, *2*, 341–375.
Rabjohn, N. "Selenium Dioxide Oxidation" **1949**, *5*, 331–386.
Rabjohn, N. "Selenium Dioxide Oxidation" **1976**, *24*, 261–415.
Sheldon, R. A.; Kochi, J. K. "Oxidative Decarboxylation of Acids by Lead Tetraacetate" **1972**, *19*, 279–421.
Swern, D. "Epoxidation and Hydroxylation of Ethylenic Compounds with Organic Peracids" **1953**, *7*, 378–433.

Monographs on Oxidation

Augustine, R. L. *Oxidation I*; Dekker: New York, 1969.
Augustine, R. L.; Trecker, D. J. *Oxidation II*; Dekker: New York, 1971.
Benson, D. *Mechanisms of Oxidation by Metal Ions*; Elsevier: Amsterdam, 1976.
Cainelli, G.; Cardillo, G. *Chromium Oxidations in Organic Chemistry*; Springer: Berlin, 1984.
Chinn, L. J. *Selection of Oxidants in Synthesis*; Dekker: New York, 1971.
Dailey, P. S. *Ozonation in Organic Chemistry*; Academic: New York, Vol. 1, 1978; Vol. 2, 1982.
Denisov, E. T.; Mitskevich, N. I.; Agabekov, V. E. *Liquid Phase Oxidation of Oxygen Containing Compounds*; Plenum: New York, 1977.
Dryhurst, G. *Periodate Oxidation of Diol and Other Functional Groups*; Pergamon: Oxford, 1970.
Fonken, G. S.; Johnson, R. A. *Chemical Oxidations with Microorganisms*; Dekker: New York, 1972.
Frimer, A. A. *Singlet Oxygen*; Chemical Rubber Company (CRC): Cleveland, OH, 1985; Vols. 1–4.
Haines, A. H. *Oxidation of Organic Compounds: Alkanes, Alkenes, Alkynes, and Arynes*; Academic: London, 1985.
Houben-Weyl *Methoden der Organischen Chemie (in German). Oxidation I, Vol. IV/1a Oxidation II, Vol. IV/1b*; Thieme: Stuttgart, New York, 1975, 1981.
House, H. O. *Modern Synthetic Reactions*; Benjamin: New York, 1972.

Lee, D. G. *The Oxidation of Organic Compounds by Permanganate Ion and Hexavalent Chromium*; Open Court Publishing: LaSalle, IL, 1980.

Lundberg, W. O. *Autoxidation and Antioxidants*; Interscience: New York, Vol. 1, 1961; Vol. 2, 1962.

Mijs, W. J.; De Jonge, G. R. H. I. *Organic Synthesis by Oxidation with Metal Compounds*; Plenum: New York, 1986.

Oxidation of Organic Compounds, I–III; Mayo, F. R., Ed.; Advances in Chemistry 75–77; American Chemical Society: Washington, DC, 1968.

Reich, L.; Stirala, S. S. *Autoxidation of Hydrocarbons and Polyolefins*; Dekker: New York, 1969.

Schaap, A. P. *Singlet Molecular Oxygen*; Dowden, Hutchinson & Ross: Stroudsburg, PA, 1976.

Schöllner, R. *Die Oxydation Organischer Verbindungen mit Sauerstoff*; Akademie Verlag: Berlin, 1964.

Stewart, R. *Oxidation Mechanisms. Applications to Organic Chemistry*; Benjamin: New York, 1964.

Trahanovsky, W. S. *Oxidation in Organic Chemistry*, Part B; Academic: New York, 1973.

Trahanovsky, W. S. *Oxidation in Organic Chemistry*, Part C; Academic: New York, 1978.

Trahanovsky, W. S. *Oxidation in Organic Chemistry*, Part D; Academic: New York, 1982.

Turney, T. A. *Oxidation Mechanism*; Butterworths: London, 1965.

Wasserman, H. H.; Murray, R. W. *Singlet Oxygen*; Academic: New York, 1979.

Wiberg, K. B. *Oxidation in Organic Chemistry*, Part A; Academic: New York, 1965.

INDEXES

AUTHOR INDEX

The numbers indicate references on pp 321–357.

Abdel-Wahab, M. F., 813
Aberhart, D. J., 1073
Abrams, C., 729
Abushanab, E., 1068
Accountius, O. E., 591
Achiwa, K., 965
Ackermann, F., 115
Adam, W., 20
Adams, R., 441
Adinolfi, M., 751
Adkins, H., 108, 242, 244
Adler, E., 822
Aebi, A., 846
Afshari, G. M., 660
Ahlers, K., 1133
Ahrens, H., 1141
Ainsworth, C., 948
Akashi, K., 674
Akulenko, N. V., 273
Akutagawa, K., 1200
Alary, J., 159
Albert, H. E., 1095
Albizati, K. F., 1152
Albright, J. D., 1008, 1009
Alcalay, W., 907
Alder, K., 928
Alfonso, L. M., 264
Allen, C. F. H., 560, 625
Allen, D. S., Jr., 765
Allphin, N. L., 632
Amin, J. H., 472
Amon, C. M., 1013
Anders, D. E., 81
Anderson, C. B., 411
Anderson, G. J., 1000, 1003
Anderson, H. V., 1060
Ando, W., 13, 14
Andrejević, V., 443, 444
Andrews, S. D., 392
Angyal, S. J., 1132
Appel, M., 476
Arbusow, B. A., 1198
Archer, G. A., 375
Arens, J. F., 337
Arkell, A., 184, 406
Armsen, R., 1195
Arnold, M. R., 354
Arora, S. K., 790
Arrias, E., 265
Arth, G. E., 592
Ascherl, A., 1104
Ashley, K. R., 761

Asinger, F., 77
Askenasy, P., 792
Atkinson, C. M., 157, 202
Attenburrow, J., 805
Audette, R. J., 917
Auret, B. J., 1046, 1047
Awasthi, A. K., 797
Awasthy, A. K., 562
Axelrod, L. R., 852
Axen, R., 979
Azadani, M. N., 778

Bacchetti, T., 384
Bach, R. D., 1099
Bachman, G. B., 862
Baciocchi, E., 416
Back, T. G., 525
Bäckvall, J.-E., 669
Bacon, R. G. R., 198
Badovskaya, L. A., 182
Bailey, J. H., 308
Bailey, J. R., 350, 514
Bailey, P. S., , 93, 98, 107, 1110, 1123, 1124
Baker, B. R., 1168
Baker, R. H., 642
Baker, W., 197, 408
Bal, B., 755
Bal, B. S., 586
Balandin, A. A., 478
Baldwin, J., 726
Ball, R. G., 1024
Ball, S., 810
Ballard, D. A., 740
Ballard, S. A., 749
Ballistreri, F. P., 534
Baltzly, R., 388
Bamberger, E., 204, 556, 657
Banay, M., 756
Banks, C. V., 508
Banwell, M. G., 1013
Barakat, M. Z., 745, 813
Baran, J. S., 950
Barbachyn, M. R., 1103
Barber, G. W., 1155
Barbier, P., 1150
Barbieri, G., 92
Barclay, B. M., 969
Bardwell, J., , 750, 1173
Barger, G., 171
Barnes, B. A., 780
Barone, G., 751

Barrett, J. H., 316
Barrette, E.-P., 276
Bartlett, P. D., 36, 40, 43, 296, 1101
Barton, D. H. R., 525, 526, 527, 798, 799, 1021
Bass, L. S., 1102
Bassignani, L., 717
Basu, S. K., 409
Basyouni, M. N., 566
Bath, S. S., 1110
Bauer, D. P., 1005
Bauer, L., 474
Baumgarten, H. E., 1182
Bavley, A., 296, 806
Bayer, R. P., 153
Beal, D. A., 722
Beard, H. G., 990
Becker, H.-D., 44, 822, 971, 972
Bednarski, T. M., 412
Beebe, T. R., 780
Beguin, C. G., 939
Behrens, C. H., 1026
Belew, J. S., 929, 1136
Bell, E. W., 706
Bell, H. M., 473, 1156
Belli, A., 360
Bender, K. A., 780
Bender, M. L., 987
Bengtsson, E. B., 462
Benz, M., 368
Berenbaum, M. B., 733
Berends, W., 30
Berger, J., 1050
Berger, M., 91
Bergh, C. L., 713
Bergmann, E., 346
Berkowitz, L. M., 940
Berndt, A., 923
Bernstein, S., 524
Beroza, M., 71
Berthelot, J., 38
Bertrand, M., 1137
Bestman, H.-J., 1195
Beucker, H., 789
Beyler, R. E., 592
Bhatt, M. V., 178, 429, 802
Bierl, B. A., 71
Bierman, M. H., 424
Bigelow, L. A., 880
Billen, G. N., 240, 241

Billimoria, J. D., 724
Billman, J. H., 884
Bingham, R. C., 545
Birch, S. F., 834
Bird, J. W., 427
Black, D. J., 1042
Blackburn, T. F., 70
Blackwood, A. C., 1089
Blair, L. K., 726
Blanchard, H. S., 68
Blangey, L., 915
Blank, R. H., 1052
Blazejewicz, L., 56
Blomquist, A. T., 359, 393, 1167
Bloom, B. M., 1063
Blumberg, R., 137
Blust, G., 75
Boatright, L. G., 465
Bocarsly, A. B., 57
Bockstahler, E. R., 517
Boedtker, E., 471
Boekelheide, V., 1188
Böeseken, J., 259, 268, 270, 271, 272
Böeseken, M. J., 265
Bogdanova, O. K., 478
Bohlmann, F., 1015
Böhme, H., 331
Booth, D. L., 760
Borger, J. P., 588
Borgers, R., 371
Börner, E., 985
Bosshard, W., 501
Boulette, B., 710
Bousquet, E. W., 737
Bowden, K., 2, 578
Bower, J. D., 920
Bowers, A., 579, 585
Boyd, D. R., 1046, 1047
Boyer, J. H., 371
Brackman, W., 61
Branchaud, B. P., 1034
Brandt, A., 717
Braslau, R., 209
Braude, E. A., 970, 973, 983, 1121
Braun, G., 293, 310
Braun, v. J., 1147
Brenner, J., 306
Breuer, S. W., 183
Brewster, A. G., 525
Brice, T. J., 455
Brill, W. F., 217
Brittain, M. J., 1120
Brockman, H., 879
Brook, A. G., 457, 970, 983
Brooks, L. A., 497
Brooks, R. E., 388
Brossi, A., 1037
Brough, J. N., 312
Brown, B. R., 440
Brown, C. W., 202
Brown, G. L., 420

Brown, H. C., 51, 317, 536, 611, 641, 646, 1201, 1202
Brown, N. C., 197
Brown, R. W., 981
Brownfain, D. S., 1149
Bruce, J. M., 754
Bruce, W. F., 1146
Brunel, Y., 939
Brunelet, T., 617
Brunken, J., 35
Brunner, M. P., 1060
Brust, B., 375
Brutcher, F. V., Jr., 781
Buc, S. R., 135
Büchi, G., 201
Büchi, J., 846
Büchner, E., 1154
Buck, A., 1111
Buck, J. S., 189
Budde, W. M., 505
Budeěšínský, Z., 695
Buehler, C. A., 1177
Burdon, J., 838
Burgdorf, C., 556
Burge, R. E., Jr., 393
Burgher, A., 515
Burkhard, J., 546
Burkow, I. C., 238
Burn, D., 975
Burt, J. G., 23
Burt, P., 300
Buse, C. T., 755
Bush, J. B., Jr., 803
Bush, J. D., 591
Buu-Hoi, N. H., 589
Bye, T. S., 494
Byers, D. J., 559
Byrd, N. R., 397
Byun, H.-S., 999

Cacchi, S., 618, 690, 974
Caciagli, V., 717
Cain, E. N., 930
Cainelli, G., 571
Calder, A., 738
Cambie, R. C., 410, 779
Cameron, A. F. B., 805
Cameron, P. J., 671
Cameron, R. E., 57
Campaigne, E., 367, 505
Campbell, I. D., 357
Campbell, J. R., 449, 701
Campbell, N., 969
Campbell, W. P., 558
Camps, F., 315
Canonica, L., 823
Caputo, J. A., 941
Cardillo, G., 540, 543, 571, 614, 616
Carey, F. A., 323
Carlsen, P. H. J., 774
Carlsmith, A., 263
Carmack, M., 498, 502, 1161
Carpino, L. A., 1185

Carr, R. V. C., 324
Cartier, X., 1127
Casati, R., 619
Cason, J., 434
Castle, J. E., 874
Castrillón, J. P. A., 795
Cavalieri, L., 1161
Cavallito, C. J., 308
Cavill, G. W. K., 437, 438
Čech, J., 1197
Cella, J. A., 326
Cha, D. Y., 952
Chamberlain, P., 997
Chambers, D., 779
Chan, K. K., 953
Chandrasekaran, S., 607
Chang, F. N., 1070
Chang, V. S., 848
Chao, T. H., 187
Chapdelaine, M. J., 958, 960
Chapman, J. H., 805
Chapman, K., 693, 1152
Chargaff, E., 759
Charibi, H., 659
Charney, W., 1051, 1056, 1057
Chaux, R., 470
Chavan, S. P., 1190
Chaykovsky, M., 979
Chechina, O. N., 123
Chemerda, J. M., 1054
Chen, S.-H., 604
Chen, Y.-Y., 279
Cheng, Y.-S., 604
Cheradame, H., 939
Chern, C.-I., 675
Chhabra, B. R., 664
Chiang, Y. H., 1002
Chitrakorn, S., 606
Chivers, G. E., 158
Chow, M.-F., 20
Church, J. M., 4, 72, 137, 1135
Ciganek, E., 161
Cinnamon, M., 1002
Cinquini, M., 773
Clardy, J. C., 1102
Clarke, H. T., 62, 635, 730, 882, 1129
Clarke, T. G., 381
Claus, A., 888, 889, 890
Claus, R. E., 1108
Cleland, J. H., 937
Clement, W. H., 60
Cliff, I. S., 622
Clutter, R. J., 892
Coates, R. R., 466
Cocker, J. D., 439
Coeur, A., 159
Cohen, M., 561
Cohen, W. D., 259
Cohen, Z., 105, 106
Cole, A. G., 104
Coleman, G. H., 645
Coll, J., 315

Collins, G. R., 361
Collins, J. C., 594, 595
Colonna, S., 773
Conant, J. B., 643
Connor, R., 547
Cook, C. D., 922
Cook, J. W., 466
Cook, M. J., 769
Cook, N. C., 85
Cook, W. J., 651
Cookson, P. G., 127
Cope, A. C., 156, 161, 396, 1187
Corey, E. J., 276, 382, 518, 602, 603, 605, 661, 720, 721, 965, 1019, 1020, 1138, 1183
Cornsworth, J. W., 839
Corson, B. B., 887
Coryell, C. D., 379
Cossió, F. P., 663
Costa, A., 487
Cotter, R. J., 396
Cowan, J. C., 81
Coxon, J. M., 512
Craig, J. C., 320
Crandall, J., 584
Crich, D., 799
Criegee, R., 48, 75, 78, 84, 109, 140, 436, 446, 789, 1154
Crook, K. E., 877
Crooks, W. J., III, 927
Crouse, D. J., 409
Crovetti, A. J., 162
Crowley, P. J., 1185
Cullis, C. F., 750, 1173
Curci, R., 534
Curragh, E. F., 818
Curran, D. P., 207
Curtin, D. Y., 821
Cutting, J. D., 328

D'Auria, M., 608
Dabovich, T. C., 554
Dahmen, K., 796
Dailey, O. D., Jr., 323
Dakin, H. D., 172, 188
Dalcanale, E., 712
Dansted, E., 512
Darling, S. D., 185
Dauben, W. G., 593
Dave, P., 999
Davey, W., 842
Davies, A. G., 127
Davies, D. D., 957
Davies, R. R., 345, 962
Davis, F. A., 1032
Davis, J. B., 1071
Day, A. C., 386, 392
Dayan, J. E., 1158
De Boer, T. J., 378
de Vries, G., 69
Deahl, T. J., 379
Dean, F. H., 472

Dean, P. M., 18, 503
Deebel, G. F., 1128
Degering, E. F., 465
Dehmlow, E. V., 192
Dehn, W. M., 740, 753
deJonge, C. R. H. I., 430
Demerson, C., 854, 953
Deming, P. H., 129
den Hollander, A. J., 615
Denis, J.-N., 325
Depke, F., 49
DePuy, C. H., 582
Deshmukh, M. N., 224
Deslongchamps, P., 111
Dess, D. B., 801
Dessau, R. M., 804
Dessy, R. E., 405
DeTar, D. F., 502
Dev, S., 597
DeVault, D., 379
Di Furia, F., 1025
Di Nunno, L., 1186
Diaper, D. G. M., 96, 100, 427
Dick, C. R., 112
Dick, K. F., 572
Dickey, J. B., 351, 694
Dickinson, J. D., 452
DiCosimo, R., 292
Diehl, H., 508
Dimant, E., 756
Dimroth, K., 923
Dittmann, W., 250
Dittmer, K., 955
Dixon, J. A., 696
Djerassi, C., 243, 579, 942, 1148
Dobrowsky, A., 170
Dodge, R. A., 887
Dodson, R. M., 1048
Doering, v. W. E., 261, 1018, 1166
Dolby, L. J., 760
Donin, M. N., 1045
Dorfman, R. I., 482
Dorsey, J. E., 313, 1180
Dost, N., 967
Douglass, I. B., 683, 684, 687
Doumaux, A. R., Jr., 225, 255
Douros, J. D., Jr., 1053
Downs, C. R., 479
Dox, A. W., 450
Doyle, M. P., 231
Drabowicz, J., 521
Dreger, E. E., 62
Drozd, J. C., 563
Du, M. T., 1001
Dübendorfer, H., 915
DuBois, R. H., 668, 671, 672
Dufraisse, C., 18
Düll, H., 80
Dulou, R., 1117
Dumont, W., 325

Duñach, E., 224
Dunbar, R. E., 354
Dunn, F. W., 955
Dunstan, W. R., 154
Dupont, G., 1117
Durham, D. L., 1101
Durland, J. R., 108
Dzhemilev, U. M., 342

Eaborn, C., 452
Eadon, G., 349
Eastman, J. F., 963
Edwards, D., 541
Egami, F., 1085
Egan, R., 1111
Eglinton, G., 58, 357
Ehrhart, O., 714
Eichelberger, J. L., 99
Einhorn, A., 840
Eisch, J., 623
Eisenbraun, E. J., 581, 1171
El-Sadr, M. M., 813
El-Wahab, M. F. A., 745
Eldridge, J. M., 843
Elix, J. A., 318
Ellenberger, S. R., 353, 1102
Ellington, P. S., 783
Ellis, B. A., 468
Emerson, W. S., 355, 1128
Emmons, W. D., 253, 257, 281, 282, 283, 284, 285, 288, 289, 290, 291, 301, 338, 1178
Endo, T., 995
Engel, R., 999
Engle, R. R., 579, 942
Engler, D. A., 531
English, J., Jr., 1153, 1155, 1158
Enomoto, S., 139
Eppstein, S. H., 1076, 1077, 1078, 1079
Erickson, R. E., 93, 114, 1123, 1124
Ertel, L., 80
Eschenmoser, A., 151
Eustance, J. W., 68
Evans, R. H., Jr., 1052
Evans, R. L., 155
Evans, R. M., 805
Evans, T. L., 208
Evans, T. W., 753
Evans, W. L., 871
Evrard, G., 325

Fabio, P. F., 1001
Failla, S., 534
Falk, K.-H., 136, 340
Faller, P., 705
Farachi, C., 776
Farkas, L., 742
Farmar, J. G., 90
Farmer, E. H., 22
Farnham, N., 304

Farrall, M. J., 612
Fatiadi, A. J., 757, 758
Fauser, K., 747
Fawcett, J. S., 1121
Fazal, N., 127
Federbush, C., 1056, 1057
Felix, D., 151
Fenical, W., 21, 39
Ferraboschi, P., 778
Ferris, A. F., 285
Fetizon, M., 377
Fichter, F., 115
Fickert, E., 890
Field, B. O., 454
Fields, E. K., 97
Fierz-David, H. E., 915
Fieser, L. F., 491, 499, 558, 627, 628, 909, 1191, 1192, 1193
Fiesselmann, H., 365
Filler, R., 629
Findlay, J. W., 968
Findley, T. W., 241, 295
Finkbeiner, H., 803
Finn, M. G., 1027
Finnegan, R. A., 220
Firouzabadi, H., 659, 660, 832, 833, 895, 896, 897
Fischer, F. G., 80, 110, 1106
Fischer, H. O. L., 336
Fishel, D. L., 633
Flatt, S. J., 538
Fleet, G. W. J., 538, 602
Fleisher, G. A., 1179
Florio, S., 1186
Floyd, M. B., 1001, 1109
Fonken, G. S., 708, 1080, 1081, 1082, 1083, 1084
Foote, C. S., 12, 13, 14, 15, 37, 41
Forrest, J., 855
Forrester, A. R., 738
Foster, G., 620, 666
Föttinger, W., 680
Francis, P. S., 1144, 1145
Frank, F. J., 595
Frankenfeld, J. W., 1053
Franzen, V., 993
Fray, M. J., 1014
Frazier, H. W., 1004
Frazier, R. H., Jr., 911
Fréchet, J. M. J., 612
Freedman, H. H., 692
Freeman, F., 668, 671, 672
Freeman, J. P., 288
Fremery, M. I., 97
Freyermuth, H. B., 135
Fried, J., 552, 1045, 1065
Friedman, L., 633, 1130
Friedman, N., 11
Friend, N. A. C., 510, 519
Friess, S. L., 303, 304, 305
Frimer, A. A., 314
Frisell, C., 235
Frohn, H.-J., 1133

Fry, E. M., 558
Fryer, R. I., 375
Fuchs, R., 941
Füeg, H. R., 846
Fujii, K., 1044
Fukunaga, T., 184
Fukuyama, M., 309
Fullerton, D. S., 593
Fulmer, R. W., 401
Furukawa, N., 1200
Fuson, R. C., 516, 775, 1151

Galbraith, A. R., 58
Gale, D. M., 445
Gall, R. J., 174
Gancher, E., 91
Ganem, B., 287
Ganem, B. E., 382
Gannon, W. F., 333
Gardella, L. A., Jr., 474
Gardner, J. H., 875, 1125
Gardner, T. S., 352
Garg, C. P., 536, 641, 646
Garner, B. J., 1021
Garrard, T. F., 997
Garrett, J. W., 656
Garrett, R. D., 1049
Garza, C., 1136
Gash, V. W., 401
Gates, M. D., Jr., 558
Geistlich, P., 400
Gelbard, G., 617
Genke, J. R., 1045
Gensler, W. J., 101
Gentles, M. J., 1051, 1056, 1057
Ghaderi, E., 832
Gibson, D. T., 1074
Gibson, R. E., 956
Giddings, S., 1207
Giersch, W., 817
Gil, G., 1137
Gilliard, X., 1127
Gilman, H., 165, 623
Gilman, N. W., 382
Gilman, S., 343
Giordano, C., 360
Giovannini, E., 910
Giovini, R., 773
Glenn, R. A., 514
Glockling, F., 408
Godfrey, C. R. A., 798
Godfrey, I. M., 318
Godoy, J., 786, 787, 788
Goemann, M., 409
Gogins, K. A. Z., 610
Goldman, I. M., 809
Goldman, L., 1008, 1009
Goldschmidt, S., 245, 982
Goldstein, A., 359
Goldstein, H., 475, 551
Golfier, M., 377
Gollnick, K., 26, 46
González, A., 663
Goodwin, R. W., 810

Gopal, H., 938
Gordon, A. J., 938
Gorodetsky, M., 11
Goulding, E., 154
Grabowich, P., 552
Grade, M. M., 208
Gravatt, G. L., 1013
Graves, D. M., 800
Gray, A. R., 516, 520
Graymore, J., 957
Greenspan, F. P., 124, 174, 239, 249
Greenwood, F. L., 73, 94
Grieco, P. A., 327, 343
Griffith, W. P., 196
Griswold, P. H., Jr., 1153
Gritter, R. J., 808
Grivsky, E. M., 388
Grob, C. A., 447, 709
Gross, B., 682
Gross, O., 921
Grosse, A. V., 1096
Gruber, J. M., 1159
Grummitt, O., 1111
Grundon, M. F., 1091
Grundy, J., 454, 456
Grunwald, J., 1036
Gubler, A., 1066
Guizard, C., 939
Gunstone, F. D., 782
Günthard, H. H., 394
Gupta, S. K., 1201
Guss, C. O., 746
Gut, M., 482
Gutsche, C. D., 736
Guziec, F. S., Jr., 530
Gwilt, J. R., 842

Haab, F., 334
Haarmann, R., 86
Hach, C. C., 508
Hadler, H. I., 1122
Halbert, A. D., 780
Hall, D. T., 937
Halsall, T. G., 585
Hamilton, G. A., 689
Hampson, N. A., 381
Han, I.-S., 590
Hanna, R. F., 112
Hansen, S. H., 238
Hanson, F. R., 1060, 1061
Hanze, A. R., 1059
Haque, M. S., 1032
Hardegger, E., 160
Harding, K. E., 572
Hardy, W. B., 187
Harfenist, M., 806
Harlow, R. L., 911
Harman, R. E., 637
Harrell, L. L., Jr., 330
Harries, C., 86
Harris, J. O., 1177
Harris, S. A., 887
Harrison, I. T., 539
Harrison, S., 539

Author Index

Harrisson, R. J., 63
Hartley, W. M., 952
Hartman, J. L., 203
Hartman, R. E., 1052
Hartman, W. W., 351, 635, 694, 1129
Hartmann, C., 893
Hartmann, M., 501
Hartshorn, M. P., 512
Hartzell, G. E., 768
Hasan, F., 570, 621
Hasan, S. K., 701
Hashimoto, J., 119
Hass, H., 871
Hass, H. B., 862, 987
Hasselbach, M., 490
Hauck, F. P., Jr., 402
Hauser, F. M., 353, 1102
Hawkins, E. G. E., 180, 931, 1163
Hawthorne, M. F., 280, 282
Hay, A. S., 59, 68, 401
Hayakawa, Y., 228
Hayano, M., 1063
Hayashi, J., 121
Hayashi, N., 1203
Hayashi, S., 553
Hazen, R. K., 887
Heathcock, C. H., 755
Hedgecock, H. C., Jr., 992
Heer, J., 752
Heiba, E. I., 804
Heider, R. L., 1098
Heiduschka, A., 484
Heilbron, I., 2
Heilbron, I. M., 578
Heinemann, R., 6
Heller, G., 886
Heller, M., 524
Helmy, E., 311
Hems, B. A., 805
Henbest, H. B., 10, 298, 439, 812, 818, 826, 827, 828, 837, 1046, 1047
Henne, A. L., 869
Hensley, M., 1074
Hentges, S. G., 951
Hepburn, S. P., 738
Hernandez, O., 323
Hernestam, S., 728
Herr, M. E., 1080, 1081, 1082, 1083, 1084
Herranz, E., 723, 1105
Herriott, A. W., 845
Hershberg, E. B., 469, 491, 622, 1051, 1056, 1057
Herz, J. E., 1045
Herzog, H. L., 1051, 1056, 1057
Hess, K., 980
Hess, W. W., 594, 595
Hesse, E., 195
Heusser, H., 400
Hey, D. G., 783
Heymann, H., 1126

Heyns, K., 5, 6, 56
Heywood, D. L., 277
Hibbert, H., 300
Hickinbottom, W. J., 564
Higgins, R., 13
Highet, R. J., 811
Hill, A., 204
Hill, J. G., 213, 1028
Hill, J. W., 873
Hirano, S., 1006, 1017
Hiroi, K., 321
Hiskey, R. G., 398
Hixon, R. M., 451
Ho, H. C., 428
Ho, T.-L., 190, 421, 422, 425, 426, 428, 528
Hock, H., 27, 28, 29, 47, 49
Hodgson, H. H., 345, 962, 990
Hoehn, H. H., 391
Hofer, P., 319
Hofmann, K. A., 714
Holloway, F., 561
Holm, R. T., 125
Holmes, A. B., 66
Holmes, H. L., 698
Holmes, R. R., 153
Holmgren, A. V., 555
Holmlund, C. E., 1052
Holum, J. R., 599
Hönig, M., 1165
Hopff, H., 460, 461
Horan, J. E., 246
Hou, K.-C., 790
House, H. O., 1, 149, 251, 333, 582
Höver, H., 109
Howard, K. L., 390
Howe, R., 1086, 1087
Hu, H., 794
Huang, P.-t., 733
Huang, S. L., 1012, 1022, 1139
Hudlický, M., 103, 191, 473, 477, 1156
Hudlický, T., 92
Hudrlik, O., 73
Hughes, M. T., 1057
Hui, R. A. H. F., 525
Huntress, E. H., 469, 622
Hurd, C. D., 653, 656
Hussey, A. S., 642
Hussmann, G. P., 279
Hutchins, R. O., 651
Hüttel, R., 177
Hymans, W. E., 584

Ichimoto, I., 1044
Ide, W. S., 189
Imamoto, T., 791
Imanishi, T., 996
Imoto, T., 1203
Indictor, N., 217
Ingold, C. F., 898
Inoue, M., 139

Ip, D., 587
Issidorides, C. H., 816
Ito, S., 67
Ito, Y., 20
Itoh, T., 216
Iwata, C., 996
Izukawa, H., 1204

Jacini, G., 384
Jacob, L. A., 1001
Jacob, T. A., 862
Jacobs, J., 270
Jacques, J., 523
Jakobson, P., 234
Jakovac, I. J., 1035
Jallabert, C., 347
Jansen, A. B. A., 805
Jarrar, A., 816
Jasiczak, J., 976
Jaunin, R., 475
Jawanda, G. S., 664
Jay, E. W. K., 854
Jay, L., 854
Jeger, O., 201, 400
Jensen, H. P., 861
Jeyaraman, R., 210, 211
Jitsukawa, K., 216
Joffe, V. A., 924
Johnson, A. P., 998
Johnson, B. D., 1001
Johnson, C. R., 767, 770, 772, 1103, 1199
Johnson, E. A., 834
Johnson, H. W., Jr., 821
Johnson, J. R., 548
Johnson, R. A., 1081, 1082, 1083, 1084
Johnson, T. B., 683, 684
Johnson, W. S., 736, 765
Jones, D. F., 1086, 1087
Jones, E. R. H., 2, 578, 585
Jones, G. E., 66
Jones, J. B., 1035
Jones, R. E., 743
Jones, R. G., 143, 876
Jones, W. J., 859, 1000, 1003
Jorissen, W. P., 1164
Joseph, J. P., 1168
Joshi, H. K., 1135
Jouitteau, C., 617
Juenge, E. C., 722
Juhasz, G., 1066
Jung, M. E., 981
Junker, P., 385
Juve, H. D., Jr., 1159

Kabalka, G. W., 317, 992
Kabasakalian, P., 1051
Kaeding, W. W., 361
Kagan, H. B., 224, 1029
Kageyama, T., 739
Kaiser, E. M., 870
Kalb, L., 921
Kalman, J. R., 449
Kalsi, P. S., 664

Kalyugina, T. Y., 182
Kamm, O., 634
Kampe, W., 1189
Kaneda, K., 216
Kaneko, C., 33
Kanemitsu, K., 734
Kanemoto, S., 236, 741
Kanno, T., 854
Kaplan, H., 513
Kaplan, M. L., 19
Karabinos, J. V., 576
Kariyone, K., 511
Karnemaat, J. N., 1060
Karrer, P., 332, 334
Kashimura, N., 1006, 1017
Katsuki, T., 221, 222, 223, 774, 1026
Kauer, J. C., 1184
Kaulen, J., 116
Kawabata, N., 119
Kawai, S., 1085
Kawai, T., 944, 1205
Kawamura, S., 934
Kazandjian, R. Z., 1039
Kearns, D. R., 21, 39
Keaveney, W. P., 91
Kehrmann, F., 648
Keinan, E., 105, 106
Keiser, J. E., 772
Kendall, E. C., 1107, 1179
Kendrick, D. A., 66
Kennedy, R. J., 205
Kent, W. H., 467
Kenyon, J., 830, 831
Kesztler, F., 784
Khan, N. A., 864
Khan, S. A., 837
Kharasch, M. S., 23, 181, 233, 1113
Kidwell, R. L., 185
Kierstead, R. W., 1050
Kihara, K., 677, 678
Kilmer, G. W., 499
Kim, C. U., 720, 721, 1019, 1020, 1138
Kim, H.-j., 535, 568
King, J. A., 500, 1169
King, S. W., 82
Kinkel, K. G., 1115
Kip, C. J., 259
Kleiderer, E. C., 885, 961
Klein, E., 858
Kleinfelter, D. C., 583
Klibanov, A. M., 1036, 1039
Klingsberg, A., 1065
Kluge, F., 879
Klyne, W., 482
Knauel, G., 49
Knight, H. B., 50
Knight, J. W., 1099
Knilling, v. W., 1104
Knowles, W. S., 89
Kobayashi, H., 128
Kobayashi, M., 711
Kobrina, L. S., 273

Koch, E., 31
Kocher, F. W., 743
Kochhar, K. S., 586
Kochi, J. K., 679
Koelsch, C. F., 891
Komppa, G., 735
Komzak, A., 493
Konaka, R., 932
Kondo, K., 32, 309
König, K.-H., 369
Korach, M., 252
Kornblum, N., 859, 892, 1000, 1003, 1004
Kornfeld, E. C., 961
Korte, F., 163
Korytnyk, W., 1141
Koteel, C., 847, 863
Koto, H., 791
Kovacic, P., 232, 1119, 1120
Kozlowski, M. A., 1054
Kozuka, S., 727
Kraft, M. E., 927
Krapcho, A. P., 843
Krapivin, G. D., 182
Krauch, C. H., 44
Krawczyk, E., 1194
Krebs, E., 229
Kremers, E., 451
Křepinský, J., 854
Krief, A., 325
Krishman, S., 689
Kröhnke, F., 984, 985
Kruse, W., 412
Kubli, H., 373
Kubota, I., 200
Kuffner, F., 784
Kuhn, D. G., 689
Kuhn, H. J., 46
Kuhn, R., 1181
Kuhnen, L., 226
Kul'nevich, V. G., 182
Kulka, M., 117
Kulkarni, S. U., 611
Kumada, M., 1203
Küng, W., 395
Kurosawa, Y., 1063
Kurz, M. E., 232, 1119
Kuwajima, I., 1140
Kwart, H., 1144, 1145
Kwart, L. D., 92

La Torre, F., 690
Laatsch, H., 423
Laba, V. I., 166
Ladner, D. W., 933
Lajis, N. H., 744
LaLancette, E. A., 1167
LaLonde, J., 650
Landa, S., 546
Lane, A. G., 107
Lang, S., 27, 28, 29, 47
Langbein, G., 575
Langheld, K., 613, 700
Langley, W. D., 199, 658
Large, R., 180, 931

Larsen, S., 532
Larson, H. O., 1003
Larson, J. R., 843
Lassak, E. V., 275
Lauer, R. F., 167, 168, 860
Lawesson, S.-O., 234, 235
Lazier, W. A., 806
Leatham, M. J., 183
Lecocq, J. J., 589
Lederer, M., 84
Lee, A. W. M., 1026
Lee, C., 835, 844
Lee, D. G., 639, 848, 849, 856, 857, 894, 937, 943
Lee, G. A., 692
Lee, H.-H., 156
Lee, J., 352
Lee, J. B., 381
Lee, S. S., 1070
Lee, T. V., 676
Leenay, T. L., 1024
Leeper, C. G., 748
Leffler, J. E., 269
Lefort, D., 1117
Lehner, A., 686
Leigh, H. M., 1076, 1077, 1078
LeMahieu, R. A., 1050
Lemieux, R. U., 763, 764, 765
Lemin, A. J., 585
Lemke, G., 1147
Leonard, N. J., 401, 402, 403, 767, 770
Leopold, B., 988
Lester, D. J., 525, 527
LeSuer, W. M., 367
Levand, O., 1003
Levin, A. I., 123
Levin, R. H., 708
Levine, A. A., 104
Levine, P., 649
Levine, R., 1172
Lewcock, W., 1094
Lewis, A. D., 299
Ley, S. V., 169, 525, 527
Li, W.-b., 1043
Lieberman, S. V., 547
Lin, Y. Y., 1062
Linderman, R. J., 800
Lindgren, B. O., 1162
Lindner, J. H. E., 46
Linn, W. J., 150, 1188
Linsert, O., 399
Linsker, F., 155
Linskeseder, M., 785
Linstead, R. P., 496, 649, 945, 946, 947, 970, 973, 983, 1093
Lissel, M., 192
Liu, H.-J., 590, 1024
Liu, K.-T., 9, 536, 771
Liu, W.-L., 1024
Livingston, J. R., Jr., 506
Lochte, H. L., 350
Lockwood, L. B., 1040

Locquin, R., 1150
Löhmer, K. H., 163
Lombardino, J. G., 398
Longley, R. I., Jr., 355
Looft, E., 1174
López, C., 663
Lorber, M., 593
Lorenz, O., 88
Louis, J.-M., 377
Löwenberg, K., 1106
Lucas, G. B., 284, 289
Luczak, J., 1007
Lugowkin, B. P., 509
Lukeš, R., 1206
Luna, H., 92
Luzzio, F. A., 530, 1183
Lynn, L., 4
Lyons, J. F., 836
Lythgoe, B., 312
Lyzwa, P., 521

Maassen, J. A., 378
Mabry, T., 1074
MacDonald, D. L., 336
MacKellar, D. G., 174, 239
Maclagan, N. F., 724
Macomber, R. S., 1005
Magasanik, B., 759
Mageli, O. L., 179
Magriotis, P. A., 276
Malan, R. L., 503
Maloney, L. S., 152
Malte, A. M., 954
Mamoli, L., 1088
Mancera, O., 243, 807
Mancuso, A. J., 1022, 1149
Mandolini, L., 416
Mangold, D., 260
Mangoni, L., 751
Manley, N. C., 294
Mann, K. M., 1060, 1061
Manzocchi, A., 537, 776, 777
Mareš, F., 103, 1097
Markees, D. G., 731
Marks, B. S., 397
Marmor, S., 691
Marsh, P. G., 114
Marshall, B., 666
Martell, A. E., 432
Martin, E. L., 1193
Martin, J. C., 801
Martin, V. S., 774, 1026
Marukyan, G. M., 478
Marvel, C. S., 464, 497
Mashiko, T., 1030
Mathieson, J. W., 1136
Matsubara, S., 236
Matsuhira, N., 734
Matsumoto, M., 32, 67, 128
Matsumura, Y., 121
Matsushita, Y., 553
Matsuura, T., 126, 677, 678
Matthews, A. O., 634
Matthews, W. S., 954
Mattox, V. R., 1107

Mattox, W. J., 1096
Mauthner, F., 913
May, L. M., 572
Maynert, E. W., 1151
Mazur, S., 37, 41
Mazur, Y., 11, 105, 106
McAleer, W. J., 1054
McCants, D., Jr., 1199
McCarty, C. G., 748
McCloskey, C. M., 645
McClure, J. D., 1118
McCurdy, J. T., 1049
McDermott, J. X., 565
McDonald, R. N., 313, 1180
McElvain, S. M., 458, 877, 883
McEvoy, F. J., 1168
McEwen, W. L., 873
McGovern, T. P., 825
McGrath, J. P., 326
McGuckin, W. F., 1107
McIntosh, A. V., 569
McIntyre, J., 202
McKeon, J. E., 255
McKeown, E., 16
McKillop, A., 194, 278, 383, 413, 414
McKittrick, B. A., 287
McLaughlin, K. C., 876
McLaughlin, T. G., 672
McLeod, G. L., 699
McMillan, F. H., 500, 1169
McOmie, J. F. W., 408
Meakins, G. D., 783
Mehta, N. B., 388
Meinert, R. N., 653
Meinwald, J., 584
Meisenheimer, J., 195
Meister, P. D., 1076, 1077, 1078, 1079
Mendenhall, G. D., 1101
Menger, F. M., 835, 844
Mertens, H.-J., 1115
Merz, V., 919
Messeguer, A., 315
Meyer, V., 792, 893
Meyers, C. Y., 702, 954
Meystre, C., 1064
Michaelis, K. O. A., 496, 947, 1093
Michaelson, R. C., 215, 328, 1031
Michel, G., 925
Middleton, W. J., 445
Midland, M. M., 51
Miescher, K., 752
Miethe, R., 1015
Mignonac, G., 7
Mihailović, M. L., 443, 444
Mikolajczyk, M., 521, 1007
Milani, F., 619
Milas, N. A., 130, 134, 152, 179, 212, 715, 716, 718
Milczanowski, S. E., 927
Milewich, L., 319, 852

Miller, J. B., 387
Miller, L. L., 610
Miller, R. D., 780
Miller, v. W., 673
Millidge, A. F., 945
Mills, A., 1207
Mills, D. C., 901
Mimoun, H., 140
Minami, K., 935
Minato, M., 120
Mineo, S., 934
Misco, P. F., 721
Mislow, K., 306
Mitchell, D. L., 96
Mitchell, P. W. D., 431
Miyaura, N., 679
Miyazawa, T., 995
Mizoguchi, M., 121
Mochizuki, K., 900
Modena, G., 1025
Modlin, L. R., Jr., 515
Moffatt, J. G., 1010, 1016
Moffett, R. B., 569
Mohan, L., 211
Moiseenkov, A. M., 342
Möller, G., 369
Moniz, W. B., 696
Monnet, P., 1127
Montanari, F., 712
Moor, K., 493
Moosavipour, H., 660
Moreau, C., 111
Morey, J., 487
Morgan, K. J., 750, 1173
Morgan, M. S., 480
Moriarty, R. M., 790, 794, 797
Moriconi, E. J., 256
Morley, J. F., 519
Morley, J. R., 381
Morrell, J. C., 1096
Morris, J. A., 901
Morris, L. J., 782
Morrow, D. F., 403
Morton, R. A., 810
Morzycki, J. W., 526, 798
Mosher, H. S., 263
Mostafavipoor, Z., 833
Motherwell, W. B., 526, 798
Moureu, C., 7, 18, 470
Moussa, G. E. M., 564, 566
Mowry, D. T., 688
Moyer, J. R., 294
Moyle, M., 63
Mugdan, M., 131
Müller, A., 925
Müller, P., 786, 787, 788
Müller, W., 329
Murphy, M., 185
Murray, H. C., 1059, 1075, 1076, 1077, 1078, 1079, 1080, 1081, 1082, 1083, 1084
Murray, P. J., 169
Murray, R. W., 19, 76, 210, 211, 820

Muxfeldt, H., 879
Muzychenko, G. F., 182

Nabih, I., 311
Naderi, M., 895, 896, 897
Nagarkatti, A. S., 441
Nagarkatti, J. P., 761
Nagashima, H., 65
Nagata, R., 126
Najafi, M. R., 991
Nakagawa, K., 932, 934, 935, 936
Nakajima, S., 918
Nakamura, A., 996
Nakata, T., 932
Nakayama, M., 553
Narain, N. K., 906
Narasaka, K., 1033
Natale, N. R., 651
Nathan, A. H., 1077
Naves, Y.-R., 299
Naylor, C. A., Jr., 1125
Neave, G. B., 344
Negishi, A., 309
Neirabeyeh, M. A., 682
Neiswender, D. D., Jr., 696
Nelson, L. A. K., 1024
Nelson, O. A., 3
Nemoto, H., 65
Nenitzescu, C. D., 389
Nes, W. R., 356
Neumüller, O. A., 26
Newman, M. S., 184, 406, 494, 504, 598, 698, 864, 1092
Newman, N., 1048
Nielsen, B. E., 1052
Nielsen, D. R., 252
Nielson, E. D., 1061
Nightingale, D. V., 624
Nikishin, G. I., 924
Nikles, E., 160
Nilsson, T., 1162
Nishihara, A., 200
Nishimura, T., 549
Nitta, Y., 944, 977, 1205
Nobile, A., 1056, 1057
Nobis, J. F., 165
Nohara, F., 307
Nonomura, S., 1044
Norberg, B., 325
Norton, D. G., 749
Norton, R. V., 687
Noureldin, N. A., 849, 894
Noyes, A. A., 379
Noyes, W. A., 350
Noyori, R., 228, 237
Nozaki, H., 236, 328, 741
Numata, T., 132

O'Brian, J. L., 82
O'Brien, J., 1037
O'Connell, P. W., 1060, 1061
O'Connor, W. F., 256
Oae, S., 132, 727

Ochrymowycz, L. A., 322
Odinokov, V. N., 83
Offenhauer, R. D., 736
Ogata, Y., 302
Ogg, C. L., 341
Ogino, T., 900
Oguri, T., 327
Ohfune, Y., 343
Ohira, S., 553
Ohkubo, H., 711
Ohloff, G., 817
Ohme, R., 372
Ohnishi, Y., 727
Ohta, H., 1055, 1058
Okamoto, Y., 1058
Okawara, M., 739
Olah, G. A., 190, 528, 529
Oldenziel, O. H., 413
Oldham, W. J., 834
Omura, K., 1011, 1012, 1023, 1139
Onodera, K., 1006, 1017
Onoue, H., 935
Oppenauer, R. V., 964
Orena, M., 540, 543, 571, 614
Oroshnik, W., 1002
Osborn, H. M. L., 1079
Osborne, E. N., 656
Oshima, K., 236, 741
Oshima, T., 1085
Ouannes, C., 1116
Ouchi, M. D., 24
Ourisson, G., 523
Overberger, C. G., 397, 398, 733
Overend, W. G., 87

Pacaud, R. A., 560
Pagano, A. S., 283, 288, 290, 301
Page, G. A., 274
Paige, J. N., 768
Palermo, R. E., 1031
Palmer, B. D., 169
Palmer, L. C., 397
Palomo, C., 663
Pamfilov, A. W., 1131
Papa, D., 1170
Papadopoulos, E. P., 816
Pappas, J. J., 91
Pappo, R., 765
Paquette, L. A., 316, 324
Parikh, J. R., 1018
Parish, E. J., 606
Park, J. R., 453
Parker, E. E., 884
Parker, K. A., 90
Parker, W. E., 341
Parmar, S. S., 906
Parrilli, M., 751
Partch, R. E., 435, 442
Partington, J. R., 453
Pasetti, A., 1100
Patrick, J. B., 95
Patrie, W. J., 231

Pattison, D. B., 1161
Pattison, F. L. M., 118, 472, 574
Pausacker, K. H., 793
Payne, C. C., 1056, 1057
Payne, G. B., 129, 138, 144, 145, 146, 147, 148, 175, 219, 254
Pearl, I. A., 206, 362, 363, 364, 366, 989, 1160
Pedersen, K., 234
Pelter, A., 998
Penmasta, R., 797
Perfetti, R. B., 959
Periasamy, M., 429, 802
Pericàs, M. A., 315
Perlman, D., 1045
Perlman, P. L., 1056, 1057
Perret, A., 230
Perrine, T. D., 926
Perrot, R., 230
Perry, M. A., 956
Perst, H., 923
Peterson, D. H., 1075, 1076, 1077, 1078, 1079
Petrow, V., 975
Pfitzner, K. E., 1016
Phillips, B., 262, 277
Phillips, G. H., 439
Piancatelli, G., 608
Picker, D., 845
Pigott, F., 489
Pike, P. E., 114
Pike, R. A., 1187
Pilgrim, W. R., 997
Pinhey, J. T., 275, 449
Pinnick, H. W., 586, 744
Pinson, R., Jr., 305
Pitchen, P., 224, 1029
Pletcher, D., 630, 652
Pobiner, H., 52
Poisel, H., 707
Pollak, A., 339
Ponti, F., 537, 777
Poos, G. I., 592
Portmann, R., 910
Posner, G. H., 958, 959, 960
Postowsky, J. J., 509
Powell, S. G., 469
Powers, J. W., 1003
Prairie, R. L., 1072
Prakash, G. K. S., 528
Prakash, I., 790, 797
Pratt, E. F., 814, 815, 825
Prelog, V., 394, 395, 493
Price, C. C., 186, 229, 576
Prichard, W. W., 703
Přikryl, J., 695
Prilezhaeva, E. N., 166
Pryde, E. H., 81
Purushothaman, K. K., 320

Quail, J. W., 917
Quayle, O. R., 643
Quici, S., 865

Author Index 371

Raasch, M. S., 874
Radkowsky, A. E., 577, 655
Radlick, P., 21, 39
Raha, C. R., 1122
Rahman, M., 654, 1143
Ramage, G. R., 920
Ramel, A., 1037
Ramsay, M. L., 780
Rao, A. V. R., 1190
Rao, C. G., 611
Rao, G. V., 178
Rao, K. S. R., 178
Rao, Y. S., 629
Rapala, R. T., 463
Rapoport, H., 376
Rappoport, Z., 404
Rascher, L., 1002
Rasmusson, G. H., 582
Rassack, R. C., 1132
Ratcliffe, R., 596
Ratcliffe, R. W., 600
Rathke, M. W., 317
Rathore, R., 607
Rawalay, S. S., 850, 851
Ray, F. E., 626
Raymond, P., 386, 392
Raymond, R. L., 1071
Re, L., 717
Reed, D., 1068
Regen, S. L., 326, 847, 863, 865
Reichardt, C., 923
Reichstein, T., 1041
Reid, C. G., 1119, 1120
Reif, D. J., 251
Reimlinger, H., 370
Reineke, L. M., 1076, 1077, 1078, 1079, 1080, 1082, 1083
Reischl, W., 905
Reissenweber, G., 260
Reist, H. N., 376
Reitsema, R. H., 632
Remenyi, M., 433
Renn, K., 982
Repič, O., 860
Reynolds, B. E., 1091
Ricciuti, C., 341
Richter, v. V., 667
Ridenour, M. W., 780
Rideout, W. H., 252
Ried, W., 385
Riegel, B., 569
Riehl, Sr. A., 544, 1175
Rieveschl, G., Jr., 626
Rigby, W., 481, 483, 829
Riley, H. A., 520
Riley, H. L., 510, 519
Ringold, H. J., 966
Ringwald, E. L., 688
Rinzema, L. C., 337
Rio, G., 38
Ripper, J., 484
Ris, C., 919
Riviere, H., 347
Robbins, F., 193

Roberts, S. M., 247, 248
Robertson, G. R., 640
Robey, R. L., 413
Robinson, C. H., 319
Roček, J., 544, 562, 563, 570, 577, 587, 621, 654, 655, 949, 1097, 1143, 1175
Rodehorst, C. H., 596
Rodewald, P. G., 804
Rodriquez, O., 20
Roe, E. T., 1040
Roebuck, A., 244
Roebuck, A. K., 242
Rogers, H. R., 565
Rohde, G., 673
Röhmer, A., 986
Rohrmann, E., 143
Rojahn, W., 858
Rol, C., 416
Rollin, P., 609
Ronzio, A. R., 507
Rosenkranz, G., 243, 807
Rosenthal, D., 552
Rosenthal, R., 746
Ross, H., 177
Ross, S., 1046
Rossiter, B. E., 213, 218, 222
Roth, H., 1181
Roth, K., 374
Roussel, M., 140
Royals, E. E., 330
Rubin, M., 258
Rubottom, G. M., 1159
Rudloff, v. E., 763, 764
Ruggli, P., 358
Ruhoff, R., 868
Ruholl, H., 122
Runquist, A. W., 959
Russel, G. A., 322
Russell, P. B., 388
Rust, F. F., 54
Rutledge, P. S., 410, 779
Ruzicka, L., 495
Ruziczka, W., 1165
Rylander, P. N., 940

Saá, J. M., 487
Sabo, E. F., 552
Sachs, F., 986
Saegebarth, K. A., 853
Saiki, M., 1204
Saito, I., 126
Sakurai, H., 1203
Sala, T., 899
Salmon, J., 662
Saltzman, H., 266, 267
Salzmann, T. N., 321
Sam, D. J., 841
Sammes, P. G., 994
Samuelsson, B., 1183
San Filippo, J., Jr., 675
Sandborn, L. T., 644, 737
Sandri, S., 540, 543, 571, 614
Sandris, C., 523

Santaniello, E., 537, 619, 776, 777, 778
Santurri, P., 193
Sardarian, A., 659, 896
Sardarian, A. R., 660, 895
Sarel, S., 567, 598
Sarett, L. H., 592
Sargent, K. H., 2
Sargent, M. V., 318, 899
Sartori, P., 796, 1133
Sauer, J. C., 573
Sauers, R. R., 286
Saunders, B. C., 1038
Sawaki, Y., 302
Sax, K. J., 1052
Saxena, N., 607
Scanlan, J. T., 240, 241, 295
Scanlon, B., 381
Scettri, A., 608
Schächter, O., 742
Schäfer, H., 1142
Schäfer, H. J., 116, 122, 902, 904
Schaap, A. P., 36, 40
Schaefer, J. P., 518
Schaeffer, G. F., 505
Schaeffer, J. R., 580
Schaub, R. E., 1168
Scheffer, A., 142
Scheffer, J. R., 17, 24
Scheit, K. H., 1189
Schenck, G. O., 26, 31, 44, 1112, 1115
Schenker, K., 394, 395
Schiess, P. W., 447
Schiessler, R. W., 246
Schildknecht, H., 680
Schipper, E., 1002
Schlecht, M. F., 535, 568
Schlein, H. N., 101
Schlessinger, R. H., 25, 34
Schleyer, v. P. R., 545, 583
Schmeisser, M., 796
Schmid, H. J., 709
Schmidt, E., 1104
Schmidt, G., 603, 661
Schmidt, H.-J., 902, 904
Schmidt, U., 707
Schmitz, E., 372
Schneider, H.-J., 329
Schneider, M., 705
Schneider, O., 714
Schniter, K., 647
Scholl, R., 407
Scholz, D., 903
Schors, A., 69
Schreiber, S. L., 1108
Schriesheim, A., 52, 53
Schröder, G., 78, 79
Schröder, M., 196
Schroeder, E. F., 871
Schubert, A., 575
Schulte-Elte, K.-H., 44, 1112
Schultz, A. G., 25
Schultz, H. S., 135

Schulze, W., 369
Schwab, J. M., 1043
Schwartz, J., 70
Schwarzer, G., 407
Schwarzkopf, O., 299
Schwenk, E., 1170
Scollar, M. P., 1036
Sebek, O. K., 1059
Sekido, F., 1044
Selwitz, C. M., 60
Senseman, C. E., 3
Seraglia, R., 1025
Sergeev, P. G., 8
Sergienko, S. R., 1090
Serra-Errante, G., 994
Severson, W. A., 455
Shaban, M. E., 566
Shanley, E. S., 124
Sharefkin, J. G., 266, 267
Sharma, A. K., 1011
Sharpless, K. B., 167, 168, 213, 214, 215, 218, 221, 222, 223, 328, 669, 670, 674, 723, 774, 860, 861, 951, 1026, 1027, 1028, 1031, 1105
Shaw, A., 916
Shaw, J. E., 601
Shechter, H., 633, 850, 851, 867
Shei, J. C., 279
Sheikh, M. Y., 349
Sheng, M. N., 227
Sherry, J. J., 601
Shibahara, M., 1062
Shieh, H. S., 1089
Shimizu, I., 348
Shimizu, M., 616, 1140
Shimizu, N., 43
Shimizu, T., 711
Shinke, S., 553
Shonle, H. A., 143
Shono, T., 121
Shriner, R. L., 102, 885, 1196
Shteingarts, V. D., 1134
Sianesi, D., 1100
Siebert, R., 575
Siegel, E., 176
Siehr, D. J., 1069
Sih, C. J., 1068, 1070
Silbert, L. S., 176
Simes, J. J. H., 275
Simmons, H. F., 841
Simpson, J. C. E., 157, 202
Sinaÿ, P., 609
Singer, A., 458
Singer, A. W., 883
Singh, J., 664
Singh, R. P., 597
Singh, S. P., 906
Sisler, H. H., 591, 665
Sivadasan, L., 1190
Skattebøl, L., 710
Sket, B., 339
Skold, C. N., 34
Skowronska, A., 1194

Sladkov, A. M., 8
Slater, G. P., 439
Slooff, G., 268, 271, 272
Small, L. D., 308
Smith, A. B., III, 1024
Smith, B., 728
Smith, B. F., 279
Smith, C., 1094
Smith, C. W., 125, 749
Smith, D. S., 396
Smith, K., 755
Smith, L. I., 73, 74, 390, 391, 557, 559, 703, 908, 914
Smith, L. L., 1062
Smith, P. J., 917
Smith, W. C., Jr., 726
Smith, W. T., 699
Smoczkiewicz, M. A., 976
Snatzke, G., 542
Sneeden, R. P. A., 55
Snoddy, A. D., 580
Snyder, H. R., 189
Sobolov, S. V., 123
Soderquist, J. A., 991
Solomon, D. H., 437, 438
Solomon, S., 710
Solomonica, E., 389
Sondheimer, F., 807
Sosnovsky, G., 181, 233, 1113
Southam, R. M., 386
Späth, E., 784
Speer, J. H., 554
Speers, L., 261, 1166
Spencer, C. F., 1187
Spielman, M. A., 498
Spillane, L. J., 703
Spitzer, U. A., 639, 943
Squires, T. G., 279
Srinivasan, N. S., 856, 857
Stacey, M., 87
Stacy, G. W., 186
Staklis, A., 1182
Stampfli, J. G., 694
Starcher, P. S., 262
Stark, B. P., 1038
Starks, C. M., 872
Staunton, J., 1171
Steinberg, A. M., 388
Steiner, E. G., 821
Stenberg, V. I., 906
Stenlake, J. B., 541
Stephens, J. R., 1172
Steppel, R. N., 313, 1180
Sternbach, L. H., 375
Sternhell, S., 449
Stevens, C. L., 297
Stevens, R. V., 693, 1152
Stewart, D., 198
Stille, J. K., 99
Stiller, K., 1109
Stiller, T. E., 648
Stobie, A., 798
Stock, A. M., 205
Stöckel, O., 5
Stoffelsma, J., 337

Stojiljković, A., 443, 444
Stork, G., 824
Story, P. R., 76
Stothers, J. B., 118, 574
Stoudt, T. H., 1054
Stratford, M. J. W., 812, 827
Strating, J., 335
Stuart, F. A., 645
Stubbings, R., 193
Stubbs, C. A., 1152
Stubbs, J. J., 1040
Stürzenhofecker, F., 250
Stürzinger, H., 332
Subbarao, H. N., 597
Sucharda, E., 459
Sucsy, A. C., 393
Suga, T., 677, 678
Suggs, J. W., 605
Sugimoto, A., 33
Sulzbacher, M., 346
Sundralingam, A., 22
Surgenor, D. M., 212
Surrey, A. R., 173
Suschitzky, H., 158, 247, 248
Suskind, S. P., 814
Sussman, S., 130
Sutter, D., 1051, 1056, 1057
Suzuki, F., 1068
Suzuki, K., 977
Suzuki, M., 237
Svátek, E., 695
Sviridova, A. V., 166
Swann, B. P., 413, 414
Swern, D., 50, 176, 240, 241, 295, 341, 729, 1011, 1012, 1022, 1023, 1139, 1149
Sydnes, L. K., 238
Symons, M. C. R., 830, 831
Syper, L., 380, 417
Szabo, H.-C., 292
Szmant, H. H., 264, 795
Szwarc, M., 916

Tabenkin, B., 1040, 1050
Tagaki, W., 727
Tait, S. J. D., 630, 652
Takada, H., 237
Takai, K., 236
Takatani, M., 1026
Takayanagi, H., 64
Takemoto, Y., 996
Talalay, P., 1072
Tam, W. W., 1152
Tamm, C., 1066
Tan, T.-L., 775, 1151
Tanaka, K.-I., 944, 1205
Tanaka, S., 33, 328
Tarbell, D. S., 164, 274
Tarbin, J. A., 194
Tarka, S. M., 506
Taschner, M. J., 1042
Tatlow, J. C., 838
Tatsumi, C., 1044
Tavares, D. F., 588
Taylor, B. J., 538

Taylor, E. C., 162, 278, 413, 414, 978
Taylor, E. R., 882
Tazuma, J., 297
Teeter, H. M., 81, 706
Teitel, S., 1037
Tek-Ling, C., 929
Telschow, J. E., 531
Teranishi, A. Y., 167, 669, 670, 860
Teranishi, R., 963
Teranishi, S., 216
Terao, S., 42
Terashima, S., 1030
Terry, E. M., 715
Tetsukawa, H., 1055
Teuber, H.-J., 488, 490
Theilacker, W., 747
Theis, R. J., 405
Thesing, J., 925
Thijs, L., 335
Thoma, R. W., 1045, 1065
Thomas, A., 818, 826, 828
Thomas, D. A., 439, 819
Thomas, L. P., 1043
Thomas, S. L. S., 945, 946, 947, 1093
Thompson, J. L., 708
Thompson, Q. E., 89, 113
Thorpe, F. G., 183
Tipson, R. S., 480
Tobin, P. S., 409
Toczek, J., 676
Todd, A. R., 440
Todd, D., 432
Todesco, P. E., 1186
Toennies, G., 1157
Tollens, B., 467
Tolstikov, G. A., 83, 342
Tomaselli, G. A., 534
Tomasz, M., 824
Tomioka, H., 741
Tong, Y.-C., 9, 771
Tonge, A. P., 769
Townley, E., 1051
Towson, J. C., 1032
Trachtenberg, E. N., 482
Trahanovsky, W. S., 415, 418, 419, 420, 424
Trave, S., 778
Traynelis, V. J., 506
Trecker, D. J., 225, 255
Treibs, W., 133
Trocha-Grimshaw, J., 10
Trost, B. M., 207, 209, 321, 1014
Trott, P., 869
Trowbridge, L., 1126
Troyanskii, E. I., 924
Trozzolo, A. M., 820
Truce, W. E., 836
Trumbull, E. R., 1151
Tsai, T. Y. R., 854
Tsang, S. M., 548
Tschitschibabin, A. E., 486

Tsong, Y. Y., 1070
Tsuchihashi, G.-I., 1058
Tsuchiya, H. M., 1048
Tsuda, Y., 734, 918
Tsuji, J., 64, 65, 120, 348
Tsuji, T., 936
Tucker, J. R., 323
Tucker, S. H., 855
Tuley, W. F., 464
Tully, M. E., 1056
Tung, C. C., 956
Turner, A. B., 966, 968
Turner, L., 263
Turner, R. B., 55, 1107, 1176
Turro, N. J., 20

Ubersax, R. W., 286
Uchiyama, M., 228
Ulatowski, T. G., 1032
Ullmann, F., 686, 878, 913
Underwood, H. M., Jr., 719
Urabe, H., 1140
Urech, J., 1067
Uskoković, M., 482
Uzbachian, J. B., 878

Vais, J., 546
Valkanas, G., 460, 461
Van Allan, J. A., 625
van Beek, H. C. A., 30
Van de Castle, J. F., 815
Van Der Beek, P. A. A., 1164
van Dort, H. M., 430
Van Nes, K., 967
Van Oot, J. G., 631
Van Rheenen, V., 952
Van Verth, J. E., 45
Vargha, L., 433
Varkony, H., 106
Varkony, T. H., 105
Vaughan, W. E., 54
Vedejs, E., 531, 532, 533
Veer, W. L. C., 245
Venier, C. G., 279
Verbeek, J., 30
Vercellone, A., 1088
Verhoeven, T. R., 214, 218
Vessal, B., 895, 896, 897
Viala, J., 1137
Viet, E. B., 636
Vilim, A., 854
Vingiello, F. A., 631
Vischer, E., 1064, 1067
Viti, S. M., 1026
Voegeli, R., 551
Volger, H. C., 61
Vollbracht, L., 430
Vorozhtsov, N. N., 1134
Vostrikov, N. S., 342

Wächter, R., 638
Wagner, H., 1195
Wagner, W. S., 624
Wakeman, N., 451
Walder, H., 912

Waldmann, H., 495
Walker, F. J., 1026
Walker, T., 805
Wallace, T. J., 52, 53, 808
Walpole, A. L., 945
Walsh, C. T., 1034
Walsh, W. L., 719
Walter, W., 732
Walther, W., 1154
Wang, K. C., 1070
Warburton, W. K., 819
Ward, G. E., 1040
Warnell, J. L., 102
Warnock, J., 612
Warrener, R. N., 930
Wasserman, H. H., 17, 42, 45, 1109
Wassmundt, F. W., 115i
Wasson, P. C., 149
Watanabe, H., 522
Watanabe, Y., 132
Waterhouse, P., 312
Waters, W. A., 16
Waters, W. L., 114
Waugh, T. D., 507, 955
Weaver, C., 164
Weaver, W. M., 1003
Weber, J.-V., 705
Webster, I. M., 557
Weedon, B. C. L., 578
Wehrli, P. A., 489
Wei, T.-Y., 606
Weijlard, J., 448
Weintraub, A., 1076, 1077, 1078, 1079
Weiss, M., 476
Weiss, R., 492
Weiss-Berg, E., 1066
Weitz, E., 142
Weizmann, C., 346
Welch, J., 528, 529
Weller, H. N., 693
Wendband, R., 650
Wenis, E., 352
Wenkert, E., 258
Wenner, W., 555
Wertheim, E., 550, 685
Westheimer, F. H., 561, 949
Weston, G. O., 975
Wettstein, A., 1064, 1067
Wexler, S., 13, 14, 15
Whalley, M., 855
Wheeler, D. M. S., 409
Wheeler, M. M., 409
White, R. W., 82, 338
White, T., 697
Whitesides, G. M., 565
Whitham, G. H., 1114
Whiting, M. C., 386, 392
Whitmore, F. C., 72, 85, 881
Wiberg, K. B., 620, 666, 853, 1142
Wicha, J., 681
Wieland, H., 374
Wiesner, K., 854, 953

Wiggins, L. F., 87
Wightman, R. H., 1021
Wildman, W. C., 811
Wiley, P. F., 914
Wiley, R. H., 203
Williams, A. L., 804
Williams, D. M., 725
Williams, D. R., 860
Williams, F. T., Jr., 867
Williams, J. H., 1168
Williams, P. H., 129, 138, 144, 147, 254, 1118
Williams, S. B., 231
Willis, J. P., 610
Willstätter, R., 368, 373
Wilson, R. A. L., 298
Wilson, T., 1116
Windaus, A., 35, 399
Wineburg, J. P., 729
Winkler, W., 866
Winstein, S., 404, 411
Wintner, C., 151
Wirtz, H., 928
Wirz, B., 1036
Witkop, B., 95, 979
Witzemann, E. J., 871
Wohl, A., 485
Wolf, C. N., 1196
Wolf, F. J., 448
Wolfe, S., 701, 898, 997
Wolford, L. T., 371
Wolfrom, M. L., 762
Wollner, R., 888
Wong, C. M., 426, 428

Wood, E. H., 548
Wood, J. H., 956
Woodard, S. S., 1026, 1027
Woodgate, P. D., 779
Woodward, G. E., 881
Woodward, R. B., 781
Woodworth, R. C., 922
Wooldridge, K. R., 973
Woolford, R. G., 118, 574
Wright, D. L., 517
Wu, C. S., 854
Wuts, P. G. M., 713

Yakobson, G. G., 273, 1134
Yamachika, N. J., 668
Yamada, K., 1204
Yamada, S.-i., 1030
Yamada, T., 1033
Yamagishi, A., 766
Yamaguchi, S., 139
Yamamoto, H., 328
Yamamoto, K., 348
Yamamoto, M., 1204
Yang, N. C., 220, 233
Yanuka, Y., 567
Yany, F., 20
Yazawa, H., 511
Yokoyama, T., 307
Yokoyama, Y., 327, 343
Yoneda, F., 977, 978
Yoshida, J., 119
Yoshifuji, S., 944, 1205
Yoshikawa, Y., 506

Yoshioka, H., 1074
Yosizawa, Z., 762
Young, D. P., 131
Young, D. W., 383
Young, L. B., 415, 418, 419, 420
Young, L. H., 424
Young, W. G., 404
Youssef, A. M., 566
Youssefyeh, R. D., 76
Yuba, K., 126
Yudis, M., 1051

Zahm, H. V., 1092
Zajacek, J. G., 227
Zally, W., 375
Zarecki, A., 681
Zaugg, H. E., 463
Zaweski, E. F., 582
Zbiral, E., 785, 905
Zelinskii, N. D., 478
Zeller, P., 358
Zhemaiduk, L. P., 83
Ziegler, J.-C., 682
Zincke, T., 704, 732
Zinke, H., 75
Zinner, H., 136, 340
Zorc, B., 927
Zupan, M., 339
Zürcher, A., 400
Zürcher, W., 1066
Zwanenburg, B., 335
Zweifel, G., 1202

SUBJECT INDEX

Numbers in **boldface** type refer to Preparative Procedures (pp 273–294). The index lists reagents, types of reactions, and types of compounds. Of individual compounds, only those mentioned in Preparative Procedures are listed.

A

Acetals, oxidation to esters, 184
Acetals, cyclic,
  oxidation to lactones, 184, 185
Acetals, unsaturated
  epoxidation, 184–185
  hydroxylation, 185
  oxidation to unsaturated esters, 184–185
Acetates, by oxidation of stannanes, 270
Acetic acid, solvent
  for ozonization, 5
  for peroxyacetic acid, 12
Acetic anhydride
  catalyst in oxidation with dimethyl sulfoxide, 43
  in preparation of peroxyacetic acid, 273
Acetobacter suboxydans
  in oxidation
    of secondary alcohols to ketones, 146
    of sugars to aldonic acids, 177, 183
  in selective oxidation of diols, 157
Acetone, oxidant, 39
Acetophenone, preparation from α-phenylethanol, **284**
Acetoxylation of ketones, 199
Acetyl chromate,
  oxidant and preparation, 26
Acetyl hypoiodite
  in degradation of tertiary alcohols, 151
  in hydroxylation of alkenes, 70, 71, **286–287**
  oxidant and preparation, 31
Acetylacetone,
  oxidation to triketopentane, **292**
Acetylenes, See *Alkynes*
Acetylenes, macrocyclic, from dihydrazones of α-diketones, 221–222, **280**
Acetylenic alcohols, See *Alcohols, acetylenic*
Acetylenic amines, oxidation to dicarbonyl compounds, 245
Acetylenic carboxylic acids, See *Carboxylic acids, acetylenic*
Acetylenic ethers, oxidation, 172
Acetylenic ketones, oxidation, 140, **283**
Acetylenic sulfides, oxidation, 261
Acetylenic sulfoxides, oxidation, 261
Acetylenic tertiary alcohols, oxidation, 151
2-Acetylmesitylene, oxidation to mesitylglyoxal, **283**

Acid anhydrides, by oxidation of α-diketones, 193, 218
Acids, nitronic, oxidation, 230–231
Acinetobacter sp., use in oxidation
  of phenylboronic acids to phenols, 270
  of selenides to selenoxides, 265
Acinetobacter strain NCIB 9871
  in Baeyer–Villiger reaction, 190, 195
  in oxidation of ketones to esters and lactones, 190, 195
Activators of dimethyl sulfoxide, 122, 144–145, 147–149
Acyl chlorides, intermediates in cleavage of alkenes to carboxylic acids, 83
Acyl esters of vicinal diols, formation from alkenes and cycloalkenes, 74–75
Acyloins, See *Hydroxy ketones*
Acyloins, ethers of,
  formation from alkynes, 91
Acyloxylation
  of alkenes and cycloalkenes, 74–75
  of aromatic rings
    with lead tetraacetate, 93
    with lead tetrakis(trifluoroacetate), 93–94
    with organic peroxides, 93
    with peroxy acids, 93, 94
Adipic acid, by oxidation of cyclohexanol
  with nitric acid, 150, **281**
  with potassium permanganate, 150, **289**
Adipic aldehyde, preparation from cis-1,2-cyclohexanediol, **281**
Adkins catalyst, for dehydrogenation, 15
Adogen 464 (methyltrialkylammonium chloride), phase-transfer reagent, 116, 137, 149
Air
  applications as oxidant, 1
  in oxidation
    of acyloins to α-diketones, 217
    of aldehydes to carboxylic acids, 174
    of aldoses, 182–183
    of alkenes and cycloalkenes to ketones, 75
    of aromatic compounds to quinones, 95
    of benzylic alcohols to aldehydes, 124
    of esters to keto esters, 223
    of hydrazo compounds, 233
    of mercaptans (thiols) to disulfides, 250
    of primary alcohols to carboxylic acids, 127

375

Air, in oxidation—*Continued*
  of primary allylic alcohols to aldehydes, 126
  of secondary alcohols to ketones, 132, 147
  of sulfoxides to sulfones, 262
  use with catalysts (catalytic oxidation), 1
  See also *Catalysts,* for oxidation with air
Alcohols
  by biochemical hydroxylation, 58
  by hydroboration of alkenes, 268
  by hydrolysis of boranes, 267
  by oxidation
    of alkenes, **277**
    of alkenes and cycloalkenes at allylic positions, 84, 85, 86
    of aralkyl allyl ethers, 171–172
    of aromatic side chains, 100–101
    of aryl ethers, 171–172
  by oxidative cleavage of alkenes and cycloalkenes, 77–78
  by reduction of ozonides, 5, 77, 78
  dehydrogenation catalyzed
    by copper, 114, 115, 178
    by metal oxides, 114, 115
    by platinum, 115
    by Raney nickel, 39, 143
  formation from alkyl hydroperoxides, 84–85
  oxidation
    to carbonyl compounds, See specific types of alcohols
    to carboxylic acids, 150, 151
  reactions during aromatization, 50
  See also specific types of alcohols
Alcohols, acetylenic
  oxidation
    to acetylenic aldehydes, 122, **293**
    to acetylenic ketones, 140
    to diacetylenic diols, 155
  oxidative coupling, 91, 155
Alcohols, aliphatic,
  electrooxidation to esters, 131
Alcohols, allylic
  epoxidation, 152–154
    preparation of enantiomers, 153
    stereoselectivity, 152
    with *tert*-butyl hydroperoxide, 152
    with chiral oxidants, 153
    with peroxy acids, 152, 153
  formation
    from alkenes, 86
    from allylic hydroperoxides, 85
  oxidation
    examples, 124, 126, 127
    to α,β-unsaturated aldehydes, 118, 120, 146, **288**
    to α,β-unsaturated ketones, 140, 141, 144
    with copper permanganate, 141
    with 2,3-dichloro-5,6-dicyano-*p*-benzoquinone, 143–144
    with manganese dioxide, 140
    with permanganates, 119

Alcohols, allylic, oxidation—*Continued*
    with potassium ferrate, 120
    with sodium permanganate, 140
  selective oxidation, 156, 157
Alcohols, benzylic
  by oxidation of arylalkanes, 100–101
  oxidation
    to aldehydes
      examples, 124–126
      with *tert*-butyl hypochlorite, 119
      with carboxypyridinium dichromates, 116
      with ceric ammonium nitrate, 115
      with *N*-chlorosuccinimide, 119
      with chromium trioxide–3,5-dimethylpyrazole complex, 118, 125
      with dinitrogen tetroxide, 115
      with permanganates, 119, 120, 125
      with potassium ferrate, 120
      with potassium hypochlorite, 119
    to ketones, 144, 149
Alcohols, homoallylic
  epoxidation, 154
  oxidation, 143
Alcohols, primary
  biochemical oxidation to carboxylic acids, 130
  decarbonylation with Raney nickel, 132
  dehydrogenation to aldehydes, 114–115
  electrolytic oxidation to carboxylic acids, 127, 139
  oxidation
    examples, 123–127
    in preference to secondary alcohols, 143
    to aldehydes
      preparative procedures, **283, 284**
      with air, 123, 124, 126
      with benzeneseleninic anhydride, 122–124, 126
      with *tert*-butyl chromate, 119, 124, 126, 128
      with *tert*-butyl hypochlorite, 119, 125
      with *N*-chlorosuccinimide, 119, 125
      with chromium trioxide–3,5-dimethylpyrazole complex, 117–118, 125, 126
      with chromium trioxide–pyridine complex (Collins reagent), 117–118, 123–126
      with chromyl chloride, 119, 123–126
      with di-*tert*-butyl chromate, 119, 124, 126
      with 2,3-dichloro-5,6-dicyano-*p*-benzoquinone, 120
      with diethyl azodicarboxylate, 126
      with dimethyl sulfide and chlorine, 120
      with dimethyl sulfide and *N*-chlorosuccinimide, 120, 123, 124
      with dimethyl sulfoxide, 121–123, 125, 126
      with 4-dimethylaminopyridinium chlorochromate, 118, 124, 125

Alcohols, primary, oxidation to
  aldehydes—Continued
    with Jones reagent, 117, 124–126
    with manganese dioxide, 119, 124,
      126, 128
    with nickel peroxide, 120, 125, 127
    with permanganates, 119, 120, 125
    with poly(vinylpyridine)–chromium
      trioxide complex, 118
    with pyridinium chlorochromate,
      118, 126
    with pyridinium dichromate, 116,
      124–126, 128
    with ruthenium tetroxide, 120, 125
    with tetrabutylammonium
      chlorochromate, 118
    with tetrabutylammonium chromate,
      116, 123, 126
    with tetrabutylammonium
      dichromate, 123, 125
    with tetrachlorobenzoquinones, 120,
      121
  to amides, 132
  to carboxylic acids
    by electrolytic oxidation, 127, 130
    with air or oxygen, 127, 130
    with *tert*-butyl chromate, 128
    with carbon tetrachloride, 130
    with chromic acid, 128
    with cupric permanganate, 129, 130
    with manganese dioxide, 128
    with nickel peroxide, 120, 129, 130
    with oxygen, 130
    with potassium hydroxide, 127
    with potassium permanganate, 129, 130
    with potassium ruthenate, 130
    with pyridinium dichromate, 128
    with ruthenium tetroxide, 120
    with sodium permanganate, 129, 130
    with sodium ruthenate, 130
    with solid potassium permanganate,
      141
    with tetrabutylammonium
      permanganate, 129, 130
    with zinc dichromate, 128, 130
  to esters
    with bromine, 131
    with *tert*-butyl chromate, 128, 131
    with manganese dioxide, 131
    with Oxone, 131
    with potassium peroxymonosulfate,
      131
    with sodium dichromate, 131
  selective oxidation, 156
Alcohols, primary, allylic, oxidation
  in preference to cyclic secondary
    alcohols, 146
  preparative procedure, **284**
Alcohols, secondary
  conversion into ketones
    by biochemical oxidation, 146
    by catalytic oxidation, 133
    by dehydrogenation, 132
    by electrooxidation, 133, 147

Alcohols, secondary, conversion into
  ketones—Continued
    by Oppenauer oxidation, 149
    by singlet oxygen, 3
  mechanism of oxidation by chromium
    compounds, 134
  oxidation
    in preference to primary alcohols, 138,
      139
    to carboxylic acids
      with chromium trioxide, 150
      with nitric acid, 150, **281**
      with potassium permanganate, 150,
        289
    to α-hydroxy hydroperoxides, 150
    to ketones
      examples, 147–149
      preparative procedures, **284, 285**
      steric effects, 135
      with *Acetobacter suboxydans,* 146
      with air, 132, 147
      with argentic oxide, 148
      with barium manganate, 140, 149
      with benzeneseleninic anhydride, 149
      with bromine, 139, 147–149
      with *tert*-butyl hypochlorite, 139, 148
      with calcium hypochlorite, 139, 148
      with ceric ammonium nitrate, 133,
        139
      with ceric sulfate, 133
      with chloral, 143
      with chlorine, 138, 147–149
      with *m*-chloroperoxybenzoic acid,
        133, 147, 148
      with *N*-chlorosuccinimide, 144, 148
      with *N*-chlorosuccinimide and
        dimethyl sulfide, 144, **293**
      with chromic acid, 134
      with chromium trioxide, 135, 147,
        148
      with chromium trioxide–3,5-
        dimethylpyrazole complex, 137,
        138, 147, 149
      with chromium trioxide–pyridine
        complex, 137, 147–149
      with chromyl chloride, 147, 149
      with Collins reagent, 137
      with copper oxide, 133, 148
      with copper permanganate
        octahydrate, 141
      with *Corynebacterium simplex,* 146,
        216
      with Dess–Martin periodinane, 148
      with 2,3-dichloro-5,6-dicyano-*p*-
        benzoquinone, 143
      with diethyl azodicarboxylate, 144,
        149
      with dimethyl sulfoxide, 144–149
      with dinitrogen tetroxide, 133, 149
      with halogens, 138
      with hexavalent chromium, 133–138,
        147–149
      with horse liver alcohol
        dehydrogenase, 146

Alcohols, secondary, oxidation to ketones—*Continued*
 with hypochlorites, 139
 with Jones reagent, 135–136
 with lead tetraacetate, 133, 149
 with manganese dioxide, 140, 144, 147–149
 with Oxone, 133
 with oxygen, 147, 149
 with peroxyacetic acid 147
 with poly(vinylpyridinium) chlorochromate, 147, 148
 with potassium dichromate, 136, 149
 with potassium ferrate, 142, 148, 149
 with potassium permanganate, 140, 147, 149
 with potassium permanganate on molecular sieves, 141, **290**
 with potassium peroxymonosulfate, 133
 with potassium ruthenate, 142
 with pyridinium chlorochromate, 138, 149
 with pyridinium chromate, 137, 148
 with pyridinium dichromate, 137
 with Raney nickel, 147
 with ruthenium tetroxide, 142, 148
 with silver carbonate, 133
 with silver oxide, 133
 with sodium bromate, 139
 with sodium bromite, 148
 with sodium dichromate, 136, 148
 with sodium hypochlorite, 139, 147, 148
 with sodium permanganate, 147, 148
 with sodium ruthenate, 142
 with solid permanganates, 141
 with supported permanganates, 141
 with tetrabutylammonium chlorochromate, 138, 149
 with tetrabutylammonium chromate, 136, 148
 with tetrabutylammonium dichromate, 148
 with tetrabutylammonium hypochlorite, 139, 149
 with tetrabutylammonium permanganate, 142, 149
 with trichloroacetaldehyde, 149
 with triphenylmethylfluoroborate, 146, 147, 148
 with yeast, 146
 with zinc dichromate, 149
 photooxidation to ketones, 133
 selective oxidation, 156, **286**
Alcohols, secondary, benzylic, oxidation to ketones, 149
Alcohols, sterically hindered, oxidation with dimethyl sulfoxide, 146
Alcohols, steroidal, selective oxidation, 157
Alcohols, tertiary
 by oxidation of methine groups, 58
 degradation
  with acetyl hypoiodite, 151
  with lead tetraacetate, 151
 dehydration to alkenes, 151

Alcohols, tertiary—*Continued*
 oxidation to hydroperoxides, 150–151
 resistance to oxidation, 114, 150
Alcohols, tertiary, acetylenic,
 oxidation to hydroperoxides, 151
Alcohols, unsaturated
 cleavage of double bonds
 epoxidation, 152–154
 oxidation
  at multiple bonds, 151–155
  to keto oxides
   with Fieser reagent, 154
   with pyridinium chlorochromate, 154
   with sodium chromate, 154
  to lactones with chromium trioxide, 154
  to unsaturated ketones with manganese dioxide, 140
Alcohols, α,β-unsaturated, oxidation with manganese dioxide, 131
Alcohols, vinylic, oxidation, 151
Aldaric acids, by oxidation of aldoses, 182, 184
Aldehyde–bisulfite addition compounds, oxidation to carboxylic acids, 178
Aldehyde derivatives, oxidation, 184–185
Aldehyde diacetates, by oxidation of aromatic side chains, 103
Aldehyde epoxides, by epoxidation of unsaturated aldehydes, 182
Aldehyde groups, by oxidation of methyl groups, 101–102, 103
Aldehydes
 biochemical oxidation to carboxylic acids, 177
 by cleavage
  of ozonides, 5
  of vicinal diols, 159–161, **281, 287**
  of ylides, 250
 by degradative oxidation
  of α-amino acids, 229
  of carboxylic acids, 224
  of α-hydroxy acids, 228–229
  of unsaturated carboxylic acids, 226
 by dehydrogenation of primary amines, 239–240
 by Guerbet reaction of alcohols, 178
 by oxidation
  of alkyl and allyl halides, 109–110
  of alkyl tosylates, 109, 110
  of aralkyl allyl ethers, 171–172
  of aralkyl halides, **292–293**
  of aryl ethers, 171–172
  of boranes, 269
  of hydrazides, 230
  of primary alcohols, 114–127
   with argentic oxide, 115, 124–126
   with benzeneseleninic anhydride, 122–124, 126
   with *tert*-butyl chromate, 124–126, 128
   with *tert*-butyl hypochlorite, 119, 125
   with carboxypyridinium dichromates, 117, 126–127

Aldehydes, by oxidation of primary alcohols—*Continued*
  with ceric ammonium nitrate, 115, 124
  with *N*-chlorosuccinimide, 119, 125
  with chromic acid on silica gel, **283**
  with chromic acid or anhydride, 116, 123–126
  with chromium trioxide–3,5-dimethylpyrazole complex, 117–118, 125–126
  with chromium trioxide–pyridine complex, 117–118, 123–126
  with chromyl chloride, 119, 123–126
  with Collins reagent, 117–118, 123–126, **283, 284**
  with Dess–Martin periodinane, 123
  with di-*tert*-butyl chromate, 119, 124, 126
  with 2,3-dichloro-5,6-dicyano-*p*-benzoquinone, 120
  with dimethyl sulfide and chlorine, 120
  with dimethyl sulfide and *N*-chlorosuccinimide, 120, 123, 125
  with dimethyl sulfoxide, 121–123, 125–126
  with 4-dimethylaminopyridinium chlorochromate, 118, 124, 125
  with dinitrogen tetroxide, 115
  with Jones reagent, 117, 124–126
  with lead tetraacetate, 115, 124–126
  with manganese dioxide, 119, 124–126, 128
  with nickel peroxide, 120, 125, 127
  with poly(vinylpyridine)–chromium trioxide complex, 118
  with potassium dichromate, 116
  with pyridinium chlorochromate, 118, 126, **284**
  with pyridinium dichromate, 116
  with ruthenium tetroxide, 120, 125
  with sodium dichromate, 116, 123
  with solid potassium permanganate, 141
  with tetrabutylammonium chlorochromate, 126
  with tetrabutylammonium dichromate, 123
  with tetrachlorobenzoquinones, 120, 121, 125
  of primary nitro compounds, 230–231
  of unsaturated ethers, 171–172
by oxidative cleavage
  of alkenes and cycloalkenes, 77–80
  of unsaturated esters, 79
by ozonolysis of dienes, 90
by reduction of ozonides, 77–78, 79

Aldehydes—Continued
from primary halides, 109–112
oxidation
  to carboxylic acids, 174–180
    with air, 174
    with ammonium polysulfide, 179
    with argentic oxide, 175–176
    with benzyltriethylammonium permanganate, 176, 177, 179
    with cumene hydroperoxide, 177, 179
    with cyclohexanone oxygenase, 177
    with dimethyl sulfoxide, 178
    with fuming nitric acid, 176
    with hydrogen peroxide, 175
    with iodosobenzene, 177, 179
    with manganese dioxide, 176, 179
    with nickel peroxide, 177, 179
    with oxygen, 174–175
    with peroxysulfuric acid, 175
    with potassium dichromate, 174, 176
    with potassium hydroxide, 180
    with potassium permanganate, 176, 179, 180
    with potassium peroxymonosulfate, 175, 179
    with potassium ruthenate, 177
    with ruthenium tetroxide, 177, 179
    with selenium dioxide, 175
    with silver oxide, 175, 179, 180
    with sodium chlorite, 179, 180
    with sodium hydroxide, 174, 180
    with sodium hydroxide–potassium hydroxide mixture, 176, 180
    with tetrabutylammonium permanganate, 176, 177
  to esters with cyclohexanone oxygenase, 177
  to peroxy acids
    with oxygen, 180
    with ozone, 180
Aldehydes, acetylenic, by oxidation of acetylenic alcohols, 118, **293**
Aldehydes, aromatic
by oxidation
  of aromatic rings, 97
  of aromatic side chains, 101–103
oxidation
  to aromatic carboxylic acids, 176–177
  to aryl formates and phenols, 180–181
Aldehydes, hydroxy,
oxidation to hydroxy acids, 182–183
Aldehydes, unsaturated
by oxidation
  of aldehydo selenoxides, 265
  of allylic alcohols with manganese dioxide, 120, 121
  examples, 124, 126, 127
  procedure, **288**
epoxidation, 182
oxidation to unsaturated carboxylic acids, 130, 175–176, 182
  with air, 174
  with argentic oxide, 175–176
  with hydrogen peroxide, 175, 182

Aldehydes, unsaturated, oxidation to unsaturated carboxylic acids—*Continued*
  with oxygen, 175
  with peroxysulfuric acid, 175, 182
  with silver oxide, 175
  with sodium chlorite, 176
Aldonic acids
  by biochemical oxidation of sugars, 177, 183
  by oxidation of sugars (aldoses), 174, 182–183
Aldoses
  degradation to aldonic acids, 228
  oxidation
    to aldaric acids, 183–184
    to aldonic acids, 182–183
Aliphatic alcohols, electrooxidation to esters, 131
Aliphatic amines, oxidation, 235
Aliphatic nitro compounds,
  by oxidation of aliphatic amines, 235
Aliquat 336 (methyltrioctylammonium chloride), phase-transfer reagent, 8
Alkalies, in reduction of alkyl hydroperoxides, 86
Alkanehydroxysulfonates, oxidation to carboxylic acids, 178
Alkanes, oxidation
  with chromic acid, 57–59
  with oxygen
    to carboxylic acids, 58
    to hydroperoxides, 58, 60
    to ketones, 58
Alkenes
  acyloxylation, 74–75
  by dehydration of tertiary alcohols, 151
  cleavage at double bonds, 77–84
    by chromyl carboxylates, 78–79
    by osmium tetroxide
      and anhydrous hydrogen peroxide, 79
      and sodium metaperiodate, 79
    by potassium permanganate, 79
    by sodium periodate, 79
    to carbonyl compounds, 80, 81
    to carboxylic acids, **288, 290**
  conversion
    into alcohols, 77–78, 84–86, 268, **277**
    into aldehydes, 77–81
    into alkyl hydroperoxides, 84–85
    into allylic esters, 86
    into amino hydroxy compounds, 74
    into carbonyl compounds, 84, 85–87
    into carboxylic acids, 77, 80–84
    into esters, 74–75, 84, 86
    into halohydrins, 73
    into α-haloketones, 76
    into α-hydroxy ketones, 76
    into ketones, 77, 78, 80
    into unsaturated ketones, 86, 87
  enantioselective hydroxylation, 71
  epoxidation
    examples, 60–64
    preparative procedure, **279**

Alkenes—Continued
  hydroboration, 268, **277**
  hydroxylation
    stereochemistry, 67–68
    stereospecificity, 70–73
    syn versus *anti*, 68–73
    with hydrogen peroxide, **277**
    with osmium tetroxide, 68, 72, 73, **290–291**
  oxidation, 60–87
    at allylic positions, 84–87
    to carboxylic acids, 81–84
    to ketones, 75–77
    to vicinal dicarbonyl compounds, 76
  oxidative cleavage, 77–84
    to aldehydes, 77–81
    to carboxylic acids, 81–84
    to ketones, 80
    with ozone, 77, 81–82
  ozonization, 66–67
Alkenes, cis, trans, syn, and *anti* hydroxylation, 67–68
Alkenes, terminal, oxidation to amides and thioamides, 203
Alkyl acetylenyl ethers, oxidation, 172
Alkyl boronates,
  by oxidation of alkylboranes, 267
Alkyl bromides
  conversion into alkyl iodides, 109, 110
  oxidation to carboxylic acids, 112–113
Alkyl chlorides
  conversion into alkyl iodides, 109, 110
  oxidation to carboxylic acids, 112–113
Alkyl chloroformates, intermediates in oxidation with dimethyl sulfoxide, 122
Alkyl fluorides, oxidation to carboxylic acids, 112–113
Alkyl groups,
  oxidation to carboxyl groups, 105–109
Alkyl halides, primary, oxidation to carboxylic acids, 112–113
Alkyl hydroperoxides
  conversion into alcohols, 84–85
  formation
    by oxidation of alkenes and cycloalkenes, 84–85
    by oxidation of alkyldichloroboranes, 270
  perepoxide intermediates, 2
  reduction with alkalies and with sulfites, 85
Alkyl iodides
  from alkyl chlorides and alkyl bromides, 109, 110
  oxidation to carboxylic acids, 112–113
Alkyl ketones,
  degradative oxidation, 208–209
Alkyl nitrites
  in oxidation of methylene groups to keto groups, 103
  oxidants and preparation, 18
Alkyl phenyl selenoxides, oxidants, 43
Alkyl thiosulfinates,
  by oxidation of disulfides, 263

Subject Index

Alkyl tosylates, preparation from alkyl halides and oxidation to aldehydes, 109–110
Alkylarylacetylenes, oxidative rearrangement, 91
Alkylboranes
 oxidation to esters of boric acid, 267–268
 preparation, 267
Alkyldichloroboranes, oxidation to alkyl hydroperoxides, 270
Alkylnaphthalenes, conversion into quinones, 94–95
Alkynes
 by oxidation of dihydrazones of α-diketones, 221, 222, **280**
 oxidation, 90–92
  to acyloins, 91
  to carboxylic acids, 91–92
  to α-diketones, 91, **280–281, 289**
  oxidative coupling to di- and polyacetylenes, 90–91
Alkynes, terminal
 coupling, **279**
 oxidation to carboxamides, 203
Allyl alcohol, hydroxylation to glycerol, **291**
Allylic alcohols, See *Alcohols, allylic*
Allylic carbon, acetoxylation with mercuric acetate, 101
Allylic halides, oxidation to aldehydes and ketones, 109–111
Allylic oxidations, 84–87
Alumina, support
 for potassium permanganate, 141
 for pyridinium chlorochromate, 118, **274**
Alumina–silica gel, support for chromyl chloride, 147, 149
Aluminum alkoxides, catalysts in Oppenauer oxidation, 39
Aluminum tert-butoxide, catalyst in Oppenauer oxidation, 39, 142
Aluminum chloride
 dehydrogenation agent, 17, 52
 in conversion of lithium borohydride into borane, 268
 in formation of borane from sodium borohydride, **277**
 in intramolecular dehydrogenation, 54
Aluminum isopropoxide, catalyst in Oppenauer oxidation, 39, 142
Aluminum oxide
 in dehydrogenative aromatization, 55
 support for sodium periodate, **287**
Amberlyst A26, support for chromium trioxide, 117, 123, 147–149
Amides
 by oxidation
  of primary alcohols, 132
  of tertiary amines, 242–243
 intermediates in Willgerodt–Kindler reaction, 203, 204
 oxidation
  to carbonates, 230

Amides, oxidation—*Continued*
  to imides, 244–245, **291**
  to isocyanates, 230
  to ureas, 230
  with lead tetraacetate, 230
  with ruthenium tetroxide, 244–245, **291**
Amides, cyclic, hydroxylation, 242–243
Amides, primary, oxidation, 230
Amine oxides
 by oxidation of tertiary amines, 236–239, **277–288**
 rearrangement to hydroxylamines, 238–239
 reoxidants of osmium oxide, 69, 262
Amines,
 oxidation of carbon chains, 242–245
Amines, acetylenic, oxidation to dicarbonyl compounds, 245
Amines, aliphatic, oxidation to aliphatic nitro compounds, 235
Amines, aromatic
 dehydrogenative couplings, 53
 oxidation to quinones, 246–248, **292**
Amines, primary
 biochemical oxidation
  to carboxylic acids, 241–242
  to nitro compounds, 236
 dehydrogenation
  to aldehydes or ketones, 239–240
  to imines, 239–240
  to nitriles, 241–242
 oxidation
  to azo compounds, 234
  to azoxy compounds, 234
  to nitro compounds, 234–236
  to nitroso compounds, 234–236, **278–279**
Amines, secondary
 by demethylation of tertiary amines, **291–292**
 dehydrogenation to imines, 240–241
 oxidation to hydroxylamines, 236
Amines, tertiary
 demethylation, 243, **291–292**
 oxidation
  to amides, 242–244
  to amine oxides, 236–239
  to ketones, 244, 245
α-*Amino acids*, degradation
 to aldehydes, 229
 to carboxylic acids, 229
Amino hydroxy compounds, formation from alkenes and cycloalkenes, 74
Aminohydroxylations, 74
Aminophenols,
 oxidation to quinones, 247–248
Ammonium cerium nitrate, See *Ceric ammonium nitrate*
Ammonium cerium sulfate, See *Ceric ammonium sulfate*
Ammonium molybdate, catalyst for oxidation with hydrogen peroxide, 7, 260

382 Oxidations in Organic Chemistry

Ammonium nitrate
 in oxidation of acyloins (α-hydroxy ketones) to α-diketones, 217, **282**
 oxidant, 19
Ammonium persulfate, oxidant, 8
Ammonium polysulfide
 in oxidation of aldehydes to carboxylic acids, 179
 oxidant in Willgerodt reaction, 202, 203, 204, 205
Ammonium vanadate, catalyst for oxidation with nitric acid, 150
Amyl nitrite (pentyl nitrite), in oxidation of methylene groups, 104
Anhydrides of carboxylic acids, by oxidation of α-keto esters, 229
Anhydrous hydrogen peroxide in ether, preparation, 7, **273**
Anthracene, oxidation, 95
Anthraquinone, by oxidation of anthracene, 95
Aralkyl allyl ethers, oxidation, 171–172
Aralkyl halides, oxidation to aldehydes, 110–111, **292–293**
Aralkyl hydroperoxides, by oxidation of arylalkanes, 99
Argentic oxide
 in degradative oxidation of α-amino acids to carboxylic acids, 229
 in dehydrogenation of primary amines to nitriles, 241
 in oxidation
  of alcohols, 115, 124, 126
  of aldehydes to carboxylic acids, 175–176
  of α-hydroxy nitriles to carboxylic acids, 179
  of methyl groups to aldehyde groups, 101
  of methylene groups to keto groups, 103–105
  of phosphines to phosphine oxides, 248
  of phosphites to phosphates, 248, 249
  of secondary alcohols to ketones, 148
  of unsaturated aldehydes, 175–176
 oxidant, 16
Argentous oxide, See *Silver oxide*
Aromatic aldehydes, See *Aldehydes, aromatic*
Aromatic alkyl groups, oxidation to carboxyl groups, 105–109
Aromatic amines, See *Amines, aromatic*
Aromatic boranes, See *Arylboranes*
Aromatic carboxylic acids, See *Carboxylic acids, aromatic*
Aromatic compounds
 conversion into endoperoxides, 2–3
 epoxidation of alkene side chains, 63–64
 hydroxylation of aromatic rings, 92–94
 oxidation
  of aromatic rings, 92–98
  of side chains, 99–109
  to quinones, 94–96
Aromatic compounds, amino hydroxy, oxidation to quinones, 247–248

Aromatic compounds, diamino, oxidation to quinones, 247–248
Aromatic compounds, polyalkylated, oxidation of side chains, 106
Aromatic endoperoxides, in formation of singlet oxygen, 1, 2
Aromatic heterocycles, See *Heterocyclic aromatic compounds*
Aromatic hydrocarbons, dehydrogenative couplings, 52–53
Aromatic ketones, See *Ketones, aromatic*
Aromatic rings
 hydroxylation, 92–94, 238
 oxidative cleavage, 96–98, 168
Aromatic side chains, oxidation
 to alcohol and ester groups, 100–101
 to carbonyl groups, 101–105
 to hydroperoxide groups, 99
Aromatization
 by catalytic and chemical dehydrogenation, 50–52
 effects
  on alcohols, 50
  on ketones, 50
  of six-membered rings, 52
 with aluminum chloride, 54
 with 2,3-dichloro-5,6-dicyano-*p*-benzoquinone, 52
 with sulfur and selenium, 50, 51
Arsenic compounds, oxidation, 250
Arsenic oxide
 in dehydrogenation of heterocyclic compounds, 52
 in preparation of nitrous anhydride, 18
Arsenic pentoxide, oxidant, 19
Arsine oxides, by oxidation of tertiary arsines, 250
Arsines, tertiary,
 oxidation to arsine oxides, 250
Arthrobacter ureafaciens, in biochemical oxidation of primary alcohols to carboxylic acids, 130
Aryl alkyl ketones, oxidation, 204
Aryl ethers, oxidation, 171–172
Aryl formates, by oxidation of aromatic aldehydes, 180–181
Arylalkanes, oxidation
 to aldehydes, ketones, and derivatives, 101–105
 to aralkyl hydroperoxides, 99
 to carboxylic acids, 105–109
Arylboranes
 intermediates in synthesis, 269–270
 oxidation to phenols, 269–270
Arylmagnesium halides,
 in synthesis of arylboranes, 269
Arylmercury halides, in synthesis of arylboranes, 269–270
Ascaridole, preparation, 87–88
Aspergillus niger
 in biochemical oxidation of sulfides to sulfoxides, 258, **294**
 in hydroxylation
  of steroidal ketones, 198
  of unsaturated ketones, 214

Subject Index 383

Aspergillus tamarii, in biochemical Baeyer–Villiger reaction, 194
Autoxidation, 4
Azelaic acid, preparation from oleic acid, 227, **288**
Azides, oxidation to nitro compounds, 232
Azines, by dehydrogenation of primary amines, 240
Azo compounds
  by oxidation
    of hydrazo compounds, 233
    of primary amines, 234
  oxidation to azoxy compounds, 232
Azoxy compounds
  by oxidation of azides, 232
  by oxidation of primary amines, 234

B

Bacillus megaterium, in hydroxylation of steroidal ketones, 198
Bacterium-infested yeast, in oxidation of hydroxy ketones to diketones, 215
Baeyer–Villiger reaction, 10, 12, 13, 187–195
  biochemical, 190, 194–195
  conversion of ketones into esters or lactones, 186–190
  mechanism, 186–187
  migratory aptitudes of groups, 187
  of dicarbonyl compounds, 218
  of α-diketones and α-keto esters, 190
  of functionalized ketones, 191–194
  of hydroxy ketones, 194–195
  of α-keto esters, 229
  of steroids, 194–195
  of unsaturated diketones, 194–195
  of unsaturated ketones, 191–193, 212
  preparative procedure, **279**
Bakers' yeast,
  in selective oxidation of diols, 157
Barbier–Loquin degradation of carboxylic acids, 151
Barium chlorate
  oxidant, 28
  reoxidant of osmium oxide, 69
Barium manganate
  in oxidation
    of benzylic alcohols to aldehydes, 125
    of diols, 155
    of phenols to quinones, 166
    of phosphines to phosphine oxides, 248
    of primary alcohols to aldehydes, 120, 125, 126
    of primary amines to azo compounds, 234
    of secondary alcohols to ketones, 140, 149
  oxidant, 33
  preparation, 140
Beckmann mixture, 24–25
Bentonite, support for potassium permanganate, 34

Benzalacetophenone, oxidation to 1,3-diphenyl-2,3-epoxy-1-propanone, **285–286**
Benzeneperoxyseleninic acid,
  oxidant and preparation, 14
Benzeneseleninic acid, in oxidation of cyclic ketones to lactones, 188
Benzeneseleninic anhydride
  by oxidation of diphenyldiselenide, 21, 48
  in oxidation
    of primary alcohols to aldehydes, 122–124, 126
    of secondary alcohols to ketones, 149
  in regeneration of ketones from their derivatives, 219
  in situ preparation, 126
  oxidant and preparation, 21, 48
Benzeneselenyl chloride
  in conversion of ketones to keto selenides, 265
  in preparation of selenides, **275**
Benzhydrol, oxidation to benzophenone, 140, **285**
Benzil, preparation
  from benzoin, **282**
  from diphenylacetylene, **280–281**
Benzodioxaborole, hydroboration agent and preparation, 268
Benzoic acid,
  preparation from *trans*-stilbene, **290**
Benzoin, oxidation to benzil, **282**
Benzophenone
  oxidant, 39
  preparation from benzhydrol, **285**
  sensitizer of oxygen, 1, 150, 169
Benzophenone hydrazone, oxidation to diphenyldiazomethane, **280**
p-*Benzoquinone disulfide*, preparation from 2-mercaptohydroquinone, **290**
Benzoxylation,
  with benzoyl peroxide and iodine, 74
Benzoyl chromate,
  oxidant and preparation, 26
Benzoyl hypoiodite,
  oxidant and preparation, 31
Benzoyl peroxide
  in benzoxylation, 74
  oxidant, 10
Benzyl tert-butyl sulfide, oxidation to benzyl *tert*-butyl sulfoxide, **294**
(–)-*Benzyl tert-butyl sulfoxide*,
  preparation from benzyl *tert*-butyl sulfide, **294**
Benzylic alcohols, See *Alcohols, benzylic*
Benzylic esters, by acyloxylation of arylalkanes, 100–101
Benzylic halides, oxidation to aromatic carbonyl compounds, 110–112
Benzyltriethylammonium permanganate
  in hydroxylation of methine groups, 58, 59
  in *syn* hydroxylation of alkenes and cycloalkenes, 72

Benzyltriethylammonium permanganate—Continued
in oxidation
of aldehydes to carboxylic acids, 176, 177, 179
of ethers to esters and lactones, 169
of tertiary amines to amides, 243
oxidant and preparation, 36
safety precautions, 36
solvents, 36
Benzyltrimethylammonium chloride, phase-transfer reagent, 74
Biochemical dehydrogenations
examples, 48–49
of diols, 159
Biochemical hydroxylations
of methylene groups, 58
of steroidal ketones, 58, 196–199
of unsaturated ketones, 214
Biochemical oxidations
Baeyer–Villiger reaction, 190, 194–195
conversion of alkyl halides into carboxylic acids, 112–113
enantioselective hydroxylation, 71
hydroxylation
of steroidal ketones, 58, 196–199
of unsaturated ketones, 214
microorganisms used, 45–46
of aldehydes to carboxylic acids, 177
of aromatic side chains to carboxyl groups, 106, 107
of boranes to borates, 267
of hydroxy aldehydes, 183
of hydroxy ketones, 215, 216
of ketones to esters and lactones, 190, 195
of methyl groups to carboxylic groups, 58
of methylene groups to alcoholic groups, 58
of phenylboronic acids to phenols, 270
of primary alcohols to carboxylic acids, 130
of primary amines
to carboxylic acids, 241–242
to nitro compounds, 236
of secondary alcohols to ketones, 146, 157
of steroidal ketones, 58, 194–195
of sugars, 183
of sulfides to sulfoxides
examples, 257, 258
preparative procedure, **294**
scope, 45–46
Biooxidations, See *Biochemical oxidations*
Bis(2,2'-bipyridyl)copper(II) permanganate
in oxidation
of mercaptans (thiols) to disulfides, 251
of primary alcohols to aldehydes, 125–126
oxidant and preparation, 36
Bis(2,2'-bipyridyl)silver permanganate, oxidant, 36
2,4-Bis(chloromethyl)anisole, oxidation to 4-methoxyisophthalaldehyde, **292–293**

Bismuth sesquioxide
in oxidation of acyloins to α-diketones, 217
oxidant, 19
Bis(tributyltin) oxide, conversion into tributyltin ethers, 139
Bis(trimethylsilyl) peroxide
in Baeyer–Villiger reaction, 187, 191, 195
in oxidation
of hydroxy ketones to lactones, 194
of phosphites to phosphates, 248, 249
of unsaturated ketones to unsaturated esters and lactones, 191
oxidant and preparation, 10
Bleach, oxidant, 27
Borane, preparation, **277**
Boranes
conversion into alcohols, 267
intermediates in conversion of alkenes to alcohols, 268
oxidation
to borates, 267
to carbonyl compounds, 269
Borates, by oxidation of boranes, 267
Boron compounds, oxidation, 267–270
Boron trifluoride, in conversion of lithium borohydride into borane, 268
Bromine
addition compound with Dabco, 139
in bromination of *o*-fluorotoluene, **282**
in generation of singlet oxygen, 1, 87
in oxidation
of ethers to carbonyl compounds, 170
of hydrazo compounds, 233
of hydroxy acids to keto acids, 228
of phenols to quinones, 164
of primary alcohols to esters, 131
of secondary alcohols to ketones, 139, 147, 148, 149
in situ preparation, 28, 29
molecular complexes, 28
oxidant, 28–29
solvents, 28
Bromine complex with Dabco, oxidant, 148
Bromine water
in degradative oxidation of α-hydroxy acids to aldehydes, 228
in oxidation of aldoses, 182
α-Bromo ketones, oxidation to α-dicarbonyl compounds, 201, **293**
N-Bromoacetamide
in formation of halohydrins, 73
in oxidation of diols to keto alcohols, **286**
in selective oxidation of steroidal diols, 157
oxidant, 29
N-Bromobenzophenoneimine, oxidant and preparation, 29
Bromohydrins, See *Halohydrins*
p-Bromophenacyl bromide, oxidation to *p*-bromophenylglyoxal hydrate, **293**
p-Bromophenylglyoxal hydrate, preparation from *p*-bromophenacyl bromide, **293**

Subject Index

N-*Bromosuccinimide*
  catalyst in oxidation with dimethyl sulfoxide, 43
  in degradative oxidation of α-hydroxy acids to aldehydes and ketones, 228
  in epoxidation
    of alkenes and cycloalkenes, 61, 63
    of unsaturated carboxylic acids, 225
    of unsaturated ketones, 212
  in formation of halohydrins, 73
  in oxidation
    of alkynes to α-diketones, 91
    of hydrazo compounds, 233
    of trialkylsilyl ethers, 172
  oxidant, 29
Buffers, in oxidations with potassium permanganate, 34
tert-*Butyl chromate*
  in oxidation of primary alcohols
    to aldehydes, 125, 126, 128
    to carboxylic acids, 128
    to esters, 128, 131
  in situ preparation, 125, 126
  oxidant and preparation, 26
tert-*Butyl hydroperoxide*
  decomposition, 9
  in epoxidation
    of alkenes, 61
    of allylic alcohols, 152
    of unsaturated aldehydes, 182
    of unsaturated ketones, 212, 213
  in oxidation
    of diphenyldiselenide, 122, 123
    of diphenyldiselenide to benzeneseleninic anhydride, 124, 126
    of lactams to imides, 245
    of sulfides to sulfoxides, 254, 257, 258, **278**
    of tertiary amines to amine oxides, 237, **277–278**
  in preparation of benzeneseleninic acid, 21
  oxidant of diphenyldiselenide, 124
  oxidant, properties, and safety precautions, 9
tert-*Butyl hypochlorite*
  in dehydrogenation of amines, 49
  in oxidation
    of primary alcohols to aldehydes, 119, 125
    of secondary alcohols to ketones, 139, 148
    of selenides to selenoxides, 265
    of sulfides to sulfoxides, 254, 257
tert-*Butyl peroxyacetate*
  in formation of allylic esters, 86
  oxidant, 10
tert-*Butyl peroxybenzoate*
  in formation of allylic esters, 86
  oxidant, 10
4-tert-*Butylcyclohexanol*, oxidation to 4-*tert*-butylcyclohexanone, **293**

4-tert-*Butylcyclohexanone*, preparation from 4-*tert*-butylcyclohexanol, **293**

C

Calcium hypochlorite
  in epoxidation
    of alkenes and cycloalkenes, 61
    of unsaturated ketones, 212–213
  in formation of singlet oxygen, 1
  in oxidation of secondary alcohols to ketones, 139, 148
  oxidant and preparation, 27
Calonectria decora
  in biochemical oxidation of sulfides to sulfoxides, 258
  in hydroxylation of steroidal ketones, 197
Camphorylsulfonyloxaziridine(s), chiral oxidants, 44, 223
Cannizzaro reaction, 174, 177, 180
Carbocyclic aromatic compounds, by catalytic dehydrogenation, 50–51
Carbon–boron bonds,
  oxidative cleavage, 267
Carbon–carbon bonds, cleavage to carbonyl groups, 159–163
Carbon–carbon double bonds
  by catalytic dehydrogenation, 49
  by dehydrogenation
    with 2,3-dichloro-5,6-dicyano-*p*-benzoquinone, 47, 48
    with mercuric acetate, 47, 48
  cleavage by ozonolysis, 77, 78, **276**
  in unsaturated alcohols, 129
  to carboxylic groups, 81–84
  with osmic acid, 162
  with sodium periodate, 102–103, 162
  epoxidation, 60–64
  hydroxylation, 31, 67–73, 89
  oxidative cleavage, 77–84
  preparative procedures, **277, 286, 290**
Carbon–nitrogen double bonds, by dehydrogenation
  of cyclic amines, 47
  with *tert*-butyl hypochlorite, 49
  with mercuric acetate, 49–50
  with silver oxide, 49
Carbon–tin bonds, oxidative cleavage, 270
Carbon tetrachloride
  in degradative oxidation of methyl ketones, 207
  in oxidation
    of methyl ketones to carboxylic acids, 205, 206
    of primary alcohols to carboxylic acids, 130
  in oxidative cleavage of sulfones, 263
  solvent for ozonization, 66
Carbonates, by oxidation of primary amides, 230
Carbonyl chloride, catalyst for oxidation with dimethyl sulfoxide, 43

Carbonyl compounds, See Aldehydes, Ketones, and specific carbonyl compounds
Carbonyl groups, by oxidation
  of aromatic side chains, 101–105
  of haloalkyl groups, **282**
  of methylene groups
    activated
      by aromatic rings, 58, 102–105
      by carbonyl groups, 199, 200
      by triple bonds, 91
    with p-nitrosodimethylaniline, **292**
Carboxyl groups
  by biochemical oxidation of halomethyl groups, 58
  by oxidation of alkyl groups, 107–109
Carboxylic acids
  by biochemical oxidation
    of aldehydes, 177
    of primary alcohols, 130
  by cleavage
    of acetylenic carboxylic acids, 226
    of ozonides, 5
  by degradative oxidation
    of alkyl ketones, 208–209
    of α-amino acids, 229
    of α-keto acids, 229
    of methyl ketones, 206–210
    of tertiary alcohols, 151
  by electrolytic oxidation of primary alcohols, 127, 130
  by oxidation
    of alcohols, **281, 289**
    of aldehyde–bisulfite addition compounds, 178
    of aldehydes
      with air, 174
      with ammonium polysulfide, 179
      with argentic oxide, 175–176
      with benzyltriethylammonium permanganate, 176, 177, 179
      with cumene hydroperoxide, 177, 179
      with cyclohexanone oxygenase, 177
      with dimethyl sulfoxide, 178
      with fuming nitric acid, 176
      with hydrogen peroxide, 175
      with iodosobenzene, 177, 179
      with manganese dioxide, 176, 179
      with nickel peroxide, 177, 179
      with oxygen, 174–175
      with peroxysulfuric acid, 175
      with potassium dichromate, 174, 176
      with potassium hydroxide, 180
      with potassium permanganate, 176, 179
      with potassium peroxymonosulfate, 175, 179
      with potassium ruthenate, 177
      with ruthenium tetroxide, 177, 179
      with selenium dioxide, 175
      with silver oxide, 175, 179
      with sodium chlorite, 179, **285–286**
      with sodium hydroxide, 174, 180

Carboxylic acids, by oxidation of aldehydes—*Continued*
      with sodium hydroxide–potassium hydroxide mixture, 176
      with tetrabutylammonium permanganate, 176, 177
    of alkanes, 58
    of alkenes and cycloalkenes, 81–84, **289, 290**
    of alkynes, 91–92
    of benzene rings, 96–98
    of dicarbonyl compounds, 218–219
    of enol ethers, 171
    of esters or lactones, 224
    of ethers, 170
    of hydrazides, 230
    of α-hydroxy nitriles
      with argentic oxide, 179
      with manganese dioxide, 179
    of ketones, 140, 202–215
    of ketones with sulfur, **282–283**
    of methyl ketones
      with carbon tetrachloride, 205, 206
      with potassium permanganate, 206
      with sodium hypobromite, **286**
      with thallium triacetate, 206
    of phenols, 163, 168
    of primary alcohols
      with air or oxygen, 127
      with tert-butyl chromate, 128
      with carbon tetrachloride, 130
      with chromic acid, 128
      with cupric permanganate, 129, 130
      with manganese dioxide, 128
      with nickel peroxide, 120, 129, 130
      with oxygen, 130
      with potassium permanganate, 129, 130
      with potassium ruthenate, 130
      with pyridinium dichromate, 128
      with ruthenium tetroxide, 120
      with sodium permanganate, 129, 130
      with sodium ruthenate, 130
      with solid potassium permanganate, 125, 126
      with tetrabutylammonium permanganate, 129, 130
      with zinc dichromate, 128, 130
    of primary alkyl halides, 112–113
    of secondary alcohols
      with chromium trioxide, 150
      with nitric acid, 150, **281**
      with potassium permanganate, 150, **289**
    of tertiary alcohols, 151
    of trialkylsilyl ethers, 172
    of unsaturated aldehydes
      with hydrogen peroxide, 182
      with peroxysulfuric acid, 182
    of vicinal diols, 163
  by oxidative cleavage of alkenes and cycloalkenes, 77, 80–84
  by ozonolysis of alkenes and cycloalkenes, 81–82

Subject Index

Carboxylic acids—*Continued*
  by Willgerodt reaction of ketones, 202–205
  degradation, 151
  degradative oxidation to aldehydes, 224
  electrolytic decarboxylative coupling, **276–277**
  oxidation, 222–229
  oxidation to peroxy acids, 222
Carboxylic acids, acetylenic
  cleavage
    with ozone, 226
    with peroxyacetic acid, 226
  oxidation to diketo carboxylic acids, 226–227
Carboxylic acids and derivatives,
  oxidation, 222–230
Carboxylic acids, aromatic
  by oxidation of aromatic aldehydes, 176–177
  formation from arylalkanes, 105–109
Carboxylic acids, unsaturated
  by oxidation of unsaturated aldehydes, 175, 176
  degradative oxidation, 226
  epoxidation, 224–225
  hydroxylation, 225–226
  stereoselective hydroxylation, 225
Carboxylic groups
  by biochemical oxidation of methyl groups, 58
  by cleavage
    of double bonds, 80–84, 227
    of triple bonds, 226, 227
3-Carboxypyridinium dichromate
  in oxidation
    of aromatic compounds to quinones, 95
    of mercaptans (thiols) to disulfides, 251
    of primary alcohols to aldehydes, 116, 126, 127
  oxidant and preparation, 25
4-Carboxypyridinium dichromate
  in oxidation of primary alcohols to aldehydes, 116
  oxidant and preparation, 25
Carboxypyridinium dichromates, in
  oxidation of primary alcohols to aldehydes, 116, 117
Caro acid
  in oxidation of primary amines to nitroso compounds, 235
  in situ preparation, 235
Catalysts
  aluminum *tert*-butoxide and aluminum isoproxide for Oppenauer oxidation, 142
  ammonium molybdate for oxidation with hydrogen peroxide, 260
  ammonium vanadate for oxidation with nitric acid, 150
  chloroplatinic acid for oxidation with oxygen, 133
  chromium compounds for alkene epoxidation, 61

Catalysts—*Continued*
  chromium oxide for dehydrogenation, 114
  chromium sesquioxide for oxidation with oxygen, 103
  cobalt acetate for oxidation with ozone, 106
  cobalt benzoate for oxidation with air, 223
  cobalt oxide for dehydrogenation, 114
  cobalt peroxide for oxidation with oxygen, 234
  cobalt sesquioxide for oxidation with oxygen or air, 124, 126
  cobaltous acetate for hydroxylation of aromatic rings, 94
  copper
    for dehydrogenation, 114, 132
    for Guerbet reaction, 178
    for oxidation with air, 124
  copper oxide for dehydrogenation, 114
  copper sulfate for oxidation with peroxysulfate, 101–102
  cupric acetate for oxidation with ammonium nitrate, 19, 217, **282**
  cupric sulfate for oxidation with oxygen, 217
  cuprous bromide for oxidation with peroxy esters, 86
  cuprous chloride
    for oxidation with oxygen, 133, 241
    for oxidative coupling, 155
  cuprous oxide for oxidation of aldehydes to carboxylic acids, 174
  2,4-dimethylpentane-2,4-diol chromate for oxidation with peroxyacetic acid, 147
  ferric chloride for oxidation with oxygen, 211–212
  ferric sulfate and ferrous sulfate for oxidation with hydrogen peroxide, 228
  for Cannizzaro reaction
    silver, 177
    silver oxide, 177
  for cleavage of alkenes and cycloalkenes to aldehydes, 79
  for dehydrogenation
    Adkins catalysts, 15
    chromium oxide, 49
    copper, 14, 114, 132
    copper bronze, 15
    oxides of copper, cobalt, and chromium, 114
    palladium, 38, 50
    phenylselenyl chloride, 48, 49
    platinum, 38, 50, 51
    silver, 15, 114, 132
    vanadium pentoxide, 49
    zinc oxide, 49
  for enantioselective oxidation
    molybdenyl acetylacetonate, 44
    titanium tetraisopropoxide, 44, 154
    vanadyl acetylacetonate, 44
  for epoxidation
    molybdenum hexacarbonyl, 153

Catalysts for epoxidation—*Continued*
  molybdenyl acetylacetonate, 153
  vanadyl acetylacetonate, 153
  for formation of peroxy acids
    methanesulfonic acid, 222
    p-toluenesulfonic acid, 222
  for hydrogenation of ozonides
    Lindlar catalyst, 81
    palladium, 77
    Raney nickel, 77
  for *anti* hydroxylation with hydrogen peroxide, 69
  for Oppenauer oxidation
    aluminum alkoxides, 39, 142
    Woelm aluminum oxide, 149
  for oxidation
    of alcohols to aldehydes, 115
    of aldehydes to carboxylic acids
      cuprous oxide, 174
      platinum, 174
      silver oxide, 174
      tris(triphenylphosphine)ruthenium dichloride, 177, 179
    with air
      cobalt benzoate, 223
      copper, 124
      copper chlorides, 75
      cuprous chloride, 15
      palladium dichloride, 75
      platinum, 182–183
      rhodium trichloride, 262
      salts of copper, palladium, and mercury, 75
      silver, 124, 126
      vanadium pentoxide, 19
    with air or oxygen
      cobalt sesquioxide, 124, 126
      platinum and platinum dioxide, 127
    with bis(trimethylsilyl) peroxide
      trimethylsilyl trifluoroacetate, 194, 195
      trimethylsilyl trifluoromethane-sulfonate, 187, 191, 195
    with *tert*-butyl hydroperoxide
      molybdenyl acetylacetonate, 153
      vanadyl acetylacetonate, 153
    with dimethyl sulfoxide
      acetic anhydride, 43
      N-bromosuccinimide, 43
      carbonyl chloride, 43
      chlorine, 43
      dicyclohexylcarbodiimide, 43
      hydrobromic acid, 43
      oxalyl chloride, 43
      phosgene, 43
      phosphoric acid, 43
      phosphorus pentoxide, 43
      pyridinium trifluoroacetate, 144
      sulfur trioxide, 43
      trifluoroacetic acid, 43
      trifluoroacetic anhydride, 43
      trifluoromethanesulfonic acid, 43
    with hydrogen peroxide
      ammonium molybdate, 7, 260

Catalysts, for oxidation with hydrogen peroxide—*Continued*
  chromium trioxide, 7
  ferric chloride, 8
  ferric sulfate and ferrous sulfate, 228
  osmium tetroxide, 8
  palladium or palladium acetate, 8
  selenium dioxide, 261
  sodium tungstate, 8
  sodium vanadate, 8
  titanium trichloride, 7
  tungsten trioxide, 7
  vanadium pentoxide, 7
  with oxygen
    chloroplatinic acid, 133
    chromium sesquioxide, 103
    cobalt compounds, 4
    cobalt peroxide, 234
    copper compounds, 4
    cupric sulfate, 217
    cuprous chloride, 133, 241
    ferric chloride, 211–212
    hydrogen bromide, 4
    iridium compounds, 4
    iron compounds, 4
    metal oxides and salts, 4
    palladium chloride, 133
    palladium compounds, 4
    palladium dichloride, 147
    platinum, 4, 123, 124, 133
    platinum compounds, 4
    platinum dioxide, 127, 147
    platinum on carbon, 182–183
    platinum oxide, 133
    rhodium compounds, 4
    transition metals, 4
    vanadium pentoxide, 4
  for oxidation with bis(trimethylsilyl) peroxide
    pyridinium dichromate, 10
    trimethylsilyl trifluoromethanesulfonate, 10
    tris(triphenylphosphine)ruthenium dichloride, 10
  for oxidation with *tert*-butyl hydroperoxide
    titanium tetrakis(isopropoxide), 254
    vanadium compounds, 237
  for oxidation with peroxyacetic acid
    2,4-dimethylpentane-2,4-diol chromate, 147
    manganese diacetylacetonate, 245
    manganese dichloride, 245
  for oxidative coupling
    cuprous chloride, 155
    cuprous salts, 90
  hydrobromic acid for oxidation with dimethyl sulfoxide, 43
  iridium compounds for oxidation with oxygen, 4
  iron compounds for oxidation with oxygen, 4
  isopropyl orthotitanate for oxidation with *tert*-butyl hydroperoxide, **278**

Subject Index 389

Catalysts—Continued
lead tetraacetate for oxidation with peroxy esters, 86
Lindlar catalyst for hydrogenation of ozonides, 81
manganese diacetylacetonate for oxidation with peroxyacetic acid, 245
manganese dichloride for oxidation with peroxyacetic acid, 245
mercuric acetate for oxidation with peroxy esters, 86
mercuric dipropionate for oxidation with chromium trioxide, 75
metal oxides and salts for oxidation with oxygen, 4
metals for oxidation of aldehydes to carboxylic acids, 174
methanesulfonic acid for formation of peroxy acids, 222
molybdenum compounds for alkene epoxidation, 61
molybdenum hexacarbonyl for epoxidation, 153
molybdenum trioxide for *anti* hydroxylation, 69
molybdenyl acetylacetonate
  for enantioselective oxidations, 44
  for epoxidation, 153
  for oxidation with *tert*-butyl hydroperoxide, 153
osmic acid for oxidation with hydrogen peroxide, 7
osmium tetroxide
  for aminohydroxylations, 74
  for hydroxylations, 69, 72, 73
  for oxidation
    of aldehydes to carboxylic acids, 177
    with chlorates, 28
    with hydrogen peroxide, 7
oxalyl chloride for oxidation with dimethyl sulfoxide, 43
palladium
  for dehydrogenation, 38, 50
  for hydrogenation of ozonides, 5, 77
  for oxidation with hydrogen peroxide, 8
palladium acetate for oxidation with hydrogen peroxide, 8
palladium chloride for oxidation with oxygen, 75, 133
palladium compounds for oxidation with oxygen, 4
palladium diacetate for oxidation with hydrogen peroxide and by electrolysis, 75
palladium dichloride, for oxidation
  of terminal alkenes, 15
  with air, 75
  with cuprous chloride, 15
  with oxygen, 147
phosgene for oxidation with dimethyl sulfoxide, 43
phosphoric acid for oxidation with dimethyl sulfoxide, 43

Catalysts—Continued
phosphorus pentoxide for oxidation with dimethyl sulfoxide, 43
platinum
  for dehydrogenative aromatization, 38, 50–51
  for oxidation
    of alcohols to aldehydes, 115, 123, 124
    of aldehydes to carboxylic acids, 174
    with air, 182–183
    with air or oxygen, 127
    with oxygen, 4, 133, 182
platinum compounds for oxidation with oxygen, 4
platinum dioxide, for oxidation
  of alcohols, 115
  with air or oxygen, 127
  with oxygen, 147
platinum oxide
  for oxidation with oxygen, 133
  for reduction of ozonides, 5
pyridinium trifluoroacetate for oxidation with dimethyl sulfoxide, 144
Raney nickel
  for dehydrogenation of alcohols, 39, 143
  for hydrogenation of ozonides, 5, 77
rhodium compounds for oxidation with oxygen, 4
rhodium trichloride for oxidation with air, 262
selenium dioxide
  for *anti* hydroxylation with hydrogen peroxide, 69
  for oxidation with hydrogen peroxide, 7, 261
silver
  for Cannizzaro reaction, 177
  for dehydrogenation, 14, 114
  for oxidation
    of aromatic aldehydes, 15
    with air, 124, 126
silver oxide
  for Cannizzaro reaction, 177
  for oxidation of aldehydes to carboxylic acids, 174
sodium orthovanadate for oxidation of sulfides, 261
sodium tungstate for oxidation with hydrogen peroxide, 8, 225
sodium vanadate for oxidation with hydrogen peroxide, 7
sulfur trioxide for oxidation with dimethyl sulfoxide, 43
titanium compounds for alkene epoxidation with *tert*-butyl hydroperoxide, 61
titanium tetraisopropoxide
  for enantioselective oxidation, 44, 154
  for oxidation with *tert*-butyl hydroperoxide, 254
titanium trichloride for oxidation with hydrogen peroxide, 7

Catalysts—Continued
  p-toluenesulfonic acid for formation of peroxy acids, 222
  transition metals
    for decomposition of hydrogen peroxide, 7
    for oxidation
      with hydrogen peroxide, 9
      with oxygen, 4
  trifluoroacetic acid for oxidation with dimethyl sulfoxide, 43
  trifluoroacetic anhydride for oxidation with dimethyl sulfoxide, 43
  trifluoromethanesulfonic acid for oxidation with dimethyl sulfoxide, 43
  trimethylsilyl trifluoroacetate for oxidation with bis(trimethylsilyl) peroxide, 194, 195
  trimethylsilyl trifluoromethanesulfonate for oxidation with bis(trimethylsilyl) peroxide, 10, 187, 191, 195
  tris(triphenylphosphine)ruthenium dichloride, for oxidation
    of aldehydes to carboxylic acids, 177, 179
    with bis(trimethylsilyl) peroxide, 10
    with iodosobenzene, 91, 92, 172, 245
  tungsten trioxide (tungstic acid)
    for *anti* hydroxylation with hydrogen peroxide, 69, **277**
    for oxidation
      of sulfides, 261
      with hydrogen peroxide, 7
  vanadium compounds
    for alkene epoxidation with *tert*-butyl hydroperoxide, 61
    for oxidation with *tert*-butyl hydroperoxide, 237
  vanadium pentoxide
    for cleavage of alkenes and cycloalkenes to aldehydes, 79
    for dehydrogenation, 49
    for *anti* hydroxylation with hydrogen peroxide, 69
    for oxidation
      of benzene rings to carboxylic acids, 97
      with air or oxygen, 19
      with *tert*-butyl hydroperoxide, 237
      with chlorates, 28
      with hydrogen peroxide, 7
      with oxygen, 4
      with sodium chlorate, 166
  vanadyl acetylacetonate
    for enantioselective oxidation, 44
    for epoxidation, 153
    for oxidation with *tert*-butyl hydroperoxide, 153
    for stereoselective epoxidation, 19
  Woelm alumina
    for dehydrogenation of hydroxy sulfides, 261
    for Oppenauer oxidation, 149
  zinc oxide for dehydrogenation, 49

Catalytic dehydrogenations, See Dehydrogenations, catalytic
Catalytic hydrogenations
  cleavage of ozonides, 5
  reduction of ozonides to aldehydes, 77–78
Catecholborane, hydroboration agent and preparation, 268
Celite, support for chromium trioxide, 117
Cellulomonas galba, in oxidation of aromatic side chains to carboxyl groups, 107
Ceric ammonium nitrate
  in acyloxylations, 100
  in cleavage of vicinal diols, 159, 161
  in conversion of hydrazides into carboxylic acids, 230
  in oxidation
    of acyloins to α-diketones, 217
    of aromatic side chains, 101, 102
    of benzylic alcohols, 115, 124
    of diols, 156
    of ketones to esters and lactones, 190
    of methylene groups to keto groups, 103, 104
    of phenols to quinones, 165, 166
    of secondary alcohols to ketones, 133, 139
    of sulfides to sulfoxides, 255, **281**
  in regeneration of ketones from their derivatives, 219
  oxidant, 17
  reoxidation by sodium bromate, 115
Ceric ammonium sulfate
  in oxidation of aromatic compounds to quinones, 95, 96
  oxidant, 17
Ceric sulfate
  in oxidation
    of diols, 156
    of secondary alcohols to ketones, 133
  reoxidation by sodium bromate, 29, 133, 139, 156
Chemical dehydrogenations, 50–52
Chiral compounds, in enantioselective hydroxylations, 71, 73
Chiral oxidants
  camphorylsulfonyloxaziridines, 223
  (R,R)-diethyl tartrate, 61, 154, 254, **278**
  hydroxamic acids, 153
  Sharpless reagent, 44, 154, 254
Chloral
  in dehydrogenation of hydroxy sulfides to keto sulfides, 261
  in oxidation of secondary alcohols to ketones, 143
  oxidant, 39
Chloramine-T *(sodium N-chloro-p-toluenesulfonamide)*
  in aminohydroxylations, 74
  oxidant, 28
Chloranil
  dehydrogenating agent, 52
  in oxidation of benzylic alcohols to aldehydes, 125
  oxidant, 40

Chlorates, oxidants of osmium tetroxide and vanadium pentoxide, 28
Chlorine
  catalyst in oxidation with dimethyl sulfoxide, 43, 122
  in oxidation
    of diols, 156
    of disulfides
      to sulfinyl chlorides, 263–264
      to sulfonyl chlorides, 264
    of isothiocyanates to sulfonyl chlorides, 252
    of isothioureas to sulfonyl chlorides, 252
    of mercaptans (thiols) to sulfonyl chlorides, 252
    of primary alcohols to aldehydes, 120
    of secondary alcohols to ketones, 138, 147, 148, 149
    of sulfinic acids to sulfonyl chlorides, 264
  oxidant and properties, 26–27
  solvents, 27
  toxicity, 26
α-*Chloro ketones,* formation from alkenes and cycloalkenes, 76
N-*Chloroamides,* in formation of amino hydroxy compounds, 74
m-*Chlorobenzoyl chloride,* in preparation of *m*-chloroperoxybenzoic acid, 222
Chlorochromates, in oxidation of mercaptans (thiols) to disulfides, 250
Chloroform, solvent
  for ozonization, 5, 66
  for peroxyacetic acid, 12
  for peroxybenzoic acid, 13
m-*Chloroperoxybenzoic acid* (m-chloroperbenzoic acid)
  in epoxidation
    of alkenes and cycloalkenes, 61
    of allylic alcohols, 152
    of allylic ethers, 171
  in oxidation
    of ketones to esters and lactones, 190
    of lactols to lactones, 184
    of mercaptals to sulfoxides, 259
    of primary amines to nitro compounds, 235
    of secondary alcohols to ketones, 133, 147, 148
    of selenides to selenones, 265
    of sulfides to sulfoxides, 253, 257
    of tertiary amines to amine oxides, 237
    of trialkylsilyl enol ethers, 172–173
  oxidant, 13–14
  preparation, 13, 222
  solvents, 13
Chlorophyll, oxygen sensitizer, 1, 88
Chloroplatinic acid, catalyst for oxidation with oxygen, 133
N-*Chlorosuccinimide*
  in oxidation
    of primary alcohols to aldehydes, 119, 120

N-*Chlorosuccinimide,* in oxidation—*Continued*
    of secondary alcohols to ketones, 144, 148, **293**
  oxidant, 28
N-*Chlorosuccinimide and dimethyl sulfide,* in oxidation of secondary alcohols to ketones, 144
3-*Cholestanone,* dehydrogenation via selenoxides, **275**
1-Cholesten-3-one, preparation from 3-cholestanone by dehydrogenation via selenoxides, **275**
Chromic acid
  generation
    from chromic oxide and water, 117
    from potassium dichromate and sodium dichromate, 24
  in degradation of carboxylic acids, 151
  in dehydrogenative coupling of phenols, 53
  in destructive oxidation of methyl-group-containing compounds to acetic acid (Kuhn–Roth method), 224
  in oxidation
    of alkanes and cycloalkanes, 57–58
    of alkyl groups, 57–58
    of allylic halides, 110–111
    of aromatic compounds to quinones, 94, 95
    of aromatic side chains
      to carbonyl groups, 103
      to carboxyl groups, 105, 106
    of primary alcohols
      to aldehydes, 116, 124–126
      to carboxylic acids, 128
    of secondary alcohols to ketones, 134
    of sulfides to sulfoxides, 253, 257
  in oxidative cleavage of ketones to carboxylic acids, 211–212
  oxidant, 21
  See also *Chromic anhydride, Chromic oxide,* and *Chromium trioxide*
Chromic acid adsorbed on silica gel, preparation, **283**
Chromic anhydride, oxidant, 21
Chromic oxide
  complexes with pyridine, 86–87
  in epoxidation of alkenes and cycloalkenes, 61, 62
  in oxidation of alkenes and cycloalkenes to unsaturated ketones, 86–87
Chromium(VI) compounds
  catalysts for alkene epoxidation with *tert*-butyl hydroperoxide, 61
  in oxidation
    of alkenes and cycloalkenes to ketones, 75
    of aromatic compounds to quinones, 94, 95
    of primary alcohols to aldehydes, 115, 123–126
    of secondary alcohols to ketones, 133, 147–149

Chromium(VI) compounds—Continued
  mechanism of oxidation of secondary
    alcohols, 134
Chromium(III) oxide
  dehydrogenation catalyst, 49, 119
  in dehydrogenative aromatization, 55
Chromium(VI) oxide, oxidant, 21
Chromium, pentavalent, in oxidation of
    secondary alcohols to ketones, 134
Chromium–pyrazole complex, See
    *Chromium trioxide–3,5-
    dimethylpyrazole complex*
Chromium–pyridine complex, See
    Chromium trioxide–pyridine complex
Chromium sesquioxide, catalyst for
    oxidation with oxygen, 103
Chromium, tetravalent, in oxidation of
    secondary alcohols to ketones, 134
Chromium trioxide
  catalyst for oxidation with hydrogen
    peroxide, 7
  conversion into tetrabutylammonium
    chromate, **285**
  in cleavage
    of alkenes and cycloalkenes, 80, 82
    of vicinal diols, 159, 162
  in hydroxylation of methine groups, 59
  in oxidation
    of aromatic compounds to quinones,
      96
    of aromatic side chains, 103
    of benzylic alcohols to aldehydes,
      125
    of benzylic halides to ketones, 112
    of diols, 155
    of ethers
      to carboxylic acids, 170
      to esters, 170
      to esters and lactones, 169
    of hydroxy ketones to diketones,
      215
    of ketones to α-dicarbonyl compounds,
      200
    of methylene groups to carbonyl
      groups, 200
    of phenols to quinones, 164
    of primary alcohols to aldehydes, 117–
      118, 123–126
    of secondary alcohols
      to carboxylic acids, 150
      to ketones, 135, 147–149
    of unsaturated alcohols to lactones,
      154
    of unsaturated esters to unsaturated
      keto esters, 226
  in preparation
    of Collins reagent, **274**
    of Jones reagent, **273, 283**
    of pyridinium chlorochromate, **274**
    of pyridinium dichromate, 25
    of tetrabutylammonium chromate, **274,
      285**
  oxidant, 21
  preparation of complex with 3,5-
    dimethylpyrazole, **284**

*Chromium trioxide–3,5-dimethylpyrazole
  complex*
  in oxidation
    of primary alcohols to aldehydes, 117–
      118, 125–126
    of secondary alcohols to ketones, 137,
      147, 149, **284**
  oxidant, 23
  preparation, 23, 137
Chromium trioxide on silica gel, in
  oxidation of alcohols to aldehydes, **283**
*Chromium trioxide–pyridine complex
  (Collins reagent)*
  in oxidation
    of primary alcohols to aldehydes, 117,
      118, 123–126, **283–284**
    of secondary alcohols to ketones, 137,
      147, 148, 149
  oxidant, 22–23
  preparation, 22, **274**
  See also *Collins reagent*
Chromyl acetate, in cleavage of alkenes and
  cycloalkenes, 78–79
Chromyl benzoate, in cleavage of alkenes
  and cycloalkenes, 78–79
Chromyl chloride
  in hydroxylation of ketones to α-hydroxy
    ketones, 196
  in oxidation
    of alkenes and cycloalkenes to
      ketones, 75–76
    of aromatic side chains, 101, 102, 105
    of primary alcohols to aldehydes, 119,
      123
    of secondary alcohols to ketones, 147,
      149
    of trialkylsilyl enol ethers, 172–173,
      196
  in preparation of α-chloro ketones, 76
  oxidant and preparation, 26
Chromyl compounds, in epoxidation of
  alkenes and cycloalkenes, 61
Chromyl nitrate, in epoxidation of alkenes
  and cycloalkenes, 62, 63
Chromyl trichloroacetate, in cleavage of
  alkenes and cycloalkenes, 78–79
Chrysene, oxidation, 95
5,6-Chrysenequinone, by oxidation of
  chrysene, 96
Cinnamaldehyde
  oxidant, 39
  preparation from cinnamyl alcohol, **284**
Cinnamaldehyde semicarbazone,
  preparation from cinnamyl alcohol,
    288
Cinnamyl alcohol, oxidation to
  cinnamaldehyde, **284, 288**
Clorox, oxidant, 27, **285**
Cobalt acetate, catalyst for oxidation with
  ozone, 106
Cobalt benzoate, catalyst for oxidation with
  air, 223
Cobalt compounds, catalysts for oxidation
  with oxygen, 4
Cobalt oxide, dehydrogenation catalyst, 114

Cobalt peroxide
  catalyst for oxidation with oxygen, 234
  in dehydrogenation of primary amines to nitriles, 241
Cobalt sesquioxide, catalyst for oxidation with oxygen or air, 124, 126
Cobaltous acetate, catalyst for hydroxylation of aromatic rings, 94
Collins reagent
  ignition during preparation, 117
  in oxidation
    of primary alcohols to aldehydes, 117–118, 124–126, **283–284**
    of secondary alcohols to ketones, 137, 149
  oxidant, 22–23
  preparation, 22, 117, 137, **274**
  solvents, 22
  See also *Chromium trioxide–pyridine complex*
Cope elimination, 259
Cope rearrangement, 237
Copper
  catalyst
    for dehydrogenation, 14, 114, 132
    for oxidation with air, 124
    in Guerbet reaction, 178
  in dehydrogenation of secondary alcohols to ketones, 132
Copper acetate, oxidant, 15
Copper bronze, dehydrogenation catalyst, 15
Copper carbonate, oxidant, 15
Copper chlorides, catalysts for oxidation with air, 75
Copper chromite
  dehydrogenation agent, 15
  in dehydrogenation
    of primary alcohols to aldehydes, 114
    of secondary alcohols to ketones, 132
Copper compounds, catalysts for oxidation with oxygen, 4
Copper, divalent, in hydroxylation of aromatic rings, 93
Copper nitrate, in oxidation of benzylic halides to aldehydes or ketones, 111
Copper oxide(s)
  dehydrogenation catalyst, 114
  in oxidation
    of primary alcohols to aldehydes, 123, 124
    of secondary alcohols to ketones, 133, 148
  oxidant, 15
Copper permanganate, in oxidation
  of mercaptans (thiols) to disulfides, 250
  of primary alcohols to carboxylic acids, 130
Copper permanganate octahydrate, in oxidation of secondary alcohols, 141
Copper sulfate
  catalyst for oxidation with peroxysulfate, 101–102

Copper sulfate—Continued
  in oxidative coupling of acetylenic alcohols, 155
Copper sulfate pentahydrate
  oxidant, 15
  support for potassium permanganate, 34, 141, 147, 149
Corynebacterium equii
  in epoxidation of alkenes, 64
  in oxidation
    of sulfides to sulfones, 254
    of sulfides to sulfoxides, 254
Corynebacterium simplex
  in dehydrogenation of carbon–carbon bonds, 48–49
  in oxidation
    of hydroxy ketones to diketones, 215, 216
    of secondary alcohols to ketones, 146, 216
Coupling, oxidative, See *Oxidative coupling*
Crown ethers
  for solubilization of potassium permanganate, 34, 129, 142, 149
  in oxidation of primary alcohols to carboxylic acids, 129
Cumene hydroperoxide
  by oxidation of cumene, 99
  in oxidation
    of aldehydes to carboxylic acids, 177, 179
    of phosphites to phosphates, 248, 249
Cunninghamella blakesleeana
  in hydroxylation of steroidal ketones, 198
  in oxidation of hydroxy ketones to diketones, 216
  in selective oxidation of diols, 157
Cupric acetate
  catalyst for oxidation with ammonium nitrate, 19, 217, **282**
  in coupling of terminal alkynes, **279**
  in oxidation of acyloins to α-diketones, 217
  in oxidative coupling of acetylenic alcohols, 155
Cupric chloride, in acyloxylation of aromatic rings, 94
Cupric nitrate,
  in oxidation of alkenes and cycloalkenes to unsaturated ketones, 86
Cupric oxide, in oxidation
  of hydrazo compounds, 233
  of primary alcohols, 115
Cupric permanganate
  in oxidation of primary alcohols to carboxylic acids, 129
  oxidant, 35
Cupric salts, in oxidative coupling of alkynes, 90
Cupric sulfate
  catalyst for oxidation with oxygen, 217
  in oxidation
    of acyloins to α-diketones, 217
    of alkenes and cycloalkenes to unsaturated ketones, 86

Cuprous bromide, catalyst for oxidation
    with peroxy esters, 86
Cuprous chloride, catalyst
    for oxidation
        with air, 15
        with oxygen, 133, 241
    for oxidative coupling of acetylenic
        alcohols, 155
Cuprous oxide, catalyst for oxidation of
    aldehydes to carboxylic acids, 174
Cuprous salts, catalysts for oxidative
    coupling by oxygen, 90
Curvularia sp., in hydroxylation of steroidal
    ketones, 198
*20-Cyano-3α,21-dihydroxy-17-pregnen-11-
    one*, oxidation to 20-cyano-21-hydroxy-
    17-pregnene-3,11-dione, **286**
Cyanohydrins,
    oxidation to α-keto nitriles, 179
*20-Cyano-21-hydroxy-17-pregnene-3,11-
    dione*, preparation from 20-cyano-
    3α,21-dihydroxy-17-pregnen-11-one,
    286
Cyclic amides, hydroxylation, 242
Cyclic ethers, oxidation to lactones, 169
Cyclic ketones,
    oxidation to dicarboxylic acids, 209
Cyclic peroxides, formation from
    conjugated dienes, 87
Cyclic sulfoxides
    by oxidation of thiiranes, 255
    of thianes, 257
Cycloalkanes, oxidation
    with chromic acid, 57–59
    with oxygen, 60
Cycloalkenes
    acyloxylation, 74–75
    cleavage
        at double bonds, 77–84
        by chromyl carboxylates, 78–79
        by osmium tetroxide
            and anhydrous hydrogen peroxide,
            79
            and sodium metaperiodate, 79
        by ozone, 77–82
        by potassium permanganate, 79
        by sodium periodate, 79
        by sodium permanganate, 80
        to dialdehydes, 79–80
        to dicarboxylic acids, 80–81
        to diols, 77
        to ketones, 80
    conversion
        into alcohols, 77–78, 84, 85, 86
        into aldehydes, 77–80
        into alkyl hydroperoxides, 84–85
        into allylic esters, 86
        into amino hydroxy compounds, 74
        into carbonyl compounds, 84, 85–87
        into carboxylic acids, 77, 80–84
        into esters, 74–75, 84, 86
        into halohydrins, 73
        into α-haloketones, 76
        into α-hydroxy ketones, 76
        into ketones, 77, 80
        into unsaturated ketones, 86

Cycloalkenes—Continued
    enantioselective hydroxylation, 71
    epoxidation, 60–64
    hydroxylation, 67–73
    oxidation
        at allylic positions, 84–87
        to ketones, 75–77
        to vicinal dicarbonyl compounds, 76
    oxidative cleavage, 77–84
    ozonization, 66, 77–82
    stereochemistry of hydroxylation, 67–68
1,6-Cyclodecanedione, preparation from *cis*-
    9,10-decalindiol, **288–289**
1,2-Cyclodecanedione dihydrazone,
    oxidation to cyclodecyne, **280**
Cyclodecyne, preparation from 1,2-
    cyclodecanedione dihydrazone, **280**
Cyclododecanol, oxidation to
    cyclododecanone, **290**
Cyclododecanone, preparation from
    cyclododecanol, **290**
1,3-Cyclohexadiene, conversion into 3,6-
    endoperoxocyclohexene, **276**
Cyclohexane, solvent for ozonization, 5
cis-1-Cyclohexanediol
    cleavage to adipic aldehyde, **281**
    preparation from cyclohexene, **286–287**
trans-1,2-Cyclohexanediol, preparation
    from cyclohexene, **277**
Cyclohexanol, oxidation to adipic acid, **281,
    289**
Cyclohexanone, oxidant, 39
Cyclohexanone oxygenase, in oxidation
    of aldehydes to carboxylic acids and
        esters, 177
    of ketones to esters and lactones
        (Baeyer–Villiger reaction), 190, 195
    of phenylboronic acids to phenols, 270
    of selenides to selenoxides, 265
Cyclohexene, hydroxylation
    to 1,2-cyclohexanediol, **286–287**
    with hydrogen peroxide, **277**
3-Cyclohexenecarboxaldehyde, oxidation to
    3-cyclohexenecarboxylic acid, **279–280**
3-Cyclohexenecarboxylic acid, preparation
    from 3-cyclohexenecarboxaldehyde,
    279–280
Cyclohexyl acetate, preparation from
    cyclohexyl methyl ketone, **279**
Cyclohexyl methyl ketone, oxidation to
    cyclohexyl acetate, **279**
Cyclopentadiene, conversion into *cis*-2-
    cyclopentene-1,4-diol, **275**
cis-2-Cyclopentene-1,4-diol, preparation
    from cyclopentadiene, **275**
Cylindrocarpon radicicola, in biochemical
    Baeyer–Villiger reaction of steroidal
    ketones, 194–195
Czapek Dox liquid medium, **294**

D

Dabco (1,4-diazabicyclo[2,2,2]octane)
    addition compound with bromine, 28,
    139, 148

Dabco (1,4-diazabicyclo[2,2,2]octane)—
    Continued
  adduct with hydrogen peroxide, 7
Dakin reaction
  mechanism, 181, 186
  oxidation of aromatic aldehydes, 180–181
DDQ, See *2,3-Dichloro-5,6-dicyano-p-benzoquinone*
cis-*9,10-Decalindiol*, oxidation to 1,6-cyclodecanedione, **288–289**
Decanal, preparation from 1-decanol, **283**
1-Decanol, oxidation to decanal with chromium trioxide, **283**
Decarbonylation, of primary alcohols with Raney nickel, 132
Degradative oxidations, See *Oxidations, degradative oxidation*
Dehydrogenating agents
  aluminum chloride, 52, 54
  aluminum oxide, 55
  arsenic oxide, 52
  chloranil, 52
  chromic acid, 53
  chromium oxide, 55
  Corynebacterium simplex, 48–49
  2,3-dichloro-5,6-dicyano-*p*-benzoquinone, 52
  ferric chloride, 53
  ferric oxide, 53
  horseradish peroxidase, 55
  hydrogen peroxide, 55
  manganese dioxide, 52, 53
  nitric acid, 52
  nitrobenzene, 52
  oxygen, 53
  periodic acid, 53
  potassium ferricyanide, 52, 55
  potassium permanganate, 53
  sulfuric acid, 53
Dehydrogenation catalysts,
  See *Catalysts*, for dehydrogenation
Dehydrogenations
  applications, 47
  aromatization, 55
  biochemical, 48–49
  by 2,3-dichloro-5,6-dicyano-*p*-benzoquinone, **275**
  catalytic
    of alcohols to aldehydes, 114, 115, 123
    of alcohols to ketones, 132
    of diols, 158
    of hydroaromatic compounds, 50–51
    with oxides of copper, cobalt, and chromium, 114, 115
    with palladium, 50–51
    with platinum, 50, 51
  chemical
    with selenium, 50, 52
    with sulfur, 50–51
  coupling, 52–53
  cyclization, 54, 55
  effects
    on alcohols, 50
    on ketones, 50
    on quaternary groups, 50
    on unsaturated side chains, 50

Dehydrogenations—Continued
  elimination of quaternary groups, 50
  intramolecular, 47, 50, 54
  introduction of double bonds, 47–50
  of alcohols to aldehydes, 114–127
  of dihydropyridines to pyridines, 241
  of diols
    biochemical, 159
    catalytic, 158
  of hydroquinones to quinones, **290**
  of hydroxy sulfides to keto sulfides, 261
  of primary amines
    to aldehydes or ketones, 239–240
    to imines, 239–240
    to nitriles, 241–242
  of secondary amines to imines, 240–241
  of thiols to disulfides, **290**
  via selenoxides, **275**
  with aluminum chloride, 17
  with copper acetate, 15
  with copper chromite, 15
  with 2,3-dichloro-5,6-dicyano-*p*-benzoquinone, 47–49
  with diphenylpicrylhydrazyl, 41
  with hydrogen peroxide, 47
  with mercuric acetate, 47, 48
  with palladium, 38
  with platinum, 38
  with quinones, 49
  with Raney nickel, 37, 39
  with selenium, 20
Dehydrogenative aromatizations
  examples, 50–52
  with aluminum oxide and chromium oxide, 55
Dehydrogenative couplings
  of aromatic amines, 53, 240
  of aromatic hydrocarbons, 52–53
  of phenols, 53
  with chromic acid, 53
  with ferric chloride, 53
  with ferric oxide, 53
  with manganese dioxide, 53
  with oxygen, 53
  with periodic acid, 53
  with potassium permanganate, 53
  with sulfuric acid, 53
Dehydrogenative cyclizations
  with horseradish peroxidase, 54, 55
  with hydrogen peroxide, 55
  with potassium ferricyanide, 55
Demethylation of tertiary amines, 243, **291–292**
Dess–Martin reagent (periodinane)
  in oxidation of alcohols to aldehydes or ketones, 123, 148, 267
  oxidant, 32
  preparation, 13, 32, 267
Destructive oxidation of methyl-group-containing compounds to acetic acid (Kuhn–Roth method), 224
Diacetylenes, formation from alkynes, 90
Diacetylenic diols,
  by coupling of acetylenic alcohols, 155

Dialdehydes
  by cleavage
    of cyclic vicinal diols, 161
    of cycloalkenes, 79–80
  by oxidation of diols, 155
Dialkyl ethers, formation of
    hydroperoxides, 168
Dialkyl tartrates, chiral reagents in
    enantioselective oxidations, 154
Dialkylalkynes, oxidation to acyloins, 91
Dialkylhydroxylamines, by oxidation of
    secondary amines, 236
Diamino aromatic compounds,
    oxidation to quinones, 247–248
Diarylacetylenes,
    oxidation to α-diketones, 91
Diarylmercury compounds, in synthesis of
    arylboranes, 269–270
1,4-Diazabicyclo[2,2,2]octane, See *Dabco*
Diazo compounds, by oxidation of
    hydrazones, 219–220, **280**
α-*Diazoketones*, by oxidation of
    monohydrazones of α-diketones, 221
Dibenzoyl peroxide
    in acyloxylation of aromatic rings, 94
    oxidant, 10
Dibenzoyl peroxide and iodine, in
    acyloxylation of aromatic rings, 94
cis-*Dibenzoylstilbene*, preparation from
    tetraphenylcyclopentadienone, **276**
Di-tert-butyl chromate
  in oxidation of primary alcohols to
    aldehydes, 119, 124, 126
  in situ preparation, 119, 124, 126
3,5-Di-tert-butyl-o-benzoquinone,
    oxidant for amines, 40
2,2-Di-tert-butylethanol, preparation from
    1,1-di-*tert*-butylethylene, **277**
1,1-Di-tert-butylethylene,
    hydroboration and oxidation, **277**
Dicarbonyl compounds
  by oxidation
    of acetylenic amines, 245
    of α-bromo ketones, **293**
    of enamines, 244, 245
    of epoxides, 173–174
    of ketones, 199–202
  formation from oxetanes, 3
  oxidation
    to acid anhydrides, 193
    to carboxylic acids, 218–219
  See also α-*Diketones* and β-*Diketones*
Dicarboxylic acids
  by cleavage of cycloalkenes, 80
  by degradative oxidation of unsaturated
    carboxylic acids, 226
  by oxidation
    of cyclic ketones, 209
    of α-diketones and o-quinones, 218
Dicarboxylic (aldaric) acids, by oxidation of
    aldoses with nitric acid, 128
2,6-Dichloroaniline, oxidation to 2,6-
    dichloronitrosobenzene, **278–279**
2,3-Dichloro-5,6-dicyano-p-benzoquinone
  dehydrogenating agent, 47–49, 52

*2,3-Dichloro-5,6-dicyano-p-benzoquinone—
    Continued*
  hydrogen acceptor in dehydrogenations,
    40
  in aromatization, 52
  in oxidation
    of alkenes and cycloalkenes to
      ketones, 75
    of diols, 156
    of phenols to quinones, 164, 165
    of primary alcohols to aldehydes, 120
    of secondary alcohols to ketones, 143
  periodic acid for reoxidation, 143
  preparation, 166
  use in dehydrogenation, **275**
2,3-Dichlorohexafluoro-2-butene,
    oxidation to trifluoroacetic acid, **289**
N-*(3,5-Dichloro-4-hydroxyphenyl)-p-
    benzoquinone imine*, oxidant for α-
    hydroxy ketones, 40
Dichloromethane, solvent
  for ozonization, 5
  for peroxyacetic acid, 12
2,6-Dichloronitrosobenzene, preparation
    from 2,6-dichloroaniline, **278–279**
Dichloroperoxymaleic acid
  in oxidation of tertiary amines to amine
    oxides, 239
  in situ preparation, 239
2,6-Dichlorophenolindophenol, oxidant, 40
Dichromates, in oxidation of mercaptans
    (thiols) to disulfides, 250
Dicyclohexylcarbodiimide, activator of
    dimethyl sulfoxide, 43, 122, 144–145,
    293
Dicyclohexyl-18-crown-6 ether,
    for solubilization of potassium
    permanganate, 129
Dicyclohexylurea, byproduct in oxidation
    with dicyclohexylcarbodiimide, 145
Diels–Alder reaction,
    formation of endoperoxides, 2–3
Dienes
  conversion
    into carbonyl compounds by
      ozonolysis, 90
    into endoperoxides, 2–3
  hydroxylation, 89
  oxidation, 87–90
  ozonolysis, 90
Diethyl azodicarboxylate
  hydrogen acceptor in dehydrogenations,
    40
  in demethylation of tertiary amines, 243,
    291–292
  in oxidation
    of benzylic alcohols to aldehydes, 126
    of hydrazo compounds, 233
    of hydroxylamines, 231–232
    of mercaptans (thiols) to disulfides,
      250–251
    of secondary alcohols to ketones, 144,
      149
    of sulfides to sulfoxides, 261
    of tertiary amines to amides, 243

Subject Index   397

Diethyl ether,
    formation of hydroperoxides, 168
Diethyl tartrate,
    chiral reagent, 44, 254, 258, **278**
Diethylmethylacetaldehyde, preparation
    from 2-ethyl-2-methyl-1-butanol, **283–284**
1,8-Difluorooctane, preparation from 5-fluorovaleric acid, **276–277**
Dihydrazones of α-diketones, oxidation to acetylenes, 221
Dihydropyridines, dehydrogenation, 241
Dihydroquinidine acetate, chiral reagent, 73
Diisopropyl peroxydicarbonate
    in acyloxylation of aromatic rings, 94
    oxidant, 10
Diisopropylamine, base in oxidation with dimethyl sulfoxide, 145
Diketo carboxylic acids, by oxidation of acetylenic carboxylic acids, 226–227
Diketo esters, by oxidation of esters of acetylenic alcohols, 226–227
Diketones
    by cleavage of cyclic vicinal diols, 161–162
    by oxidation
        of diols, 155
        of hydroxy ketones, 215–218
α-Diketones
    by oxidation
        of acetylenes
            with potassium permanganate, 91, **289**
            with thallium trinitrate, 91, **280–281**
        of acyloins (α-hydroxy ketones), 217–218, **282**
        of ketones, 200, **283**
        of vicinal diols, **288–289**
    oxidation
        to acid anhydrides, 193, 218
        to dicarboxylic acids, 218
    See also *Dicarbonyl compounds*
α-Diketones, dihydrazones, oxidation to acetylenes, 221
α-Diketones, monohydrazones, oxidation
    to acetylenes, 221
    to α-diazoketones, 221
β-Diketones
    Baeyer–Villiger reaction, 190
    oxidation, 218–219
    See also *Dicarbonyl compounds*
Diketones, unsaturated,
    Baeyer–Villiger reaction, 194–195
9,10-Diketostearic acid,
    preparation from stearolic acid, **289**
Dimethyl cellosolve (dimethyl ether of ethylene glycol), solvent for peroxyacetic acid, 12
Dimethyl sulfide
    in oxidation
        of diols, 156
        of primary alcohols to aldehydes, 120
        of secondary alcohols to ketones, 144, **293**
    in reduction of ozonides, 5, 78–79, 90

Dimethyl sulfide and chlorine
    in oxidation
        of diols, 156
        of primary alcohols to aldehydes, 120
    in selective dehydrogenation of alcohols, 42
Dimethyl sulfide and N-chlorosuccinimide
    in oxidation of primary alcohols to aldehydes, 120, 123, 125
    in selective dehydrogenation of alcohols, 42
Dimethyl sulfoxide
    activators, 122, 144–145, 147–149
    in dehydrogenation of primary amines to aldehydes, 240
    in oxidation
        of alcohols to carbonyl compounds, 43, **293**
            of aldehyde–bisulfite addition compounds to carboxylic acids, 178
            of aldehydes to carboxylic acids, 178
            of alkynes to α-diketones, 91
            of benzylic alcohols to aldehydes, 125
            of benzylic halides to aldehydes or ketones, 111
            of α-bromo ketones to α-dicarbonyl compounds, 201–202, **293**
            of diols, 155
            of epoxides (oxiranes), 173–174
            of hydroxy ketones to diketones, 215, 216
            of ketones to α-dicarbonyl compounds, 200–201
            of α-methyl and α-methylene groups, 200, 201
            of primary alcohols to aldehydes, 121–123, 125, 126
            of primary halides to aldehydes, 109, 110
            of secondary alcohols to ketones, 144–149
            of secondary halides to ketones, 109, 110
    mechanism of oxidation reaction, 43, 121–122, 144–145
    role of activator, 121
    scavenger in oxidation with sodium chlorite, 28
    solvent in oxidation with chromium trioxide, 136
Dimethyl sulfoxide and N-bromosuccinimide, in oxidation of alkynes to α-diketones, 91
4-Dimethylaminopyridinium chlorochromate
    in oxidation of primary alcohols to aldehydes, 118, 124, 125
    oxidant and preparation, 23
2,6-Dimethylaniline, oxidation to 2,6-dimethyl-*p*-benzoquinone, **292**
2,6-Dimethyl-p-benzoquinone, preparation from 2,6-dimethylaniline, **292**
cis-1,2-Dimethyl-1,2-cyclopentanediol,
    preparation from 1,2-dimethylcyclopentene, **290–291**

1,2-Dimethylcyclopentene, hydroxylation to *cis*-1,2-dimethyl-1,2-cyclopentanediol, **290–291**
Dimethyldioxirane
  in epoxidation
    of alkenes and cycloalkenes, 61, 63
    of unsaturated esters, 225
  in oxidation
    of primary amines to nitro compounds, 235
    of sulfides to sulfoxides, 254
    of tertiary amines to amine oxides, 237
  in situ formation, 63
Dimethyldodecylamine, oxidation to dimethyldodecylamine oxide, **277–278**
Dimethyldodecylamine oxide, preparation from dimethyldodecylamine, **277–278**
Dimethylhydrazones, oxidation to parent ketones, 219
2,7-Dimethylocta-3,5-diyn-2,7-diol, preparation from 2-ethyl-3-butyn-2-ol, **279**
2,4-Dimethylpentane-2,4-diol chromate, catalyst in oxidation with peroxyacetic acid, 147
3,5-Dimethylpyrazole, preparation of complex with chromium trioxide, **284**
Dinitrogen tetroxide
  in oxidation
    of phenols to quinones, 166
    of primary alcohols to aldehydes, 115
    of secondary alcohols to ketones, 133, 149
    of sulfides to sulfoxides, 257
  oxidant and preparation, 18
3,5-Dinitromesitylglyoxal, oxidant, 40
2,4-Dinitrophenylhydrazones, conversion into parent ketones, 219
Diols
  biochemical dehydrogenation, 159
  by hydroxylation of alkenes and cycloalkenes, 68–73, **290–291**
  by oxidative cleavage of cycloalkenes, 77
  catalytic dehydrogenation, 158
  oxidation
    to dialdehydes, 155, 161
    to diketones, 155, 162
    to esters, 131
    to lactones, 131
    with barium manganate, 155
    with ceric ammonium nitrate, 156
    with ceric sulfate, 156
    with chlorine, 156
    with chromium trioxide, 155
    with 2,3-dichloro-5,6-dicyano-*p*-benzoquinone, 156
    with dimethyl sulfide and chlorine, 156
    with dimethyl sulfoxide 155
    with methyl phenyl sulfide and chlorine, 156
    with silver carbonate, 157
    with sodium hypochlorite, 156
  selective oxidation
    with *Acetobacter suboxydans*, 157
    with bakers' yeast, 157

Diols, selective oxidation—*Continued*
    with *Cunninghamella blakesleeana*, 157
    with *Helicostylum piriforme*, 157
    with 4-methoxy-2,2,6,6-tetramethyl-1-oxopiperidinium chloride, 157
    with *Rhizopus arrhizus*, 157
Diols, diacetylenic, by coupling of acetylenic alcohols, 155
Diols, steroidal
  cleavage to ketones, 162
  selective oxidation, 156–157
Diols, vicinal
  by hydroxylation of double bonds, 67–73
  by oxidation of alkenes, **286–287**
  cleavage
    to aldehydes and ketones, 159–163, **287**
    with ceric ammonium nitrate, 159, 161
    with chromium trioxide, 159, 162
    with iodobenzene diacetate, 160
    with iodosobenzene or iodoxybenzene, 160
    with lead tetraacetate, 159, 160, 161, **281**
    with manganese dioxide, 159, 161
    with periodic acid, 159, 160, **287**
    with potassium dichromate, 159, 161
    with potassium periodate, 159
    with sodium bismuthate, 159, 161, 162
    with sodium periodate, 160
    with trivalent iodine compounds, 159
  mechanism of cleavage with periodic acid, 159
  oxidation to carboxylic acids, 163
Diols with primary and secondary hydroxyls, oxidation to lactones, 157, 158
Dioxane
  formation of hydroperoxides, 169
  molecular complex with bromine, 28
Dioxetanes
  by oxidation
    with singlet oxygen, 2, 3, 64–65
    with triphenyl phosphite ozonide, 64, 65
  formation, 64–65
  intermediates in alkene oxidation, 84–85
Dioxiranes
  oxidants and preparation, 9
  See also *Dimethyldioxirane*
Diphenyl diselenide, in conversion of epoxides into allylic alcohols, 266
Diphenyl sulfide, oxidation to diphenyl sulfoxide, **281**
Diphenyl sulfoxide, preparation from diphenyl sulfide, **281**
Diphenylacetylene, oxidation to benzil, **280–281**
Diphenyldiazomethane, preparation from benzophenone hydrazone, **280**
Diphenyldiselenide
  oxidant, 48
  oxidation
    to benzeneseleninic acid, 21

Diphenyldiselenide, oxidation—*Continued*
  to benzeneseleninic anhydride
    in dehydrogenation of steroids, 48
    in oxidation of alcohols to
      aldehydes, 122–124, 126
1,3-Diphenyl-2,3-epoxy-1-propanone,
  preparation from benzalacetophenone,
    285
Diphenylpicrylhydrazyl
  dehydrogenating agent, 41
  in oxidation of hydrazo compounds, 233
  preparation, 41
Disiamylborane
  hydroborating agent, 268
  preparation, 268
Disulfides
  by oxidation of mercaptans (thiols), 250–
    251, **290**
  oxidation
    to monosulfoxides (alkyl
      thiosulfinates), 263
    to sulfinyl chlorides, 263–264
    to sulfonic acids, 264
    to sulfonyl chlorides, 264
Disulfones, by oxidation of mercaptals or
  mercaptoles, 260
Divalent copper, in hydroxylation of
  aromatic rings, 93
DMS, See *Dimethyl sulfide*
DMSO, See *Dimethyl sulfoxide*
Double bonds, See *Carbon–carbon double
  bonds* and *Carbon–nitrogen double
  bonds*

E

Elbs hydroxylation (persulfate oxidation), 8,
  163
Electrolysis, anodes, electrolytes, and
  electrolytic cells, 6, **276**
Electrolytic decarboxylative coupling,
  of sodium carboxylates, 224, **276–277**
Electrolytic oxidations
  of aliphatic alcohols to esters, 131
  of alkenes and cycloalkenes
    to epoxides, 61
    to ketones, 75
  of aromatic side chains to carboxyl
    groups, 105
  of benzene rings to carboxylic acids, 98
  of primary alcohols to carboxylic acids,
    127, 130
  of pyridine homologues, 108, 109
  of secondary alcohols to ketones, 133,
    147
  of vicinal diols to carboxylic acids, 163
Electrophilic substitutions in pyridine ring,
  238
cis *Elimination,* via sulfoxides, 259
Enamines, oxidation to dicarbonyl
  compounds, 244, 245
Enantiomeric excess in chiral oxidations,
  154, 258, **278**

Enantiomers, by epoxidation of allylic
  alcohols, 153
Enantioselective hydroxylations
  biochemical, 71
  of alkenes and cycloalkenes, 71
  of sulfones and sulfoxides, 73
  use of chiral compounds, 71, 73
  with *Pseudomonas putida,* 71
Enantioselective oxidations
  applications, 45
  catalyst, 154
  chiral reagents
    camphorylsulfonyloxaziridines, 44
    dialkyl tartrates, 154
    N-phenylcamphoryl hydroxamic acid,
      44
    Sharpless reagent, 44, 154
  of esters, 223
  of sulfides to sulfoxides, 257–258
  requirements, 44
Endooxides (endoperoxides)
  conversion into unsaturated *cis* diols, 87–
    89
  from aromatic compounds, 2–3, 88, 89
  from conjugated dienes, 2, 3, 87–89
  generation of singlet oxygen, 88, 89
  reduction with thiourea, 87–88
3,6-Endoperoxocyclohexene, preparation
  from 1,3-cyclohexadiene, **276**
Ene mechanism, 2–3, 84
Enol ethers, See *Ethers, enol*
Enolates,
  oxidation to α-hydroxy ketones, 196
Eosin, oxygen sensitizer, 88
Epoxidations
  catalysts, 61
  chiral reagents, 61
  of acyclic allylic alcohols, 153
  of alkenes and cycloalkenes
    by electrolysis, 61
    with *N*-bromosuccinimide, 61, 63
    with *tert*-butyl hydroperoxide, 61, 62
    with calcium hypochlorite, 61
    with *m*-chloroperoxybenzoic acid, 61,
      63
    with chromic oxide, 61, 62
    with chromyl compounds, 61, 62,
      63
    with *Corynebacterium equii,* 62
    with dimethyldioxirane, 61, 63
    with hydrogen peroxide, 60, 62, 63
    with iodine and silver oxide, 61, 63
    with iodine triacetate, 61
    with iodine tris(trifluoroacetate), 62
    with microorganisms, 61
    with Oxone, 63
    with oxygen, 60, 62
    with ozone, 61, 62
    with peroxy acids, 61
    with peroxyacetic acid, 62
    with peroxybenzoic acid, 63
    with peroxymaleic acid, 62
    with peroxyphthalic acid, 63
    with peroxytrifluoroacetic acid, 61, **279**
    with sodium hypochlorite, 61, 63

Epoxidations, of alkenes and
cycloalkenes—*Continued*
with tetrabutylammonium persulfate,
63
with thallium triacetate, 61
of allylic alcohols, 152–154
of allylic ethers, 171
of enol ethers
with peroxy acids, 170
with pyridinium chlorochromate, 171
of homoallylic alcohols, 154
of unsaturated acetals, 184–185
of unsaturated alcohols, 152–154
of unsaturated aldehydes, 182
of unsaturated carboxylic acids, 224–225
of unsaturated esters, 225
of unsaturated ethers, 170–171
of unsaturated ketones, 191, 192, 212–214, **285**
of unsaturated nitriles, 230
of vinyl ethers, 170
Epoxides
conversion into allylic alcohols, 266
from alkenes and cycloalkenes, 61–64
from halohydrins, 73
oxidation
to dicarbonyl compounds, 173–174
to hydroxy hydroperoxides, 173–174
to α-hydroxy ketones, 173–174
with dimethyl sulfoxide, 173–174
with hydrogen peroxide, 173–174
with periodic acid, 173–174
Epoxy acetals,
from unsaturated acetals, 184–185
Epoxy aldehydes,
from unsaturated aldehydes, 182
Epoxy esters,
from unsaturated ketones, 191, 192
Epoxy ketones,
from unsaturated ketones, 192, **285**
1,2-Epoxypentane,
preparation from 1-pentene, **279**
Ester groups, by oxidation of aromatic side chains, 101–105
Esters
by electrooxidation of aliphatic alcohols, 131
by oxidation
of acetals, 184
of aldehydes with cyclohexanone oxygenase, 177
of alkenes and cycloalkenes at allylic positions, 84, 86
of aromatic aldehydes, 181
of diols, 131
of enol ethers, 171
of ethers
with benzyltriethylammonium permanganate, 169
with chromium trioxide, 169
with ozone, 169
with potassium permanganate, 169
with ruthenium tetroxide, 169, 170
with zinc dichromate, 169
with zinc permanganate, 169

Esters, by oxidation—*Continued*
of glycosides, 184
of ketones
with ceric ammonium nitrate, 190
with *m*-chloroperoxybenzoic acid, 190
with cyclohexanone oxygenase, 190, 195
with hydrogen peroxide, 187–188
with peroxy acids, 186–187, 188–190, 195
with peroxyacetic acid, 188
with peroxybenzoic acid, 189
with peroxymaleic acid, 190
with peroxyphthalic acid, 190
with peroxytrifluoroacetic acid, 188
with potassium peroxymonosulfate, 188, 195
of primary alcohols
with bromine, 131
with *tert*-butyl chromate, 128, 131
with manganese dioxide, 131
with Oxone, 131
with potassium peroxymonosulfate, 131
with sodium dichromate, 131
of trialkylsilyl ethers, 172
by Willgerodt–Kindler reaction of ketones, 204
enantioselective oxidation, 223
hydroxylation, 223
oxidation, 222–229
oxidation of alcohol part to carboxylic groups, 224
oxidation to α-keto esters, 223
oxidative coupling in α positions, 223
Esters, benzylic, by acyloxylation of arylalkanes, 100–101
Esters of acetylenic alcohols,
oxidation to diketo esters, 226–227
Esters of hydroxy ketones,
from ketones, 199
Esters, unsaturated
by decomposition of alkoxycarbonyl selenoxides, 265
by oxidation
of unsaturated acetals, 184–185
of unsaturated alcohols, 131
of unsaturated ketones, 191, 192
epoxidation, 225
oxidation, 226
oxidative cleavage to aldehydes, 79
Etard reaction, 105
Ethers
formation of hydroperoxides, 168–169
oxidation
to carbonyl compounds with bromine, 170
to carboxylic acids with chromium trioxide, 170
to esters and lactones
with benzyltriethylammonium permanganate, 169
with chromium trioxide, 169
with ozone, 169

Subject Index 401

Ethers, oxidation to esters and lactones—*Continued*
    with potassium permanganate, 169
    with pyridinium chlorochromate, 171
    with ruthenium tetroxide, 169, 170
    with zinc dichromate, 169
    with zinc permanganate, 169
Ethers, acetylenic
  oxidation to keto esters, 172
  oxidation with iodosobenzene, 172
Ethers, allylic, epoxidation, 171
Ethers, aralkyl, allyl, and aryl, oxidation to alcohols and aldehydes, 171–172
Ethers, cyclic, oxidation to lactones, 169
Ethers, enol
  epoxidation
    with peroxy acids, 170
    with pyridinium chlorochromate, 171
  oxidation
    to carboxylic acids with nitric acid, 171
    to esters or lactones, 171
    to α-hydroxy ketones, 196
Ethers, trialkylsilyl
  by oxidation of silanes, 270
  oxidation
    to carboxylic acids, 172
    to esters, 172
    to α-hydroxy ketones, 172–173
    to ketones, 172–173
    to quinones, 173
    with *N*-bromosuccinimide, 172
    with *m*-chloroperoxybenzoic acid, 172–173
    with chromyl chloride, 172–173
    with Jones reagent, 172
    with periodic acid, 172, 173
    with pyridinium chlorochromate, 173
    with triphenyl phosphite ozonide, 173
Ethers, unsaturated
  epoxidation
    with peroxy acids, 170–171
    with pyridinium chlorochromate, 171
  oxidation
    of multiple bonds, 170–173
    to aldehydes, 171–172
Ethers, vinyl,
  epoxidation with peroxy acids, 170
Ethyl acetate, solvent
  for ozonization, 5
  for peroxyacetic acid, 12
Ethyl 1-acetyl-6-oxopiperidine-2-carboxylate, preparation from ethyl 1-acetylpiperidine-2-carboxylate, **291**
Ethyl 1-acetylpiperidine-2-carboxylate, oxidation to ethyl 1-acetyl-6-oxopiperidine-2-carboxylate, **291**
2-Ethyl-3-butyn-2-ol, oxidative coupling, **279**
4-Ethylcyclohexanol, oxidation to 4-ethylcyclohexanone, **285**
4-Ethylcyclohexanone, preparation from 4-ethylcyclohexanol, **285**
Ethylene glycol dimethyl ether, solvent for peroxyacetic acid, 12

N-*Ethylephedrine*, chiral component in enantioselective oxidations, 44
2-Ethyl-2-methyl-1-butanol, oxidation to diethylmethylacetaldehyde, **283**
Excited-state oxygen, definition, 1
Explosions
  in handling triphenylmethylphosphonium permanganate, 36
  in oxidations with peroxymonosulfuric acid, 8
  in reactions with pyridinium dichromate, 25
  of α-hydroxy hydroperoxides, 150

F

Ferric chloride
  as catalyst for oxidation
    with hydrogen peroxide, 8
    with oxygen, 211–212
  in dehydrogenative coupling of phenols, 53
  in oxidation
    of aromatic amines to quinones, 246–247
    of hydroquinones to quinones, **290**
    of mercaptans (thiols) to disulfides, 250, 251
    of phenols to quinones, 166, 251
    of thiols to disulfides, **290**
  in oxidative coupling of esters, 223
  oxidant, 37
Ferric oxide
  conversion into ferric sulfate, 37
  in dehydrogenative coupling, 53
Ferric sulfate
  catalyst for oxidation with hydrogen peroxide, 228
  in oxidation of phenols to quinones, 166
  oxidant and preparation, 37
Ferrous sulfate, catalyst for oxidation with hydrogen peroxide, 228
Fieser reagent
  in oxidation of unsaturated alcohols to keto oxides, 154
  oxidant, 21
o-*Fluorobenzaldehyde*,
  preparation from o-fluorotoluene, **282**
o-*Fluorobenzaldehyde*,
  preparation from o-fluorotoluene, **282**
o-*Fluorobenzyl bromide*,
  preparation from o-fluorotoluene, **282**
  conversion into o-fluorobenzaldehyde, **282**
5-Fluorovaleric acid, conversion into 1,8-difluorooctane, **276–277**
Formamides, by oxidation of tertiary amines, 242–243
Formic acid, in *anti* hydroxylation of alkenes and cycloalkenes, 69
Fremy salt
  in oxidation
    of aromatic amines to quinones, 246, **292**

Fremy salt, in oxidation—Continued
    of phenols to quinones, 164
    oxidant, 20
Friedel–Crafts reaction,
    for ketone synthesis, 204
Fuming nitric acid, See *Nitric acid, fuming*
Functionalized ketones, oxidation by
    Baeyer–Villiger reaction, 191–194
Fusarium lini, in hydroxylation of steroidal
    ketones, 199
Fusarium solani, in substitutive
    hydroxylation of unsaturated
    carboxylic acids, 225–226

G

Gaseous oxygen,
    conversion into singlet oxygen, 1
Gibberella saubinetii, in hydroxylation of
    steroidal ketones, 199
Glucose phenyl osazone, cleavage to
    mesoxaldehyde osazone, **287**
Glycerol,
    preparation from allyl alcohol, **291**
Glycosides, oxidation to esters, 184
Glycosides, unprotected, oxidation, 139
Ground-state oxygen
    applications as oxidant, 4
    definition, 1
Group 1 elements, derivatives, 14–16
Group 2 elements, derivatives, 16
Group 3 elements, derivatives, 17
Group 4 elements, derivatives, 17–18
Group 5 elements, derivatives, 18–20
Group 6 elements, derivatives, 20–26
Group 7 elements, derivatives, 26–37
Group 8 elements, derivatives, 37–39
Guerbet reaction, 177–178

H

Halides, allylic, oxidation to aldehydes and
    ketones, 109–111
Halides, benzylic, oxidation to aromatic
    carbonyl compounds, 110–112
Halides, primary, conversion into
    aldehydes, 109–112
Halides, secondary, oxidation to ketones,
    109, 110
Haloalkyl groups, conversion into carbonyl
    groups, **282**
Haloforms, byproducts of hypohalite
    oxidation of methyl ketones, 207
Halogen compounds, oxidation
    to acids, 112–113
    to aldehydes, 109–112
    to ketones, 112
Halogenated alkenes, epoxidation, 62
Halogens, in oxidation of secondary
    alcohols to ketones, 138, 148
Halogens in alkaline media, in degradation
    of methyl ketones to carboxylic acids,
    206

Halohydrins
    formation from alkenes and cycloalkenes,
        73
    intermediates in synthesis of epoxides, 73
α-*Haloketones,* formation from alkenes and
    cycloalkenes, 76
Halomethyl groups, biochemical oxidation
    to carboxyl groups, 58
Hantzsch synthesis, 241
Helicostylum piriforme
    in oxidation of hydroxy ketones to
        diketones, 216
    in selective oxidation of diols, 157
Helminthosporium sp., in biochemical
    oxidation of sulfides to sulfoxides, 258
Heptanal, preparation from 1-heptanol, **284**
1-Heptanol, oxidation to heptanal, **284**
Heterocyclic aromatic compounds
    by catalytic dehydrogenation, 50, 52
    oxidation of side chains, 102–103
Heterocyclic compounds, dehydrogenation
    with arsenic oxide, 52
    with manganese dioxide, 52
    with mercuric acetate, 49, 50
    with nitric acid, 52
    with nitrobenzene, 52
    with potassium ferricyanide, 52
Hexacyanoferrate(III), See *Potassium
    ferricyanide*
Hexadecyltributylphosphonium periodate,
    in degradation of α-hydroxy acids to
    aldehydes, 228
Hexamethylenetetramine (hexamine)
    in conversion of aralkyl halides to
        aldehydes, **292–293**
    in dehydrogenation of primary amines to
        aldehydes, 240
    in oxidation
        of allylic halides to carbonyl
            compounds, 110
        of benzylic halides to aldehydes or
            ketones, 111–112
    oxidant, 39
Hexane, solvent for ozonization, 66
Homoallylic alcohols, epoxidation, 154
Horse liver alcohol dehydrogenase, in
    oxidation of secondary alcohols to
    ketones, 146
Horseradish peroxidase
    in dehydrogenative cyclizations, 55
    in oxidation of phenols to quinones, 164,
        165
Hydrazides, oxidation, 230
Hydrazo compounds,
    oxidation to azo compounds, 233
Hydrazones, oxidation to diazo
    compounds, 219–220, **280**
Hydroaromatic compounds,
    dehydrogenation to aromatic
    compounds, 50, 52, 54
Hydroboration, of alkenes, 268, **277**
Hydrobromic acid, catalyst in oxidation
    with dimethyl sulfoxide, 43
Hydrocarbons, by electrochemical coupling
    of carboxylic acid salts, 224

Hydrocarbons, aromatic,
  dehydrogenative couplings, 52–54
Hydrogen bromide, catalyst for oxidation
  with oxygen, 4, 58
Hydrogen peroxide
  adduct with Dabco, 7
  anhydrous
    in cleavage of alkenes and
      cycloalkenes, 79
    preparation, 7, **273**
  as oxidant of phenylselenyl halides, 43
  as reoxidant, 187–188
  as reoxidant of osmium tetroxide, 69, 225
  decomposition, 7
  dehydrogenating agent, 47
  in degradative oxidation
    of α-amino acids to carboxylic acids,
      229
    of α-hydroxy acids to aldehydes, 228
    of α-keto acids to carboxylic acids, 229
  in dehydrogenation of carbon–carbon
    bond, 48, 49
  in dehydrogenative cyclizations, 55
  in epoxidation
    of alkenes and cycloalkenes, 60, 63
    of halogenated alkenes, 62
    of unsaturated carboxylic acids, 225
    of unsaturated ketones, 191, 212, 213
    of unsaturated nitriles, 230
  in formation of singlet oxygen, 1, 2
  in hydroxylation of alkenes and
    cycloalkenes, 69, 72, **277**
  in oxidation
    of aldehydes to carboxylic acids, 175
    of alkenes and cycloalkenes to
      ketones, 75
    of alkylboranes to alcohols, **277**
    of aromatic aldehydes, 180–181
    of boranes to borates, 267
    of carboxylic acids to peroxy acids, 222
    of cyclic ketones to lactones, 187
    of dicarbonyl compounds to carboxylic
      acids, 218
    of α-diketones to acid anhydrides, 193
    of epoxides (oxiranes), 173–174
    of ketones to peroxides, 186
    of mercaptans (thiols)
      to disulfides, 250, 252
      to sulfinic acids, 252
    of nitroso compounds, 231
    of phenols to quinones, 164
    of primary amines to azoxy
      compounds, 234
    of secondary amines to
      hydroxylamines, 236
    of selenides to selenoxides, 265
    of sulfides
      to sulfones, 253–256
      to sulfoxides, 253–256, 257
    of tertiary alcohols to hydroperoxides,
      151
    of tertiary amines to amine oxides, 236
    of tertiary arsines to arsine oxides, 250
    of unsaturated aldehydes
      to epoxy aldehydes, 182

Hydrogen peroxide, in oxidation of
  unsaturated aldehydes—*Continued*
    to unsaturated carboxylic acids, 175,
      182
  of unsaturated sulfides to unsaturated
    sulfones, 261
  in oxidative cleavage of ozonides, 5,
    226
  in preparation of bis(trimethylsilyl)
    peroxide, 10
  molecular complexes, 7
  oxidations
    catalysts, 7–8
    scope, 8
  physical properties, 6–7
  preparation, 7
  reoxidant of osmium oxide, 69
  scavenger in oxidation with sodium
    chlorite, 28
Hydrogen peroxide, alkaline, in epoxidation
  of unsaturated aldehydes, 182
  of unsaturated ketones, 213
Hydrogen peroxide and derivatives, 6–10
Hydrogen peroxide, anhydrous,
  in oxidation of phosphites to
    phosphates, 248
Hydrogen peroxide, 30% concentration
  in oxidation of primary amines to nitroso
    compounds, **278–279**
  in oxidation of selenides to selenoxides,
    275
Hydrogen peroxide, 90% concentration
  decomposition, 7
  in preparation of peroxytrifluoroacetic
    acid, **279**
  physical properties, 7
  preparation, **273**
  safety precautions in preparation, 7
Hydrogen sulfide, in preparation of
  ammonium sulfide, 203
*Hydrogenations, catalytic, See Catalytic
  hydrogenations*
Hydrogens, tertiary,
  oxidation with oxygen, 60
Hydroperoxides
  by oxidation
    of alkanes and cycloalkanes, 58, 60
    of aromatic side chains, 99
    of ethers, 168–169
    of tertiary alcohols, 151
  formation with singlet oxygen, 2
Hydroperoxides, tertiary, by oxidation
  of methine groups, 58, 60
  of tertiary alcohols, 150–151
Hydroquinones,
  dehydrogenation to quinones, **290**
Hydroxy acetals, by oxidation of
  unsaturated acetals, 185
α-*Hydroxy carboxylic acids,* degradative
  oxidation
    to aldehydes, 228
    to keto acids, 227–228
    to ketones, 228–229
Hydroxy compounds, conversion into
  tributyltin ethers, 139

Hydroxy esters
  by oxidation of cyclic ketones, 188
  oxidation to keto esters, 227–228
Hydroxy hydroperoxides
  by oxidation
    of epoxides, 173–174
    of secondary alcohols, 150
  safety precautions in handling, 150
Hydroxy ketones
  Baeyer–Villiger reaction, 194–195
  biochemical oxidation, 215, 216
  by hydroxylation of ketones, 196–199
  by oxidation
    of alkynes, 91
    of diols, 157
    of enolates and enol ethers, 196
    of epoxides, 173–174
    of trialkylsilyl enol ethers, 172–173
  oxidation to diketones, 215–218, **282**
  formation from alkenes and cycloalkenes, 76
α-*Hydroxy nitriles*, oxidation to carboxylic acids
  with argentic oxide, 179
  with manganese dioxide, 179
Hydroxy steroids,
  by biochemical hydroxylation, 196–199
  oxidation to keto steroids, 143
Hydroxy sulfides,
  dehydrogenation to keto sulfides, 261
Hydroxyl groups
  addition across double bonds, 67–73
  stereochemistry of addition across double bonds, 67–68
Hydroxyl groups, tertiary,
  by oxidation of methine groups, 58
Hydroxylamines, oxidation
  to nitroso compounds, 231–232
  to nitroxyls, 232
Hydroxylations
  at benzylic positions, 100–101
  biochemical, See *Biochemical hydroxylations*
  enantioselective, of sulfones and sulfoxides, 73
  of alkenes and cycloalkenes
    stereochemistry, 67–68
    with iodine and silver acetate, 70–71
    with iodine and silver benzoate, 70–71
    with osmium tetroxide, 72–73, **290–291**
  of aromatic rings
    in phenols, 163, 164
    with *Pseudomonas putida*, 93
    with pyridine oxide, 238
  of cyclic amides, 242–243
  of double bonds, 67–63, **277, 286–287, 290–291**
  of esters, 223
  of ketones, 196–199
  of phenols, 163, 164
  of steroids, 71, 196–199
  of sulfoxides, 73
  of unsaturated acetals, 185
  of unsaturated carboxylic acids, 225–226

Hydroxylations—Continued
  of unsaturated ketones, 214
  of unsaturated steroidal ketones, 196–199
  of unsaturated sulfoxides, 262
  stereospecific, 70
  syn versus *anti*, 67–68
anti *Hydroxylations*
  catalyst, 7
  of alkenes and cycloalkenes
    with hydrogen peroxide
      in the presence of formic acid, 69, 72, 226
      in the presence of trifluoroacetic acid, 69
      in the presence of tungsten trioxide, 72
    with iodine and silver acetate or thallium acetate, 70–72
syn *Hydroxylations*
  catalyst, 7
  of alkenes and cycloalkenes
    with benzyltriethylammonium permanganate, 72
    with iodine and silver acetate, 70, 72
    with iodine and thallium acetate, 71, 72
    with osmic acid, 68
    with osmium tetroxide, 68, 71–73
    with potassium permanganate, 68, 226
    with triphenylmethylphosphonium permanganate, 72
  of aromatic rings, 93
Hydroxylations, enantioselective, 73
Hydroxylations, stereospecific, 67–73, 262
Hydroxylations, substitutive,
  by *Fusarium solani*, 225–226
Hydroxylations, transannular,
  of cyclic amides, 242
Hygrophorus conicus, in biochemical oxidation of primary amines to carboxylic acids, 241–242
Hypochlorites, in oxidation
  of primary alcohols to aldehydes, 119, 126
  of secondary alcohols to ketones, 139, 147–149
Hypohalites, in oxidation
  of ketones, 205
  of side chains to carboxylic groups, 210
Hypohalites, alkaline
  in degradative oxidation of methyl ketones to carboxylic acids, 206–207
  side reactions in oxidations, 208
Hypoiodites
  in degradation of methyl ketones, 209
  in tests for methyl ketones, 210
  preparation, 31

I

Ignition
  during preparation of Collins reagent, 117

Ignition—Continued
  in handling tetrabutylammonium
    permanganate, 36, 142
  in reactions of zinc and magnesium
    permanganate, 36
Imides, by oxidation
  of amides, 244, **291**
  of lactams, 245
Imines
  by dehydrogenation of secondary amines,
    240–241
  intermediates, 239
Intermolecular coupling,
  by dehydrogenation, 47
Intramolecular rearrangements,
  during dehydrogenation, 50
Iodine
  addition product
    with silver acetate, 31
    with silver benzoate, 31
  in benzoxylation, 74
  in epoxidation of alkenes and
    cycloalkenes, 61, 63
  in hydroxylation of alkenes and
    cycloalkenes, 70–72, **286**
  in preparation of α-iodo ketones, 76
  oxidant and properties, 29
Iodine acetate, oxidant and properties, 31
Iodine and silver acetate or silver benzoate,
  in hydroxylation of alkenes and
    cycloalkenes, 70–72
Iodine and silver chromate, oxidant, 29
Iodine and silver oxide
  in epoxidation of alkenes and
    cycloalkenes, 61, 63
  oxidant, 29
Iodine and thallium acetate,
  in hydroxylation of alkenes and
    cycloalkenes, 71, 72
Iodine compounds, trivalent,
  in cleavage of vicinal diols, 159
Iodine triacetate
  in epoxidation of alkenes and
    cycloalkenes, 61
  oxidant and preparation, 31
Iodine tris(trifluoroacetate)
  in epoxidation of alkenes and
    cycloalkenes, 62
  oxidant and preparation, 31
Iodo compounds, oxidation
  to iodoso compounds, 266–267
  to iodoxy compounds, 266–267
α-*Iodo ketones,* formation from alkenes
    and cycloalkenes, 76
Iodobenzene diacetate
  in cleavage of vicinal diols, 160
  in hydroxylation of esters, 223
  in oxidation of sulfides to sulfoxides, 256
  oxidant and preparation, 31–32
Iodobenzene dichloride
  in oxidation of selenides to selenoxides,
    265
  oxidant, 31
o-*Iodobenzoic acid,* conversion into
    periodinane (Dess–Martin reagent), 13

Iodoform test, for methyl ketones, 210
Iodopentafluorobenzene, conversion into
    pentafluoroiodobenzene
    bis(trifluoroacetate), 32
Iodoso compounds, by oxidation of iodo
    compounds, 266–267
Iodosobenzene
  in cleavage of vicinal diols, 160
  in oxidation
    of acetylenic amines to dicarbonyl
      compounds, 245
    of acetylenic ethers, 172
    of aldehydes to carboxylic acids, 177,
      179
    of sulfides to sulfoxides, 257
  oxidant and preparation, 31
Iodosobenzene diacetate,
  by oxidation of iodobenzene, 266
o-*Iodosobenzoic acid*
  in hydroxylation of ketones to α-hydroxy
    ketones, 196
  oxidant, 31
p-*Iodosotoluene,* oxidant, 31
Iodoxy compounds, by oxidation of iodo
    compounds, 266–267
Iodoxybenzene
  applications as oxidant, 32
  in cleavage of vicinal diols, 160
  oxidant of diphenyldiselenide, 48
  preparation, 32
m-*Iodoxybenzoic acid*
  oxidant, 32
  oxidant of diphenyldiselenide, 48
Ion exchanger,
  support for chromium trioxide, 117
Ion exchanger IRA 900,
  support of calcium hypochlorite, 148
Iridium compounds,
  catalysts for oxidation with oxygen, 4
Iron, in reduction of pyridine oxide to
    pyridine, 238
Iron compounds,
  catalysts for oxidation with oxygen, 4
Isocyanates,
  by oxidation of primary amides, 230
Isopelletierine, preparation from *N*-
    methylisopelletierine, 243, **291–292**
Isopropyl orthotitanate,
  catalyst for oxidation with *tert*-butyl
    hydroperoxide, 44, 153–154, 258, **278**
Isothiocyanates,
  oxidation to sulfonyl chlorides, 252
Isothioureas,
  oxidation to sulfonyl chlorides, 252

J

Jones reagent
  in oxidation
    of acetylenic alcohols to ketones, **283**
    of primary alcohols to aldehydes, 117,
      124–126

Jones reagent, in oxidation—*Continued*
  of secondary alcohols to ketones, 135–136
  of trialkylsilyl ethers, 172
  in regiospecific oxidation of ethers, 172
  oxidant, 21, 22
  preparation, 22, 135–136, **273**

K

Ketimines, oxidation to oxaziridines, 219
Keto acids, by oxidation
  of hydroxy aldehydes, 182–183
  of hydroxy carboxylic acids, 227–228
  of sugars, 182–183
α-*Keto acids,* degradative oxidation to carboxylic acids, 229
Keto alcohols, steroidal, by oxidation of diols with *N*-bromoacetamide, **286**
α-*Keto aldehydes,*
  by oxidation of ketones, 200
Keto esters,
  by oxidation of hydroxy esters, 227–228
α-*Keto esters*
  Baeyer–Villiger reaction, 190, 229
  by oxidation
    of acetylenic ethers, 172
    of esters, 223
  oxidation to acid anhydrides, 193, 229
Keto esters, unsaturated, by oxidation of unsaturated esters, 226
α-*Keto nitriles,*
  by oxidation of cyanohydrins, 179
Keto oxides, by oxidation of unsaturated alcohols
  with Fieser reagent, 154
  with pyridinum chlorochromate, 154
  with sodium chromate, 154
Keto steroids,
  by oxidation of hydroxy steroids, 143
Keto sulfides, by dehydrogenation of hydroxy sulfides, 261
Ketone derivatives, oxidation, 219–222
Ketones
  by biochemical oxidation of secondary alcohols, 146
  by catalytic oxidation of secondary alcohols, 132, 133, 147, 149
  by degradative oxidation of α-hydroxy acids, 228–229
  by dehydrogenation
    of primary amines, 239–240
    of secondary alcohols, 132, 148
  by electrooxidation of secondary alcohols, 133, 147
  by Oppenauer oxidation of secondary alcohols, 149
  by oxidation
    of alkanes, 58
    of alkenes and cycloalkenes, 75–77, 85–87
    of allylic alcohols, with sodium permanganate, 140

Ketones, by oxidation—*Continued*
  of benzylic alcohols, 144
  of boranes, 269
  of ethers, 170
  of organomercury compounds, 271
  of secondary alcohols
    steric effects, 135
    with *Acetobacter suboxydans,* 146
    with air, 132, 147, 149
    with argentic oxide, 148
    with barium manganate, 140, 149
    with benzeneseleninic anhydride, 149
    with bromine, 139, 147, 148, 149
    with *tert*-butyl hypochlorite, 139, 148
    with calcium hypochlorite, 139, 148
    with ceric ammonium nitrate, 133, 139
    with ceric sulfate, 133
    with chloral, 143, 149
    with chlorine, 138, 147–149
    with *m*-chloroperoxybenzoic acid, 133, 147, 148
    with *N*-chlorosuccinimide, 144, 148
    with *N*-chlorosuccinimide and dimethyl sulfide, 144
    with chromic acid, 134
    with chromium trioxide, 135, 147–149
    with chromium trioxide–3,5-dimethylpyrazole complex, 137, 147, 149, **284**
    with chromium trioxide–pyridine complex, 137, 147–149
    with chromyl chloride, 147, 149
    with Collins reagent, 137, 147–149
    with copper oxide, 133, 148
    with copper permanganate octahydrate, 141
    with *Corynebacterium simplex,* 146, 216
    with Dess–Martin periodinane, 148
    with 2,3-dichloro-5,6-dicyano-*p*-benzoquinone, 143
    with diethyl azodicarboxylate, 144, 149
    with dimethyl sulfide and *N*-chlorosuccinimide, **293**
    with dimethyl sulfoxide, 144–149
    with dinitrogen tetroxide, 133, 149
    with halogens, 138, 147–149
    with hexavalent chromium, 133, 147–149
    with horse liver alcohol dehydrogenase, 146
    with hypochlorites, 139, 147, 149
    with Jones reagent, 135–136, 148, 149
    with lead tetraacetate, 133, 149
    with manganese dioxide, 140, 144, 147–149
    with Oxone, 133
    with oxygen, 147, 149
    with peroxyacetic acid, 147

Subject Index

Ketones, by oxidation of secondary alcohols—*Continued*
  with poly(vinylpyridinium) chlorochromate, 147, 148
  with potassium dichromate, 136, 149
  with potassium ferrate, 142, 148, 149
  with potassium permanganate, 140, 147, 149
  with potassium permanganate on molecular sieves, 149, **290**
  with potassium peroxymonosulfate, 133
  with potassium ruthenate, 142
  with pyridinium chlorochromate, 138, 149
  with pyridinium chromate, 137, 148
  with pyridinium dichromate, 137
  with Raney nickel, 147
  with ruthenium tetroxide in situ, 142, 148
  with silver carbonate, 133
  with silver oxide, 133
  with sodium bromate, 139
  with sodium bromite, 148
  with sodium dichromate, 136, 148, **285**
  with sodium hypochlorite, 139, 147, 148
  with sodium permanganate, 147, 148
  with sodium ruthenate, 142
  with solid permanganates, 141
  with supported permanganates, 141
  with tetrabutylammonium chlorochromate, 138, 149
  with tetrabutylammonium chromate, 136, 148, **285**
  with tetrabutylammonium dichromate, 148
  with tetrabutylammonium hypochlorite, 139, 149
  with tetrabutylammonium permanganate, 142, 149
  with trichloroacetaldehyde, 149
  with triphenylmethylfluoroborate, 146–148
  with yeast, 146
  with zinc dichromate, 149
  of secondary benzylic alcohols, 149
  of secondary halides, 109, 110
  of secondary nitro compounds, 230–231
  of tertiary alcohols, 151
  of tertiary amines, 244
  of trialkylsilyl ethers, 172–173
by oxidative cleavage of alkenes and cycloalkenes, 77, 80
by ozonolysis of alkenes and cycloalkenes, 80
cleavage of carbon chain, 134
conversion
  into esters of hydroxy ketones, 199

Ketones, conversion—*Continued*
  into phenols during aromatization, 50–51
degradative oxidations to carboxylic acids with one less carbon, 206–210
from alkenes, 75–76
hydroxylation
  with chromyl chloride, 196
  with *o*-iodosobenzoic acid, 196
  with oxodiperoxy molybdenum complex with pyridine and hexamethylphosphoramide, 196
oxidation
  by Willgerodt reaction and Willgerodt–Kindler reaction, 202–205, **282–283**
  to carboxylic acids, 202–215
  to α-dicarbonyl compounds, 199–202
  to α-diketones, 200, **283**
  to esters and lactones
    with ceric ammonium nitrate, 190
    with *m*-chloroperoxybenzoic acid, 190
    with cyclohexanone oxygenase, 190, 195
    with peroxy acids, 188–190, 195
    with peroxybenzoic acid, 189
    with peroxymaleic acid, 190
    with peroxyphthalic acid, 190
    with peroxytrifluoroacetic acid, 188
    with potassium peroxymonosulfate, 188, 195
  to α-hydroxy ketones, 196
  to α-keto aldehydes, 200
  with hypohalites, 205
  with potassium permanganate, 140
  with sulfur, **282–283**
oxidative cleavage
  with chromic acid, 211–212
  with nitric acid, 211
  with oxygen, 211
  with potassium permanganate, 211–212
  with potassium superoxide, 211
reduction during aromatization, 50
regeneration from their derivatives by oxidation, 219–220
Ketones, acetylenic, by oxidation of acetylenic alcohols
  with Jones reagent, **283**
  with manganese dioxide, 140
Ketones, aromatic, by oxidation
  of arylalkanes, 101–105
  of benzylic halides, 112
Ketones, aryl alkyl,
  oxidation to aryl alkanoic acids, 204
Ketones, cyclic,
  oxidation to dicarboxylic acids, 209
Ketones, derivatives, oxidation, 219–222
Ketones, functionalized
  Baeyer–Villiger reaction, 191–194
  oxidation, 191–194
Ketones, heterocyclic,
  Willgerodt reaction, 205
Ketones, hydroxy, See *Hydroxy ketones*

Ketones, methyl
  degradative oxidation to carboxylic acids
    with carbon tetrachloride, 205-206
    with halogens, 206
    with hypohalites, 206-210, **286**
    with hypoiodites, 209-210
    with nitric acid, 210
    with potassium permanganate, 210
    with sodium dichromate, 210
  oxidation
    to carboxylic acids
      with carbon tetrachloride, 205-206
      with thallium trinitrate, 205-206
    to α-keto carboxylic acids, 206
  qualitative test, 210
Ketones, steroidal
  biochemical hydroxylation, 58
  oxidation to lactones, 194-195
Ketones, trihalomethyl
  hydrolysis to carboxylic acids, 206
  intermediates in oxidation of methyl ketones, 206
Ketones, unsaturated
  Baeyer-Villiger reaction, 191-193, 212
  biochemical hydroxylation, 214
  by oxidation
    of allylic alcohols, 140, 143, 144
    of keto selenoxides, 265
    of unsaturated alcohols, 140
  cleavage, 215
  epoxidation
    preparative procedure, **285**
    with alkaline hydrogen peroxide, 213
    with *N*-bromosuccinimide, 212
    with *tert*-butyl hydroperoxide, 212, 213
    with calcium hypochlorite, 212-213
    with hydrogen peroxide, 191, 212, 213
    with peroxy acids, 191, 192, 212-214
    with sodium or calcium hypochlorite, 212, 213
  hydroxylation, 214
  mechanism of oxidation, 192
  oxidation
    to epoxides, 191, 192, 212-214
    to unsaturated esters and lactones, 191, 192
  oxidative cleavage by ozone, 215
Ketones, unsaturated, steroidal,
  hydroxylation, 196-199
Kiliani mixture, 24-25
Kindler modification of Willgerodt reaction, 20, 203
Kolbe electrosynthesis, 224, **276-277**
Kuhn-Roth determination of methyl groups, 224

L

Lactams, conversion into imides, 245
Lactols, oxidation to lactones, 184, 185
Lactones
  by oxidation
    of cyclic acetals, 184, 185

Lactones, by oxidation—Continued
    of cyclic ethers
      with ruthenium tetroxide, 169
      with zinc dichromate, 169
      with zinc permanganate, 169
    of cyclic ketones
      with benzeneseleninic acid, 188
      with hydrogen peroxide, 187
      with peroxy acids, 186-190, 195
      with potassium peroxymonosulfate, 188, 195
    of diols, 131
    of diols with primary and secondary hydroxyls, 157, 158
    of enol ethers, 171
    of ethers, 169-170
    of ketones
      with ceric ammonium nitrate, 190
      with *m*-chloroperoxybenzoic acid, 190
      with cyclohexanone oxygenase, 190, 195
      with peroxy acids, 188-190, 195
      with peroxyacetic acid, 188
      with peroxybenzoic acid, 189
      with peroxymaleic acid, 190
      with peroxyphthalic acid, 190
      with peroxytrifluoroacetic acid, 188
      with potassium peroxymonosulfate, 188, 195
    of lactols, 184, 185
    of unsaturated alcohols with chromium trioxide, 154
  formation from alkenes and cycloalkenes, 74
  oxidation of alcohol part to carboxylic groups, 224
Lactones, steroidal, by oxidation of steroidal ketones, 194-195
Lactones, unsaturated, by oxidation of unsaturated ketones, 191
Lead, anode in electrolytic cells, 6
Lead dioxide
  in oxidation
    of aromatic amines to quinones, 247-248
    of phenols to quinones, 164, 165, 166, 167
  in preparation of diphenylpicrylhydrazyl, 41
  oxidant, 17
Lead nitrate
  in oxidation of benzylic halides to aldehydes or ketones, 111, **282**
  oxidant, 19
Lead oxide, red
  in degradative oxidation of α-hydroxy acids to ketones, 229
  in situ conversion into lead tetraacetate, 160
Lead oxides, in oxidation of phenols to quinones, 166
Lead superoxide, See *Lead dioxide*
Lead superoxide (red), in generation of lead tetraacetate in situ, 101

Subject Index

Lead tetraacetate
  applications as oxidant, 18
  in acetoxylation
    at benzylic positions, 100, 101
    of ketones, 199
  in cleavage
    of carbon-tin bonds, 270
    of vicinal diols, 159, 160, 161, **281**
  in degradation of tertiary alcohols, 151
  in dehydrogenation of primary amines to nitriles, 241
  in formation of vicinal diacetates, 74
  in oxidation
    of aromatic amines to quinones, 247–248
    of benzylic alcohols, 115, 124
    of hydrazones to diazo compounds, 220
    of phenols to quinones, 167
    of primary allylic alcohols to aldehydes, 124
    of primary amides, 230
    of secondary alcohols to ketones, 133, 149
    of stannanes into acetates, 270
    of vinylic alcohols, 151
  in situ generation, 101
  in situ preparation from red lead oxide, 160
Lead tetrakis(trifluoroacetate)
  applications as oxidant, 18
  in acyloxylation at benzylic positions, 100
  preparation, 18
Lemieux–Rudloff reagents,
  in cleavage of alkenes, **288**
Lieben test, 210
Lithium aluminum hydride,
  in reduction of ozonides, 5, 78
Lithium borohydride, source of borane for hydroboration, 268

M

Magnesium permanganate
  applications as oxidant, 36
  ignition with solvents, 36
  silica gel support, 36
  solvents, 36
Manganese diacetylacetonate, catalyst for oxidation with peroxyacetic acid, 245
Manganese dichloride, catalyst for oxidation with peroxyacetic acid, 245
Manganese dioxide
  activity, 33
  applications as oxidant, 33
  byproduct of oxidation with potassium permanganate, 34
  in cleavage of vicinal diols, 159, 161
  in dehydrogenation
    of heterocyclic compounds, 52
    of secondary amines to imines, 240–241
  in dehydrogenative couplings, 53
  in demethylation of tertiary amines, 243

Manganese dioxide—Continued
  in oxidation
    of aldehydes to carboxylic acids, 176, 179
    of allylic alcohols to aldehydes, 126, **288**
    of aromatic side chains, 101
    of benzylic alcohols to aldehydes, 124
    of hydrazo compounds, 233
    of hydrazones to diazo compounds, 221
    of α-hydroxy nitriles to carboxylic acids, 179
    of hydroxylamines to nitroso compounds, 231–232
    of mercaptans (thiols) to disulfides, 250–251
    of methylene groups to keto groups, 103, 104
    of oximes to nitro compounds, 219
    of phosphines to phosphine oxides, 248
    of primary alcohols
      to aldehydes, 119, 124, 128
      to carboxylic acids, 128
      to esters, 131
    of primary amines to azo compounds, 234
    of secondary alcohols to ketones, 140, 144, 147–149
    of sulfides to sulfoxides, 253
    of tertiary amines to amides, 242
    of unsaturated and acetylenic alcohols, 140
    of α,β-unsaturated primary alcohols to esters, 131
    of vicinal diols to ketones, 161, 162, **288**
  in oxidative cleavage of tertiary amines to carbonyl compounds, 244
  preparation, 32–33, **274–275**
  reduction to manganese sulfate by sulfur dioxide, 35
  solvents, 33
Manganese sulfate tetrahydrate, in preparation of manganese dioxide, **274**
Manganese triacetate
  applications as oxidant, 32
  in oxidation of alkenes and cycloalkenes to lactones, 74
Manganic acetate,
  applications as oxidant, 32
Mechanisms
  of Baeyer–Villiger reaction, 186–187
  of cleavage of vicinal diols with periodic acid, 159
  of Dakin reaction, 181
  of ene reaction, 2–3
  of Oppenauer oxidation, 143
  of oxidation
    of secondary alcohols with chromium compounds, 134
    of unsaturated ketones with peroxy acids, 191, 192
    with dimethyl sulfoxide, 43, 121–122, 144–145

Mechanisms, of oxidation—*Continued*
  with singlet oxygen, 2–3
  of ozonization, 65–66
  of Willgerodt reaction, 203
Mercaptals, oxidation to sulfones and
  sulfoxides, 259–261
Mercaptans, oxidation
  to disulfides, 250–251
  to sulfinic acids, 252
  to sulfonic acids, 251–252
  to sulfonyl chlorides, 252
Mercapto group, replacement with
  hydrogen, 252
2-Mercaptohydroquinone, oxidation to
  disulfide of *p*-benzoquinone, **290**
Mercaptoles, oxidation to sulfones and
  sulfoxides, 259–261
Mercuric acetate
  applications as oxidant, 16
  dehydrogenating agent, 16, 47, 48
  in acetoxylation of allylic carbon, 101
  in dehydrogenation
    of amines, 49–50
    of secondary amines to imines, 240–241
  in demethylation of tertiary amines, 243
Mercuric bromide
  applications as oxidant, 16
  in oxidation of methylene groups to keto
    groups, 103–105
Mercuric dipropionate, catalyst for
  oxidations with chromium trioxide, 75
Mercuric oxide
  applications as oxidant, 16
  in oxidation
    of α-diketone dihydrazones to
      acetylenes, 221, 222, **280**
    of hydrazones to diazo compounds,
      220, 221, **280**
    of phenols to quinones, 164, 166
Mercuric trifluoroacetate
  applications as oxidant, 16
  in oxidation of phenols to quinones, 164–166
Mercurous trifluoroacetate
  applications as oxidant, 16
  in oxidation of α-diketone
    monohydrazones to acetylenes, 221
Mercury,
  in removal of colloidal selenium, 199
Mercury immersion lamp, in photochemical
  generation of singlet oxygen, **275**
Mesitylglyoxal
  applications as oxidant, 40
  preparation from 2-acetylmesitylene, **283**
Mesoxaldehyde osazone, preparation from
  glucose phenyl osazone, **287**
Metal oxides and salts, catalysts for
  oxidation with oxygen, 4
Metals, catalysts for oxidation of aldehydes
  to carboxylic acids, 174
Metaperiodic acid, oxidant, 30
Methanesulfonic acid, catalyst for formation
  of peroxy acids, 222

Methanol, solvent for ozonization, 5
Methenamine (hexamethylenetetramine),
  oxidant, 39
Methine groups
  hydroxylation
    retention of configuration, 58
    with benzyltriethylammonium
      permanganate, 58, 59
    with chromic acid, 57–58
    with chromium trioxide, 59
    with *p*-nitroperoxybenzoic acid, 58–59
    with ozone, 59
  oxidation to hydroperoxides, 60
4-Methoxyisophthalaldehyde, preparation
  from 2,4-bis(chloromethyl)anisole,
  292–293
4-(p-Methoxyphenyl)acetophenone,
  oxidation to 4-(*p*-
  methoxyphenyl)benzoic acid, **286**
4-(p-Methoxyphenyl)benzoic acid,
  preparation from 4-(*p*-
  methoxyphenyl)acetophenone, **286**
*4-Methoxy-2,2,6,6-tetramethyl-1-
  oxopiperidinium chloride*, in selective
  oxidation of diols, 157
Methyl azelaldehyde, preparation from
  methyl oleate by ozonolysis, **276**
Methyl chloride, solvent for ozonization, 66
Methyl groups
  biochemical oxidation to carboxylic
    groups, 58
  oxidation
    to aldehyde groups, 101–102, 103
    with chromic acid, 57–58
Methyl ketones, See *Ketones, methyl*
Methyl 2-naphthyl ketone, oxidation to 2-
  naphthylacetic acid, **282–283**
Methyl oleate, ozonolysis, **276**
Methyl phenyl sulfide (thioanisole)
  in oxidation of diols, 156
  oxidation
    to methyl phenyl sulfone, **278**
    to methyl phenyl sulfoxide, **288**
Methyl phenyl sulfide and chlorine,
  in oxidation of diols, 156
Methyl phenyl sulfone, preparation from
  methyl phenyl sulfide, **278**
Methyl phenyl sulfoxide, by oxidation
  with sodium periodate, **287**
  with sodium periodate on alumina, **287–288**
Methyl p-tolyl sulfide, oxidation to methyl
  p-tolyl sulfoxide, **278**
Methyl p-tolyl sulfoxide, preparation from
  methyl *p*-tolyl sulfide by oxidation with
  Sharpless reagent, **278**
Methylene blue, oxygen sensitizer
  in cleavage of double bonds, 80
  in formation
    of dioxetanes, 64, 65
    of singlet oxygen, 1
  in oxidation of enamines, 245
Methylene chloride, solvent for
  benzeneseleninic acid, 14

Subject Index 411

Methylene groups
  biochemical hydroxylation to alcoholic groups, 58
  oxidation
    to carbonyl groups, 58, 103–105, 223, **292**
    with chromic acid, 57–58
    with chromium trioxide, 200
α-*Methylene groups in ketones* oxidation
  with dimethyl sulfoxide, 200, 201
  with selenium dioxide, 199, 200
N-*Methylephedrine*, chiral component in enantioselective oxidation, 44
N-*Methylisopelletierine*, demethylation to isopelletierine, 243, **291–292**
N-*Methylmorpholine* N-*oxide*
  in oxidation of phosphites to phosphates, 248, 249
  oxidant, 42
  reoxidant of osmium tetroxide, 72
Methyltrioctylammonium trichloride (Aliquat 336), phase-transfer reagent, 8
Microbial oxidations,
  See *Biochemical oxidations*
Microorganisms
  cultivation, 197
  in biochemical oxidations, 45–46
  in epoxidation of alkenes and cycloalkenes, 61
  in hydroxylation of steroidal ketones, 196–199
  See also specific microorganisms
Migratory aptitudes in Baeyer-Villiger reaction, 187
Molecular complexes
  bromine–Dabco, 28
  bromine–dioxane, 28
  bromine–pyridine, 28
  hydrogen peroxide–Dabco, 7
Molecular sieves, support for potassium permanganate
  in oxidation
    of alcohols
      to aldehydes, 119, 125, 126
      to ketones, 147, 149
    in organic solvents, 34
  preparation, 141, **290**
Molozonides,
  intermediates in ozonization, 5, 65
Molybdenum acetylacetonate, oxidant, 21
Molybdenum compounds,
  catalysts for alkene epoxidation, 61
Molybdenum hexacarbonyl
  catalyst for epoxidation, 153
  oxidant, 21
Molybdenum hexafluoride
  in regeneration of ketones from their derivatives, 220
  oxidant, 21
Molybdenum oxide,
  in hydroxylation of esters, 223
Molybdenum oxychloride
  in regeneration of ketones from their derivatives, 220

Molybdenum oxychloride—Continued
  oxidant, 21
Molybdenum trioxide,
  catalyst for *anti* hydroxylation, 69
Molybdenyl acetylacetonate, catalyst
  for enantioselective oxidation, 44
  for epoxidation, 153
  for oxidation, 153
Monohydrazones of α-diketones,
  oxidation to diazo ketones, 221
Monosulfoxides,
  by oxidation of disulfides, 263
Montmorillonite, support for potassium permanganate, 34
Morpholine, in Willgerodt–Kindler reaction, 203, **282**
Mortierella isabellina, in biochemical oxidation of sulfides to sulfoxides, 258
Multiple bonds, See *Carbon–carbon double bonds* and *Carbon–nitrogen double bonds*

N

Naphthalene, preparation from tetralin, **275**
Naphthalene homologues, oxidation to carboxylic acids, 107
2-*Naphthylacetic acid*, preparation from methyl 2-naphthyl ketone, **282–283**
2-*Naphthylacetic acid thiomorpholide*, intermediate in preparation of 2-naphthylacetic acid, **282**
NBA, See N-*Bromoacetamide*
NBS, See N-*Bromosuccinimide*
NCS, See N-*Chlorosuccinimide*
Nef reaction, 230
Nickel, anode in electrolytic cells, 6
Nickel dioxide (nickel peroxide)
  in oxidation
    of aldehydes to carboxylic acids, 177, 179
    of aromatic side chains to carboxyl groups, 105, 108
    of hydrazones to diazo compounds, 220
    of primary alcohols
      to aldehydes, 125, 127
      to amides, 132
      to carboxylic acids, 120, 129, 130
    of primary amines
      to azo compounds, 234
      to nitriles, 241
  oxidant, 37
  preparation, 37, 129
Nickel sulfate hexahydrate,
  conversion into nickel peroxide, 37
Nitrates of α-hydroxy ketones,
  oxidation, 201
Nitric acid
  in degradation of methyl ketones to carboxylic acids, 210
  in dehydrogenation
    of dihydroxypyridines to pyridines, 241
    of heterocyclic compounds, 52

Nitric acid—Continued
in oxidation
of aldehydes to carboxylic acids, 176, 183–184
of aromatic amines to quinones, 247
of aromatic side chains to carboxyl groups, 105, 106, 108
of enol ethers to carboxylic acids, 171
of esters to carboxylic acids, 224
of hydrazo compounds to azo compounds, 233
of iodo compounds to iodoxy compounds, 266
of mercaptans (thiols) to sulfonic acids, 252
of perfluorinated aromatic compounds to quinones, 113–114
of primary alcohols to carboxylic acids, 128
of pyridine homologues to carboxylic acids, 108–109
of secondary alcohols to carboxylic acids, 150, 211, **281**
of sulfides to sulfoxides, 257
in oxidative cleavage
of ketones to carboxylic acids, 211
of ozonides, 5
oxidant and properties, 19
Nitric acid, dilute,
in oxidation of sugars, 183–184
Nitric acid, fuming, in oxidation
of aldehydes to carboxylic acids, 176
of iodopentafluorobenzene, 32
Nitriles
by dehydrogenation of primary amines, 241–242
oxidation, 230
Nitriles, unsaturated, epoxidation, 230
Nitro compounds
applications as oxidants, 42
by biochemical oxidation of primary amines, 236
by oxidation
of azides, 232
of nitroso compounds, 231
of oximes, 219
of primary amines, 234, 235–236
oxidation, 230–231
See also specific nitro compounds
Nitro compounds, aliphatic
by oxidation of aliphatic amines, 235
oxidation, 230–231
Nitroamines,
by oxidation of nitrosoamines, 231
Nitrobenzene
in dehydrogenation of heterocyclic compounds, 52
oxidant, 42
p-*Nitrobenzoyl peroxide,* oxidant, 10
Nitrogen compounds
oxidation, 230–248
See also specific types of nitrogen-containing organic compounds
Nitrogen dioxide, in oxidation of perfluoroalkyl hydrides, 113

Nitrogen oxides, in oxidation of phenols to quinones, 166
Nitrogen sesquioxide, See *Nitrous anhydride*
3-*Nitromesitylglyoxal,* oxidant, 40
Nitronic acids, oxidation, 230–231
p-*Nitroperoxybenzoic acid*
in epoxidation
of alkenes and cycloalkenes, 61
of allylic alcohols, 153
in hydroxylation of methine groups, 58–59
oxidant and preparation, 14
p-*Nitrophenylhydrazones,* conversion into parent ketones, 219
2-*Nitropropane,* oxidant, 42
2-*Nitropropane, sodium salt,* in oxidation of benzylic halides to aldehydes or ketones, 111
Nitroso compounds
by oxidation of primary amines
with hydrogen peroxide, 235, **278–279**
with peroxy acids, 235, **278–279**
with potassium peroxysulfate (Oxone), 235
in oxidation
of benzylic halides to aldehydes or ketones, 112
of α-bromo ketones to α-dicarbonyl compounds, 201
oxidation to nitro compounds, 231
Nitrosoamines,
oxidation to nitroamines, 231
Nitrosobenzene
in oxidation of α-bromo ketones to α-dicarbonyl compounds, 201
oxidant, 41
p-*Nitrosodimethylaniline*
in oxidation of methylene groups to carbonyl groups, 201, **292**
oxidant, 41
Nitrous acid
in oxidation of amino phenols to quinones, 246
oxidant and preparation, 18
Nitrous anhydride
in oxidation of esters to keto esters, 223
oxidant and preparation, 18
Nitroxyls,
by oxidation of hydroxylamines, 232
Nocardia strain 107–332, in oxidation of aromatic side chains to carboxyl groups, 106, 107
Nucleoside phosphites,
oxidation to phosphates, 248–249

O

4,6-*Octadiynl,* preparation from 4,6-octadiyn-1-ol, **293**
4,6-*Octadiyn-1-ol,* oxidation to 4,6-octadiynal, **293**
3-*Octyn-2-ol,*
oxidation to 3-octyn-2-one, **283**

3-Octyn-2-one,
    preparation from 3-octyn-2-ol, **283**
Olefins, See *Alkenes*
Oleic acid, oxidation to pelargonic acid and azelaic acid, 227, **288**
Oppenauer oxidation
    hydrogen acceptors, 39
    mechanism, 143
    of alcohols to ketones, 142–143, 149
Orange benzene, 116
Organic oxidants, 39–44
Organic peroxy acids
    applications as oxidants, 10–14
    in formation of singlet oxygen, 1, 2
    in oxidative cleavage of phenols, 168
    See also *Peroxy acids* and specific peroxy acids
Organobromine compounds, oxidants, 29
Organochlorine compounds, oxidants, 28
Organomagnesium compounds,
    oxidation to phenols, 271
Organomercury compounds,
    oxidation to ketones, 271
Osmic acid
    catalyst for oxidation with hydrogen peroxide, 7
    in cleavage of double bonds, 162
    in syn hydroxylation of alkenes and cycloalkenes, 68, 71–73
    oxidant, 38
    See also *Osmium tetroxide*
Osmium oxide,
    reagents for reoxidation, 69, 262
Osmium tetroxide
    catalyst
        for aminohydroxylation, 74
        for oxidation
            of aldehydes to carboxylic acids, 177
            with chlorates, 28
            with hydrogen peroxide, 7
    in enantioselective hydroxylation of alkenes, 71
    in hydroxylation
        of alkenes and cycloalkenes, 68, 71–73
        of unsaturated carboxylic acids, 225
        of unsaturated ketones, 214
        of unsaturated sulfoxides, 73
    in oxidation of sulfoxides to sulfones, 262
    mixture with sodium periodate (metaperiodate), 30
    oxidant, 38
    reagents for reoxidation
        amine oxides, 69, 72, 262
        hydrogen peroxide, 69, 225
        N-methylmorpholine *N*-oxide, 72
        potassium chlorate, 69, 225
        silver chlorate, 69, 225
        sodium chlorate, 69, 225
    See also *Osmic acid*
Osmium tetroxide and sodium periodate (metaperiodate), in cleavage of alkenes and cycloalkenes, 79
Oxalyl chloride, activator of dimethyl sulfoxide, 43, 122, 147–149

Oxaziridines
    by oxidation of ketimines (Schiff bases), 219
    formation, 44
Oxidation agents
    air, oxygen, ozone, and electrolysis, 1–6
    biochemical (microbial) agents, 45–46
    chiral oxidants, 44–45
    derivatives of Group 1 elements, 14–16
    derivatives of Group 2 elements, 16
    derivatives of Group 3 elements, 17
    derivatives of Group 4 elements, 17–18
    derivatives of Group 5 elements, 18–20
    derivatives of Group 6 elements, 20–26
    derivatives of Group 7 elements, 26–37
    derivatives of Group 8 elements, 37–39
    hydrogen peroxide and derivatives, 6–20
    organic oxidants, 39–44
    organic peroxy acids, 10–14
    See also specific reagents or substrates
Oxidations
    at methine groups, 58
    biochemical hydroxylation of unsaturated ketones, 214
    degradative oxidation
        of α-amino acids to aldehydes or carboxylic acids, 229
        of carboxylic acids to aldehydes, 224
        of α-hydroxy acids to aldehydes or ketones, 228
        of α-keto acids to carboxylic acids, 229
        of methyl ketones to carboxylic acids, 206–210
        of unsaturated carboxylic acids, 226
    destructive oxidation of methyl-group-containing compounds to acetic acid (Kuhn–Roth method), 224
    enantioselective oxidation of sulfides to sulfoxides, 257–258, **294**
    hydroxylation of unsaturated carboxylic acids, 225–226
    of acetals, 184
    of acetylenes, 90–92
    of acetylenic acids, 227
    of acetylenic alcohols
        to acetylenic aldehydes, 118, 122, **293**
        to acetylenic ketones, 136, **283**
    of acetylenic amines, 245
    of acyloins, 217–218
    of alcohols, 114–163, **283**
    of aldehyde derivatives, 184–185
    of aldehydes, 174–185
    of aldehydes
        to carboxylic acids, 174–180, **285–286**
        to peroxy acids, 180
    of alkanes and cycloalkanes, 57–60
    of alkenes and cycloalkenes
        to alcohols, 77–78
        to aldehydes, 77–80
        to carboxylic acids, 77, 80–84
        to dioxetanes, 64–65
        to epoxides, 60–64
        to halohydrins, amino hydroxy compounds, and esters, 73–75
        to ketones, 75–77, 80

Oxidations, of alkenes and cycloalkenes—
    Continued
  to ozonides, 65–67
  to vicinal diols, 67–73
of alkyl acetylenyl ethers, 172
of alkyl ketones, 208–209
of alkynes, 90–92
of allylic alcohols to α,β-unsaturated
    aldehydes, 120–121, 124, 126, **288**
of amides, 229–230
of amides to imides, 244
of amino hydroxy aromatic compounds,
    247–248
of aralkyl halides
  to aldehydes, 110–112, **292–293**
  to ketones, 110, 112
of aromatic aldehydes
  to aryl formates, 180–181
  to phenols, 180–181
of aromatic amines to quinones, 246–248
of aromatic compounds, 92–109
of arsenic compounds, 250
of arylalkanes
  to aldehydes, ketones, and derivatives,
      101–105
  to aralkyl hydroperoxides, 99
  to aromatic carboxylic acids, 105–109
  to benzylic alcohols and esters, 100–
      101
of azides, 232
of azo compounds, 232
of benzylic alcohols to aldehydes, 118–
    121, 124–126
of boron compounds, 267–270
of α-bromo ketones to α-dicarbonyl
    compounds, 201
of carbon chains of amines, 242–245
of carboxylic acids and derivatives, 222–
    230
of cycloalkenes, See *Oxidations*, of
    alkenes and cycloalkenes
of diamino aromatic compounds, 247–
    248
of dicarbonyl compounds to carboxylic
    acids, 218–219
of dienes, 87–90
of dihydrazones of α-diketones, 221
of α-diketones, 218
of β-diketones, 218–219
of dimethylhydrazones, 219
of 2,4-dinitrophenylhydrazones, 219
of diols, 155–159
of disulfides, 263–264
of enamines to dicarbonyl compounds,
    244, 245
of epoxides, 173–174
of esters, 222–229
of ethers, 168–174
of functionalized ketones, 191–194
of glycosides, 184
of halogen derivatives, 109–114
of heterocyclic ketones, 205
of hydrazides, 229–230
of hydrazo compounds, 233

Oxidations—Continued
of hydrazones to diazo compounds, 219–
    220
of hydroxy aldehydes, 182
of hydroxy ketones, 194–195, 215–218,
    282
of hydroxylamines, 231–232
of iodo compounds, 266–267
of ketimines to oxaziridines, 219
of ketone derivatives, 219–222
of ketones
  to carboxylic acids, 140, 202–215
  to α-dicarbonyl compounds, 199–202
  to esters and lactones, 186–190, **279**
  to α-hydroxy ketones, 196
  to peroxides, 186
of lactams to imides, 245
of lactols to lactones, 184, 185
of mercaptals, 259–261
of mercaptans, 250–252
of mercaptoles, 259–261
of methyl groups, 58, 101–103
of methyl ketones, 205–210
of monohydrazones of α-diketones, 221
of multiple bonds in unsaturated ethers,
    170–173
of nitriles, 229–230
of nitro compounds, 230–231
of nitrogen compounds, 230–248
of *p*-nitrophenylhydrazones, 219
of nitroso compounds, 231
of organomagnesium compounds, 271
of organomercury compounds, 271
of oximes to nitro compounds, 219
of oxiranes, 173–174
of phenols to quinones, 163–168
of phenylboronic acids to phenols, 270
of phenylhydrazones, 219
of phosphine selenides, 249
of phosphine sulfides, 249
of phosphorus compounds, 248–250
of primary alcohols
  to aldehydes, 114–127, **284**
  to amides, 132
  to carboxylic acids, 127–130
  to esters, 131–132
of primary allylic alcohols to aldehydes,
    120–121, 124, 126–127
of primary amines at nitrogen, 234–236
of primary aromatic amines to nitroso
    compounds, **278–279**
of primary benzylic alcohols to
    aldehydes, 119, 120, 124–126
of Schiff bases to oxaziridines, 219
of secondary alcohols
  to carboxylic acids, 150, **289**
  to α-hydroxy hydroperoxides, 150
  to ketones, 132–149, **285, 290**
of secondary amines at nitrogen, 236
of selenides, 265–266
of selenophosphates to phosphates, 250
of semicarbazones, 219
of side chains of aromatic compounds,
    99–109

Oxidations—Continued
  of silicon compounds, 270
  of steroidal ketones, 194–195
  of sugars, 174, 182–184
  of sulfides to sulfones, 252–256, 259, **278**
  of sulfides to sulfoxides
    examples, 252–259
    with ceric ammonium nitrate, **281**
    with sodium periodate, **287**
  of sulfones, 263
  of sulfoxides, 262–263
  of sulfur compounds, 250–264
  of tertiary alcohols, 150–151
  of tertiary amines
    at nitrogen, 236–239
    to amides, 242–243
    to amine oxides, 236–239, **277–278**
    to carbonyl compounds, 244
  of thioacetals, 259–261
  of thiols, 250–252
  of thiophosphates to phosphates, 250
  of tin compounds, 270
  of tosylates, 109–114
  of tosylhydrazones, 219
  of trimethylsilyl enol ethers, 172–173
  of unprotected glycosides, 139
  of unsaturated acetals, 154–185
  of unsaturated alcohols
    at multiple bonds, 151–155
    examples, 124, 126, 127, **284**
    to aldehydes, 115, 118, 120–122
    to ketones, 136, 137, 140, 143–145
  of unsaturated carboxylic acids, 224–225
  of unsaturated diketones, 194–195
  of unsaturated esters, 79, 225, 226
  of unsaturated ethers, 170–173
  of unsaturated ketones, 212–215
  of vicinal dihydrazones to alkynes, 221, **280**
  of vicinal diols to ketones, **288–289**
  of ylides, 250
  regiospecific, 101
  selective oxidations of diols, 155–156
  transannular, 154
  Willgerodt–Kindler reaction, 203
  Willgerodt reaction, 202–205
  with acetyl hypoiodite, 31, **286–287**
  with acyl chromates, 26
  with aldehydes, 39, 149
  with alkyl nitrites, 18, 104
  with alkyl phenyl selenoxides, 43–44
  with alumina-supported sodium periodate, **287–288**
  with amine oxides, 42, 109, 267
  with ammonium nitrate, 19, **282**
  with ammonium persulfate, 8
  with argentic oxide, 16
  with arsenic pentoxide, 19
  with barium manganate, 33
  with benzeneperoxyseleninic acid, 14
  with benzeneseleninic anhydride, 21
  with benzoyl hypoiodite, 31
  with benzoyl peroxide, 10

Oxidations—Continued
  with benzyltriethylammonium permanganate, 36
  with bis(2,2′-bipyridyl)copper(II) permanganate, 36, 125, 126
  with bis(2,2′-bipyridyl)silver permanganate, 36
  with bismuth sesquioxide, 19, 217
  with bis(trimethylsilyl) peroxide, 10, 195
  with bromine, 28–29
  with bromine complexes, 28–29
  with *N*-bromoacetamide, 29, **286**
  with *N*-bromobenzophenoneimine, 29
  with *N*-bromosuccinimide, 29
  with *tert*-butyl chromate, 26
  with *tert*-butyl hydroperoxide, 9, **277–278**
  with *tert*-butyl hypochlorite, 27
  with *tert*-butyl peroxyacetate, 10
  with *tert*-butyl peroxybenzoate, 10
  with calcium hypochlorite, 27
  with carbon tetrachloride, 39
  with carboxypyridinium dichromates, 25
  with Caro acid, 8
  with ceric ammonium nitrate, **281**
  with ceric ammonium sulfate, 17
  with Chloramine-T, 28
  with chlorates, 28
  with chlorine, 26–27
  with *m*-chloroperoxybenzoic acid, 13
  with *N*-chlorosuccinimide, 28
  with chromium trioxide (chromic acid), 21, **283**
  with chromium trioxide-3,5-dimethylpyrazole complex, 23, **284**
  with chromium trioxide–pyridine complex, 22–23, **283–284**
  with chromyl chloride, 26
  with Collins reagent, 22–23, **283–284**
  with cupric acetate, 15, **279**
  with cupric permanganate, 35
  with Dess–Martin reagent (periodinane), 32
  with dibenzoyl peroxide, 10
  with diethyl azodicarboxylate, 40, **291–292**
  with diisopropyl peroxydicarbonate, 10
  with dimethyl sulfide and chlorine, 42
  with dimethyl sulfide and *N*-chlorosuccinimide, 42, **293**
  with dimethyl sulfoxide
    bases, 145
    catalysts, 144
    mechanism, 43, 144–145
    preparative procedure, **293**
    scope, 43
  with 4-dimethylaminopyridinium chlorochromate, 23, 124, 125
  with dinitrogen tetroxide, 18
  with dioxiranes, 9
  with diphenylpicrylhydrazyl, 41
  with dipyridine chromium(VI) oxide, 22–23
  with ferric chloride, **290**
  with ferric compounds, 37

Oxidations—Continued
with Fieser reagent, 21
with Fremy salt, 20, **292**
with hexamine (hexamethylenetetramine or methenamine), 39
with hydrogen peroxide, 6, 8, **277**
with iodine, 29
with iodine acetate, 31
with iodine and silver chromate, 29
with iodine and silver oxide, 29
with iodine triacetate, 31
with iodine tris(trifluoroacetate), 31
with iodobenzene diacetate, 31–32
with iodobenzene dichloride, 31
with iodoso compounds, 31
with iodosobenzene, 31
with iodoxybenzene, 32
with *m*-iodoxybenzoic acid, 32
with Jones reagent, 21, 22
with ketones, 39
with lead dioxide, 17
with lead nitrate, 19, **282**
with lead tetraacetate, 18, **281**
with lead tetrakis(trifluoroacetate), 18
with magnesium permanganate, 36
with manganese dioxide, 32, **288–289**
with manganese triacetate, 32
with manganic acetate, 32
with manganic sulfate, 32
with mercuric acetate, 16
with mercuric bromide, 16
with mercuric oxide, 16, **280**
with mercuric trifluoroacetate, 16
with mercurous trifluoroacetate, 16
with molybdenum compounds, 21
with nickel peroxide, 37
with nitric acid, 19, **281**
with nitro compounds, 42
with *p*-nitrobenzoyl peroxide, 10
with nitrogen sesquioxide, 18
with *p*-nitroperoxybenzoic acid, 14
with nitrosobenzene, 41
with *p*-nitrosodimethylaniline, 41, **292**
with nitrous acid, 18
with nitrous anhydride, 18
with organic peroxy acids, 10–14, **278–279**
with organobromine compounds, 29
with organochlorine compounds, 28
with osmium tetroxide (osmic acid), 38, **290–291**
with Oxone, 9
with 1-oxo-2,2,6,6-tetramethylpiperidinium chloride, 42
with oxygen, 1–4
with ozone, 4–6
with pentafluoroiodobenzene bis(trifluoroacetate), 32
with performic acid, 10–11
with periodic acid dihydrate (paraperiodic acid), 30, **287**
with periodic acid (metaperiodic acid), 30
with peroxyacetic acid, 11–12, **278–279**
with peroxybenzoic acid, 13, **279**
with peroxydichloromaleic acid, 14

Oxidations—Continued
with peroxyformic acid, 10–11
with peroxylauric acid, 14
with peroxymaleic acid, 14
with peroxymonosulfuric acid, 8
with peroxypentafluorobenzoic acid, 14
with peroxyphthalic acid, 14
with peroxypropionic acid, 14
with peroxytrifluoroacetic acid, 12–13, **279**
with persulfuric acid, 8
with poly(vinylpyridinium) chlorochromate, 23
with potassium-permanganate-coated molecular sieves, **290**
with potassium bromate, 29
with potassium chromate, 23
with potassium dichromate, 24
with potassium ferrate, 37
with potassium ferricyanide [hexacyanoferrate(III)], 37
with potassium hydroxide, 20
with potassium hypobromite, 29
with potassium hypochlorite, 27
with potassium hypoiodite, 29–30
with potassium manganate, 33
with potassium nitrosodisulfonate, 20, **292**
with potassium periodate, 30
with potassium permanganate, 34–35, **289**
with potassium peroxysulfate, **278**
with potassium persulfate, 8, 9
with potassium ruthenate, 38
with potassium superoxide, 8
with purple benzene, 34, **290**
with pyridinium chlorochromate, 23, **284**
with pyridinium chlorochromate adsorbed on alumina, **284**
with quinolinium chlorochromate, 25–26
with quinones, 40
with red lead, 17
with ruthenium tetroxide, 37–38, **291**
with selenious acid, 20
with selenium, 20
with selenium dioxide, 20, **283**
with Sharpless reagent, 10, **278**
with silver carbonate, 16
with silver chromate, 23
with silver oxide, 15, **279–280**
with singlet oxygen, 2–3
with sodium bismuthate, 19
with sodium bromate, 29
with sodium bromite, 29
with sodium chlorite, 27–28, **285–286**
with sodium dichromate, 24, **285**
with sodium hypobromite, 29, **286**
with sodium hypochlorite, 27, **285**
with sodium hypoiodite, 29
with sodium iodate, 30
with sodium perborate, 8
with sodium periodate (sodium metaperiodate), 30, **287**
with sodium periodate and potassium permanganate, **288**

Oxidations—Continued
  with sodium permanganate monohydrate, 34
  with sodium peroxide, 8
  with sodium persulfate, 8
  with sodium ruthenate, 38
  with sulfomonoperacid, 8
  with sulfur, 20, **282–283**
  with sulfuric acid, 20
  with sulfuryl chloride, 20
  with tetrabutylammonium chlorochromate, 24
  with tetrabutylammonium chromate, 23–24, **285**
  with tetrabutylammonium hypochlorite, 27
  with tetrabutylammonium periodate, 30–31
  with tetrabutylammonium permanganate, 36, **290**
  with thallium compounds, 17
  with thallium trinitrate, **280–281**
  with trichloroisocyanuric acid, 28
  with triphenyl phosphite ozonide, 42
  with triphenylmethyl tetrafluoroborate, 41
  with triphenylmethylphosphonium permanganate, 36
  with tungsten hexafluoride, 21
  with vanadium compounds, 19
  with zinc permanganate, 36
  See also *Dehydrogenations, Oxidative coupling,* and specific reactions
Oxidations, asymmetric, of sulfides to sulfoxides, 257–258, **278**
Oxidations, biochemical,
  See *Biochemical oxidations*
Oxidations, catalytic,
  of secondary alcohols to ketones, 133
Oxidations, enantioselective,
  See *Enantioselective oxidations*
Oxidations, stereospecific,
  See *Stereospecific oxidations*
Oxidative cleavage
  of alkenes and cycloalkenes, 77–84
  of carbon–boron bonds, 267
  of carbon–carbon double bonds, 77–84
  of carbon–tin bonds, 270
  of ketones, 211–212
  of ozonides, 5, 226
  of silicon–silicon bonds, 270
  of sulfones, 263
  of tertiary amines to carbonyl compounds, 244
  of unsaturated ketones, 215
Oxidative coupling
  of acetylenic alcohols, 155
  of esters, 223
Oximes, oxidation
  to nitro compounds, 219
  to parent ketones, 219
Oxiranes
  by epoxidation
    of alkenes and cycloalkenes, 61–64
    of unsaturated alcohols, 153–154

Oxiranes, by epoxidation—Continued
    of unsaturated aldehydes, 182
    of unsaturated carboxylic acids and esters, 225
    of unsaturated ketones, 212–214
  oxidation
    to dicarbonyl compounds, 173–174
    to hydroxy hydroperoxides, 173–174
    to α-hydroxy ketones, 173–174
    with dimethyl sulfoxide, 173–174
    with hydrogen peroxide, 173–174
    with periodic acid, 173–174
Oxodiperoxymolybdenum–pyridine–hexamethylphosphoric triamide
  in hydroxylation of ketones, 196
  oxidant, 21
Oxone
  in epoxidation
    of alkenes and cycloalkenes, 63
    of carbon–nitrogen bond, 44
  in oxidation
    of aromatic side chains to carboxyl groups, 105, 106
    of iodo compounds to iodoxy compounds, 267
    of primary alcohols to esters, 131
    of primary amines to nitroso compounds, 235
    of secondary alcohols to ketones, 133
    of sulfides to sulfones, 254, 255
    of sulfides to sulfoxides, 254, 255
  in selective oxidation of sulfides to sulfones, **278**
  oxaziridine formation, 44
  oxidant, 9
  See also *Potassium peroxymonosulfate*
1-Oxo-2,2,6,6-tetramethylpiperidinium chloride, oxidant, 42
Oxygen
  applications as oxidant, 1
  excited state, 1–3
  ground state, 1, 4
  in cleavage of alkenes and cycloalkenes to ketones, 80
  in dehydrogenation of primary amines to nitriles, 241
  in dehydrogenative couplings, 53
  in demethylation of tertiary amines, 243
  in dioxetane formation, 64–65
  in epoxidation
    of alkenes, 60
    of halogenated alkenes, 62
  in formation of hydroperoxides, 60, 169
  in oxidation
    of alcohols
      to aldehydes, 114–115, 123, 124
      to carboxylic acids, 127, 130
      to ketones, 133, 147, 149
    of aldehydes
      to carboxylic acids, 174–175
      to peroxy acids, 180
    of aldoses, 182
    of alkanes
      to carboxylic acids, 58
      to hydroperoxides, 58

Oxygen, in oxidation of alkanes—*Continued*
    to ketones, 58
  of alkyldichloroboranes to alkyl hydroperoxides, 270
  of aromatic side chains
    to carboxyl groups, 106
    to hydroperoxides, 99
  of benzylic alcohols to aldehydes, 124
  of disulfides to sulfonic acids, 264
  of mercaptans (thiols) to sulfonic acids, 251
  of methylene groups to keto groups, 103
  of primary alcohols
    to aldehydes, 123
    to carboxylic acids, 127, 130
  of primary amines to azo compounds, 234
  of secondary alcohols
    to hydroperoxides, 150
    to ketones, 133, 147, 149
  of tertiary hydrogens, 60
  of unsaturated aldehydes, 175
  of vicinal diols to carboxylic acids, 163
    to hydroperoxides, 99
  in oxidative cleavage of ketones to carboxylic acids, 211
  in oxidative coupling of acetylenic alcohols, 155
  sensitizers
    benzophenone, 1, 150, 169
    chlorophyll, 1
    methylene blue, 1, 245
    rose bengal
      in demethylation of tertiary amines, 243
      in generation of singlet oxygen, 1, **275**
      in oxidation
        at allylic positions, 84
        of dienes with oxygen, 87
        of secondary alcohols, 147, 149
      in stereospecific addition of singlet oxygen, 64
  singlet, 1–3
  triplet, 1
Oxygenations, with singlet oxygen, **275–276**
Ozone
  applications as oxidant, 5–6
  determination of excess, 5
  in cleavage
    of acetylenic carboxylic acids, 226
    of methyl oleate, **276**
    of nitrogen ylides, 232
  in epoxidation of alkenes and cycloalkenes, 61, 62
  in formation of ozonides, 67
  in hydroxylation of methine groups, 59
  in oxidation
    of acetals to esters, 184
    of aldehydes to peroxy acids, 180
    of aromatic side chains to carboxyl groups, 105, 106
    of azides
      to aldehydes, 232
      to nitro compounds, 232

Ozone, in oxidation—*Continued*
    of ethers to esters and lactones, 169
    of glycosides to esters, 184
    of nickel to nickel peroxide, 37
    of organomercury compounds to ketones, 271
    of phosphines to phosphine oxides, 248
    of phosphites to phosphates, 248, 249
  in oxidative cleavage of alkenes and cycloalkenes, 77, 78
  in preparation of triphenyl phosphite ozonide, **276**
  in regeneration of ketones from their derivatives, 219
  physical properties and preparation, 4
  reaction with double bonds, 65–66
Ozonides
  cleavage
    to alcohols, 5, 77
    to aldehydes, 5, 77–79
    to carboxylic acids, 5, 81, 82
  formation
    cis–*trans* isomerism, 66
    mechanism, 65–66
  isolation, 66–67
  oxidative cleavage
    with hydrogen peroxide, 5, 81, 82, 226
    with nitric acid, 5
    with silver oxide, 5
  reductive cleavage
    by catalytic hydrogenation, 5, 77–78
    to alcohols with lithium aluminum hydride, 78
    to aldehydes, 5, 77–78, 79
    with dimethyl sulfide, 5, 78–79
    with lithium aluminum hydride, 5, 77, 78
    with sodium borohydride, 5, 78
    with sodium iodide, 5
    with trimethyl phosphite, 5, 78–79
    with triphenylphosphine, 5
    with zinc, 5, 78
Ozonization (ozonolysis)
  intermediates, 65–66
  isolation of ozonides, 66–67
  mechanism, 65–66
  of alkenes and cycloalkenes to ketones, 80
  of esters of acetylenic alcohols to diketo esters, 227
  of unsaturated ketones, 215
  procedure, 4–5, **276**
  safety precautions, 5, 67
  solvents, 5, 66

P

Palladium, catalyst
  for dehydrogenation, 38, 50–51
  for oxidation with hydrogen peroxide, 8
  for reduction of ozonides, 5, 77
Palladium acetate, catalyst for oxidation with hydrogen peroxide, 8

Subject Index 419

Palladium chloride,
  See Palladium dichloride
Palladium compounds, catalysts for oxidation with oxygen, 4
Palladium dichloride, catalyst for oxidation
  with air, 15, 75
  with oxygen, 75, 133, 147
Paraperiodic acid, See Periodic acid
PCC, See Pyridinium chlorochromate
Pelargonaldehyde, preparation from methyl oleate by ozonolysis, 276
Pelargonic acid,
  preparation from oleic acid, **288**
Penicillium chrysogenum and Penicillium lilacinum
  in biochemical Baeyer–Villiger reaction, 194–195
  in oxidation of steroidal ketones, 194–195
Pentafluoroiodobenzene bis(trifluoroacetate), oxidant and preparation, 32
Pentane, solvent for ozonization, 5, 66
Pentanetrione (triketopentane), preparation from acetylacetone, **292**
1-Pentene, epoxidation to 1,2-epoxypentane, **279**
Pentyl nitrite, in oxidation of methylene groups to keto groups, 104
Peracetic acid, See Peroxyacetic acid
Perbenzoic acid, See Peroxybenzoic acid
Perboric acid, in oxidation of primary aromatic amines to nitro compounds, 235
Perepoxides, intermediates in alkyl hydroperoxide formation, 2
Perfluoroalkyl hydrides, oxidation to perfluorocarboxylic acids, 113
Perfluoroalkyl iodides, oxidation to perfluorocarboxylic acids, 113
Performic acid, See Peroxyformic acid
Perhydrol, 7
Periodates, in oxidation
  of α-hydroxy acids to aldehydes, 228
  of sulfides to sulfones, 254
  of sulfides to sulfoxides, 254, **287**
Periodic acid
  in cleavage
    of saccharides, 160
    of vicinal diols, 159, 160, **287**
  in dehydrogenative couplings, 53
  in oxidation
    of aromatic side chains, 101, 102–103
    of epoxides (oxiranes), 173–174
    of trialkylsilyl ethers, 172, 173
  oxidant, 30
  reoxidant
    of 2,3-dichloro-5,6-dicyano-p-benzoquinone, 143
    of ruthenium compounds, 38
Periodic acid dihydrate, oxidant, 30
Permanganates, solid or supported, in oxidation of alcohols
  to aldehydes, 125, 125
  to ketones, 141, 148, 149
Peroxides, by oxidation of ketones, 186

Peroxy acids
  by oxidation
    of aldehydes with oxygen or ozone, 180
    of carboxylic acids, 222
  in epoxidation
    of alkenes and cycloalkenes, 61–63, **279**
    of unsaturated ethers, 170
    of unsaturated ketones, 212, 213
  in hydroxylation
    of alkenes and cycloalkenes, 69
    of unsaturated carboxylic acids, 225
  in oxidation
    of acetals to esters or lactones, 184, 185
    of aromatic aldehydes, 180–181
    of dicarbonyl compounds to carboxylic acids, 218, 219
    of epoxides (oxiranes), 173–174
    of ketones to esters or lactones, 186–187, 188–190, 195, **279**
    of primary amines to azoxy compounds, 234
    of primary aromatic amines to nitro compounds, 235–236
    of selenides to selenoxides, 265
    of sulfides
      to sulfones, 253–254
      to sulfoxides, 253–254
    of trialkylsilyl ethers, 173
    of unsaturated ketones to unsaturated esters and lactones, 191, 192, 195
  preparation, 222
Peroxyacetic acid
  in cleavage of acetylenic carboxylic acids, 226
  in epoxidation of alkenes and cycloalkenes, 61–63
  in oxidation
    of α-dicarbonyl compounds, 193
    of β-diketones, 218–219
    of iodo compounds
      to iodoso diacetates, 266
      to iodoxy compounds, 266
    of ketimines to oxaziridines, 219
    of ketones to esters and lactones, 188, 189
    of lactams to imides, 245
    of primary amines
      to azoxy compounds, 234
      to nitro compounds, 235
      to nitroso compounds, 235
    of primary aromatic amines to nitro compounds, 235–236
    of o-quinones to acid anhydrides, 193
    of secondary alcohols to ketones, 147
    of sulfides
      to sulfones, 259
      to sulfoxides, 253, 259
    of tertiary amines to amine oxides, 236
    of unsaturated acetals, 184–185
    of unsaturated ketones to unsaturated esters and lactones, 191
  in oxidative cleavage of phenols, 168
  oxidant, 12
  oxidant of ruthenium compounds, 38

Peroxyacetic acid—Continued
 preparation, 11–12, **273, 278–279**
 solvents
  acetic acid, 12
  chloroform, 12
  dichloromethane, 12
  dimethyl cellosolve (dimethyl ether of ethylene glycol), 12
  ethyl acetate, 12
Peroxybenzoic acid
 analytical determination, 13
 in cleavage of silicon–silicon bonds, 270
 in epoxidation
  of alkenes and cycloalkenes, 61, 63
  of allylic alcohols, 152
  of unsaturated carboxylic acids, 225
  of unsaturated esters, 225
  of unsaturated ketones, 213
 in hydroxylation of unsaturated carboxylic acids, 225
 in oxidation
  of acetylenic sulfides to acetylenic sulfoxides, 261
  of azo compounds, 232
  of disulfides to monosulfoxides (alkyl thiosulfinates), 263
  of ketones
   to esters, **279**
   to esters and lactones, 189
  of primary amines to nitroso compounds, 235
  of silanes to silyl ethers, 270
  of tertiary amines to dicarbonyl compounds, 244
 oxidant, 13
 preparation, 13, 222
Peroxycarboximidic acid
 formation, 60
 reaction with alkenes, 60
Peroxydichloromaleic acid,
 in oxidation of tertiary amines to amine oxides, 237
Peroxyformic acid
 in hydroxylation of unsaturated carboxylic acids, 225–226
 in oxidation
  of alkenes and cycloalkenes to ketones, 75, 76
  of tertiary amines, 237–239
 in situ preparation, 225–226
 oxidant and preparation, 10–11
Peroxylauric acid
 in epoxidation
  of alkenes and cycloalkenes, 61
  of unsaturated esters, 225
 oxidant, 14
 preparation, 14, 222
Peroxymaleic acid
 in epoxidation of alkenes and cycloalkenes, 61, 62
 in oxidation
  of ketones to esters and lactones, 190
  of primary aromatic amines to nitro compounds, 236
 oxidant and preparation, 14

Peroxymonosulfuric acid
 explosion with alcohols, 8
 in oxidation
  of aldehydes to carboxylic acids, 175
  of disulfides to sulfonic acids, 264
  of primary aromatic amines to nitro compounds, 235
  of unsaturated aldehydes, 175, 182
 oxidant, 8
 preparation, 175
Peroxypentafluorobenzoic acid,
 oxidant and preparation, 14
Peroxyphthalic acid
 in epoxidation
  of alkenes and cycloalkenes, 61, 63
  of unsaturated ketones, 213–214
 in oxidation
  of β-diketones, 218
  of α-keto esters to anhydrides, 193, 229
  of ketones to esters and lactones, 190
  of mercaptoles to disulfones, 260
  of sulfides to sulfoxides, 256
 oxidant, 14
 preparation, 14, 222
Peroxypropionic acid
 in oxidation of mercaptals to disulfones, 260
 oxidant and preparation, 14
Peroxytrifluoroacetic acid
 analytical determination, 12
 in epoxidation
  of alkenes and cycloalkenes, 61, **279**
  of allylic ethers, 171
 in oxidation
  of ketones to esters and lactones, 188–189
  of nitrosoamines, 231
  of primary aromatic amines to nitro compounds, 236
  of selenides to selenones, 265
  of sulfides
   to sulfones, 253
   to sulfoxides, 253
  of tertiary amines to amine oxides, 237
 in situ preparation, **279**
 oxidant and preparation, 12–13
Persulfates, in hydroxylation of phenols, 163
Persulfuric acid, See Peroxymonosulfuric acid
Peziza sp., in hydroxylation of steroidal ketones, 198
Phase-transfer reagents
 Adogen 464, 116, 137, 149
 Aliquat 336, 8
 benzyltrimethylammonium chloride, 74
 in oxidations with potassium permanganate, 34
 quaternary ammonium salts, 34
 tetrabutylammonium bisulfate, 123, 148
 tetrabutylammonium bromide, 254, 263, **290**
 tetrabutylammonium chloride, 80, 148
 tetrabutylammonium hydrogen sulfate, 24, 27, 241

Phase-transfer reagents—Continued
  tetrabutylammonium
    trifluoromethanesulfonate, 147
Phenols
  by oxidation
    of aromatic aldehydes, 180–181
    of aromatic boranes, 269
    of phenylboronic acids, 270
  dehydrogenative coupling, 53, 163, 165
  from arylmercury bromides, 270
  hydroxylation
    with persulfate, 163–164
    with polyphenol oxidase, 163
  oxidation
    to carboxylic acids, 163, 168
    to free radicals, 163
    to quinones
      with barium manganate, 166
      with bromine, 164
      with ceric ammonium nitrate, 165, 166
      with chromium trioxide, 164
      with 2,3-dichloro-5,6-dicyano-*p*-benzoquinone, 164, 165
      with dinitrogen tetroxide, 166
      with ferric chloride, 166
      with ferric sulfate, 166
      with Fremy salt, 164
      with horseradish peroxidase, 164, 165
      with hydrogen peroxide, 164
      with lead dioxide, 164–167
      with lead oxides, 166
      with lead tetraacetate, 167
      with mercuric oxide, 164, 166
      with mercuric trifluoroacetate, 164, 165
      with potassium chromate, 166
      with silver oxide, 166, 167
      with sodium chlorate, 166
      with sodium dichromate, 166
      with sodium iodate, 167–168
  oxidative cleavage with organic peroxy acids, 168
  oxidative coupling, 163, 165
Phenoxy radicals,
  by oxidation of phenols, 163
Phenylboronic acids,
  oxidation to phenols, 270
N-Phenylcamphoryl hydroxamic acid,
  chiral component in enantioselective oxidations, 44
α-*Phenylethanol,*
  oxidation to acetophenone, **284**
Phenylhydrazones,
  oxidation to parent ketones, 219
Phenylselenyl bromide, oxidant, 43
Phenylselenyl chloride
  in dehydrogenation of carbon–carbon bond, 48, 49
  oxidant, 43
Phosgene, catalyst for oxidation with dimethyl sulfoxide, 43, 122
Phosphates, by oxidation
  of selenophosphates, 250
  of thiophosphates, 250

Phosphine oxides, by oxidation
  of phosphine selenides, 249
  of phosphine sulfides, 249
  of tertiary phosphines, 248
  of ylides, 250
Phosphine selenides,
  oxidation to phosphine oxides, 249
Phosphine sulfides,
  oxidation to phosphine oxides, 249
Phosphines, tertiary,
  oxidation to phosphine oxides, 248
Phosphites, trialkyl and triaryl,
  oxidation to phosphates, 248–249
Phosphoric acid, catalyst for oxidation with dimethyl sulfoxide, 43, 122
Phosphorus compounds,
  oxidation, 248–250
Phosphorus pentoxide, catalyst for oxidation with dimethyl sulfoxide, 43
Photooxidation,
  of secondary alcohols to ketones, 133
Photooxidation sensitizers,
  See *Oxygen,* sensitizers
Phthalic anhydride, oxidation to peroxyphthalic acid, 222
Platinum
  anodes in electrolytic cells, 6
  catalyst for oxidation
    of alcohols, 115
    of aldehydes to carboxylic acids, 174
    with air or oxygen, 127, 182–183
    with oxygen, 123, 124, 127, 133
  dehydrogenation catalyst, 38, 50, 51
Platinum and platinum compounds,
  catalysts for oxidation with oxygen, 4
Platinum dioxide, catalyst for oxidation
  of alcohols, 115
  with air or oxygen, 127
  with oxygen, 147
Platinum oxide, catalyst
  for oxidation with oxygen, 133
  for reduction of ozonides, 5
Polyphenol oxidase,
  in hydroxylation of phenols, 163
Poly(vinylpyridine), support for pyridinium chlorochromate, 138
Poly(vinylpyridine)–chromium trioxide complex, in oxidation of primary alcohols to aldehydes, 118
Poly(vinylpyridinium) chlorochromate
  in oxidation of secondary alcohols to ketones, 147, 148
  oxidant and preparation, 23
Potassium bromate
  in generation of bromine, 29
  in oxidation of *o*-iodobenzoic acid to Dess–Martin reagent (periodinane), 13
Potassium chlorate
  oxidant, 28
  reoxidant of osmium oxide, 69, 225
Potassium chromate
  in oxidation of phenols to quinones, 166
  oxidant and preparation, 23
Potassium dichromate
  applications as oxidant, 24

Potassium dichromate—Continued
   in cleavage of vicinal diols, 159, 161
   in oxidation
      of aldehydes to carboxylic acids, 174, 176
      of aromatic amines to quinones, 246, 247
      of aromatic side chains, 102
      of benzylic halides to aldehydes, 113
      of hydroxylamines, 231–232
      of nitroso compounds, 231
      of primary alcohols, 116
      of secondary alcohols to ketones, 136, 149
   in preparation of silver dichromate, 25
Potassium ferrate
   in dehydrogenation of primary amines to aldehydes, 240
   in oxidation
      of primary alcohols to aldehydes, 120, 125, 126
      of secondary alcohols to ketones, 142, 148, 149
   oxidant and preparation, 37
Potassium ferricyanide [hexacyanoferrate(III)]
   in conversion of hydrazides into aldehydes, 230
   in dehydrogenation
      of heterocyclic compounds, 52
      of primary amines to nitriles, 241
   in dehydrogenative cyclizations, 55
   in demethylation of tertiary amines, 243
   in oxidation of phenols to phenoxy radicals, 163
   oxidant, 37
Potassium hydroxide
   in hydroxylation of aromatic rings, 92
   in oxidation
      of aldehydes to carboxylic acids, 180
      of primary alcohols to carboxylic acids, 127
   oxidant, 20
Potassium hypochlorite
   in degradative oxidation of methyl ketones, 207
   in oxidation of benzylic alcohols to aldehydes, 119, 126
   oxidant and preparation, 27
Potassium hypoiodite
   in degradative oxidation of methyl ketones, 210
   in situ preparation, 210
   oxidant and preparation, 29–30
Potassium manganate, oxidant and preparation, 33
Potassium nitrosodisulfonate,
   See *Fremy salt*
Potassium periodate
   in cleavage of vicinal diols, 159
   oxidant, 30
Potassium permanganate
   in cleavage of alkenes and cycloalkenes, 79, 82–84

Potassium permanganate—Continued
   in degradative oxidation
      of methyl ketones to carboxylic acids, 210
      of unsaturated carboxylic acids
         to aldehydes, 226
         to dicarboxylic acids, 226
   in dehydrogenation of primary amines
      to azines, 240
      to ketones, 240
   in dehydrogenative couplings, 53
   in hydroxylation
      of dienes, 89
      of unsaturated acetals, 185
      of unsaturated carboxylic acids, 225–226
   in *syn* hydroxylation of alkenes and cycloalkenes, 68
   in oxidation
      of acetylenic carboxylic acids to diketo carboxylic acids, 226–227
      of aldehydes to carboxylic acids, 176, 179
      of alkenes to carboxylic acids, **289**
      of alkenes and cycloalkenes to vicinal dicarbonyl compounds, 76
      of alkynes
         to α-diketones, 91, **289**
         to carboxylic acids, 91–92
      of aromatic side chains to carboxyl groups, 105, 106, 108
      of benzene rings to carboxylic acids, 97, 98
      of benzylic alcohols to aldehydes, 125
      of esters of acetylenic alcohols to diketo esters, 227
      of ethers to esters and lactones, 169
      of hydroxy acids to keto acids, 227
      of hydroxy aldehydes to keto acids, 182–183
      of iodoso compounds to iodoxy compounds, 266
      of lactones to dicarboxylic acids, 224
      of mercaptoles to disulfones, 260
      of methyl ketones
         to carboxylic acids, 206
         to α-keto carboxylic acids, 206
      of methylene groups to keto groups, 103
      of nitro compounds, 230–231
      of primary alcohols
         to aldehydes, 119–120, 125–126
         to carboxylic acids, 129, 130
      of primary allylic alcohols to aldehydes, 126
      of primary amines to nitro compounds, 235
      of pyridine homologues, 108–109
      of secondary alcohols
         to carboxylic acids, 150, **289**
         to ketones, 140, 147, 149, **290**
      of selenides to selenones, 265
      of sulfinic acids to sulfonic acids, 264
      of tertiary amines to amides, 242, 243–244

Subject Index

Potassium permanganate, in oxidation—
  Continued
    of unsaturated carboxylic acids to diketo acids, 227
  in oxidative cleavage of ketones to carboxylic acids, 211
  in preparation
    of α-hydroxy ketones, 76–77
    of manganese dioxide, 32, **274–275**
  in situ conversion into tetrabutylammonium permanganate, **290**
  mixture with sodium periodate (metaperiodate), 30
  oxidant, 34–35
  reduction to manganate with potassium iodide, 140
  solubility, 34
  solubilization with crown ethers, 129, 142, 149
  solvents, 34
  supports
    bentonite, 34
    copper sulfate pentahydrate, 34
    molecular sieves, 34
    montmorillonite, 34
  use of buffers in oxidations, 34
Potassium permanganate adsorbed on alumina, preparation, 141
Potassium permanganate adsorbed on molecular sieves, preparation, 125, 126, 141, **290**
Potassium permanganate and sodium periodate
  in degradative oxidation of unsaturated carboxylic acids to dicarboxylic acids, 226
  in oxidation of primary alcohols, 125, 126, 141
Potassium peroxydisulfate, in oxidation of aromatic side chains, 101–102
Potassium peroxymonosulfate (Oxone)
  in Baeyer–Villiger reaction, 188, 195
  in oxidation
    of aldehydes to carboxylic acids, 175, 179
    of cyclic ketones
      to hydroxy esters, 188
      to lactones, 188, 195
    of iodo compounds to iodoxy compounds, 267
    of ketones to esters and lactones, 188, 195
    of mercaptans (thiols) to sulfonic acids, 251
    of phosphines to phosphine oxides, 248
    of primary alcohols to esters, 131
    of primary amines to nitroso compounds, 235
    of secondary alcohols to ketones, 133
    of sulfides
      to sulfones, 254, 255, **278**
      to sulfoxides, 254, 255

Potassium persulfate
  in oxidation of α-dicarbonyl compounds, 193
  in preparation of potassium ruthenate, 142
  oxidant, 8, 9
  reoxidant of ruthenium, 142
Potassium ruthenate
  in dehydrogenation of primary amines to nitriles, 241
  in oxidation
    of aldehydes to carboxylic acids, 177
    of primary alcohols to carboxylic acids, 130
    of secondary alcohols to ketones, 142
  oxidant and preparation, 38
Potassium superoxide
  in oxidative cleavage of ketones to carboxylic acids, 211
  oxidant, 8
Preparative procedures, **273–294**
Prevost reaction
  anti hydroxylation of double bonds, 31, 70–72
  Woodward modification, 31, 70–72
Primary alcohols, See *Alcohols, primary*
Primary alkyl halides, oxidation
  to aldehydes, 109–112
  to carboxylic acids, 112–113
Primary amides, oxidation, 220
Primary amines, See *Amines, primary*
Primary nitro compounds,
  oxidation, 230–231
Pseudomonas putida
  in enantioselective hydroxylation, 71
  in *syn* hydroxylation, 93
Purple benzene
  in oxidation of alkenes to carboxylic acids, **290**
  preparation, 34, **290**
  See also *Tetrabutylammonium permanganate*
Pyridine,
  molecular complex with bromine, 28
Pyridine homologues, oxidation to carboxylic acids, 108–109
Pyridine oxide, oxidant, 42, 238
Pyridine ring,
  electrophilic substitutions, 238
Pyridine-α-diselenide, oxidation to pyridine-α-seleninic anhydride, 86
Pyridine-α-seleninic anhydride
  in oxidation of alkenes and cycloalkenes, 86
  in situ formation, 86
Pyridinium chlorochromate
  adsorption on silica gel or poly(vinylpyridine), 138
  in epoxidation of enol ethers, 171
  in oxidation
    of boranes to aldehydes, 269
    of ethers to esters or lactones, 171
    of methylene groups to keto groups, 103, 104

Pyridinium chlorochromate, in oxidation—
  Continued
  of primary alcohols to aldehydes, 118, 126, **284**
  of secondary alcohols to ketones, 138, 149
  of trialkylsilyl ethers, 173
  of unsaturated alcohols to keto oxides, 154
  oxidant, 23
  preparation, 23, 138
Pyridinium chlorochromate adsorbed on alumina, preparation, **274**
Pyridinium chromate
  in oxidation
    of primary allylic alcohols to aldehydes, 124
    of secondary alcohols to ketones, 137, 148
  preparation, 137
Pyridinium dichromate
  applications, 25
  catalyst for oxidation with bis(trimethylsilyl) peroxide, 10
  in oxidation
    of primary alcohols
      to aldehydes, 116, 128
      to carboxylic acids, 128
    of secondary alcohols to ketones, 137
  oxidant, 25
  preparation, 25, 116, 137
  safety precautions, 25
Pyridinium trifluoroacetate, activator of dimethylsulfoxide, 122, 144

Q

Quaternary ammonium salts, solubilizers of potassium permanganate, 34
Quaternary groups, elimination during dehydrogenation, 50
Quinidine acetate, chiral component in enantioselective oxidation, 44
Quinine acetate, chiral component in enantioselective oxidation, 44
Quinolinium chlorochromate,
  oxidant and preparation, 25–26
Quinones
  by dehydrogenation of hydroquinones, 166, 167, **290**
  by oxidation
    of aromatic amines, 246–248, **292**
    of aromatic compounds, 94–96
    of phenols
      with barium manganate, 166
      with bromine, 164
      with ceric ammonium nitrate, 165, 166
      with chromium trioxide, 164
      with 2,3-dichloro-5,6-dicyano-*p*-benzoquinone, 164, 165
      with dinitrogen tetroxide, 166
      with ferric chloride or ferric sulfate, 166

Quinones, by oxidation of phenols—
  Continued
      with Fremy salt, 164
      with horseradish peroxidase, 164, 165
      with hydrogen peroxide, 164
      with lead dioxide, 164–167
      with lead tetraacetate, 167
      with mercuric oxide, 164, 166
      with mercuric trifluoroacetate, 164–165
      with potassium chromate, 166
      with silver oxide, 166, 167
      with sodium chlorate, 166
      with sodium dichromate, 166
      with sodium iodate, 167–168
    of trialkylsilyl ethers, 173
  dehydrogenation agents, 49
  from perfluorinated aromatic compounds, 113–114
  See also o-*Quinones*
o-*Quinones,*
  by oxidation of phenols, 164
  oxidation
    to acid anhydrides, 193
    to dicarboxylic acids, 218
  See also *Quinones*

R

Raney nickel
  catalyst
    for dehydrogenation of alcohols, 39, 143
    for reduction of ozonides, 5, 77
  dehydrogenating agent, 37
  in decarbonylation of primary alcohols, 132
  in dehydrogenation of secondary alcohols to ketones, 132
  in oxidation of secondary alcohols to ketones, 147
  in removal of colloidal selenium, 199
Red lead, oxidant, 17
Red lead oxide, See *Lead oxide, red*
Reductions,
  of pyridine oxide to pyridine, 238
Reductive cleavage, of ozonides, 5
Regiospecific oxidations, 101
Reoxidation
  of osmium compounds
    by amine oxides, 69, 72, 262
    by hydrogen peroxide, 69, 225
    by potassium chlorate, 69, 225
    by silver chlorate, 69, 225
    by sodium chlorate, 69, 225
  of ruthenium compounds
    by periodic acid, 38
    by peroxyacetic acid, 38
    by sodium bromate, 29, 38, 120
    by sodium hypochlorite, 27, 38, 142, 148
    by sodium periodate (metaperiodate), 30, 38, 148, 244–245

Resonance energy,
  of aromatic compounds, 50
Resorcinol,
  scavenger in oxidation with sodium chlorite, 28
Retention of configuration,
  in hydroxylation of methine groups, 58
Rhizopus arrhizus
  in hydroxylation of steroidal ketones, 197
  in oxidation of hydroxy ketones to diketones, 216
  in selective oxidation of diols, 157
Rhizopus nigricans, in hydroxylation of steroidal ketones, 197
Rhizopus stolonifer, in biochemical oxidation of sulfides to sulfoxides, 258
Rhodium compounds, catalysts for oxidation with oxygen, 4
Rhodium trichloride,
  catalyst for oxidation with air, 262
Rose bengal, oxygen sensitizer
  in demethylation of tertiary amines, 243
  in formation of singlet oxygen, 1, **275**
  in oxidation
    at allylic positions, 84
    of dienes with oxygen, 87
    of secondary alcohols, 147, 149
  in stereospecific addition of singlet oxygen, 64
Rubrene, in generation of singlet oxygen, 88–89
Ruff degradation, 228
Ruthenium compounds, oxidation to ruthenium tetroxide
  by periodic acid, 38
  by peroxyacetic acid, 38
  by potassium persulfate, 148
  by sodium bromate, 29, 38, 130
  by sodium hypochlorite, 27, 38, 142, 148
  by sodium periodate, 30, 38, 148, 244–245
Ruthenium dioxide
  conversion into ruthenium tetroxide, 38, 83, 148, 244–245
  in preparation of sodium ruthenate, 142
Ruthenium tetroxide
  from ruthenium dioxide, 38, 245, **291**
  from ruthenium trichloride, 38, 169
  in cleavage of alkenes and cycloalkenes to carboxylic acids, 82
  in oxidation
    of aldehydes to carboxylic acids, 177, 170
    of alkynes to α-diketones, 91
    of amides to imides, 244–245, **291**
    of benzene rings to carboxylic acids, 96, 97
    of benzylic alcohols to aldehydes, 125
    of ethers to esters and lactones, 169, 170
    of hydroxy acids to keto acids, 227–228
    of lactams to imides, 245
    of primary alcohols
      to aldehydes, 120, 125
      to carboxylic acids, 120

Ruthenium tetroxide, in oxidation—*Continued*
    of secondary alcohols to ketones, 142, 148
    of sulfides to sulfones and sulfoxides, 255
  in situ preparation, 244–245
  oxidant and preparation, 27, 38
  safety precautions, 38
  solubility, 38
Ruthenium trichloride
  conversion into ruthenium tetroxide, 82, 244
  in preparation of potassium ruthenate, 142
  oxidation to ruthenium tetroxide
    with sodium bromate, 120
    with sodium hypochlorite, 27, 38, 142
    with sodium periodate, 148

S

Saccharides, cleavage
  with periodic acid, 160
  with sodium periodate, 160
Safety advisory, 272
Safety precautions
  for handling
    benzyltriethylammonium permanganate, 36
    tert-butyl hydroperoxide, 9
    α-hydroxy peroxides, 150
    pyridinium dichromate, 25
    ruthenium tetroxide, 38
    tetrabutylammonium permanganate, 36
    thallium compounds, 17
    triphenylmethylphosphonium permanganate, 36
  for preparation of 90% hydrogen peroxide, 7
  for work with ozone, 5, 67
Schiff bases, oxidation to oxaziridines, 219
Secondary alcohols,
  See *Alcohols, secondary*
Secondary amines, See *Amines, secondary*
Secondary halides,
  oxidation to ketones, 109, 110
Secondary nitro compounds,
  oxidation, 230–231
Selective oxidations
  biochemical, 157
  of allylic alcohols, 156, 157
  of hydroxyl groups in steroids, 155–156
  of primary alcohols, 156
  of secondary alcohols, 156, **286**
  of steroidal alcohols, 157
  of steroidal diols, 156–157
Selectivity, in oxidation of primary and secondary alcohols, 138, 139
Selenides, oxidation to selenones and selenoxides, 265
Selenious acid
  in oxidation of ketones to α-dicarbonyl compounds, 199, 200
  oxidant, 20

Selenium
  dehydrogenating agent, 20
  in chemical dehydrogenation, 50, 51
  removal of colloidal form, 21, 199
Selenium dioxide
  catalyst
    for *anti* hydroxylation with hydrogen peroxide, 69
    for oxidation with hydrogen peroxide, 7, 261
  in oxidation
    of aldehydes to carboxylic acids, 175
    of alkenes and cycloalkenes
      at allylic positions, 86
      to vicinal dicarbonyl compounds, 76
    of alkynes to α-diketones, 91
    of aromatic compounds to quinones, 95
    of aromatic side chains, 101, 102
    of ethers, 171–172
    of ketones to α-dicarbonyl compounds, 199, 200, **283**
    of methylene groups to keto groups, 103, 104
  oxidant, 20
Selenones, by oxidation of selenides, 265
Selenophosphates,
  oxidation to phosphates, 250
Selenoxides
  by oxidation of selenides, 265
  decomposition to unsaturated carbonyl compounds, 265
  intermediates for unsaturated carbonyl compounds and esters, 265
  use in dehydrogenation, **275**
Semicarbazones,
  oxidation to parent ketones, 219
Sensitizers of oxygen
  methylene blue, 245
  rose bengal, 243
  for singlet-oxygen formation, 1, 65, 87–88
  zinc tetraphenylphorphyrin, 65
  See also *Oxygen,* sensitizers
Sharpless reagent
  chiral reagent in enantioselective oxidations, 154
  composition, 10, 44
  in enantioselective oxidation of sulfides to sulfoxides, 254, 257–258, **278**
  oxidant, 10
Side chains of aromatic compounds,
  oxidation, 99–109
Side reactions, in oxidations with alkaline hypohalites, 208
Silanes, oxidation to silyl ethers, 270
Silica gel, support
  for chromic acid, **283**
  for chromium trioxide, 117, **283**
  for magnesium permanganate, 36
  for ozone, 248
  for pyridinium chlorochromate, 138
  for pyridinium chromate, 148
  for zinc permanganate, 36, 169
Silica gel–alumina, support for chromyl chloride, 147, 149

Silicon–silicon bonds, oxidative cleavage, 270
Silicon compounds, oxidation, 270
Silver, catalyst
  for dehydrogenation, 15, 114, 132
  for oxidation, 15
  for oxidation with air, 124, 126
  in Cannizzaro reaction, 177
Silver acetate
  addition product with iodine, 31
  in hydroxylation of alkenes and cycloalkenes, 70–72, **286**
  in preparation of acetyl hypoiodite, 31
Silver benzoate
  addition product with iodine, 31
  in hydroxylation of alkenes and cycloalkenes, 70–72
  in preparation of benzoyl hypoiodite, 31
Silver carbonate
  in oxidation
    of diols, 157
    of hydroxylamines, 231–232
    of secondary alcohols to ketones, 133
  oxidant, 16
Silver chlorate
  oxidant, 28
  reoxidant of osmium oxide, 69, 225
Silver p-chlorodinitrobenzoate,
  in preparation of hypoiodites, 31
Silver chromate,
  oxidant and preparation, 23
Silver chromate and iodine,
  in preparation of α-iodo ketones, 76
Silver dichromate–pyridine complex
  applications, 25
  oxidant and preparation, 25
Silver 3,5-dinitrobenzoate,
  in preparation of hypoiodites, 31
Silver nitrate
  in oxidation of α-bromo ketones to nitrates of α-hydroxy ketones, 202
  in preparation of silver oxide, **279**
Silver oxide
  catalyst
    for oxidation of aldehydes to carboxylic acids, 174
  in Cannizzaro reaction, 177
  in dehydrogenation of amines, 49
  in epoxidation of alkenes and cycloalkenes, 61, 63
  in oxidation
    of aldehydes to carboxylic acids, 175, 179
    of aromatic amines to quinones, 247–248
    of hydrazo compounds, 233
    of hydrazones to diazo compounds, 221
    of hydroxylamines, 231
    of phenols to quinones, 166, 167
    of primary alcohols to aldehydes, 124
    of secondary alcohols to ketones, 133
    of unsaturated aldehydes, 175
  in oxidative cleavage of ozonides, 5, 81
  oxidant, 16
  preparation, 15–16, **279–280**

Silver oxide and nitric acid, in oxidative decomposition of ozonides, 81
Silver permanganate, in oxidation of primary amines to azo compounds, 234
Silver peroxysulfate, in dehydrogenation of primary amines to aldehydes, 240
Silver(II) oxide, See *Argentic oxide*
Silyl ethers, See *Ethers, trialkylsilyl*
Simonini complex, 31
Singlet oxygen
  applications as oxidant, 2–3
  by decomposition
    of aromatic endoperoxides, 1, 2, 88
    of triphenyl phosphite ozonide, 1, 2
  by irradiation of gaseous oxygen, 1
  definition, 1
  from endoperoxides, 88
  from hydrogen peroxide
    with bromine, 1
    with calcium hypochlorite, 1
    with sodium hypochlorite,1, 2
  from organic peroxy acids, 1, 2
  from triphenyl phosphite ozonide, 42
  in formation
    of cyclic peroxides, 87
    of dioxetanes, 64–65
    of hydroperoxides, 2
  in oxidation
    of alkenes and cycloalkenes, 84
    of tertiary amines to dicarbonyl compounds, 244, 245
  mechanism of oxidation reaction, 2–3
  photochemical generation, 87
  preparation
    from hydrogen peroxide and sodium hypochlorite, **276**
    from triphenyl phosphite ozonide, **276**
    photochemical generation, **275**
  stereospecific addition to double bonds, 64
Sodium bismuthate
  in cleavage of vicinal diols, 159, 161, 162
  oxidant, 19
Sodium borohydride
  in reduction of ozonides, 5, 78
  source of borane for hydroboration, 268, **277**
Sodium bromate
  applications as oxidant, 29
  in oxidation
    of acyloins to α-diketones, 217
    of secondary alcohols to ketones, 139
    of cerium and ruthenium compounds, 29
  oxidant
    of cerous sulfate to ceric sulfate, 156
    of ruthenium compounds, 38, 120
    of sodium bromide, 28
  reoxidant of cerium compounds, 115, 133, 139
Sodium bromite
  in oxidation
    of secondary alcohols to ketones, 139, 148
    of sulfides to sulfoxides, 255
  oxidant, 29

Sodium carboxylates, electrolytic decarboxylative coupling, 224
Sodium chlorate
  in oxidation of phenols to quinones, 166
  oxidant, 28
  reoxidant of osmium oxide, 69, 225
Sodium chlorite
  in oxidation
    of aldehydes to carboxylic acids, 179, **285–286**
    of unsaturated aldehydes, 176
  oxidant, 27–28
Sodium N-chloro-p-toluenesulfonamide, See *Chloramine-T*
Sodium chromate, in oxidation of unsaturated alcohols to keto oxides, 154
Sodium dichromate
  applications, 24
  in degradation of methyl ketones to carboxylic acids, 210
  in oxidation
    of alcohols to ketones, **285**
    of allylic halides, 110–111
    of aromatic amines to quinones, 246
    of aromatic compounds to quinones, 95
    of aromatic side chains to carboxyl groups, 106, 107, 108
    of benzene rings to carboxylic groups, 96
    of boranes to ketones, 269
    of hydroxylamines, 231–232
    of methylene groups to keto groups, 103, 104
    of phenols to quinones, 166
    of primary alcohols
      to aldehydes, 116, 123, 125, 126
      to esters, 131
    of secondary alcohols to ketones, 136, 148
  oxidant, 24
Sodium hydroxide, in oxidation
  of aldehydes to carboxylic acids, 174, 180
  of primary alcohols to carboxylic acids, 127
Sodium hydroxide and silver oxide, in oxidation of primary alcohols to carboxylic acids, 127
Sodium hydroxide–potassium hydroxide mixture, in oxidation of aldehydes to carboxylic acids, 176
Sodium hypobromite
  in degradation
    of alkyl ketones to carboxylic acids, 209, **286**
    of methyl ketones, 208
  oxidant and preparation, 29
Sodium hypochlorite
  applications as oxidant, 27
  in degradative oxidation
    of α-amino acids to aldehydes, 229
    of methyl ketones, 207
  in dehydrogenation of primary amines
    to ketones, 240
    to nitriles, 241

Sodium hypochlorite—Continued
  in epoxidation
    of alkenes and cycloalkenes, 61, 63
    of unsaturated ketones, 212, 213, **285**
  in formation of singlet oxygen, 1, 2, 87
  in oxidation
    of alkynes to α-diketones, 91
    of aromatic amines to quinones, 247
    of aromatic side chains to carboxyl groups, 105, 187
    of diols, 156
    of ruthenium dioxide to ruthenium tetroxide, 83
    of secondary alcohols to ketones, 139, 147, 148
    of sulfides to sulfoxides, 254
  preparation, 27
  reoxidant
    of ruthenium compounds, 27, 142
    of ruthenium oxide, 91, 96, 97
Sodium hypoiodite,
  oxidant and preparation, 29
Sodium iodate
  in oxidation of phenols to quinones, 167–168
  oxidant, 30
Sodium iodide
  in conversion of alkyl chlorides and alkyl bromides into alkyl iodides, 109, 110
  in reduction of ozonides, 5
Sodium metaperiodate,
  See *Sodium periodate*
Sodium nitrite, in oxidation of aromatic amines to quinones, 246
Sodium orthovanadate,
  catalyst for sulfide oxidation, 261
Sodium perborate
  in oxidation
    of primary amines to azo compounds, 234
    of sulfides to sulfones and sulfoxides, 255
  oxidant and properties, 8
Sodium periodate
  applications as oxidant, 30
  in cleavage
    of alkenes and cycloalkenes
      to aldehydes, 79–80
      to ketones, 80, **288**
    of double bonds, 102–103, 162
    of saccharides, 160
    of vicinal diols, 160
    of ylides, 250
  in degradation of unsaturated carboxylic acids to dicarboxylic acids, 226
  in oxidation
    of acetylenic sulfides to sulfoxides, 261
    of alkynes to α-diketones, 91
    of hydroxy acids to keto acids, 227–228
    of mercaptals to sulfoxides, 259
    of ruthenium dioxide to ruthenium tetroxide, 83, 148
    of ruthenium trichloride to ruthenium tetroxide, 83, 148

Sodium periodate, in oxidation—*Continued*
    of selenides to selenoxides, 265
    of sulfides to sulfoxides
      examples, 254, 255, 257, 259
      preparative procedure, **287**
  in preparation of sodium ruthenate, 142
  mixture
    with osmium tetroxide, 30
    with potassium permanganate, 30
  oxidant
    of phenylselenyl halides, 43
    of ruthenium compounds, 30, 38, 148, 244–245
    of ruthenium oxide, 91, 96, 97
Sodium periodate, alumina-supported
  preparation, **287**
  use as oxidant, **288**
Sodium permanganate
  in cleavage
    of alkenes and cycloalkenes to carboxylic acids, 82
    of cycloalkenes to dicarboxylic acids, 80
  in oxidation
    of primary alcohols to carboxylic acids, 129, 130
    of secondary alcohols to ketones, 140, 147, 148
    of sulfoxides to sulfones, 262
Sodium permanganate monohydrate,
  oxidant, 34
Sodium peroxide, oxidant and properties, 8
Sodium peroxysulfate, in dehydrogenation of primary amines to aldehydes, 240
Sodium persulfate, oxidant, 8
Sodium polysulfide, in conversion of *p*-nitrotoluene into *p*-aminobenzaldehyde, 101–102
Sodium ruthenate
  in oxidation
    of primary alcohols to carboxylic acids, 130
    of secondary alcohols to ketones, 142
  oxidant and preparation, 38
Sodium tungstate, catalyst for oxidation with hydrogen peroxide, 8, 225
Sodium vanadate, catalyst for oxidation with hydrogen peroxide, 7
Solubilization of potassium permanganate
  by crown ethers, 34
  by phase-transfer agent, 34
Solvents
  for benzeneseleninic acid, 14
  for benzyltriethylammonium permanganate, 36
  for bromine, 28
  for chlorine, 27
  for *m*-chloroperoxybenzoic acid, 13
  for Collins reagent, 22
  for magnesium permanganate, 36
  for manganese dioxide, 33
  for oxidation with chromium trioxide, 136
  for ozonization, 5, 66

Subject Index

Solvents—Continued
  for peroxyacetic acid, 12
  for peroxybenzoic acid, 13
  for potassium permanganate, 34
  for tetrabutylammonium permanganate, 36
  for zinc permanganate, 36
Sommelet oxidation, **292–293**
Sommelet reaction, 39, 110, 111
Spirocyclic amides, hydroxylation, 242–243
Sporotrichum sulfurescens,
  in hydroxylation of cyclic amides, 242
Stannanes, oxidation, 270
Stearolic acid, oxidation to 9,10-diketostearic acid, **289**
Stereochemistry,
  in cleavage of vicinal diols, 161
Stereoselective hydroxylations,
  of unsaturated carboxylic acids, 225
Stereoselectivity,
  in epoxidation of allylic alcohols, 152
Stereospecific additions,
  of singlet oxygen, 64
Stereospecific hydroxylations, 67–73, 262
  syn addition and *trans* addition, 67–73
  formation of *meso,* DL, DL-*erythro,* DL-*threo, cis,* and *trans* diols, 67, 68
Stereospecific oxidations,
  of sulfides to sulfoxides, 259
Steric effects, in oxidation of secondary alcohols to ketones, 135
Sterically hindered alcohols, oxidation with dimethyl sulfoxide, 146
Steroidal alcohols, selective oxidation, 157
Steroidal diols, oxidation, 156–157, 162
Steroidal ketones
  biochemical hydroxylation, 58, 197–199
  oxidation to lactones, 194–195
Steroidal lactones, by oxidation of unsaturated ketones, 191
Steroids
  allylic oxidation, 86–87
  Baeyer–Villiger reaction, 194–195
  biochemical hydroxylation, 196–199
  dehydrogenation, 47–49
  formation of carbon–carbon double bonds, 47
  hydroxylation, 71
  oxidation of secondary alcoholic group to keto group, 136, 143, 144
  oxidation to keto steroids, 86–87
  selective oxidation of hydroxyl groups, 155–156
Steroids, hydroxy,
  oxidation to keto steroids, 143
Steroids, keto, by oxidation of hydroxy steroids, 136, 143, 144
trans-*Stilbene,*
  oxidation to benzoic acid, **290**
Streptomyces aureofaciens, in hydroxylation of unsaturated ketones, 214
Streptomyces thioluteus, in oxidation of primary amines to nitro compounds, 236

Sugars
  biochemical oxidation to aldonic acids, 177, 183
  cleavage, 160, 161
  oxidation
    to aldaric (dicarboxylic) acids, 128, 182, 184
    to aldonic acids, 182–183
    to keto aldonic acids, 183
Sulfides
  asymmetric oxidation to sulfoxides, **278**
  biochemical oxidation to sulfoxides, **294**
  enantioselective oxidation to sulfoxides, 257–258
  oxidation
    to sulfoxides or sulfones, 252–259
    to sulfoxides
      with ceric ammonium nitrate, **281**
      with sodium periodate, **287**
  selective oxidation to sulfones with potassium peroxysulfate, **278**
Sulfides, acetylenic,
  oxidation to acetylenic sulfoxides, 261
Sulfides, unsaturated, oxidation to unsaturated sulfoxides or sulfones, 257, 259–261
Sulfilimines, oxidation to sulfoximines, 262–263
Sulfinic acids
  by oxidation of mercaptans (thiols), 252
  oxidation
    to sulfonic acids, 264
    to sulfonyl chlorides, 264
Sulfinyl chlorides,
  by oxidation of disulfides, 263–264
Sulfomonoperacid
  in oxidation of aromatic side chains to carboxyl groups, 105, 106
  oxidant, 8
Sulfones
  by oxidation
    of sulfides, 252–256, **278**
    of sulfoxides, 262
    of thioacetals (mercaptals and mercaptoles), 259–260
  oxidative cleavage, 263
Sulfones, unsaturated, by oxidation of unsaturated sulfides, 256, 259
Sulfonic acids, by oxidation
  of disulfides, 264
  of mercaptans (thiols) 251–252
  of sulfinic acids, 264
  of sulfones, 263
Sulfonyl chlorides, by oxidation
  of disulfides, 264
  of isothiocyanates and isothioureas, 252
  of mercaptans (thiols), 252
  of sulfinic acids, 264
Sulfoxides
  by asymmetric oxidation of sulfides, **278**
  by oxidation of sulfides
    examples, 252–259
    with *Aspergillus niger,* **294**
    with ceric ammonium nitrate, **281**
    with sodium periodate, **287**

Sulfoxides—Continued
  by oxidation of thioacetals (mercaptals and mercaptoles), 259–260
  optical purity, **294**
  oxidation to sulfones, 262
  thermal decomposition to α,β-unsaturated ketones or esters, 259
Sulfoxides, acetylenic, by oxidation of acetylenic sulfides, 261
Sulfoxides, cyclic,
  by oxidation of thiiranes, 255
Sulfoxides, unsaturated
  by oxidation of unsaturated or acetylenic sulfides, 256, 257, 259, 261
  hydroxylation, 73, 262
Sulfoximines, by oxidation of sulfilimines, 262–263
Sulfur
  in chemical dehydrogenation, 50–51
  in oxidation of ketones to carboxylic acids, 203, **282**
  oxidant, 20
Sulfur and ammonium sulfide, oxidant in Willgerodt reaction, 203, 204
Sulfur and morpholine, in Willgerodt–Kindler reaction, 203–205
Sulfur compounds, oxidation, 250–264
Sulfur dioxide, in reduction of manganese dioxide, 35
Sulfur trioxide,
  dimethyl sulfoxide activator, 43
Sulfur trioxide–pyridine,
  dimethyl sulfoxide activator, 122
Sulfuric acid
  in dehydrogenative coupling, 53
  in preparation of ferric sulfate, 37
  oxidant, 20
  scavenger in oxidation with sodium chlorite, 28
Sulfuryl chloride
  in oxidation of sulfides to sulfoxides, 255, 256
  oxidant, 20
Superoxol, 7
Supports
  activated charcoal or asbestos for palladium and platinum, 50
  alumina for potassium permanganate, 141
  alumina for pyridinium chlorochromate, **274**
  alumina–silica gel for chromyl chloride, 147, 149
  aluminum oxide for sodium periodate, **287**
  Amberlyst A26 for chromium trioxide, 117, 123, 147, 148
  bentonite or montmorillonite for potassium permanganate, 34
  Celite for chromium trioxide, 117
  copper sulfate pentahydrate for potassium permanganate, 34, 141, 147, 149
  ion exchanger for chromium trioxide, 117
  ion exchanger IRA 900 for calcium hypochlorite, 148

Supports—Continued
  molecular sieves for potassium permanganate
    examples, 119, 141, 147, 149
    preparative procedure, **290**
  poly(vinylpyridine) for pyridinium chlorochromate, 138
  silica gel
    for chromium trioxide, 117, **283**
    for magnesium permanganate, 36
    for ozone, 248
    for permanganates, 36
    for pyridinium chlorochromate, 138
    for pyridinium chromate, 148
    for zinc permanganate, 36, 169
  silica gel–alumina for chromyl chloride, 147, 149
  Woelm alumina for carbonyl compound oxidants, 143
Swern oxidation, 43, 145

T

Tertiary acetylenic alcohols,
  oxidation to hydroperoxides, 151
Tertiary alcohols, See *Alcohols, tertiary*
Tertiary amines, See *Amines, tertiary*
Tetraalkylammonium dichromate, in oxidation of primary alcohols, 116
Tetraalkylammonium permanganate, in cleavage of alkenes and cycloalkenes to carboxylic acids, 82
Tetrabutyl chlorochromate
  in oxidation of secondary alcohols to ketones, 138
  preparation, 138
Tetrabutylammonium bisulfate, phase-transfer reagent, 24, 139, 148, 149
Tetrabutylammonium bromide, phase-transfer reagent, 254, 263, **290**
Tetrabutylammonium chloride
  in preparation of tetrabutylammonium chromate, **274**
  phase-transfer reagent, 80, 148, **285**
Tetrabutylammonium chlorochromate
  in oxidation
    of mercaptans (thiols) to disulfides, 251
    of primary alcohols to aldehydes, 118, 126, 127
    of secondary alcohols to ketones, 149
  oxidant and preparation, 24
Tetrabutylammonium chromate
  in oxidation
    of primary alcohols to aldehydes, 123, 126
    of secondary alcohols to ketones, 136, 148
  in situ preparation from chromium trioxide, **285**
  oxidant, 23–24
  preparation, 23–24, 136, **274**
Tetrabutylammonium dichromate
  in oxidation
    of aromatic compounds to quinones, 95

Tetrabutylammonium dichromate, in oxidation—*Continued*
  of benzylic alcohols to aldehydes, 125
  of secondary alcohols to ketones, 148
  in situ preparation, 116
Tetrabutylammonium hydrogen sulfate,
  phase-transfer reagent, 27, 241
Tetrabutylammonium hypochlorite
  in oxidation
    of benzylic alcohols to aldehydes, 125
    of benzylic halides to ketones, 112
    of secondary alcohols to ketones, 139, 149
  in oxidation of sulfilimines to sulfoximines, 262–263
  in situ preparation, 139, 262–263
  oxidant and preparation, 27
Tetrabutylammonium periodate
  in degradative oxidation
    of carboxylic acids to aldehydes, 224
    of α-hydroxy acids to aldehydes, 228
  in oxidation
    of benzylic halides to aldehydes or ketones, 111
    of sulfides to sulfoxides, 255
  oxidant and preparation, 30–31
Tetrabutylammonium permanganate
  ignition, 36, 142
  in oxidation
    of aldehydes to carboxylic acids, 176, 177
    of primary alcohols to carboxylic acids, 129, 130
    of secondary alcohols to ketones, 142, 149
  oxidant, 36
  preparation, 36, 129, 142
  safety precautions, 36
  solvents, 36
  See also *Purple benzene*
Tetrabutylammonium persulfate
  in epoxidation of alkenes and cycloalkenes, 63
  in oxidation of sulfides to sulfones, 253
Tetrabutylammonium trifluoromethanesulfonate, phase-transfer reagent, 147
Tetrachloro-o-benzoquinone, in oxidation of primary allylic alcohols to aldehydes, 120, 121, 126
Tetrachloro-p-benzoquinone, oxidant, 40, 120, 125
Tetracyclone, conversion into endoperoxide, 89
Tetrahydrofuran
  formation of hydroperoxides, 169
  solvent for benzeneseleninic acid, 14
Tetraisopropyl orthotitanate
  asymmetric oxidation of sulfides to sulfoxides, **278**
  catalyst for oxidation with *tert*-butyl hydroperoxide, 44, 154, 254, 258
  for epoxidation with *tert*-butyl hydroperoxide, 61
Tetralin, dehydrogenation to naphthalene, **275**

Tetraphenylcyclopentadienone, conversion into *cis*-dibenzoylstilbene, **276**
  into endoperoxide, 89
Tetraphenylnaphthacene,
  generation of singlet oxygen, 88–89
Tetrapyridine silver dichromate, See *Silver dichromate–pyridine complex*
Thallium compounds
  safety precautions, 17
  toxicity, 17
Thallium monoacetate
  in hydroxylation of alkenes and cycloalkenes, 71, 72
  oxidant, 17
Thallium triacetate
  applications as oxidant, 17
  in acyloxylation of alkenes and cycloalkenes, 74–75
  in epoxidation of alkenes and cycloalkenes, 61
  in formation of vicinal diacetates, 74–75
Thallium trinitrate
  in oxidation
    of alkynes (acetylenes)
      to acyloins, 91
      to carboxylic acids, 91–92
      to α-diketones, 91, **280–281**
    of methyl ketones to carboxylic acids, 205, 206
  oxidant, 17
Thiiranes,
  oxidation to cyclic sulfoxides, 255
Thioacetals, oxidation
  to disulfoxides or disulfones, 260
  to sulfoxides and sulfones, 259–261
Thioamides, intermediates in Willgerodt reaction, 203, 204
Thioanisole (methyl phenyl sulfide),
  oxidation to methyl phenyl sulfoxide, **287**
Thiol group,
  replacement with hydrogen, 252
Thiols
  dehydrogenation to disulfides, **290**
  oxidation
    to disulfides, 250–251
    to sulfinic acids, 252
    to sulfonic acids, 251–252
    to sulfonyl chlorides, 252
Thiomorpholides, intermediates in Willgerodt–Kindler reaction, 203, 205
Thionyl chloride,
  activator of dimethyl sulfoxide, 122
Thiophosphates,
  oxidation to phosphates, 250
Thiourea,
  in reduction of endoperoxides, 87–88
Tillman reagent, oxidant, 40
Tin compounds, oxidation, 270
Titanium salts
  catalysts for alkene epoxidation with *tert*-butyl hydroperoxide, 61
  in reduction of pyridine oxide to pyridine, 238
Titanium tetraisopropoxide,
  See *Tetraisopropyl orthotitanate*

Titanium trichloride, catalyst for oxidation with hydrogen peroxide, 7
p-*Toluenesulfonic acid,* catalyst for formation of peroxy acids, 222
Torulopsis gropengiesseri, in oxidation
  of alkyl halides to carboxylic acids, 112–113
  of halomethyl groups to carboxyl groups, 58
Tosylates, oxidation, 109–114
Tosylhydrazones,
  oxidation to parent ketones, 219
Transannular oxidations, 154
Transition metals, catalysts
  for decomposition
    of *tert*-butyl hydroperoxide, 9
    of hydrogen peroxide, 7
  for oxidation with oxygen, 4
Trialkylsilyl ethers, See *Ethers, trialkylsilyl*
Tributyltin ethers, from hydroxy compounds and bis(tributyl)tin oxide, 139
Trichloroacetaldehyde
  in oxidation of secondary alcohols to ketones, 149
  oxidant, 39
Trichloroacetyl chromate, oxidant and preparation, 26
Trichloroisocyanuric acid, oxidant, 28
Triethylamine, base in oxidation with dimethyl sulfoxide, 145
Triethylbenzylammonium chloride, phase-transfer reagent, 89
Trifluoroacetic acid
  catalyst for oxidation with dimethyl sulfoxide, 43
  in *anti* hydroxylation of alkenes and cycloalkenes, 69
  preparation from 2,3-dichlorohexafluoro-2-butene, **289**
Trifluoroacetic anhydride
  activator of dimethyl sulfoxide, 43, 122
  in preparation of peroxytrifluoroacetic acid, **279**
Trifluoroacetyl chromate,
  oxidant and preparation, 26
Trifluoromethanesulfonic acid, catalyst for oxidation with dimethyl sulfoxide, 43
Trifluoroperacetic acid,
  See *Peroxytrifluoroacetic acid*
Trihalomethyl ketones, intermediates in oxidation of methyl ketones, 206
Triketopentane (pentanetrione),
  preparation from acetylacetone, **292**
Trimethyl phosphite,
  in reduction of ozonides, 5, 78–79
Trimethylamine oxide
  applications as oxidant, 42
  in oxidation
    of boranes to borates, 267
    of primary halides to aldehydes, 109, 110
  oxidant of osmium oxide, 73, 262
Trimethylsilyl enol ethers
  conversion into α-hydroxy ketones, 196
  preparation from ketones, 196

Trimethylsilyl ethers,
  See *Ethers, trialkylsilyl*
Trimethylsilyl peroxide, in oxidation of benzylic alcohols to aldehydes, 126
Trimethylsilyl trifluoroacetate, catalyst for oxidation with bis(trimethylsilyl) peroxide, 194, 195
Trimethylsilyl trifluoromethanesulfonate, catalyst for oxidation with bis(trimethylsilyl) peroxide, 10, 187, 191, 195
Triphenyl phosphite, conversion into triphenyl phosphite ozonide, **276**
Triphenyl phosphite ozonide
  decomposition, 42
  in formation
    of dioxetane, 64, 65
    of singlet oxygen, 1, 2, 87
  in oxidation of trialkylsilyl enol ethers, 173
  oxidant, 42
  preparation from triphenyl phosphite, **276**
Triphenylmethyl tetrafluoroborate
  oxidant, 41
  in oxidation of secondary alcohols to ketones, 146, 147, 148
Triphenylmethylphosphonium permanganate
  explosion during handling, 36
  in *syn* hydroxylation of alkenes and cycloalkenes, 72
  oxidant and preparation, 36–37
  solvents, 37
Triphenylphosphine,
  catalyst for reduction of ozonides, 5
Triple bonds,
  cleavage to carboxylic groups, 226
Triplet oxygen, definition, 1
Tris(triphenylphosphine)ruthenium dichloride, catalyst for oxidation
  of aldehydes to carboxylic acids, 177, 179
  with bis(trimethylsilyl) peroxide, 10
  with iodosobenzene, 92, 172, 245
Tungsten hexafluoride
  in regeneration of ketones from their derivatives, 220
  oxidant, 21
Tungsten trioxide (tungstic acid), catalyst
  for *anti* hydroxylation with hydrogen peroxide, 69, **277**
  for oxidation with hydrogen peroxide, 7, 261

U

Unprotected glycosides, oxidation, 139
Unsaturated alcohols, See *Alcohols, unsaturated*
Unsaturated aldehydes, See *Aldehydes, unsaturated*
Unsaturated carboxylic acids, See *Carboxylic acids, unsaturated*
Unsaturated cis diols, formation from endoperoxides, 87–88

Subject Index 433

Unsaturated esters, See *Esters, unsaturated*
α,β-*Unsaturated esters,* by oxidation of
   primary alcohols, 131
Unsaturated ethers, See *Ethers, unsaturated*
Unsaturated keto esters, by oxidation of
   unsaturated esters, 226
Unsaturated ketones,
   See *Ketones, unsaturated*
Unsaturated nitriles, epoxidation, 230
Unsaturated selenoxides, oxidation, 265
Unsaturated steroidal ketones,
   hydroxylation, 196–199
Unsaturated sulfones,
   See *Sulfones, unsaturated*
Unsaturated sulfoxides,
   See *Sulfoxides, unsaturated*
Ureas, by oxidation of primary amides, 230

V

Vanadium compounds, catalysts
   for alkene epoxidation with *tert*-butyl
      hydroperoxide, 61
   for oxidation
      with *tert*-butyl hydroperoxide, 237
      with hydrogen peroxide, 261
Vanadium pentoxide, catalyst
   for cleavage of alkenes and cycloalkenes
      to aldehydes, 79
   for dehydrogenation, 49
   for hydroxylation with hydrogen
      peroxide, 19
   for oxidation
      of benzene rings to carboxylic acids, 97
      with air, 19
      with *tert*-butyl hydroperoxide, 237,
         277–278
      with chlorates, 28
      with hydrogen peroxide, 7
      with oxygen, 4, 19
      with sodium chlorate, 166
Vanadyl acetylacetonate, catalyst
   for enantioselective oxidation, 44
   for epoxidation, 153
   for oxidation with *tert*-butyl
      hydroperoxide, 153
   for stereoselective epoxidation, 19
Vanillic acid,
   preparation from vanillin, **285–286**
Vanillin, oxidation to vanillic acid, **285–286**
Vicinal diacetates, formation from alkenes
   and cycloalkenes, 74–75
Vicinal dicarbonyl compounds,
   formation from alkenes, 76
Vicinal diols, See *Diols, vicinal*
Vicinal diols, acyl esters, formation from
   alkenes and cycloalkenes, 74–75
Vinylic alcohols, oxidation, 151

W

Wieland degradation of carboxylic acids,
   151
Willgerodt–Kindler reaction, 203–205, **282–283**
Willgerodt reaction
   Kindler modification, 20, 203–205
   of heterocyclic ketones, 205
Woelm alumina
   catalyst
      for dehydrogenation of hydroxy
         sulfides, 261
      in Oppenauer oxidation, 149
      support for carbonyl compound oxidants,
         143
Woodward modification of Prevost reaction,
   31, 70, **286–287**

Y

Yeast, in oxidation of secondary alcohols to
   ketones, 146
Ylides, cleavage by sodium periodate, 250

Z

Zinc, in reduction of ozonides, 5, 78
Zinc dichromate
   in oxidation
      of alkynes to α-diketones, 91
      of aromatic compounds to quinones,
         95
      of ethers to esters and lactones, 169
      of primary alcohols to carboxylic acids,
         128, 130
      of secondary alcohols to ketones, 149
   in regeneration of ketones from their
      derivatives, 219
Zinc dichromate dihydrate, oxidant and
   preparation, 25
Zinc dust, in reductive cleavage of
   ozonides, **276**
Zinc oxide, dehydrogenation catalyst, 49
Zinc permanganate
   ignition with solvents, 36
   in dehydrogenation of primary amines to
      aldehydes, 240
   in oxidation
      of alkynes to α-diketones, 91
      of ethers to esters and lactones, 169
   oxidant, 36
   silica gel support, 36
   solvents, 36
Zinc tetraphenylphorphyrin, sensitizer in
   oxidations with singlet oxygen, 65

Editing and indexing by Ann Maureen R. Rouhi
Production by Paula M. Bérard
Cover design by Amy Hayes

Typeset by Techna Type, Inc., York, PA
Printed and bound by Maple Press Company, York, PA

Other ACS Books

Chemical Structure Software for Personal Computers
Edited by Daniel E. Meyer, Wendy A. Warr, and Richard A. Love
ACS Professional Reference Book; 107 pp;
clothbound, ISBN 0-8412-1538-3; paperback, ISBN 0-8412-1539-1

Personal Computers for Scientists: A Byte at a Time
By Glenn I. Ouchi
276 pp; clothbound, ISBN 0-8412-1000-4; paperback, ISBN 0-8412-1001-2

Biotechnology and Materials Science: Chemistry for the Future
Edited by Mary L. Good
160 pp; clothbound, ISBN 0-8412-1472-7; paperback, ISBN 0-8412-1473-5

Polymeric Materials: Chemistry for the Future
By Joseph Alper and Gordon L. Nelson
110 pp; clothbound, ISBN 0-8412-1622-3; paperback, ISBN 0-8412-1613-4

The Language of Biotechnology: A Dictionary of Terms
By John M. Walker and Michael Cox
ACS Professional Reference Book; 256 pp;
clothbound, ISBN 0-8412-1489-1; paperback, ISBN 0-8412-1490-5

Cancer: The Outlaw Cell, Second Edition
Edited by Richard E. LaFond
274 pp; clothbound, ISBN 0-8412-1419-0; paperback, ISBN 0-8412-1420-4

Practical Statistics for the Physical Sciences
By Larry L. Havlicek
ACS Professional Reference Book; 198 pp; clothbound; ISBN 0-8412-1453-0

The Basics of Technical Communicating
By B. Edward Cain
ACS Professional Reference Book; 198 pp;
clothbound, ISBN 0-8412-1451-4; paperback, ISBN 0-8412-1452-2

The ACS Style Guide: A Manual for Authors and Editors
Edited by Janet S. Dodd
264 pp; clothbound, ISBN 0-8412-0917-0; paperback, ISBN 0-8412-0943-X

Chemistry and Crime: From Sherlock Holmes to Today's Courtroom
Edited by Samuel M. Gerber
135 pp; clothbound, ISBN 0-8412-0784-4; paperback, ISBN 0-8412-0785-2

For further information and a free catalog of ACS books, contact:
American Chemical Society
Distribution Office, Department 225
1155 16th Street, NW, Washington, DC 20036
Telephone 800-227-5558